Methods in Enzymology

Volume 140

CUMULATIVE SUBJECT INDEX

Volumes 102–119, 121–134

METHODS IN ENZYMOLOGY

EDITORS-IN-CHIEF

Sidney P. Colowick Nathan O. Kaplan

Methods in Enzymology

Volume 140

Cumulative Subject Index

Volumes 102–119, 121–134

ACADEMIC PRESS, INC.
Harcourt Brace Jovanovich, Publishers
San Diego New York Berkeley Boston
London Sydney Tokyo Toronto

ACADEMIC PRESS, INC.
1250 Sixth Avenue, San Diego, California 92101

United Kingdom Edition published by
ACADEMIC PRESS INC. (LONDON) LTD.
24–28 Oval Road, London NW1 7DX

Library of Congress Catalog Card Number: 54-9110

ISBN 0–12–182040-8

PRINTED IN THE UNITED STATES OF AMERICA

88 89 90 91 9 8 7 6 5 4 3 2 1

Table of Contents

Preface

The idea for a cumulative index was recognized by the founding editors who prepared one for Volumes I through VI of *Methods of Enzymology* by weeding and interpolating from the entries that had been indexed in the individual volumes. As the series developed in both number and complexity, different individuals with different backgrounds served as volume indexers. Subsequently, the series was fortunate in having Dr. Martha G. Dennis and Dr. Edward A. Dennis accept the challenge of computerizing the data available from these individual indexes, and this effort resulted in Volumes 33, 75, and 95, which cover Volumes 1 through 80.

Although each of the three books produced with the aid of computerization provided an appropriate cumulative index, major problems were encountered. One was time, both expensive computer time and lag time before such efforts resulted in publication. The most important difficulty was that the compilers were hampered by the lack of uniformity in the indexing of the individual volumes, resulting in the need for much hand editing to achieve a reasonable collation. The products were very decent, if uneven, indexes that also contributed to the methodology of computerized indexing, albeit with much delay and great expense.

This cumulative index has been produced by the staff of Academic Press. The indexers have gone back to the text rather than to the individual volume indexes and, using a uniform set of guidelines, have culled the major topics, leading to five to ten entries for each article. Thus, if searching by name for one of the dozen substrates of an enzyme, the isolation of which is being presented, it probably will not be found in the index, although the assay substrate may be there. Nor will the specific inhibitors of the enzyme be itemized, although the topic of the enzyme's inhibition will form an entry. The index, then, is not complete, but should lead to the broad subject headings. Since there is a tendency to identify specific topics and methods with particular individuals, a contributor index is included. Finally, the complete table of contents of each of the volumes indexes is included.

Contents of Volumes 102–119, 121–134

VOLUME 102

Hormone Action (Part G: Calmodulin and Calcium-Binding Proteins)

1. Calmodulin Purification and Fluorescent Labeling — John R. Dedman and Marcia A. Kaetzel 1
2. Purification of Plant Calmodulin — James Michael Anderson 9
3. Purification of Calmodulin by Ca^{2+}-Dependent Affinity Chromatography — Harry Charbonneau, Rita Hice, Russell C. Hart, and Milton J. Cormier 17
4. Assay of Calmodulin by Ca^{2+}-Dependent Phosphodiesterase — Robert W. Wallace, E. Ann Tallant, and Wai Yiu Cheung 39
5. The Red Blood Cell as a Model for Calmodulin-Dependent Ca^{2+} Transport — Thomas R. Hinds and Frank F. Vincenzi 47
6. Myosin Light Chain Phosphorylation in Smooth Muscle and Nonmuscle Cells as a Probe of Calmodulin Function — Paul J. Silver and James T. Stull 62
7. Assessment of Ca^{2+}–Calmodulin Formation in Intact Vertebrate Skeletal Muscle — David R. Manning and James T. Stull 74
8. Spectroscopic Analyses of Calmodulin and Its Interactions — Rachel E. Klevit 82
9. Production of Polyclonal and Monoclonal Antibodies to Calmodulin and Utilization of These Immunological Probes — James G. Chafouleas, Mary E. Riser, Lisette Lagacé, and Anthony R. Means 104
10. Localization of Calmodulin and Calmodulin Acceptor Sites by Fluorescence Methods — Michael J. Welsh 110
11. Immunofluorescence Localization of Calmodulin in Unfixed Frozen Tissue Sections — Jeffrey F. Harper and Alton L. Steiner 122
12. Ca^{2+} Binding to Calmodulin — James D. Potter, Priscilla Strang-Brown, Patricia L. Walker, and Shozo Iida 135
13. Preparation of Calmodulin Crystals — William J. Cook and John S. Sack 143
14. Preparation of Fluorescent Labeled Calmodulins — Bradley B. Olwin and Daniel R. Storm 148
15. Posttranslational Modification of Calmodulin — Timothy J. Murtaugh, Paul M. Rowe, Pamela L. Vincent, Lynda S. Wright, and Frank L. Siegel 158
16. Techniques for Measuring the Interaction of Drugs with Calmodulin — Benjamin Weiss 171
17. Naphthalenesulfonamides as Calmodulin Antagonists — Hiroyoshi Hidaka and Toshio Tanaka 185

18. Synthesis and Characterization of Calmodulin Antagonistic Drugs
RUSSELL C. HART, MICHAEL D. BATES, MILTON J. CORMIER, GERALD M. ROSEN, AND P. MICHAEL CONN 195
19. Detection of Calmodulin-Binding Polypeptides Separated in SDS–Polyacrylamide Gels by a Sensitive [^{125}I]Calmodulin Gel Overlay Assay
JOHN R. GLENNEY, JR. AND KLAUS WEBER 204
20. Use of Calmodulin Affinity Chromatography for Purification of Specific Calmodulin-Dependent Enzymes
RAJENDRA K. SHARMA, WILLIAM A. TAYLOR, AND JERRY H. WANG 210
21. Calmodulin as an Integral Subunit of Phosphorylase Kinase from Rabbit Skeletal Muscle
COLIN PICTON, SHIRISH SHENOLIKAR, ROGER GRAND, AND PHILIP COHEN 219
22. Isolation and Characterization of Bovine Brain Calcineurin: A Calmodulin-Stimulated Protein Phosphatase
C. B. KLEE, M. H. KRINKS, A. S. MANALAN, P. COHEN, AND A. A. STEWART 227
23. Purification and Radioimmunoassay of Calmodulin-Dependent Protein Phosphatase from Bovine Brain
E. ANN TALLANT, ROBERT W. WALLACE, AND WAI YIU CHEUNG 244
24. Assay of S-100 Protein by an Enzyme Immunoassay Method
HIROYOSHI HIDAKA, TOYOSHI ENDO, AND KANEFUSA KATO 256
25. Covalent Regulation of the Cardiac Sarcoplasmic Reticulum Calcium Pump: Purification and Properties of Phospholamban, a Substrate of cAMP-Dependent Protein Kinase and Ca^{2+}–Calmodulin-Dependent Phospholamban Kinase
CHRISTIAN J. LE PEUCH, DANIELLE A.-M. LE PEUCH, AND JACQUES G. DEMAILLE 261
26. Ca^{2+}Dependent Neutral Protease and Proteolytic Activation of Ca^{2+}-Activated, Phospholipid-Dependent Protein Kinase
NORIO KAJIKAWA, AKIRA KISHIMOTO, MAKOTO SHIOTA, AND YASUTOMI NISHIZUKA 279
27. Enzyme-Linked Immunoabsorbent Assay (ELISA) and Radioimmunoassay (RIA) for the Vitamin D-Dependent 28,000 Dalton Calcium-Binding Protein
BARBARA E. MILLER AND ANTHONY W. NORMAN 291
28. Chemical Approaches to the Calmodulin System THOMAS C. VANAMAN 296

VOLUME 103
HORMONE ACTION (PART H: NEUROENDOCRINE PEPTIDES)

Section I. Preparation of Chemical Probes

1. Synthesis and Use of Neuropeptides JOHN M. STEWART 3
2. Preparation of Radiolabeled Neuroendocrine Peptides RICHARD N. CLAYTON 32
3. Ligand Dimerization: A Technique for Assessing Receptor–Receptor Interactions P. MICHAEL CONN 49
4. Photoaffinity Labeling in Neuroendocrine Tissues ELI HAZUM 58
5. Microisolation of Neuropeptides
FRED S. ESCH, NICHOLAS C. LING, AND PETER BÖHLEN 72

Section II. Equipment and Technology

6. Electrophysiological Assays of Mammalian Cells Involved in Neurohormonal Communication
BERNARD DUFY, LUCE DUFY-BARBE, AND JEFFERY BARKER 93
7. Electrophysiological Techniques in Dissociated Tissue Culture
GARY L. WESTBROOK AND PHILLIP G. NELSON 111
8. Neuroendocrine Cells *in Vitro*: Electrophysiology, Triple-Labeling with Dye Marking, Immunocytochemical and Ultrastructural Analysis, and Hormone Release
JAMES N. HAYWARD, TROY A. REAVES, JR., ROBERT S. GREENWOOD, AND RICK B. MEEKER 132
9. Single-Channel Electrophysiology: Use of the Patch Clamp
FREDERICK SACHS AND ANTHONY AUERBACH 147
10. Push–Pull Perfusion Technique in the Median Eminence: A Model System for Evaluating Releasing Factor Dynamics
NORMAN W. KASTING AND JOSEPH B. MARTIN 176
11. Horseradish Peroxidase: A Tool for Study of the Neuroendocrine Cell and Other Peptide-Secreting Cells
RICHARD D. BROADWELL AND MILTON W. BRIGHTMAN 187
12. Visualization of Enkephalin Receptors by Image-Intensified Fluorescence Microscopy
STEVEN G. BLANCHARD, KWEN-JEN CHANG, AND PEDRO CUATRECASAS 219
13. Flow Cytometry as an Analytic and Preparative Tool for Studies of Neuroendocrine Function JOHN C. CAMBIER AND JOHN G. MONROE 227

Section III. Preparation and Maintenance of Biological Materials

14. Optimization of Culture Conditions for Short-Term Pituitary Cell Culture
NIRA BEN-JONATHAN, EDNA PELEG, AND MICHAEL T. HOEFER 249
15. Separation of Cells from the Rat Anterior Pituitary Gland
WESLEY C. HYMER AND J. MICHAEL HATFIELD 257
16. Culture of Dispersed Anterior Pituitary Cells on Extracellular Matrix
RICHARD I. WEINER, CYNTHIA L. BETHEA, PHILIPPE JAQUET, JOHN S. RAMSDELL, AND DENIS J. GOSPODAROWICZ 287
17. Continuous Perifusion of Dispersed Anterior Pituitary Cells: Technical Aspects
WILLIAM S. EVANS, MICHAEL J. CRONIN, AND MICHAEL O. THORNER 294
18. Preparation and Maintenance of Adrenal Medullary Chromaffin Cell Cultures
STEVEN P. WILSON AND NORMAN KIRSHNER 305
19. Techniques for Culture of Hypothalamic Neurons
CATHERINE LOUDES, ANNIE FAIVRE-BAUMAN, AND ANDRÉE TIXIER-VIDAL 313
20. Techniques in the Tissue Culture of Rat Sympathetic Neurons
MARY I. JOHNSON AND VINCENT ARGIRO 334
21. Methodological Considerations in Culturing Peptidergic Neurons
WILLIAM J. SHOEMAKER, ROBERT A. PETERFREUND, AND WYLIE VALE 347
22. Preparation and Properties of Dispersed Rat Retinal Cells
JAMES M. SCHAEFFER 362
23. Punch Sampling Biopsy Technique MIKLÓS PALKOVITS 368

Section IV. Use of Chemical Probes

24. Excitotoxic Amino Acids as Neuroendocrine Research Tools
JOHN W. OLNEY AND MADELON T. PRICE 379
25. Use of Excitatory Amino Acids to Make Axon-Sparing Lesions of Hypothalamus
J. VICTOR NADLER AND DEBRA A. EVENSON 393
26. Use of Specific Ion Channel Activating and Inhibiting Drugs in Neuroendocrine Tissue P. MICHAEL CONN 401

Section V. Quantitation of Neuroendocrine Substances

27. Development and Use of Ultrasensitive Enzyme Immunoassays
GLENNWOOD E. TRIVERS, CURTIS C. HARRIS, CATHERINE ROUGEOT, AND FERNAND DRAY 409
28. Strategies for the Preparation of Haptens for Conjugation and Substrates for Iodination for Use in Radioimmunoassay of Small Oligopeptides
WILLIAM W. YOUNGBLOOD AND JOHN STEPHEN KIZER 435
29. Preparation and Use of Specific Antibodies for Immunohistochemistry of Neuropeptides LOTHAR JENNES AND WALTER E. STUMPF 448
30. Production of Monoclonal Antibodies Reacting with the Cytoplasm and Surface of Differentiated Cells
RICHARD M. SCEARCE AND GEORGE S. EISENBARTH 459
31. Voltammetric and Radioisotopic Measurement of Catecholamines
PAUL M. PLOTSKY 469
32. Microradioenzymic Assays for the Measurement of Catecholamines and Serotonin RALPH L. COOPER AND RICHARD F. WALKER 483
33. Hypothalamic Catecholamine Biosynthesis and Neuropeptides
DAVID K. SUNDBERG, BARBARA A. BENNETT, AND MARIANA MORRIS 493
34. Methods for the Study of the Biosynthesis of Neuroendocrine Peptides *in Vivo* and *in Vitro*
JEFFREY F. MCKELVY, JAMES E. KRAUSE, AND JEFFREY D. WHITE 511
35. Methods for Investigating Peptide Precursors in the Hypothalamus
DAVID K. SUNDBERG, MARIANA MORRIS, AND KENNETH A. GRUBER 524
36. Measurement of the Degradation of Luteinizing Hormone Releasing Hormone by Hypothalamic Tissue JAMES E. KRAUSE AND JEFFREY F. MCKELVY 539
37. Measurement of β-Endorphin and Enkephalins in Biological Tissues and Fluids
JAU-SHYONG HONG, KAZUAKI YOSHIKAWA, AND R. WAYNE HENDREN 547
38. Assay of Corticotropin-Releasing Factor
WYLIE VALE, JOAN VAUGHAN, GAYLE YAMAMOTO, THOMAS BRUHN, CAROLYN DOUGLAS, DAVID DALTON, CATHERINE RIVIER, AND JEAN RIVIER 565
39. Direct Radioligand Measurement of Dopamine Receptors in the Anterior Pituitary Gland
MARC C. CARON, BRIAN F. KILPATRICK, AND DAVID R. SIBLEY 577
40. Ornithine Decarboxylase: Marker of Neuroendocrine and Neurotransmitter Actions THEODORE A. SLOTKIN AND JORGE BARTOLOME 590

Section VI. Localization of Neuroendocrine Substances

41. Secretion of Hypothalamic Dopamine into the Hypophysial Portal Vasculature: An Overview — JOHN C. PORTER, MARIANNE J. REYMOND, JUN ARITA, AND JANICE F. SISSOM 607
42. Neurotransmitter Histochemistry: Comparison of Fluorescence and Immunohistochemical Methods — ROBERT Y. MOORE AND J. PATRICK CARD 619
43. Simultaneous Localization of Steroid Hormones and Neuropeptides in the Brain by Combined Autoradiography and Immunocytochemistry — MADHABANANDA SAR AND WALTER E. STUMPF 631
44. Immunocytochemistry of Steroid Hormone Receiving Cells in the Central Nervous System — JOAN I. MORRELL AND DONALD W. PFAFF 639
45. Techniques for Tracing Peptide-Specific Pathways — LARRY W. SWANSON 663
46. Immunocytochemistry of Endorphins and Enkephalins — FLOYD E. BLOOM AND ELENA L. F. BATTENBERG 670

Section VII. Summary

47. Previously Published Articles from Methods in Enzymology Related to Neuroendocrine Peptides 691

VOLUME 104

ENZYME PURIFICATION AND RELATED TECHNIQUES (PART C)

Section I. Chromatography

1. Affinity Chromatography — MEIR WILCHEK, TALIA MIRON, AND JOACHIM KOHN 3
2. Immobilization of Ligands with Organic Sulfonyl Chlorides — KURT NILSSON AND KLAUS MOSBACH 56
3. Hydrophobic Chromatography — SHMUEL SHALTIEL 69
4. Affinity Chromatography on Immobilized Dyes — CHRISTOPHER R. LOWE AND JAMES C. PEARSON 97
5. Displacement Chromatography of Proteins — ELBERT A. PETERSON AND ANTHONY R. TORRES 113
6. High-Performance Liquid Chromatography: Care of Columns — C. TIMOTHY WEHR 133
7. High-Performance Size-Exclusion Chromatography — KLAUS UNGER 154
8. High-Performance Ion-Exchange Chromatography — FRED E. REGNIER 170
9. Reversed-Phase High-Performance Liquid Chromatography — MILTON T. W. HEARN 190
10. High-Performance Liquid Affinity Chromatography — PER-OLOF LARSSON 212
11. Optimal pH Conditions for Ion Exchangers on Macroporous Supports — JAMES S. RICHEY 223

Section II. Electrophoresis

12. Systems for Polyacrylamide Gel Electrophoresis — PERRY J. BLACKSHEAR 237
13. High-Resolution Preparative Isoelectric Focusing — BERTOLD J. RADOLA 256
14. Affinity Electrophoresis — VÁCLAV HOŘEJŠÍ 275
15. Preparative Isotachophoresis — CHRISTOPHER J. HOLLOWAY AND RÜDIGER V. BATTERSBY 281

Section III. Techniques for Membrane Proteins

16. Solubilization of Functional Membrane Proteins — LEONARD M. HJELMELAND AND ANDREAS CHRAMBACH 305
17. Separating Detergent from Proteins — ANNA J. FURTH, HILARY BOLTON, JENNIFER POTTER, AND JOHN D. PRIDDLE 318
18. Purification of Integral Membrane Proteins — JOS VAN RENSWOUDE AND CHRISTOPH KEMPF 329
19. Reconstitution of Membrane Proteins — RICHARD D. KLAUSNER, JOS VAN RENSWOUDE, AND BENJAMIN RIVNAY 340

Section IV. Other Separation Systems

20. Protein Precipitation with Polyethylene Glycol — KENNETH C. INGHAM 351
21. Affinity Partitioning — GÖTE JOHANSSON 356
22. Affinity Precipitation of Dehydrogenases — PER-OLOF LARSSON, SUSANNE FLYGARE, AND KLAUS MOSBACH 364
23. Protein Crystallization: The Growth of Large-Scale Single Crystals — GARY L. GILLILAND AND DAVID R. DAVIES 370
24. Immunosorbent Separations — GARY J. CALTON 381

Section V. Related Techniques

25. Preparation of High-Purity Laboratory Water — GARY C. GANZI 391
26. Buffers for Enzymes — JOHN S. BLANCHARD 404
27. Protein Assay for Dilute Solutions — ENRICO CABIB AND ITZHACK POLACHECK 415
28. Enzyme Localization in Gels — MARY J. HEEB AND OTHMAR GABRIEL 416
29. Gel Protein Stains: A Rapid Procedure — A. H. REISNER 439
30. Gel Protein Stains: Silver Stain — CARL R. MERRIL, DAVID GOLDMAN, AND MARGARET L. VAN KEUREN 441
31. Gel Protein Stains: Glycoproteins — JOHN E. GANDER 447
32. Gel Protein Stains: Phosphoproteins — JOHN A. CUTTING 451
33. Western Blots — JAIME RENART AND IGNACIO V. SANDOVAL 455
34. Fluorography for the Detection of Radioactivity in Gels — WILLIAM M. BONNER 460

VOLUME 105
OXYGEN RADICALS IN BIOLOGICAL SYSTEMS

Section I. Chemistry and Biochemistry of Oxygen and Intermediate States of Its Reduction

1. Chemistry of Dioxygen MONIKA J. GREEN AND H. ALLEN O. HILL 3
2. Biochemistry of Oxygen Radical Species M. BRUNORI AND G. ROTILIO 22
3. Characterization of Singlet Oxygen CHRISTOPHER S. FOOTE, F. C. SHOOK, AND R. B. ABAKERLI 36
4. Role of Iron in Oxygen Radical Reactions BARRY HALLIWELL AND JOHN M. C. GUTTERIDGE 47

Section II. Isolation and Assays of Enzymes or Substances Resulting in Formation or Removal of Oxygen Radicals

A. One-Electron Transfer Enzyme Reactions Resulting in O_2^- Production

5. Overview: Biological Sources of O_2^- IRWIN FRIDOVICH 59
6. Overview: Superoxygenase RYOTARO YOSHIDA AND OSAMU HAYAISHI 61
7. Methods for the Study of Superoxide Chemistry in Nonaqueous Solutions JOAN S. VALENTINE, ANDREW R. MIKSZTAL, AND DONALD T. SAWYER 71
8. Generation of Superoxide Radicals in Aqueous and Ethanolic Solutions by Vacuum-UV Photolysis BENON H. J. BIELSKI 81

B. Isolation, Purification, Characterization, and Assay of Antioxygenic Enzymes

9. Isolation and Characterization of Superoxide Dismutase J. V. BANNISTER AND W. H. BANNISTER 88
10. Superoxide Dismutase Assays L. FLOHÉ AND F. ÖTTING 93
11. Subcellular Distribution of Superoxide Dismutases in Rat Liver BRUCE L. GELLER AND DENNIS R. WINGE 105
12. Assays of Glutathione Peroxidase LEOPOLD FLOHÉ AND WOLFGANG A. GÜNZLER 114
13. Catalase *in Vitro* HUGO AEBI 121
14. Assays of Lipoxygenase, 1,4-Pentadiene Fatty Acids, and O_2 Concentrations: Chemiluminescence Methods SIMO LAAKSO, ESA-MATTI LILIUS, AND PEKKA TURUNEN 126

C. Antioxidants/Prooxidants

15. High-Performance Liquid Chromatography Methods for Vitamin E in Tissues JUDITH L. BUTTRISS AND ANTHONY T. DIPLOCK 131
16. Vitamin E Analysis Methods for Animal Tissues INDRAJIT D. DESAI 138

17. Simultaneous Determination of Reduced and Oxidized Ubiquinones
MASAHIRO TAKADA, SATORU IKENOYA, TERUAKI YUZURIHA, AND KOUICHI KATAYAMA 147
18. Assay of Carotenoids NORMAN I. KRINSKY AND SUDHAKAR WELANKIWAR 155
19. Uric Acid: Functions and Determination
PAUL HOCHSTEIN, LINDA HATCH, AND ALEX SEVANIAN 162

D. Detection and Characterization of Oxygen Radicals

20. Pulse Radiolysis Methodology KLAUS-DIETER ASMUS 167
21. Electron and Hydrogen Atom Transfer Reactions: Determination of Free Radical Redox Potentials by Pulse Radiolysis L. G. FORNI AND R. L. WILLSON 179
22. Spin Trapping EDWARD G. JANZEN 188
23. Spin Trapping of Superoxide and Hydroxyl Radicals
GERALD M. ROSEN AND ELMER J. RAUCKMAN 198
24. Reaction of ·OH GIDON CZAPSKI 209
25. Electron Spin Resonance Spin Destruction Methods for Radical Detection
ROLF J. MEHLHORN AND LESTER PACKER 215
26. Low-Level Chemiluminescence as an Indicator of Singlet Molecular Oxygen in Biological Systems ENRIQUE CADENAS AND HELMUT SIES 221
27. High-Pressure Liquid Chromatography-Electrochemical Detection of Oxygen Free Radicals ROBERT A. FLOYD, C. ANN LEWIS, AND PETER K. WONG 231
28. Survey of the Methodology for Evaluating Negative Air Ions: Relevance to Biological Studies RONALD PETHIG 238

E. Genetic Methods for Detection/Assay of Oxygen Radical Species

29. Detection of Oxidative Mutagens with a New *Salmonella* Tester Strain (TA102)
DAVID E. LEVIN, MONICA HOLLSTEIN, MICHAEL F. CHRISTMAN, AND BRUCE N. AMES 249
30. Determination of the Mutagenicity of Oxygen Free Radicals Using Microbial Systems HOSNI M. HASSAN AND CARMELLA S. MOODY 254
31. Assay of Rate of Aging of Conidia of *Neurospora crassa*
KENNETH D. MUNKRES AND CHERYL A. FURTEK 263

Section III. Assay of Modes of Biological Damage Imposed by O_2 and Reduced Species

A. Lipid Peroxidation

32. Chemistry of Lipid Peroxidation NED A. PORTER 273
33. Overview of Methods Used for Detecting Lipid Peroxidation T. F. SLATER 283
34. Chemical Methods for the Detection of Lipid Hydroperoxides
WILLIAM A. PRYOR AND LAWRENCE CASTLE 293
35. Comparative Studies on Different Methods of Malonaldehyde Determination
R. P. BIRD AND H. H. DRAPER 299
36. Concentrating Ethane from Breath to Monitor Lipid Peroxidation *in Vivo*
GLEN D. LAWRENCE AND GERALD COHEN 305

37. Assay of Ethane and Pentane from Isolated Organs and Cells
ARMIN MÜLLER AND HELMUT SIES 311
38. Detection of Malonaldehyde by High-Performance Liquid Chromatography
HERMANN ESTERBAUER, JOHANNA LANG, SYLVIA ZADRAVEC, AND TREVOR F. SLATER 319
39. Assay for Blood Plasma or Serum KUNIO YAGI 328
40. Spectrophotometric Detection of Lipid Conjugated Dienes
RICHARD O. RECKNAGEL AND ERIC A. GLENDE, JR. 331
41. Fluorescence Damage Products of Lipid Peroxidation
CORA J. DILLARD AND AL L. TAPPEL 337
42. Calibration of Microspectrophotometers as It Applies to the Detection of Lipofuscin and the Blue- and Yellow-Emitting Fluorophores *in Situ*
WILLIAM S. STARK, GREGORY V. MILLER, AND KENDALL A. ITOKU 341
43. Detection of the Metabolism of Polycyclic Aromatic Hydrocarbon Derivatives to Ultimate Carcinogens during Lipid Peroxidation
THOMAS A. DIX AND LAWRENCE J. MARNETT 347
44. Detection of Picomole Levels of Lipid Hydroperoxides Using a Dichlorofluorescein Fluorescent Assay
RICHARD CATHCART, ELIZABETH SCHWIERS, AND BRUCE N. AMES 352

B. Plasma Membrane

45. Measurement of O_2^- Production by Human Neutrophils. The Preparation and Assay of NADPH Oxidase-Containing Particles from Human Neutrophils
MICHÈLE MARKERT, PATRICIA C. ANDREWS, AND BERNARD M. BABIOR 358
46. Measurement of O_2^- Secreted by Monocytes and Macrophages
RICHARD B. JOHNSTON, JR. 365
47. Superoxide Production PETER J. O'BRIEN 370
48. Endothelial Culture, Neutrophil or Enzymic Generation of Free Radicals: *In Vitro* Methods for the Study of Endothelial Injury
CHARLES F. MOLDOW AND HARRY S. JACOB 378
49. Systemic Consequences of O_2^- Production ROLANDO DEL MAESTRO 386
50. Calcium and Calmodulin in Neutrophil Activation
HAROLD P. JONES AND JOE M. MCCORD 389
51. Measurement of Oxidizing Radicals by Polymorphonuclear Leukocytes
DENNIS P. CLIFFORD AND JOHN E. REPINE 393
52. Antimicrobial Activity of Myeloperoxidase
SEYMOUR J. KLEBANOFF, ANN M. WALTERSDORPH, AND HENRY ROSEN 399
53. Determination of Microbial Damage Caused by Oxygen Free Radicals, and the Protective Role of Superoxide Dismutase HOSNI M. HASSAN 404

C. Microsomal Membranes

54. Oxidizing Radical Generation by Prostaglandin H Synthase
PAUL H. SIEDLIK AND LAWRENCE J. MARNETT 412
55. Assay of *in Situ* Radicals by Electron Spin Resonance RONALD P. MASON 416

D. Organelles

56. Chloroplasts: Formation of Active Oxygen and Its Scavenging KOZI ASADA 422
57. Determination of the Production of Superoxide Radicals and Hydrogen Peroxide in Mitochondria ALBERTO BOVERIS 429

58. Hydroperoxide Effects on Redox State of Pyridine Nucleotides and Ca^{2+} Retention by Mitochondria — CHRISTOPH RICHTER 435

Section IV. Pathology, Cancer, Aging

59. Glutathione Disulfide (GSSG) Efflux from Cells and Tissues — HELMUT SIES AND THEO P. M. AKERBOOM 445
60. Stable Tissue Free Radicals — W. LOHMANN AND H. NEUBACHER 451
61. Assay of Superoxide Dismutase Activity in Tumor Tissue — LARRY W. OBERLEY AND DOUGLAS R. SPITZ 457
62. *In Vitro* Cell Cultures as Tools in the Study of Free Radicals and Free Radical Modifiers in Carcinogenesis — CARMIA BOREK 464
63. Radicals in Melanin Biochemistry — ROGER C. SEALY 479
64. Assay of Lipofuscin/Ceroid Pigment *in Vivo* during Aging — R. S. SOHAL 484

Section V. Enzymes, Viral Activity, and Cell Viability as End Points for Study of Free Radical Damage

65. The Use of Selective Free Radical Probes to Study Active Sites in Enzymes and Viruses — J. LESLIE REDPATH 491

Section VI. Drugs: Environmental Induction of Radical Formation and Radical Species

66. Detection and Measurement of Drug-Induced Oxygen Radical Formation — MARTYN T. SMITH, HJÖRDIS THOR, AND STEN ORRENIUS 505
67. Alloxan and 6-Hydroxydopamine: Cellular Toxins — GERALD COHEN AND RICHARD E. HEIKKILA 510
68. Microsomal Oxidant Radical Production and Ethanol Oxidation — ARTHUR I. CEDERBAUM AND GERALD COHEN 516
69. Exacerbation of Superoxide Radical Formation by Paraquat — HOSNI M. HASSAN 523
70. Ethidium Binding Assay for Reactive Oxygen Species Generated from Reductively Activated Adriamycin (Doxorubicin) — J. WILLIAM LOWN 532
71. Chemical Carcinogenesis: Benzopyrene System — STEPHEN A. LESKO 539

VOLUME 106
POSTTRANSLATIONAL MODIFICATIONS (PART A)

Section I. General Aspects

1. Nomenclature and Symbolism of Amino Acids — WALDO E. COHN 3
2. Ion-Exchange Chromatography of Some Amino Acid Derivatives Found in Proteins — CHRISTOPHER C. Q. CHIN 17

3. High-Performance Liquid Chromatography Probes for Posttranslationally Modified Amino Acids KENNETH A. WALSH AND TATSURU SASAGAWA 22
4. Identification of Posttranslationally Modified Amino Acids in Proteins by Mass Spectrometry STEVEN A. CARR AND KLAUS BIEMANN 29
5. Complete Enzymatic Hydrolysis of Wool and Its Morphological Components KARIN RÖPER, J. FÖHLES, AND H. KLOSTERMEYER 58
6. Enzymatic Hydrolysis of Native, Soluble Proteins and Glycoproteins CHRISTOPHER C. Q. CHIN AND FINN WOLD 69

Section II. Nonenzymatic Modifications

7. Measurement of Nonenzymatic Protein Glycosylation RUDOLF FLÜCKIGER AND PAUL M. GALLOP 77
8. Nonenzymatic Glucosylation of Lysine Residues in Albumin JOHN W. BAYNES, SUZANNE R. THORPE, AND MARTHA H. MURTIASHAW 88
9. *In Vivo* Racemization in Mammalian Proteins JEFFREY L. BADA 98
10. Pyridoxal Phosphate-Binding Site in Enzymes; Reduction and Comparison of Sequences YOSHIMASA MORINO AND FUJIO NAGASHIMA 116

Section III. Modifications at the Level of Aminoacyl-tRNA

11. Properties and Specificity of Methionyl-$tRNA_f^{Met}$ Formyltransferase from *Escherichia coli* SYLVAIN BLANQUET, PHILIPPE DESSEN, AND DANIEL KAHN 141
12. Glu-$tRNA^{Gln}$: An Intermediate in Yeast Mitochondrial Protein Synthesis NANCY C. MARTIN AND MURRAY RABINOWITZ 152
13. Characterizing the Function of O^{β}-Phosphoseryl-tRNA THOMAS S. STEWART AND STEPHEN SHARP 157

Section IV. Covalent Modification of the α-Amino and α-Carboxyl Groups of Proteins

14. Amino-Terminal Acetylation of Proteins: An Overview SUSUMU TSUNASAWA AND FUMIO SAKIYAMA 165
15. Rat Pituitary N^{α}-Acetyltransferase JACK E. DIXON AND TERRY A. WOODFORD 170
16. Actin Amino-Terminal Acetylation and Processing in a Rabbit Reticulocyte Lysate KENT L. REDMAN AND PETER A. RUBENSTEIN 179
17. Amino-Terminal Protein Transacetylase from Wheat Germ HIROSHI KIDO, ALBERTO VITA, AND B. L HORECKER 193
18. Aminoacyl-tRNA: Protein Transferases CHARLES E. DEUTCH 198
19. Identification of N-Terminal Myristyl Blocking Groups in Proteins ALASTAIR AITKEN AND PHILIP COHEN 205
20. Detection of an N-Terminal Glucuronamide Linkage in Proteins P. E. KOLATTUKUDY 210
21. Occurrence, Detection, and Biosynthesis of Carboxy-Terminal Amides GÜNTHER KREIL 218
22. Tyrosine Incorporation in Tubulin MARTIN FLAVIN AND HIROMU MUROFUSHI 223
23. Ubiquitination of Proteins HARRIS BUSCH 238

Section V. Protein Alkylations/Dealkylations (Arylations)

24. A Simple Analysis of Methylated Proteins WOON KI PAIK 265
25. *S*-Adenosylmethionine: Protein (Arginine) *N*-Methyltransferase (Protein Methylase I) (Wheat Germ) MARTIN TUCK AND WOON KI PAIK 268
26. Enzymatic Methylation and Demethylation of Protein-Bound Lysine Residues WOON KI PAIK AND PETER DIMARIA 274
27. Methylated Lysines and 3-Methylhistidine in Myosin: Tissue and Developmental Differences GABOR HUSZAR 287
28. *S*-Adenosylmethionine:Protein-carboxyl *O*-Methyltransferase (Protein Methylase II) SANGDUK KIM 295
29. The Protein Carboxylmethyltransferase Involved in *Escherichia coli* and *Salmonella typhimurium* Chemotaxis JEFFREY B. STOCK, STEVEN CLARKE, AND DANIEL E. KOSHLAND, JR. 310
30. Carboxylmethyl Esterase of Bacterial Chemotaxis MARK A. SNYDER, JEFFREY B. STOCK, AND DANIEL E. KOSHLAND, JR. 321
31. Isolation of D-Aspartic Acid β-Methyl Ester from Erythrocyte Carboxyl Methylated Proteins STEVEN CLARKE, PHILIP N. MCFADDEN, CLARE M. O'CONNOR, AND LILLIAN L. LOU 330
32. Hypusine Formation: A Unique Posttranslational Modification in Translation Initiation Factor eIF-4D HERBERT L. COOPER, MYUNG HEE PARK, AND J. E. FOLK 344
33. Histidinoalanine: A Naturally Occurring Cross-Linking Amino Acid RONALD L. SASS AND MARY E. MARSH 351
34. S^{β}-(2-Histidyl)cysteine: Properties, Assay, and Occurrence KONRAD LERCH 355
35. S^{β}-(Bilin)cysteine Derivatives: Structures, Spectroscopic Properties, and Quantitation ALEXANDER N. GLAZER 359
36. N-Terminal Glyceride-Cysteine Modification of Membrane Penicillinases in Gram-Positive Bacteria J. OLIVER LAMPEN AND JENNIFER B. K. NIELSEN 365
37. Covalent Attachment of Flavin to Flavoproteins: Occurrence, Assay, and Synthesis THOMAS P. SINGER AND WILLIAM S. MCINTIRE 369
38. Diphthamide in Elongation Factor 2: ADP-Ribosylation, Purification, and Properties JAMES W. BODLEY, PATRICIA C. DUNLOP, AND BRIAN G. VANNESS 378
39. Diphthamide: *In Vitro* Biosynthesis J. M. MOEHRING AND T. J. MOEHRING 388

Section VI. Protein Glycosylations; ADP-Ribosylation

40. ADP-Ribosylarginine NORMAN J. OPPENHEIMER 399
41. Use of Guanylhydrazones as Substrates for Guanidine-Specific Mono-ADP-Ribosyltransferases GOPALAN SOMAN, JOYCE F. MILLER, AND DONALD J. GRAVES 403
42. Toxin ADP-Ribosyltransferases That Act on Adenylate Cyclase Systems JOEL MOSS AND MARTHA VAUGHAN 411
43. Coliphage-Induced ADP-Ribosylation of *Escherichia coli* RNA Polymerase C. G. GOFF 418
44. NAD:Arginine Mono-ADP-Ribosyltransferases from Animal Cells JOEL MOSS AND MARTHA VAUGHAN 430
45. Overview of Poly(ADP-ribosyl)ation MARK E. SMULSON AND TAKASHI SUGIMURA 438

46. Structure of Poly(ADP-ribose) MASANAO MIWA AND TAKASHI SUGIMURA 441
47. Evidence for Poly(ADP-ribosyl) Derivatives of Carboxylates in Histone
KUNIHIRO UEDA AND OSAMU HAYAISHI 450
48. Purification and Characterization of (ADP-Riobosyl)$_n$ Proteins
PETER ADAMIETZ AND HELMUTH HILZ 461
49. Quantification of Protein-Bound ADP-Ribosyl and (ADP-Ribosyl)$_n$ Residues
KLAUS WIELCKENS, REINHARD BREDEHORST, AND HELMUTH HILZ 472
50. Determination of *in Vivo* Levels of Polymeric and Monomeric ADP-Ribose by Fluorescence Methods
MYRON K. JACOBSON, D. MICHAEL PAYNE, RAFAEL ALVAREZ-GONZALES, HECTOR JUAREZ-SALINAS, JAMES L. SIMS, AND ELAINE L. JACOBSON 483
51. Quantification of *in Vivo* Levels of Poly(ADP-ribose): Tritium Labeling Method and Radioimmunoassay MASANAO MIWA AND TAKASHI SUGIMURA 495
52. Poly(ADP-ribose) Synthetase
KUNIHIRO UEDA, JINGYUAN ZHANG, AND OSAMU HAYAISHI 500
53. Poly(ADP-ribose) Glycohydrolase and Phosphodiesterases in the Analysis of Poly(ADP-ribosyl)ated Proteins
WILLIAM R. KIDWELL AND MICHAEL R. PURNELL 505
54. Poly(ADP-ribosyl)ation of Nucleosomal Chromatin: Electrophoretic and Immunofractionation Methods MARK E. SMULSON 512
55. Hydroxyproline Glycosides in the Plant Kingdom DEREK T. A. LAMPORT 523

VOLUME 107
POSTTRANSLATIONAL MODIFICATIONS (PART B)

Section I. Protein Acylations/Deacylations

1. Chemical Properties, Isolation, and Analysis of *O*-Phosphates in Proteins
TODD M. MARTENSEN 3
2. Techniques in the Detection and Characterization of Phosphoramidate-Containing Proteins JAMES M. FUJITAKI AND ROBERTS A. SMITH 23
3. Phosphorus-31 Nuclear Magnetic Resonance Studies of Phosphorylated Proteins
MANFRED BRAUER AND BRIAN D. SYKES 36
4. Protein Kinases PETER J. ROACH 81
5. Protein (Serine and Threonine) Phosphate Phosphatases
S. SHENOLIKAR AND T. S. INGEBRITSEN 102
6. Detection and Identification of Substrates for Protein Kinases: Use of Proteins and Synthetic Peptides
LORENTZ ENGSTRÖM, PIA EKMAN, ELISABET HUMBLE, ULF RAGNARSSON, AND ÖRJAN ZETTERQVIST 130
7. Assays for Regulatory Properties of Polyamine-Dependent Protein Kinase
VAITHILINGAM SEKAR, VALERIE J. ATMAR, AND GLENN D. KUEHN 154
8. UDP-*N*-acetylglucosamine:Lysosomal Enzyme *N*-Acetylglucosamine-1-phosphotransferase MARC L. REITMAN, LES LANG, AND STUART KORNFELD 163
9. O^{β}-(*N*-Acetyl-α-glucosamine-1-phosphoryl)serine in Proteinase I from *Dictyostelium discoideum* GARY L. GUSTAFSON AND JOHN E. GANDER 172

10. 5′-Nucleotidyl-*O*-tyrosine Bond in Glutamine Synthetase — SUE GOO RHEE 183
11. Determination and Occurrence of Tyrosine *O*-Sulfate in Proteins — WIELAND B. HUTTNER 200
12. Protein Side-Chain Acetylations — VINCENT G. ALLFREY, EUGENE A. DI PAOLA, AND RICHARD STERNER 224
13. The $N\varepsilon$-(γ-Glutamic)lysine Cross-Link: Method of Analysis, Occurrence in Extracellular and Cellular Proteins — ARIEL G. LOEWY 241
14. $N\varepsilon$-(β-Aspartyl)lysine — H. KLOSTERMEYER 258
15. Formation of $N\varepsilon$-(Biotinyl)lysine in Biotin Enzymes — NEIL H. GOSS AND HARLAND G. WOOD 261

Section II. Oxidations, Hydroxylations, and Halogenations

16. Formation and Isomerization of Disulfide Bonds in Proteins: Protein Disulfide-Isomerase — DAVID A. HILLSON, NIGEL LAMBERT, AND ROBERT B. FREEDMAN 281
17. Enzymatic Reduction–Oxidation of Protein Disulfides by Thioredoxin — ARNE HOLMGREN 295
18. Nonenzymatic Formation and Isomerization of Protein Disulfides — DONALD B. WETLAUFER 301
19. Disulfide Bond Formation in Proteins — THOMAS E. CREIGHTON 305
20. Redox Control of Enzyme Activities by Thiol/Disulfide Exchange — HIRAM F. GILBERT 330
21. Enzymatic Reduction of Methionine Sulfoxide Residues in Proteins and Peptides — NATHAN BROT, HENRY FLISS, TIMOTHY COLEMAN, AND HERBERT WEISSBACH 352
22. Prolyl Hydroxylase in Plants — MAARTEN J. CHRISPEELS 361
23. Mixed-Function Oxidation of Histidine Residues — RODNEY L. LEVINE 370
24. Dityrosine: *In Vitro* Production and Characterization — RENATO AMADÒ, ROBERT AESCHBACH, AND HANS NEUKOM 377
25. Isodityrosine, a Diphenyl Ether Cross-Link in Plant Cell Wall Glycoprotein: Identification, Assay, and Chemical Synthesis — STEPHEN C. FRY 388
26. Assay of Dihydroxyphenylalanine (Dopa) in Invertebrate Structural Proteins — J. HERBERT WAITE AND CHRISTINE V. BENEDICT 397
27. Halogenated Tyrosine Derivatives in Invertebrate Scleroproteins: Isolation and Identification — S. HUNT 413
28. Protein Bromination by Bromoperoxidase from *Penicillus capitatus* — J. A. MANTHEY, L. P. HAGER, AND K. D. MCELVANY 439
29. Iodination by Thyroid Peroxidase — JOSEPH T. NEARY, MORRIS SOODAK, AND FARAHE MALOOF 445
30. Thyroid Hormones: Mechanism of Phenoxy Ether Formation — J. NUNEZ 476
31. Iodotyrosine Deiodinase from Bovine Thyroid — ISADORE N. ROSENBERG AND AJIT GOSWAMI 488

Section III. Miscellaneous Derivatives

32. γ-Carboxyglutamic Acid: Assay Methods — GARY L. NELSESTUEN 503
33. Vitamin K-Dependent Plasma Proteins — GARY L. NELSESTUEN 507

34. Osteocalcin: Isolation, Characterization, and Detection
C. M. GUNDBERG, P. V. HAUSCHKA, J. B. LIAN, AND P. M. GALLOP 516
35. Specific Tritium Labeling of γ-Carboxyglutamic Acid Residues in Proteins
PAUL A. PRICE 544
36. Decarboxylation of γ-Carboxyglutamic Acid Residues in Proteins
PAUL A. PRICE 548
37. Glutamate Carboxylase: Assays, Occurrence, and Specificity
TOM BRODY AND J. W. SUTTIE 552
38. β-Carboxyaspartic Acid
TAD H. KOCH, M. ROBERT CHRISTY, ROBERT M. BARKLEY, RICHARD SLUSKI, DENISE BOHEMIER, JOHN J. VAN BUSKIRK, AND WOLFF M. KIRSCH 563
39. Occurrence and Characterization of Selenocysteine in Proteins
THRESSA C. STADTMAN 576
40. Identification of Selenocysteine by High-Performance Liquid Chromatography and Mass Spectrometry H. E. GANTHER, R. J. KRAUS, AND S. J. FOSTER 582
41. Oxidation States of Glutathione Peroxidase
H. E. GANTHER AND R. J. KRAUS 593
42. Selenocysteine-Containing Proteins and Glutathione Peroxidase
AL L. TAPPEL, WAYNE C. HAWKES, ERIC C. WILHELMSEN, AND MARVIN A. MOTSENBOCKER 602
43. Characterization of Selenomethionine in Proteins MARK X. SLIWKOWSKI 620
44. Citrulline in Proteins from the Enzymatic Deimination of Arginine Residues
J. A. ROTHNAGEL AND G. E. ROGERS 624

VOLUME 108

IMMUNOCHEMICAL TECHNIQUES (PART G: SEPARATION AND CHARACTERIZATION OF LYMPHOID CELLS)

Section I. Surgical Techniques in Immunology

1. Methods of Bursectomy BRUCE GLICK AND IMRE OLÁH 3
2. Thymectomy and Engraftment of Thymic Tissues PIERSON J. VAN ALTEN 10
3. Skin Grafting DAVID STEINMULLER 20
4. Graft-versus-Host Reactions DAVID STEINMULLER 28
5. Adoptive Transfer DAVID STEINMULLER 35

Section II. Methods for the Separation and Purification of Populations and Subpopulations of Lymphoreticular Cells

6. Preparation of Single-Cell Suspensions from Lymphoid Organs
NORMAN J. KLEIMAN, DANIEL L. FRIEDMAN, AND GIOVANNI DI SABATO 43
7. Rosetting Techniques to Detect Cell Surface Markers on Mouse and Human Lymphoreticular Cells BRUCE E. ELLIOTT AND HUGH F. PROSS 49
8. Separation of Cells by Velocity Sedimentation RICHARD G. MILLER 64
9. Separation of Lymphocytes, Granulocytes, and Monocytes from Human Blood Using Iodinated Density Gradient Media ARNE BØYUM 88

10. Analytical and Preparative Equilibrium Density Separation of Lymphoid Cells on Albumin and Metrizamide — KEN SHORTMAN 102
11. Separation of Lymphocytes on Antibody-Coated Plates — MICHAEL G. MAGE 118
12. Cell Separation on Cellular Monolayers — MICHAEL G. MAGE 125
13. Separation of Cells by Affinity Chromatography on Protein A Gels — VICTOR GHETIE AND JOHN SJÖQUIST 132
14. Separation of Lymphoid Cells Using Immunoadsorbent Affinity Chromatography — ROGER A. HUBBARD, SAMUEL F. SCHLUTER, AND JOHN J. MARCHALONIS 139
15. Use of Anti-Fab Columns for the Isolation of Human Lymphocyte Populations — PAUL L. ROMAIN AND STUART F. SCHLOSSMAN 148
16. Fractionation of Human Lymphocytes on *Helix pomatia* A Hemagglutinin-Sepharose and Wheat Germ Agglutinin-Sepharose — U. HELLSTRÖM, M.-L. HAMMARSTRÖM, S. HAMMARSTRÖM, AND P. PERLMANN 153
17. Fractionation of Subpopulations of Mouse and Human Lymphocytes by Peanut Agglutinin or Soybean Agglutinin — YAIR REISNER AND NATHAN SHARON 168
18. Electrophoretic Separation of Lymphoid Cells — ERNIL HANSEN AND KURT HANNIG 180
19. Fluorescence-Activated Cell Sorting: Theory, Experimental Optimization, and Applications in Lymphoid Cell Biology — D. R. PARKS AND L. A. HERZENBERG 197
20. Isolation of Human Mononuclear Leukocyte Subsets by Countercurrent Centrifugal Elutriation — HENRY C. STEVENSON 242
21. Use of Antisera and Complement for the Purification of Lymphocyte Subpopulations — FUNG-WIN SHEN 249
22. Suicide of Lymphoid Cells — A. BASTEN AND P. CRESWICK 254
23. Elimination of Specific Immunoreactive T Lymphocytes with 5-Bromodeoxyuridine and Light — ETHAN M. SHEVACH 262
24. Collection of Mouse Thoracic Duct Lymphocytes — ROBERT KORNGOLD AND JACK R. BENNINK 270
25. Methods for the Collection of Peritoneal and Alveolar Macrophages — RICHARD M. MCCARRON, DIANA K. GOROFF, JORDAN E. LUHR, MARY A. MURPHY, AND HERBERT B. HERSCOWITZ 274
26. Separation and Collection of Rat Liver Macrophages — JUSTINE S. GARVEY AND MATTHEW F. HEIL 285
27. Separation of Macrophages on Plastic and Glass Surfaces — D. E. MOSIER 294
28. Separation of Lymphoid Cells on Nylon Wool Columns — DAVID A. LITVIN AND DAVID L. ROSENSTREICH 298
29. Use of Sephadex G-10 to Separate Macrophages and Lymphocytes — ROBERT I. MISHELL AND BARBARA B. MISHELL 303
30. Depletion of Macrophages from Heterogeneous Cell Populations by the Use of Carbonyl Iron — DENNIS M. WONG AND LUIGI VARESIO 307
31. Preparation and Use of Monoclonal Antimacrophage Antibodies — MAY-KIN HO AND TIMOTHY A. SPRINGER 313
32. Elimination of Macrophages with Silica and Asbestos — ELLIOTT KAGAN AND DAN-PAUL HARTMANN 325
33. Use of Horseradish Peroxidase and Fluorescent Dextrans to Study Fluid Pinocytosis in Leukocytes — JANET M. OLIVER, RICHARD D. BERLIN, AND BRUCE H. DAVIS 336

34. Use of the Nude Mouse in Immunological Research
BEPPINO C. GIOVANELLA AND JAMES T. HUNTER 348
35. Isolation of Epidermal Langerhans Cells
VERA B. MORHENN AND EVA A. PFENDT 357
36. Cryopreservation of Lymphoid Cells J. DIXON GRAY AND S. H. GOLUB 363

Section III. Methods for the Study of Surface Immunoglobulin

37. Ligand-Induced Patching and Capping of Surface Immunoglobulins F. LOOR 371
38. Direct and Mixed Antiglobulin Rosetting Reaction DAVID G. HAEGERT 386
39. Ultrastructural Localization of Surface and Intracellular Immunoglobulin in Human Lymphoid Cell Suspensions D. G. NEWELL 393
40. Use of Protein A in the Detection and Quantitation of Immunoglobulin G on the Surface of Lymphocytes VICTOR GHETIE AND JOHN SJÖQUIST 405
41. Use of Fluorescent Antibodies in the Study of Lymphoid Cell Membrane Molecules LUCIANA FORNI AND STEFANELLO DE PETRIS 413
42. Biosynthetic, Surface Labeling, and Isolation of Membrane Immunoglobulin
DOROTHY YUAN AND ELLEN S. VITETTA 426

Section IV. Methods for the Detection, Purification, and Biochemical Characterization of Lymphoid Cell Surface Antigens

43. Molecules Encoded with the Murine Major Histocompatibility Complex: An Overview W. LEE MALOY AND JOHN E. COLIGAN 437
44. Analysis of H-2 and Ia Antigens by Two-Dimensional Polyacrylamide Gel Electrophoresis PATRICIA P. JONES 452
45. Isolation of Antigenic Proteins from Lymphocytes by Binding Immune Complexes to Staphylococcal Protein A SUSAN E. CULLEN 467
46. Use of Monoclonal Antibodies in the Study of the Fine Specificity of Antigens Coded for by the Major Histocompatibility Complex in Mice
LORRAINE FLAHERTY AND MICHAEL A. LYNES 481
47. β_2-Microglobulin ETTORE APPELLA AND JANET A. SAWICKI 494
48. Structural Analysis of Murine Class I Major Histocompatibility Complex Antigens RODERICK NAIRN 505
49. Isolation and Analysis of the Murine Ia Molecular Complex
SUSAN E. CULLEN 523
50. Immunochemical Purification and Analysis of Qa and TL Antigens
MARK J. SOLOSKI AND ELLEN S. VITETTA 549
51. Purification and Molecular Cloning of Rat Ia Antigens
W. ROBERT MCMASTER 558
52. Purification of Human HLA-A and HLA-B Class I Histocompatibility Antigens
JOSÉ A. LÓPEZ DE CASTRO 582
53. Use of Monoclonal Antibody Immunoaffinity Columns to Purify Subsets of Human HLA-DR Antigens
DAVID W. ANDREWS, M. ROSA BONO, JAMES F. KAUFMAN, PETER KNUDSEN, AND JACK L. STROMINGER 600

54. Use of the HLA-DR Antigens Incorporated into Liposomes to Generate HLA-DR Specific Cytotoxic T Lymphocytes
JOAN C. GORGA, JUDITH FORAN, STEVEN J. BURAKOFF, AND JACK L. STROMINGER 607
55. HLA Antigens in Serum M. A. PELLEGRINO, C. RUSSO, AND J. P. ALLISON 614
56. Identification and Characterization of Human T Lymphocyte Antigens
PAUL L. ROMAIN, ORESTE ACUTO, AND STUART F. SCHLOSSMAN 624
57. Thy-1.1 and Thy-1.2 Alloantigens: An Overview MICHELLE LETARTE 642
58. Isolation and Characterization of the Thy-1 Glycoproteins of Rat Brain and Thymus MICHELLE LETARTE 654
59. Lyt-1, Lyt-2, and Lyt-3 Antigens
PAUL D. GOTTLIEB, EDWARD B. REILLY, PAUL J. DURDA, AND MALGORZATA NIEZGODKA 666
60. Detection by Immunochemical Techniques of Cell Surface Markers on Epidermal Langerhans Cells PAUL R. BERGSTRESSER AND DOLORES V. JUAREZ 683

VOLUME 109
HORMONE ACTION (PART I: PEPTIDE HORMONES)

Section I. Receptor Assays

1. Assay for the Glucagon Receptor FRANCISCO J. ROJAS AND L. BIRNBAUMER 1
2. Wheat Germ Lectin-Sepharose Affinity Adsorption Assay for the Soluble Glucagon Receptor RAVI IYENGAR AND JOHN T. HERBERG 13
3. Assaying Binding of Nerve Growth Factor to Cell Surface Receptors
RONALD D. VALE AND ERIC M. SHOOTER 21
4. Assays for Calcitonin Receptors
ANNE P. TEITELBAUM, ROBERT A. NISSENSON, AND CLAUDE D. ARNAUD 40
5. Assay for Parathyroid Hormone Receptors
ROBERT A. NISSENSON, ANNE P. TEITELBAUM, AND CLAUDE D. ARNAUD 48
6. Gastrin Receptor Assay LEONARD R. JOHNSON 56
7. Assay for the Cholecystokinin Receptor on Pancreatic Acini
ROBERT T. JENSEN AND JERRY D. GARDNER 64
8. Methods for Studying the Platelet-Derived Growth Factor Receptor
DANIEL F. BOWEN-POPE AND RUSSELL ROSS 69
9. Binding Assays for Epidermal Growth Factor GRAHAM CARPENTER 101
10. Radioligand Assay for Angiotensin II Receptors
HARTMUT GLOSSMANN, ALBERT BAUKAL, GRETI AGUILERA, AND KEVIN J. CATT 110

Section II. Identification of Receptor Proteins

11. General Principles for Photoaffinity Labeling of Peptide Hormone Receptors
ALEX N. EBERLE AND PIERRE N. E. DE GRAAN 129
12. Photoaffinity Labeling of Prolactin Receptors
PAUL A. KELLY, MASAO KATOH, JEAN DJIANE, AND SENKITI SAKAI 156

13. Photoaffinity Labeling of the Insulin Receptor
CECIL C. YIP AND CLEMENT W. T. YEUNG 170
14. Affinity Cross-Linking of Receptors for Insulin and the Insulin-Like Growth Factors I and II JOAN MASSAGUÉ AND MICHAEL P. CZECH 179
15. The Preparation of Biologically Active Monomeric Ferritin–Insulin and Its Use as a High Resolution Electron Microscopic Marker of Occupied Insulin Receptors ROBERT M. SMITH AND LEONARD JARETT 187
16. Cross-Linking of hCG to Luteal Receptors TAE H. JI AND INHAE JI 203
17. Covalent Labeling of the Hepatic Glucagon Receptor
JOHN T. HERBERG AND RAVI IYENGAR 207

Section III. Methods for Identification of Internalized Hormones and Hormone Receptors

18. Peptide Hormone Receptors in Intracellular Structures from Rat Liver
BARRY I. POSNER, MASOOD N. KHAN, AND JOHN J. M. BERGERON 219
19. Assessment of Receptor Recycling in Mammalian Hepatocytes: Perspectives Based on Current Techniques JOE HARFORD AND GILBERT ASHWELL 232
20. Preparation of Low-Density "Endosome" and "Endosome"-Depleted Golgi Fractions from Rat Liver W. H. EVANS 246
21. Isolation of Receptosomes (Endosomes) from Human KB Cells
ROBERT B. DICKSON, JOHN A. HANOVER, IRA PASTAN, AND MARK C. WILLINGHAM 257

Section IV. Preparation of Hormonally Responsive Cells and Cell Hybrids

22. Preparation of Isolated Leydig Cells
F. F. G. ROMMERTS, R. MOLENAAR, AND H. J. VAN DER MOLEN 275
23. Preparation of Dispersed Pancreatic Acinar Cells and Dispersed Pancreatic Acini
JERRY D. GARDNER AND ROBERT T. JENSEN 288
24. The Preparation, Culture, and Incubation of Rat Anterior Pituitary Cells for Static and Dynamic Studies of Secretion
PAULINE R. M. DOBSON AND BARRY L. BROWN 293
25. Isolation and Functional Aspects of Free Luteal Cells
JUDITH L. LUBORSKY AND HAROLD R. BEHRMAN 298
26. Culture and Characteristics of Hormone-Responsive Neuroblastoma × Glioma Hybrid Cells
BERND HAMPRECHT, THOMAS GLASER, GEORG REISER, ERNST BAYER, AND FRIEDRICH PROPST 316
27. Primary Glial Cultures as a Model for Studying Hormone Action
BERND HAMPRECHT AND FRIDOLIN LÖFFLER 341
28. Study of Receptor Function by Membrane Fusion: The Glucagon Receptor in Liver Membranes Fused to a Foreign Adenylate Cyclase
SONIA STEINER AND MICHAEL SCHRAMM 346
29. Isolation of ACTH-Resistant Y1 Adrenal Tumor Cells
BERNARD P. SCHIMMER 350

30. Induction of Glucagon Responsiveness in Transformed MDCK Cells Unresponsive to Glucagon
SUZANNE K. BECKNER, FREDERICK J. DARFLER, AND MICHAEL C. LIN 356
31. Characterization of Hormone-Sensitive Madin–Darby Canine Kidney Cells
MICHAEL C. LIN, SUZANNE K. BECKNER, AND FREDERICK J. DARFLER 360
32. Construction of Hormonally Responsive Intact Cell Hybrids by Cell Fusion: Transfer of β-Adrenergic Receptor and Nucleotide Regulatory Protein(s) in Normal and Desensitized Cells D. SCHULSTER AND D. M. SALMON 366
33. Establishment and Characterization of Fibroblast-Like Cell Lines Derived from Adipocytes with the Capacity to Redifferentiate into Adipocyte-Like Cells
R. NEGREL, P. GRIMALDI, C. FOREST, AND G. AILHAUD 377
34. Establishment of Rat Fetal Liver Lines and Characterization of Their Metabolic and Hormonal Properties: Use of Temperature-Sensitive SV40 Virus
JANICE YANG CHOU 385

Section V. Purification of Membrane Receptors and Related Techniques

35. Purification of Insulin Receptor from Human Placenta
STEVEN JACOBS AND PEDRO CUATRECASAS 399
36. Purification of the LDL Receptor
WOLFGANG J. SCHNEIDER, JOSEPH L. GOLDSTEIN, AND MICHAEL S. BROWN 405
37. Synthesis of Biotinyl Derivatives of Peptide Hormones and Other Biological Materials FRANCES M. FINN AND KLAUS H. HOFMANN 418
38. Purification of N_S and N_i, the Coupling Proteins of Hormone-Sensitive Adenylyl Cyclases without Intervention of Activating Regulatory Ligands
JUAN CODINA, WALTER ROSENTHAL, JOHN D. HILDEBRANDT, LUTZ BIRNBAUMER, AND RONALD D. SEKURA 446

Section VI. Assays of Hormonal Effects and Related Functions

39. Analysis of Hormone-Induced Changes of Phosphoinositide Metabolism in Rat Liver M. A. WALLACE AND J. N. FAIN 469
40. Effect of Bradykinin on Prostaglandin Production by Human Skin Fibroblasts in Culture
CAROLE L. JELSEMA, JOEL MOSS, AND VINCENT C. MANGANIELLO 480
41. Direct Chemical Measurement of Receptor-Mediated Changes in Phosphatidylinositol Levels in Isolated Rat Liver Plasma Membranes
JOSEPH EICHBERG AND CHARLES A. HARRINGTON 504
42. Assay for Calcium Channels HARTMUT GLOSSMANN AND DAVID R. FERRY 513
43. Assessment of Effects of Vasopressin, Angiotensin II, and Glucagon on Ca^{2+} Fluxes and Phosphorylase Activity in Liver
P. F. BLACKMORE AND J. H. EXTON 550
44. Pertussis Toxin: A Tool for Studying the Regulation of Adenylate Cyclase
RONALD D. SEKURA 558
45. ADP-Ribosylation of Membrane Components by Pertussis and Cholera Toxin
FERNANDO A. P. RIBEIRO-NETO, RAFAEL MATTERA, JOHN D. HILDEBRANDT, JUAN CODINA, JAMES B. FIELD, LUTZ BIRNBAUMER, AND RONALD D. SEKURA 566

46. Measurement of mRNA Concentration of mRNA Half-Life as a Function of Hormonal Treatment
JOHN R. RODGERS, MARK L. JOHNSON, AND JEFFREY M. ROSEN 572
47. Radioactive Labeling and Turnover Studies of the Insulin Receptor Subunits
JOSE A. HEDO AND C. RONALD KAHN 593
48. Phosphorylation of the Insulin Receptor in Cultured Hepatoma Cells and a Solubilized System
MASATO KASUGA, MORRIS F. WHITE, AND C. RONALD KAHN 609

Section VII. Antibodies in Hormone Action

49. Development of Monoclonal Antibodies against Parathyroid Hormone: Genetic Control of the Immune Response to Human PTH
SAMUEL R. NUSSBAUM, C. SHIRLEY LIN, JOHN T. POTTS, JR., ALAN S. ROSENTHAL, AND MICHAEL ROSENBLATT 625
50. Monoclonal Antibodies to Gonadotropin Subunits
PAUL H. EHRLICH, WILLIAM R. MOYLE, AND ROBERT E. CANFIELD 638
51. Assay of Antibodies Directed against Cell Surface Receptors
SIMEON I. TAYLOR, LISA H. UNDERHILL, AND BERNICE MARCUS-SAMUELS 656
52. Characterization of Antisera to Prolactin Receptors
PAUL A. KELLY, MASAO KATOH, JEAN DJIANE, LOUIS-MARIE HOUDEBINE, AND ISABELLE DUSANTER-FOURT 667
53. Assays of Thyroid-Stimulating Antibody J. M. MCKENZIE AND M. ZAKARIJA 677
54. A Monoclonal Antibody to Growth Hormone Receptors
J. S. A. SIMPSON AND H. G. FRIESEN 692
55. Site Specific Monoclonal Antibodies to Insulin
TIMOTHY P. BENDER AND JOYCE A. SCHROER 704

Section VIII. General Methods

56. Chemical Deglycosylation of Glycoprotein Hormones
P. MANJUNATH AND M. R. SAIRAM 725
57. Disasssembly and Assembly of Glycoprotein Hormones
THOMAS F. PARSONS, THOMAS W. STRICKLAND, AND JOHN G. PIERCE 736
58. Purification of Human Platelet-Derived Growth Factor
ELAINE W. RAINES AND RUSSELL ROSS 749
59. Isolation of Rat Somatomedin WILLIAM H. DAUGHADAY AND IDA K. MARIZ 773
60. Purification of Human Insulin-Like Growth Factors I and II
PETER P. ZUMSTEIN AND RENÉ E. HUMBEL 782
61. Purification of Somatomedin-C/Insulin-Like Growth Factor I
MARJORIE E. SVOBODA AND JUDSON J. VAN WYK 798
62. 5′-*p*-Fluorosulfonylbenzoyl Adenosine as a Probe of ATP-Binding Sites in Hormone Receptor-Associated Kinases
SUSAN A. BUHROW AND JAMES V. STAROS 816
63. A Radioimmunoassay for Cyclic GMP with Femtomole Sensitivity Using Tritiated Label and Acetylated Ligands
PAULINE R. M. DOBSON AND PHILIP G. STRANGE 827

VOLUME 110
STEROIDS AND ISOPRENOIDS (PART A)

Section I. Early Stages in Isoprenoid Biosynthesis

1. Rat Liver Acetoacetyl-CoA Synthetase
JAMES D. BERGSTROM AND JOHN EDMOND 3
2. Use of Oxygenated Sterols to Probe the Regulation of 3-Hydroxy-3-methylglutaryl-CoA Reductase and Sterologenesis
FREDERICK R. TAYLOR AND ANDREW A. KANDUTSCH 9
3. 3-Hydroxy-3-methylglutaryl-CoA Synthase from Chicken Liver
HENRY M. MIZIORKO 19
4. 3-Hydroxy-3-methylglutaryl-CoA Reductases from Pea Seedlings
DAVID WARWICK RUSSELL 26
5. 3-Hydroxy-3-methylglutaryl-CoA Reductase in the Latex of *Hevea brasiliensis*
ABDULLAH SIPAT 40
6. Insect 3-Hydroxy-3-methylglutaryl-CoA Reductase DANIEL J. MONGER 51
7. Radioenzymatic Assay of Plasma Mevalonate
THOMAS S. PARKER, RON R. KOPITO, AND HENRI BRUNENGRABER 58
8. Mevalonate Kinase JOHN W. PORTER 71
9. Phosphomevalonate Kinase from Pig Liver
JAIME EYZAGUIRRE AND SERGIO BAZAES 78
10. Mevalonate 5-Pyrophosphate Decarboxylase from Chicken Liver
EMILIO CARDEMIL AND ANA MARÍA JABALQUINTO 86
11. Isopentenyldiphosphate Δ-Isomerase DENNIS M. SATTERWHITE 92
12. Shunt Pathway of Mevalonate Metabolism
BERNARD R. LANDAU AND HENRI BRUNENGRABER 100

Section II. Linear Condensations of Isoprenoids

13. Photolabile Analogs of the Allylic Pyrophosphate Substrate of Prenyltransferases
CHARLES M. ALLEN AND TSUNEO BABA 117
14. Photoaffinity Substrate Analogs for Eukaryotic Prenyltransferase
HANS C. RILLING 125
15. Synthesis of Allylic and Homoallylic Isoprenoid Pyrophosphates
V. JO DAVISSON, A. B. WOODSIDE, AND C. DALE POULTER 130
16. Eukaryotic Prenyltransferases HANS C. RILLING 145
17. Enzymatic Hydrolysis of Polyprenyl Pyrophosphates
TANETOSHI KOYAMA, HIROSHI FUJII, AND KYOZO OGURA 153
18. Prenyltransferase from Human Liver GRAHAM F. BARNARD 155
19. Prenyltransferases of Pumpkin Fruit
KYOZO OGURA, TOKUZO NISHINO, TOSHIHIRO SHINKA, AND SHUICHI SETO 167
20. Purification of Farnesylpyrophosphate Synthetase by Affinity Chromatography
DESIREE L. BARTLETT, CHI-HSIN RICHARD KING, AND C. DALE POULTER 171
21. Geranylgeranylpyrophosphate Synthetase of Pig Liver
HIROSHI SAGAMI, KOICHI ISHII, AND KYOZO OGURA 184

22. Geranylpyrophosphate Synthetase–Geranylgeranylpyrophosphate Synthetase from *Micrococcus luteus* HIROSHI SAGAMI AND KYOZO OGURA 188
23. Hexaprenylpyrophosphate Synthetase of *Micrococcus luteus* B-P 26 HIROSHI FUJII, TANETOSHI KOYAMA, AND KYOZO OGURA 192
24. Heptaprenylpyrophosphate Synthetase from *Bacillus subtilis* IKUKO SAGAMI, HIROSHI FUJII, TANETOSHI KOYAMA, AND KYOZO OGURA 199
25. Nonaprenylpyrophosphate Synthetase from *Micrococcus luteus* HIROSHI SAGAMI AND KYOZO OGURA 206
26. Enzymatic Synthesis of Phytoene BENJAMIN L. JONES AND JOHN W. PORTER 209
27. Carotene Mutants of *Phycomyces* ENRIQUE CERDÁ-OLMEDO 220
28. Carotene Synthesis in *Capsicum* Chromoplasts BILAL CAMARA 244
29. Carotenoid Synthesis in *Neurospora crassa* WERNER RAU AND URSULA MITZKA-SCHNABEL 253
30. Carotene Synthesis in Spinach (*Spinacia oleracea* L.) Chloroplasts and Daffodil (*Narcissus pseudonarcissus* L.) Chromoplasts HANS KLEINIG AND PETER BEYER 267
31. Prenylation of Chlorophyllide *a* in *Capsicum* Plastids BILAL CAMARA 274
32. Purification and Characterization of Undecaprenylpyrophosphate Synthetase CHARLES M. ALLEN 281
33. *In Vitro* Synthesis of C_{15}–C_{60} Polyprenols in a Cell-Free System of *Myxococcus fulvus* and Determination of Chain Length by High-Performance Liquid Chromatography PETER BEYER AND HANS KLEINIG 299
34. Xanthophyll Cycles HARRY Y. YAMAMOTO 303
35. Biosynthesis of Plastoquinone J. F. PENNOCK 313
36. Aspulvinone Dimethylallyltransferase IKUKO SAGAMI, NOBUTOSHI OJIMA, KYOZO OGURA, AND SHUICHI SETO 320
37. 4-Hydroxybenzoate Polyprenyltransferase from Rat Liver ARUN GUPTA AND HARRY RUDNEY 327
38. Dimethylallylpyrophosphate:L-Tryptophan Dimethylallyltransferase HANS C. RILLING 335
39. Assay and Partial Purification of the Cytokinin Biosynthetic Enzyme Dimethylallylpyrophosphate:5′-AMP Transferase N. G. HOMMES, D. E. AKIYOSHI, AND R. O. MORRIS 340
40. Isolation of Cytokinins by Immunoaffinity Chromatography and Analysis by High-Performance Liquid Chromatography Radioimmunoassay E. M. S. MACDONALD AND R. O. MORRIS 347
41. Squalene Synthetase WILLIAM S. AGNEW 359
42. Formation of 12-*cis*-Dehydrosqualene Catalyzed by Squalene Synthetase TOKUZO NISHINO AND HIROHIKO KATSUKI 373
43. Squalene Epoxidase from Rat Liver Microsomes TERUO ONO AND YOH IMAI 375

Section III. Cyclization Reactions

44. Monoterpene and Sesquiterpene Cyclases RODNEY CROTEAU AND DAVID E. CANE 383
45. Carbocyclases from *Citrus limonum* OSVALDO CORI AND MARÍA CECILIA ROJAS 406
46. Isolation and Identification of Diterpenes from Termite Soldiers GLENN D. PRESTWICH 417

VOLUME 111
STEROIDS AND ISOPRENOIDS (PART B)

Section I. Methodology

1. A Comparison of Methods for the Identification of Sterols — WILLIAM R. NES — 3
2. High-Performance Liquid Chromatography of Sterols: Yeast Sterols — RUSSELL J. RODRIGUEZ AND LEO W. PARKS — 37
3. High-Performance Liquid Chromatography of Bile Acids — WILLIAM H. ELLIOTT AND ROGER SHAW — 51
4. Mass Spectrometry of Bile Acids — JAN SJÖVALL, ALEXANDER M. LAWSON, AND KENNETH D. R. SETCHELL — 63
5. General Carotenoid Methods — GEORGE BRITTON — 113
6. Gas Chromatography of Isoprenoids — WALTER JENNINGS AND GARY TAKEOKA — 149
7. High-Performance Liquid Chromatography of Carotenoids — MANFRED RUDDAT AND OSCAR H. WILL, III — 189
8. Isolation and Assay of Dolichol and Dolichyl Phosphate — W. LEE ADAIR AND R. KENNEDY KELLER — 201
9. Invertebrate Carotenoproteins — P. F. ZAGALSKY — 216
10. Separation of Mevalonate Phosphates and Isopentenyl Pyrophosphate by Thin-Layer Chromatography and of Short-Chain Prenyl Phosphates by Ion-Pair Chromatography on a High-Performance Liquid Chromatography Column — PETER BEYER, KLAUS KREUZ, AND H. KLEINIG — 248
11. Reversed-Phase High-Performance Liquid Chromatography of C_5 to C_{20} Isoprenoid Benzoates and Naphthoates — TSAN-HSI YANG, T. MARK ZABRISKIE, AND C. DALE POULTER — 252

Section II. Sterol Metabolism

12. Lecithin–Cholesterol Acyltransferase and Cholesterol Transport — CHRISTOPHER J. FIELDING — 267
13. Assay of Cholesteryl Ester Transfer Activity and Purification of a Cholesteryl Ester Transfer Protein — YASUSHI OGAWA AND CHRISTOPHER J. FIELDING — 274
14. Cholesterol Acyltransferase — JEFFREY T. BILLHEIMER — 286
15. Sterol Carrier Protein — MARY E. DEMPSEY — 293
16. Epoxide Hydrolases in the Catabolism of Sterols and Isoprenoids — BRUCE D. HAMMOCK, DAVID E. MOODY, AND ALEX SEVANIAN — 303
17. Biosynthesis and Interconversion of Sterols in Plants and Marine Invertebrates — L. J. GOAD — 311
18. Yeast Sterols: Yeast Mutants as Tools for the Study of Sterol Metabolism — LEO W. PARKS, CYNTHIA D. K. BOTTEMA, RUSSELL J. RODRIGUEZ, AND THOMAS A. LEWIS — 333
19. Enzymatic Dealkylation of Phytosterols in Insects — YOSHINORI FUJIMOTO, MASUO MORISAKI, AND NOBUO IKEKAWA — 346
20. Side-Chain Cleavage of Cholesterol by Gas Chromatography–Mass Spectrometry in a Selected Ion Monitoring Mode — MASUO MORISAKI, MIKIO SHIKITA, AND NOBUO IKEKAWA — 352

21. Purification for Rabbit and Rat Liver of Cytochromes *P*-450 Involved in Bile Acid Biosynthesis
STEFAN ANDERSSON, HANS BOSTRÖM, HENRY DANIELSSON, AND KJELL WIKVALL 364
22. Biosynthesis and Metabolism of Ecdysteroids and Methods of Isolation and Identification of the Free and Conjugated Compounds
H. H. REES AND R. E. ISAAC 377
23. Ecdysone Conjugates: Isolation and Identification
C. HETRU, B. LUU, AND J. A. HOFFMANN 411
24. Ecdysone Oxidase J. KOOLMAN 419
25. Ecdysteroid Carrier Proteins J. KOOLMAN 429
26. Ecdysone 3-Epimerase
MALCOLM J. THOMPSON, GUNTER F. WEIRICH, AND JAMES A. SVOBODA 437
27. 20-Hydroxyecdysone Binding Protein from Locust Hemolymph
R. FEYEREISEN 442
28. Ecdysone 20-Monooxygenase GUNTER F. WEIRICH 454
29. Measurement and Characterization of Ecdysteroid Receptors
BECKY A. SAGE AND JOHN D. O'CONNOR 458

Section III. Metabolism of Other Isoprenoids

30. Dolichol Kinase, Phosphatase, and Esterase Activity in Calf Brain
CARLOTA SUMBILLA AND CHARLES J. WAECHTER 471
31. Purification and Properties of the Juvenile Hormone Carrier Protein from the Hemolymph of *Manduca sexta* RONALD C. PETERSON 482
32. Analysis of Juvenile Hormone Esterase Activity
BRUCE D. HAMMOCK AND RICHARD M. ROE 487
33. Cellular Juvenile Hormone Binding Proteins ERNEST S. CHANG 494
34. Experimental Techniques for Photoaffinity Labeling of Juvenile Hormone Binding Proteins of Insects with Epoxyfarnesyl Diazoacetate
GLENN D. PRESTWICH, JOHN K. KOEPPE, GAE E. KOVALICK, JOHN J. BROWN, ERNEST S. CHANG, AND AMBARISH K. SINGH 509
35. Radiochemical Assay for Juvenile Hormone III Biosynthesis *in Vitro*
R. FEYEREISEN 530
36. Epoxyfarnesoic Acid Methyltransferase GUNTER F. WEIRICH 540
37. *S*-Adenosylmethionine:γ-Tocopherol Methyltransferase (*Capsicum* Chromoplasts) BILAL CAMARA 544
38. Dolichyl Pyrophosphate Phosphatase in Brain
MALKA G. SCHER AND CHARLES J. WAECHTER 547

VOLUME 112
DRUG AND ENZYME TARGETING (PART A)

Section I. Microencapsulation Techniques

1. Albumin Microspheres: Heat and Chemical Stabilization
ANTHONY F. YAPEL, JR. 3

2. Hydrophilic Albumin Microspheres
WILLIAM E. LONGO AND EUGENE P. GOLDBERG 18
3. Incorporation of Water-Soluble Drugs in Albumin Microspheres
E. TOMLINSON AND J. J. BURGER 27
4. Technetium-99m Labeling of Albumin Microspheres Intended for Drug Targeting
J. J. BURGER, E. TOMLINSON, J. E. DE ROO, AND J. PALMER 43
5. Biophysical Drug Targeting: Magnetically Responsive Albumin Microspheres
ANDREW E. SENYEI, CHARLES F. DRISCOLL, AND KENNETH J. WIDDER 56
6. Attachment of Monoclonal Antibodies to Microspheres
LISBETH ILLUM AND PHILIP D. E. JONES 67
7. Nylon-Encapsulated Pharmaceuticals
JAMES W. MCGINITY AND GEORGE W. CUFF 84
8. Poly(lactide-*co*-glycolide) Microcapsules for Controlled Release of Steroids
DONALD R. COWSAR, THOMAS R. TICE, RICHARD M. GILLEY, AND
JAMES P. ENGLISH 101
9. Crystallized Carbohydrate Spheres for Slow Release and Targeting
ULF SCHRÖDER 116
10. Poly(alkyl acrylate) Nanoparticles JÖRG KREUTER 129
11. Ethylcellulose Microcapsules for Selective Drug Delivery
TETSURO KATO, KATSUO UNNO, AND AKIO GOTO 139
12. Polyacrolein Microspheres: Preparation and Characteristics
M. CHANG, G. RICHARDS, AND A. REMBAUM 150
13. Polyacrolein Microspheres SHLOMO MARGEL 164
14. Poly(vinylpyridine) Microspheres A. SCHWARTZ AND A. REMBAUM 175
15. Artificial Cells Containing Multienzyme Systems THOMAS MING SWI CHANG 195

Section II. Drug Conjugates

16. Preparation of Antibody–Toxin Conjugates
A. J. CUMBER, J. A. FORRESTER, B. M. J. FOXWELL, W. C. J. ROSS, AND
P. E. THORPE 207
17. Protein Conjugates of Fungal Toxins HEINZ FAULSTICH AND LUIGI FIUME 225
18. Transferrin Receptor as a Target for Antibody–Drug Conjugates
DERRICK L. DOMINGO AND IAN S. TROWBRIDGE 238
19. Lectins as Carriers: Preparation and Purification of a Concanavalin A–Trypsin Conjugate W. THOMAS SHIER 248
20. Hormone–Drug Conjugates J. M. VARGA 259
21. Polylysine–Drug Conjugates LYLE J. ARNOLD, JR. 270
22. Dextran and Inulin Conjugates as Drug Carriers LUIGI MOLTENI 285
23. Carrier Potential of Glycoproteins JUDITH L. BODMER AND ROGER T. DEAN 298

Section III. Prodrugs

24. Theory and Practice of Prodrug Kinetics ROBERT E. NOTARI 309
25. Prodrug Lability Prediction through the Use of Substituent Effects
M. CHARTON 323
26. Alteration of Drug Metabolism by the Use of Prodrugs SIDNEY D. NELSON 340
27. Formation of Prodrugs of Amines, Amides, Ureides, and Imides
HANS BUNDGAARD 347

28. Design of Prodrugs for Improved Gastrointestinal Absorption by Intestinal Enzyme Targeting
DAVID FLEISHER, BARBRA H. STEWART, AND GORDON L. AMIDON 360
29. Targeting of Drugs to the Brain NICHOLAS BODOR 381

Section IV. Polymer Systems

30. Controlled Release and Magnetically Modulated Release Systems for Macromolecules ROBERT LANGER, LARRY BROWN, AND ELAZER EDELMAN 399
31. Poly(ortho ester) Biodegradable Polymer Systems
J. HELLER AND K. J. HIMMELSTEIN 422
32. Poly(lactic/glycolic acid) Biodegradable Drug–Polymer Matrix Systems
JUDITH P. KITCHELL AND DONALD L. WISE 436
33. Transdermal Dosage Forms JANE E. SHAW 448
34. Microsealed Drug Delivery Systems: Fabrications and Performance
YIE W. CHIEN 461
35. Osmotic Delivery Systems for Research JOHN W. FARA 470
36. Enzymatically Controlled Drug Release Systems
B. D. RATNER AND T. A. HORBETT 484
37. Membrane Systems: Theoretical Aspects
K. L. SMITH AND H. K. LONSDALE 495
38. Membrane Systems: Practical Applications K. L. SMITH 504
39. Implantable Infusion Pumps: Clinical Applications PERRY J. BLACKSHEAR 520
40. Implantable Infusion Pumps: Practical Aspects
PERRY J. BLACKSHEAR, BRUCE D. WIGNESS, ANNE M. ROUSSELL, AND ALFRED M. COHEN 530

VOLUME 113
GLUTAMATE, GLUTAMINE, GLUTATHIONE, AND RELATED COMPOUNDS

Section I. Glutamate

1. L-Glutamate Decarboxylase from Brain
JANG-YEN WU, LARRY DENNER, CHIN-TARNG LIN, AND GUOXING SONG 3
2. L-Glutamate Decarboxylase from Bacteria MARGARET L. FONDA 11
3. L-Glutamate Dehydrogenase from Bovine Liver HARVEY F. FISHER 16
4. *N*-Acetylglutamate Synthase SUSAN G. POWERS-LEE 27
5. *N*-Methyl-L-glutamate Synthase LOUIS B. HERSH 36
6. Glutamyl-tRNA Synthetase from *Escherichia coli*
JACQUES LAPOINTE, SYLVAIN LEVASSEUR, AND DANIEL KERN 42
7. Purification of Glutamyl-tRNA Synthetase from *Bacillus subtilis*
MARIO PROULX AND JACQUES LAPOINTE 50
8. Glutaminyl-tRNA Synthetase of *Escherichia coli*
PATRICIA HOBEN AND DIETER SÖLL 55

9. Preparation of Optically Pure Isotopically Labeled L-Glutamate
TINGSEN XU AND ALTON MEISTER 59
10. L-Glutamate-L-Amino Acid Transaminases ARTHUR J. L. COOPER 63
11. Glutamate-Aspartate Transaminase ARTHUR J. L. COOPER 66
12. Glutamate-Alanine Transaminase ARTHUR J. L. COOPER 69
13. Glutamate-Branched-Chain Amino Acid Transaminase ARTHUR J. L. COOPER 71
14. Glutamate-Aromatic Amino Acid Transaminase ARTHUR J. L. COOPER 73
15. Glutamate-Ornithine Transaminases ARTHUR J. L. COOPER 76
16. Glutamate γ-Aminobutyrate Transaminase ARTHUR J. L. COOPER 80
17. Glutamate-Aspartate Transaminase from Microorganisms
TOSHIHARU YAGI, HIROYUKI KAGAMIYAMA, MITSUHIRO NOZAKI, AND KENJI SODA 83
18. L-Kynurenine Transaminase from *Hansenula schneggii*
KATSUYUKI TANIZAWA, YASUHIKO ASADA, AND KENJI SODA 90
19. L-Lysine Transaminase from *Flavobacterium lutescens*
KATSUYUKI TANIZAWA, TOHRU YOSHIMURA, AND KENJI SODA 96
20. Taurine-Glutamate Transaminase
KAZUO YONAHA, SEIZEN TOYAMA, AND KENJI SODA 102
21. D-Glutamate-D-Amino Acid Transaminase from Bacteria
WANDA M. JONES, THOMAS S. SOPER, HIROSHI UENO, AND JAMES M. MANNING 108
22. Δ^1-Pyrroline-5-carboxylate Synthase from Rat Intestinal Mucosa
JAMES J. KRAMER, JERRY G. HENSLEE, YASUO WAKABAYASHI, AND MARY ELLEN JONES 113
23. β-Methylaspartate-Glutamate Mutase from *Clostridium tetanomorphum*
H. A. BARKER 121
24. γ-Carboxyglutamic Acid
JANE B. LIAN, CAREN M. GUNDBERG, PETER V. HAUSCHKA, AND PAUL M. GALLOP 133
25. Capsular Poly-γ-D-glutamate Synthesis in *Bacillus licheniformis*
FREDERIC A. TROY, II 146
26. Synthesis of Folylpolyglutamates LINDA D'ARI AND JESSE C. RABINOWITZ 169

Section II. Glutamine

27. Glutamine Synthetase from Mammalian Tissues ALTON MEISTER 185
28. Glutamine Synthetase from Muscle W. BRUCE ROWE 199
29. Glutamine Synthetase from *Escherichia coli*
SUE GOO RHEE, P. BOON CHOCK, AND EARL R. STADTMAN 213
30. Glutaminase from Mammalian Tissues
ELLING KVAMME, INGEBORG AA. TORGNER, AND GERD SVENNEBY 241
31. Glutaminase from *Acinetobacter glutaminsificans* JOHN S. HOLCENBERG 257
32. Glutamine Amidotransferases HOWARD ZALKIN 263
33. Amidophosphoribosyltransferase HOWARD ZALKIN 264
34. GMP Synthetase HOWARD ZALKIN 273
35. Glucosamine-6-Phosphate Synthase HOWARD ZALKIN 278
36. CTP Synthetase HOWARD ZALKIN 282
37. Anthranilate Synthase HOWARD ZALKIN 287
38. *p*-Aminobenzoate Synthase HOWARD ZALKIN 293

39. NAD Synthetase — HOWARD ZALKIN 297
40. Glu-tRNA^{Gln} Amidotransferase — HOWARD ZALKIN 303
41. Carbamyl Phosphate Synthetase (Glutamine-Utilizing) from *Escherichia coli* — DEBORAH S. KASEMAN AND ALTON MEISTER 305
42. Glutamate Synthase from *Escherichia coli*, *Klebsiella aerogenes*, and *Saccharomyces cerevisiae* — ALTON MEISTER 327
43. Glutamine Transaminase L from Rat Liver — ARTHUR J. L. COOPER AND ALTON MEISTER 338
44. Glutamine Transaminase K from Rat Kidney — ARTHUR J. L. COOPER AND ALTON MEISTER 344
45. α-Keto Acid ω-Amidase from Rat Liver — ARTHUR J. L. COOPER, THOMAS E. DUFFY, AND ALTON MEISTER 350
46. Transglutaminases — J. E. FOLK AND SOO IL CHUNG 358

Section III. Glutathione

47. Glutathione Biosynthesis; γ-Glutamylcysteine Synthetase from Rat Kidney — GAIL FOURÉ SEELIG AND ALTON MEISTER 379
48. γ-Glutamylcysteine Synthetase from Erythrocytes — GAIL FOURÉ SEELIG AND ALTON MEISTER 390
49. Glutathione Synthetase from Rat Kidney — ALTON MEISTER 393
50. γ-Glutamyl Transpeptidase from Kidney — SURESH S. TATE AND ALTON MEISTER 400
51. γ-Glutamyl Transpeptidase: Kinetics and Mechanism — R. DONALD ALLISON 419
52. γ-Glutamylcyclotransferase from Rat Kidney — ALTON MEISTER 438
53. 5-Oxo-L-prolinase from Rat Kidney — ALTON MEISTER, OWEN W. GRIFFITH, AND JOANNE M. WILLIAMSON 445
54. 5-Oxo-L-prolinase from *Pseudomonas putida* — ANDREW P. SEDDON, LUYUAN LI, AND ALTON MEISTER 451
55. Synthesis of L-2-Oxothiazolidine-4-carboxylic Acid — BRIAN BOETTCHER AND ALTON MEISTER 458
56. Formation of Respiratory $^{14}CO_2$ from Radiolabeled Substrates — OWEN W. GRIFFITH 461
57. Metabolism *in Vivo* of 5-Oxo-L-proline and Its Inhibition by Analogs of 5-Oxo-L-proline — TINGSEN XU AND ALTON MEISTER 468
58. Microvillus Membrane Peptidases That Catalyze Hydrolysis of Cysteinylglycine and Its Derivatives — SURESH S. TATE 471
59. Glutathione Reductase — INGER CARLBERG AND BENGT MANNERVIK 484
60. Glutathione Peroxidase — BENGT MANNERVIK 490
61. Glutathione Transferases: An Overview — WILLIAM B. JAKOBY 495
62. Glutathione Transferase from Human Liver — MARGARETA WARHOLM, CLAES GUTHENBERG, CHRISTER VON BAHR, AND BENGT MANNERVIK 499
63. Glutathione Transferase Isoenzymes from Rat Liver Cytosol — HELGI JENSSON, PER ÅLIN, AND BENGT MANNERVIK 504
64. Glutathione Transferase from Rat Testis — CLAES GUTHENBERG, PER ÅLIN, AND BENGT MANNERVIK 507
65. Cysteine Conjugate β-Lyase — JAMES L. STEVENS AND WILLIAM B. JAKOBY 510
66. Cysteine *S*-Conjugate *N*-Acetyltransferase — MICHAEL W. DUFFEL AND WILLIAM B. JAKOBY 516

67. Thioltransferase from Human Placenta
KERSTIN LARSON, VÉRONIQUE ERIKSSON, AND BENGT MANNERVIK 520
68. Glutaredoxin from *Escherichia coli* and Calf Thymus ARNE HOLMGREN 525
69. Thiol:Protein Disulfide Exchange Enzymes
JOHN E. MORIN AND JACK E. DIXON 541
70. Determination of Glutathione and Glutathione Disulfide in Biological Samples
MARY E. ANDERSON 548
71. Preparation of γ-Glutamyl Amino Acids by Chemical and Enzymatic Methods
MARY E. ANDERSON AND ALTON MEISTER 555
72. Separation of γ-Glutamyl Amino Acids by Ion-Exchange Chromatography
DANIEL WELLNER 564
73. Enzymatic Synthesis of Isotopically Labeled Glutathione by Means of an Immobilized Cell Matrix
RICHARD J. BRIDGES, KOUSAKU MURATA, WILLIAM ABBOTT, AND ALTON MEISTER 567
74. Methods for the Selective Modification of Glutathione Metabolism and Study of Glutathione Transport ALTON MEISTER 571

Section IV. Aspartate and Asparagine

75. L-Aspartate α-Decarboxylase JOANNE M. WILLIAMSON 589
76. Aspartate Kinases I, II, and III from *Escherichia coli* GEORGES N. COHEN 596
77. Aspartate-semialdehyde Dehydrogenase from *Escherichia coli*
GEORGES N. COHEN 600
78. Asparagine Transaminase from Rat Liver
ARTHUR J. L. COOPER AND ALTON MEISTER 602
79. Asparaginase JOHN C. WRISTON, JR. 608
80. Aspartate Ammonia-lyase MASANOBU TOKUSHIGE 618
81. Aspartate Transcarbamylase from *Bacillus subtilis*
JOHN S. BRABSON, MICHAEL R. MAURIZI, AND ROBERT L. SWITZER 627

Section V. α-Aminoadipate

82. α-Aminoadipate RICHARD A. HARTLINE 639
83. Kynurenine Aminotransferase from Kidney Supernatant
RICHARD A. HARTLINE 664

VOLUME 114

DIFFRACTION METHODS FOR BIOLOGICAL MACROMOLECULES (PART A)

Section I. Introduction

1. Early Days of Protein Crystallography MAX PERUTZ 3
2. Introduction to Crystallography ROBERT M. SWEET 19

Section II. Crystallization and Treatment of Crystals

3. Theory of Protein Solubility TSUTOMU ARAKAWA AND SERGE N. TIMASHEFF 49
4. Nucleation and Growth of Protein Crystals: General Principles and Assays G. FEHER AND Z. KAM 77
5. Crystallization of Macromolecules: General Principles ALEXANDER MCPHERSON 112
6. Use of Polyethylene Glycol in the Crystallization of Macromolecules ALEXANDER MCPHERSON 120
7. Crystallization of Proteins by Variation of pH or Temperature ALEXANDER MCPHERSON 125
8. Crystallization in Capillary Tubes GEORGE N. PHILLIPS, JR. 128
9. Seed Enlargement and Repeated Seeding C. THALLER, G. EICHELE, L. H. WEAVER, E. WILSON, R. KARLSSON, AND J. N. JANSONIUS 132
10. Treatment and Manipulation of Crystals IVAN RAYMENT 136
11. Protein Crystallization by Free Interface Diffusion F. R. SALEMME 140
12. Flow Cell Construction and Use GREGORY A. PETSKO 141
13. Preparation of Isomorphous Heavy-Atom Derivatives GREGORY A. PETSKO 147
14. Heavy Metal Ion–Nucleic Acid Interaction SUNG-HOU KIM, WHAN-CHUL SHIN, AND R. W. WARRANT 156
15. Crystallization and Heavy-Atom Derivatives of Polynucleotides STEPHEN R. HOLBROOK AND SUNG-HOU KIM 167
16. Determination of Protein Molecular Weight, Hydration, and Packing from Crystal Density B. W. MATTHEWS 176
17. Crystal Density Measurements Using Aqueous Ficoll Solutions EDWIN M. WESTBROOK 187

Section III. Data Collection

A. Photographic Techniques

18. Photographic Science and Microdensitometry in X-Ray Diffraction Data Collection MICHAEL ELDER 199
19. Oscillation Method with Large Unit Cells STEPHEN C. HARRISON, FRITZ K. WINKLER, CLARENCE E. SCHUTT, AND RICHARD DURBIN 211
20. Determining the Intensity of Bragg Reflections from Oscillation Photographs MICHAEL G. ROSSMANN 237
21. A Rotation Camera Used with a Synchrotron Radiation Source R. FOURME AND R. KAHN 281
22. X-Ray Sources WALTER C. PHILLIPS 300
23. A Systematic Method for Aligning Double-Focusing Mirrors WALTER C. PHILLIPS AND IVAN RAYMENT 316

B. Diffractometry

24. Diffractometry HAROLD W. WYCKOFF 330
25. Measuring X-Ray Diffraction Data from Large Proteins with X-Ray Diffractometry ROBERT J. FLETTERICK AND JURGEN SYGUSCH 386

26. On the Design of Diffractometers to Measure a Number of Reflections Simultaneously P. J. ARTYMIUK AND D. C. PHILLIPS 397
27. Multiwire Area X-Ray Diffractometers RONALD HAMLIN 416
28. Software for a Diffractometer with Multiwire Area Detector A. J. HOWARD, C. NIELSEN, AND NG H. XUONG 452
29. Television Area Detector Diffractometers U. W. ARNDT 472
30. A General Purpose, Computer-Configurable Television Area Detector for X-Ray Diffraction Applications KENNETH KALATA 486
31. Experimental Neutron Protein Crystallography B. P. SCHOENBORN 510
32. Automated Peak Fitting Procedure for Processing Data from an Area Detector and Its Application to the Neutron Structure of Trypsin ANTHONY A. KOSSIAKOFF AND STEVEN A. SPENCER 530
33. Neutron Diffraction: A Facility for Data Collection and Processing at the National Bureau of Standards Reactor ALEXANDER WLODAWER 551

VOLUME 115
DIFFRACTION METHODS FOR BIOLOGICAL MACROMOLECULES (PART B)

Section I. Primary Phasing

1. Overview of Phasing by Isomorphous Replacement KEITH D. WATENPAUGH 3
2. Least-Squares MIR Calculations: Minimum Variance Fourier Coefficient (MVFC) Refinement J. SYGUSCH 15
3. Least-Squares Refinement ROBERT A. SPARKS 23
4. Direct Phase Determination Based on Anomalous Scattering WAYNE A. HENDRICKSON, JANET L. SMITH, AND STEVEN SHERIFF 41
5. Use of the Rotation and Translation Functions EATON LATTMAN 55
6. Phase Refinement/Extension by Density Modification A. TULINSKY 77
7. Resolution of Phase Ambiguity in Macromolecular Crystallography BI-CHENG WANG 90
8. Free Atom Insertion and Refinement as a Means of Extending and Refining Phases N. W. ISAACS AND R. C. AGARWAL 112
9. On the Derivation of Dynamic Information from Diffraction Data D. I. STUART AND D. C. PHILLIPS 117

Section II. Modeling

10. Optical Matching of Physical Models and Electron Density Maps: Early Developments FREDERIC M. RICHARDS 145
11. Some Minor Refinements on the Richards Optical Comparator and Methods for Model Coordinate Measurement F. R. SALEMME 154
12. Interactive Computer Graphics: FRODO T. ALWYN JONES 157
13. Rationalization of Molecular Models JAN HERMANS 171
14. Interpretation of Electron Density Maps JANE S. RICHARDSON AND DAVID C. RICHARDSON 189

15. Computer Skeletonization and Automatic Electron Density Map Analysis
JONATHAN GREER 206

Section III. Combined Model and Phase Refinement

16. Overview of Refinement in Macromolecular Structure Analysis
LYLE H. JENSEN 227
17. Classic (F_o–F_c) Fourier Refinement STEPHAN T. FREER 235
18. Real Space Refinement R. DIAMOND 237
19. Stereochemically Restrained Refinement of Macromolecular Structures
WAYNE A. HENDRICKSON 252
20. Constrained–Restrained Least-Squares (CORELS) Refinement of Proteins and Nucleic Acids JOEL L. SUSSMAN 271
21. Experience with Various Techniques for the Refinement of Protein Structures
J. DEISENHOFER, S. J. REMINGTON, AND W. STEIGEMANN 303
22. Fast Fourier Transform Calculation of Electron Density Maps
LYNN F. TEN EYCK 324

Section IV. Presentation and Analysis of Structure

23. Describing Patterns of Protein Tertiary Structure JANE S. RICHARDSON 341
24. Schematic Drawings of Protein Structures JANE S. RICHARDSON 359
25. Computer-Generated Pictures of Proteins
ARTHUR M. LESK AND KARL D. HARDMAN 381
26. Macromolecule Backbone Models BYRON RUBIN 391
27. Comparison of Protein Structures
BRIAN W. MATTHEWS AND MICHAEL G. ROSSMANN 397
28. Domains in Proteins: Definitions, Location, and Structural Principles
JOËL JANIN AND CYRUS CHOTHIA 420
29. Automatic Recognition of Domains in Globular Proteins GEORGE D. ROSE 430
30. Calculation of Molecular Volumes and Areas for Structures of Known Geometry
FREDERIC M. RICHARDS 440

VOLUME 116

IMMUNOCHEMICAL TECHNIQUES (PART H: EFFECTORS AND MEDIATORS OF LYMPHOID CELL FUNCTIONS)

Section I. Purification and Characterization of Serum Immunoglobulins

1. Immunoglobulin G (IgG)
PETER D. GOREVIC, FRANCES C. PRELLI, AND BLAS FRANGIONE 3
2. Immunoglobulin M (IgM) DAVID J. HAYZER AND JEAN-CLAUDE JATON 26
3. Immunoglobulin A (IgA) JIRI MESTECKY AND MOGENS KILIAN 37
4. Immunoglobulin E (IgE) KIMISHIGE ISHIZAKA 76
5. Immunoglobulin D (IgD) HANS L. SPIEGELBERG 95
6. Light Chains of Human Immunoglobulins ALAN SOLOMON 101

7. Plasmacytomas and Murine Immunoglobulins
SANDRA J. SMITH-GILL, FRED D. FINKELMAN, AND MICHAEL POTTER 121
8. Antiisotypic Antibodies DAVID M. SPALDING AND WILLIAM J. KOOPMAN 146
9. Antiallotypic Antibodies SUSAN JACKSON AND THOMAS J. KINDT 157
10. Antiidiotypic Antibodies DAVID E. BRILES AND JOHN F. KEARNEY 174
11. Crystallization of Immunoglobulins and Their Fragments for X-Ray Diffraction Studies ROBERTO J. POLJAK 190
12. Methods for the Study of the Metabolism of Immunoglobulins
THOMAS A. WALDMANN 201

Section II. Preparation, Purification, and Characterization of Thymic Hormones and Peptides of Thymic Origin

13. Thymic Hormones: An Overview
TERESA L. K. LOW AND ALLAN L. GOLDSTEIN 213
14. Thymosin fraction 5 and 5A TERESA L. K. LOW AND ALLAN L. GOLDSTEIN 219
15. Thymosin α_1 and Polypeptide β_1
TERESA L. K. LOW AND ALLAN L. GOLDSTEIN 233
16. Thymosin β_4 TERESA L. K. LOW AND ALLAN L. GOLDSTEIN 248
17. Prothymosin α and α_1-like Peptides
A. A. HARITOS, GREGORY J. GOODALL, AND B. L. HORECKER 255
18. Thymosin β_4-like Peptides
B. L. HORECKER, SUSAN ERICKSON-VIITANEN, AND EWALD HANNAPPEL 265
19. Thymulin
JEAN-MARIE PLEAU, LOUIS NOEL GASTINEL, AND JEAN-FRANÇOIS BACH 269
20. Thymopoietin and Ubiquitin TAPAN AUDHYA AND GIDEON GOLDSTEIN 279

Section III. Preparation, Purification, and Characterization of Antigen-Specific Lymphokines

21. Antigen-Specific Factors: An Overview
D. R. WEBB, J. A. KAPP, AND C. W. PIERCE 295
22. Antigen-Specific Suppressor Factors from Hybridoma Cell Lines
JUDITH A. KAPP, CRAIG M. SORENSEN, AND CARL W. PIERCE 303
23. Antigen-Specific Suppressor T Cells and Their Soluble Products
MASARU TANIGUCHI, MASAMOTO KANNO, AND TAKASHI SAITO 311
24. GAT Antigen-Specific Suppressor Factors
K. KRUPEN, C. TURCK, S. STEIN, J. A. KAPP, AND D. R. WEBB 325
25. Antigen-Specific Helper Factor to Poly(Tyr,Glu)-poly(DLAla)-poly(Lys), TGAL
MICHAEL J. TAUSSIG 340

Section IV. Preparation, Purification, and Characterization of Antigen-Nonspecific Lymphokines

26. Antigen-Nonspecific Lymphokines: An Overview JOOST J. OPPENHEIM 357
27. B Cell Growth and Differentiation Factors
KATHRYN BROOKS, RAFAEL FERNANDEZ, AND ELLEN S. VITETTA 372
28. Migration Inhibitory Factor HEINZ G. REMOLD AND ALMA D. MEDNIS 379
29. Soluble Immune Response Suppressor (SIRS)
THOMAS M. AUNE, H. WILLIAM SCHNAPER, AND CARL W. PIERCE 395

30. IgG-Binding Factors
W. H. FRIDMAN, M. J. GELABERT, M. DAËRON, J. MONCUIT, I. LÖWY, J. THÈZE, AND C. NÉAUPORT-SAUTÈS 403
31. A Soluble T Cell Suppressor Factor for the Mixed Leukocyte Response (MLR-TsF) MARGARET BECKWITH, CHRIS ARHELGER, AND SUSAN RICH 416
32. Allogeneic Effect Factor DAVID H. KATZ AND RIAZ I. ZUBERI 428
33. Human Lymphotoxin BHARAT B. AGGARWAL 441
34. Human Tumor Necrosis Factor
BHARAT B. AGGARWAL AND WILLIAM J. KOHR 448
35. Human Interleukin 1 IGAL GERY AND JOHN A. SCHMIDT 456
36. Interleukin 1 from Human Leukemic Monocytes
LAWRENCE B. LACHMAN, LI-CHEN NANCY SHIH, AND DALE C. BROWN 467
37. Murine Interleukin 1 MITCHELL DUKOVICH AND STEVEN B. MIZEL 480
38. Human Interleukin 2 RICHARD J. ROBB 493
39. Murine Interleukin 2
VERNER PAETKAU, DENIS RIENDEAU, AND R. CHRIS BLEACKLEY 526
40. Interleukin 3
JAMES N. IHLE, YACOB WEINSTEIN, JONATHAN KELLER, LOUIS HENDERSON, AND EDMUND PALASZYNSKI 540
41. Murine Mast Cell Growth Factor YEE-PANG YUNG AND KARL WELTE 553
42. The Macrophage Colony-Stimulating Factor, CSF-1 E. RICHARD STANLEY 564
43. Murine Granulocyte-Macrophage Colony-Stimulating Factor
ANTONY W. BURGESS AND EDOUARD C. NICE 588
44. Granulocyte Colony-Stimulating Factor NICOS A. NICOLA 600

VOLUME 117
ENZYME STRUCTURE (PART J)

Section I. Size, Shape, and Polydispersity of Macromolecules

1. Characterization of Macromolecules by Sedimentation Equilibrium in the Air-Turbine Ultracentrifuge ROBERT J. POLLET 3
2. Measurement of Changes of Hydrodynamic Properties by Sedimentation
ROBERT W. OBERFELDER, THOMAS G. CONSLER, AND JAMES C. LEE 27
3. Determination of Protein Molecular Weight in Complexes with Detergent without Knowledge of Binding
JACQUELINE A. REYNOLDS AND DARRELL R. MCCASLIN 41
4. Calculation of Partial Specific Volumes of Proteins in 8 *M* Urea Solution
V. PRAKASH AND SERGE N. TIMASHEFF 53
5. Calculation of the Partial Specific Volume of Proteins in Concentrated Salt and Amino Acid Solutions T. ARAKAWA AND SERGE N. TIMASHEFF 60
6. Molecular Weight Determinations from Radiation Inactivation
J. T. HARMON, T. B. NIELSEN, AND E. S. KEMPNER 65
7. Measurement of Molecular Weights of Elongated Molecules by Viscoelastometry
BRUNO H. ZIMM 94
8. Measurement of Hydrodynamic Properties of Active Enzyme by Sedimentation
LYNDAL K. HESTERBERG AND JAMES C. LEE 97

9. Computer-Controlled Scanning Gel Chromatography
LAWRENCE C. DAVIS, THEODORE J. SOCOLOFSKY, AND GARY A. RADKE 116
10. Hydrodynamic Characterization of Random Coil Polymers by Size Exclusion Chromatography PHIL G. SQUIRE 142
11. Structural Interpretation of Hydrodynamic Measurements of Proteins in Solution through Correlations with X-Ray Data
THOMAS F. KUMOSINSKI AND HELMUT PESSEN 154
12. The Use of Covolume in the Estimation of Protein Axial Ratios
LAWRENCE W. NICHOL AND DONALD J. WINZOR 182
13. Application of Transient Electric Birefringence to the Study of Biopolymer Structure PAUL J. HAGERMAN 198
14. Measurements of Protein Hydration by Various Techniques
HELMUT PESSEN AND THOMAS F. KUMOSINSKI 219
15. Correlation Function Profile Analysis of Polydisperse Macromolecular Solutions and Colloidal Suspensions B. CHU, J. R. FORD, AND H. S. DHADWAL 256

Section II. Interaction of Macromolecules with Ligands and Linkages

16. Nonlinear Least-Squares Analysis
MICHAEL L. JOHNSON AND SUSAN G. FRASIER 301
17. Ligand Binding to Proteins by Equilibrium Gel Penetration
HIROSHI KIDO, ALBERTO VITA, AND B. L. HORECKER 342
18. Measurement of Protein–Ligand Interactions by Gel Chromatography
JOSÉ MANUEL ANDREU 346
19. Ultrafiltration in Ligand-Binding Studies
ALKIS J. SOPHIANOPOULOS AND JUDITH A. SOPHIANOPOULOS 354
20. Protein Chromatography on Hydroxyapatite Columns
MARINA J. GORBUNOFF 370
21. Measurement of Ligand–Protein Interaction by Electrophoretic and Spectroscopic Techniques ROBERT W. OBERFELDER AND JAMES C. LEE 381
22. Measurement of Ligand Binding to Proteins by Fluorescence Spectroscopy
LARRY D. WARD 400
23. Measurement of Metal–Ligand Distances by EXAFS ROBERT A. SCOTT 414
24. Velocity Sedimentation Study of Ligand-Induced Protein Self-Association
GEORGE C. NA AND SERGE N. TIMASHEFF 459
25. Measurement and Analysis of Ligand-Binding Isotherms Linked to Protein Self-Associations GEORGE C. NA AND SERGE N. TIMASHEFF 496

VOLUME 118
PLANT MOLECULAR BIOLOGY

Section I. Cell Wall and Membrane

1. Isolation and Characterization of Plant Cell Walls and Cell Wall Components
WILLIAM S. YORK, ALAN G. DARVILL, MICHAEL MCNEIL, THOMAS T. STEVENSON, AND PETER ALBERSHEIM 3
2. Isolation of the Plasma Membrane THOMAS K. HODGES AND DAVID MILLS 41

Section II. Nucleus

3. Purification and Restriction Endonuclease Analysis of Plant Nuclear DNA
JOHN C. WATSON AND WILLIAM F. THOMPSON 57
4. Analyzing Genome Variation in Plants CAROL RIVIN 75
5. Synchronization of Plant Cells in Culture and in Meristems by Aphidicolin
FRANCESCO SALA, MARIA GRAZIA GALLI, GUIDO PEDRALI-NOY, AND SILVIO SPADARI 87
6. Isolation and Characterization of DNA Polymerase α from Spinach
CHERYL WARD AND ARTHUR WEISSBACH 97

Section III. Cytoplasmic Protein Synthesis

7. Purification and Properties of Protein Synthesis Initiation and Elongation Factors from Wheat Germ
SANDRA R. LAX, STEPHEN J. LAUER, KAREN S. BROWNING, AND JOANNE M. RAVEL 109
8. The Wheat Germ Protein Synthesis System
S. N. SEAL, A. SCHMIDT, AND A. MARCUS 128
9. Purification and Characterization of Polypeptide Chain Elongation Factor 1 from Plants SHIN-ICHIRO EJIRI 140
10. The Synthesis of High-Molecular-Weight Proteins in the Wheat Germ Translation System
M. D. MORCH, G. DRUGEON, W. ZAGORSKI, AND A. L. HAENNI 154

Section IV. The Chloroplast

A. Genome Structure

11. Isolation and Structural Analysis of Chloroplast DNA JEFFREY D. PALMER 167
12. Purification and Properties of Chloroplast DNA Polymerase
KRISHNA K. TEWARI 186

B. Transcription–Translation

13. Improved Preparative Methods for Isolation and Purification of Tobacco Chloroplast Ribosomes, Ribosomal Proteins, and rRNA
N. GAIL HEWLETT AND DON P. BOURQUE 201
14. Isolation and Characterization of Chloroplast and Cytoplasmic Transfer RNAs
ANDRÉ STEINMETZ AND JACQUES-HENRY WEIL 212
15. *In Vitro* Transcription of Chloroplast Protein Genes
EMIL M. OROZCO, JR., JOHN E. MULLET, LINDA HANLEY-BOWDOIN, AND NAM-HAI CHUA 232
16. Chloroplast Gene Expression and Promoter Identification in Chloroplast Extracts
WILHELM GRUISSEM, BRUCE M. GREENBERG, GERARD ZURAWSKI, AND RICHARD B. HALLICK 253
17. Coupled Transcription–Translation in Chloroplast Lysates
J. D. J. BARD, D. P. BOURQUE, AND D. ZAITLIN 270

18. Translation by Isolated Pea Chloroplasts
HELEN T. NIVISON, LEONARD E. FISH, AND ANDRE T. JAGENDORF 282
19. Purification of Chloroplast Elongation Factors
ORSOLA TIBONI AND ORIO CIFERRI 296
20. Use of an *in Vitro* Dipeptide System to Determine the Translation Initiation Sites of Chloroplast Genes
MARK BLOOM, NATHAN BROT, BENNETT N. COHEN, AND HERBERT WEISSBACH 309

C. Photosynthetic Systems

21. Isolation and Characterization of Chloroplast Envelope Membranes
KENNETH KEEGSTRA AND ABDULLAH E. YOUSIF 316
22. Characterization of the Thylakoid Membrane by Subfractionation Analyses
BERTIL ANDERSSON 325
23. Isolation of Chlorophyll-Binding Proteins of Green Plants
RICHARD S. WILLIAMS, ELIZABETH K. SHAW, LESLIE E. SIEBURTH, AND JOHN BENNETT 338
24. Subunit Structure and Biogenesis of ATP Synthase and Photosystem I Reaction Center NATHAN NELSON 352
25. Identification of cDNA Clones Representing Phytochrome and Other Low Abundance Red-Light Regulated Sequences
HOWARD P. HERSHEY AND PETER H. QUAIL 369
26. Identification and Characterization of the *psbA* Gene Product: The 32-kDa Chloroplast Membrane Protein
JONATHAN B. MARDER, AUTAR K. MATTOO, AND MARVIN EDELMAN 384

D. RuBP Carboxylase

27. Cloning and Expression of Genes for the Small Subunit of Ribulosebisphosphate Carboxylase C. F. WIMPEE AND E. M. TOBIN 396
28. Separation and Reassembly of the Subunits of Ribulosebisphosphate Carboxylase T. JOHN ANDREWS 410
29. The Cloning and Expression in *Escherichia coli* of RuBP Carboxylase/Oxygenase Large Subunit Genes
C. R. SOMERVILLE, L. MCINTOSH, J. FITCHEN, AND M. GUREVITZ 419

Section V. Mitochondria

30. The Isolation of Mitochondria and Mitochondrial DNA
MAUREEN R. HANSON, MAURY L. BOESHORE, PHILLIP E. MCCLEAN, MARY A. O'CONNELL, AND HELEN T. NIVISON 437
31. Analysis of the Genome Structure of Plant Mitochondria
DAVID M. LONSDALE, TONY P. HODGE, AND PETER J. STOEHR 453
32. Strategies for the Identification and Analysis of Higher Plant Mitochondrial Genes
ADAM J. DAWSON, VALERIE P. JONES, AND CHRISTOPHER J. LEAVER 470
33. Preparation of Maize Mitochondrial DNA–Protein Complex for Electron Microscopy C. S. LEVINGS III AND R. H. MEINTS 485

34. Isolation of Plant Mitochondrial RNA
DAVID B. STERN AND KATHLEEN J. NEWTON 488
35. Isolation and Characterization of Single-Stranded and Double-Stranded RNAs in Mitochondria ANNE M. SCHUSTER AND PAUL H. SISCO 497

Section VI. Nitrogen Metabolism

36. Isolation and Characterization of Nitrogenase from *Klebsiella pneumoniae*
VINOD K. SHAH 511
37. Tn5 Mapping of Rhizobium Nitrogen Fixation Genes G. DITTA 519
38. H_2 Uptake Negative (Hup^-) Mutants of *Rhizobium* and Their Use in the Isolation of *hup* Genes R. J. MAIER AND S. S. M. HOM 528

Section VII. Cell Culture and Transformation

39. Establishment of Calli and Suspension Cultures ROBERTA H. SMITH 539
40. Protoplasts: Isolation, Culture, Plant Regeneration
INGO POTRYKUS AND RAYMOND D. SHILLITO 549
41. Fusion and Transformation of Plant Protoplasts
J. B. POWER, M. R. DAVEY, J. P. FREEMAN, B. J. MULLIGAN, AND E. C. COCKING 578
42. Organelle Transfer ESRA GALUN AND DVORA AVIV 595
43. Induction and Selection of Chloroplast-Coded Mutations in *Nicotiana*
ROBERT FLUHR AND AGNES CSÉPLO 611

Section VIII. Gene Transfer

44. Gene Transfer in Plants: Production of Transformed Plants Using Ti Plasmid Vectors
STEPHEN G. ROGERS, ROBERT B. HORSCH, AND ROBERT T. FRALEY 627
45. Plasmid Vectors for the Genetic Analysis and Manipulation of Rhizobia and Other Gram-Negative Bacteria
R. SIMON, M. O'CONNELL, M. LABES, AND A. PÜHLER 640
46. Plant Virus Vectors: Cauliflower Mosaic Virus N. BRISSON AND T. HOHN 659
47. Direct Gene Transfer to Plants JERZY PASZKOWSKI AND MICHAEL W. SAUL 668

Section IX. Virology

48. Propagation and Purification of RNA Plant Viruses LESLIE C. LANE 687
49. Propagation of DNA Viruses TOM J. GUILFOYLE 696
50. *In Vitro* Transcription of Infectious Viral RNA from Cloned cDNA
PAUL AHLQUIST 704
51. Use of RNA Probes to Detect Plant RNA Viruses
S. J. GARGER AND T. H. TURPEN 717
52. Preparation and Use of cDNA Probes for Detection of Viral Genomes
PETER PALUKAITIS 723

53. ELISA Techniques
MICHAEL F. CLARK, RICHARD M. LISTER, AND MOSHE BAR-JOSEPH 742
54. Serotyping Plant Viruses with Monoclonal Antibodies EDWARD L. HALK 766

VOLUME 119
INTERFERONS (PART C)

Section I. Introduction

1. Interferon from 1981 to 1986 SIDNEY PESTKA 3
2. Interferon Standards and General Abbreviations SIDNEY PESTKA 14

Section II. Induction of Interferons

A. Human Interferons

3. Effect of Purification Procedures on the Composition of Human Leukocyte Interferon Preparations
HANNA-LEENA KAUPPINEN, SINIKKA HIRVONEN, AND KARI CANTELL 27
4. Large-Scale Production of Human Interferon from Lymphoblastoid Cells
A. W. PHILLIPS, N. B. FINTER, C. J. BURMAN, AND G. D. BALL 35
5. Large-Scale Production and Recovery of Human Leukocyte Interferon from Peripheral Blood Leukocytes BERNARD HOROWITZ 39
6. Induction of Human Interferon Gamma with Phorbol Esters and Phytohemagglutinin JAN VILČEK, JUNMING LE, AND Y. K. YIP 48
7. Production and Partial Purification of Human Immune Interferon
KARI CANTELL, SINIKKA HIRVONEN, AND HANNA-LEENA KAUPPINEN 54
8. Production of Human Immune Interferon from Leukocytes Cocultured with Exogenous Cells PHILIP C. FAMILLETTI AND DONNA STREMLO 63
9. Induction of Human Immune Interferon with Ionophores
FERDINANDO DIANZANI, GUIDO ANTONELLI, AND MARIA R. CAPOBIANCHI 69
10. Production of Human Immune Interferon with Normal Human Leukocytes by the Inducers A-23187 and Mezerein IRWIN A. BRAUDE AND CRISS TARR 72
11. Large-Scale Production and Recovery of Human Immune Interferon
ROBERT VAN REIS, STUART E. BUILDER, AND ANTHONY S. LUBINIECKI 77
12. Large-Scale Production, Concentration, and Partial Purification of Human Immune Interferon JERZY A. GEORGIADES 83
13. Production and Partial Purification of Human Immune Interferon Induced with Concanavalin A, Staphylococcal Enterotoxin A, and OK-432
MARC DE LEY, JOZEF VAN DAMME, AND ALFONS BILLIAU 88
14. Induction of Human Immune Interferon from Peripheral Blood Mononuclear Cells Induced with Combinations of T Cell Mitogens
MASAFUMI TSUJIMOTO AND NAOKI HIGASHI 93

15. Preparation and Partial Purification of Human Interferon δ
MILES F. WILKINSON AND ALAN G. MORRIS 96
16. Preparation and Assay of Poly ICL-CM Dextran, An Interferon Inducer of Reduced Toxicity
JAKE BELLO, EDWARD N. GRANADOS, MICHAEL MCGARRY, AND JUDITH A. O'MALLEY 103

B. Non-Human Interferons

17. Interferon Induction: Dose–Response Curves PHILIP I. MARCUS 106
18. Induction of High Titer Chicken Interferon
MARGARET J. SEKELLICK AND PHILIP I. MARCUS 115
19. *In Vitro* Induction of Interferon from Guinea Pig Fibroblasts
TIMOTHY R. WINSHIP 125
20. Induction of Equine Interferons TILAHUN YILMA 130
21. Induction and Characterization of Bovine Leukocyte Interferon
SARA COHEN, BARUCH VELAN, TAMAR BINO, HAGAI ROSENBERG, AND AVIGDOR SHAFFERMAN 136
22. Production and Purification of Mouse Immune Interferon
MARC DE LEY, HUBERTINE HEREMANS, AND ALFONS BILLIAU 145

Section III. Purification of Interferons

23. Large-Scale Purification of Recombinant Human Leukocyte Interferons
S. JOSEPH TARNOWSKI, SWAPAN K. ROY, RAYMOND A. LIPTAK, DAVID K. LEE, AND ROBERT Y. NING 153
24. Purification of Recombinant Human IFN-α2
DAVID R. THATCHER AND NIKOS PANAYOTATOS 166
25. Purification of Recombinant Human Fibroblast Interferon Produced in *Escherichia coli*
JOHN A. MOSCHERA, DIANA WOEHLE, KELLY PENG TSAI, CHIEN-HWA CHEN, AND S. JOSEPH TARNOWSKI 177
26. Purification of Recombinant Human Interferon β Expressed in *Escherichia coli*
LEO S. LIN, RALPH YAMAMOTO, AND ROBERT J. DRUMMOND 183
27. Purification of Natural Human Immune Interferon Induced by A-23187 and Mezerein IRWIN A. BRAUDE 193
28. Nickel Chelate Chromatography of Human Immune Interferon
DORIAN H. COPPENHAVER 199
29. Purification of Recombinant Human Immune Interferon
HSIANG-FU KUNG, YU-CHING E. PAN,. JOHN MOSCHERA, KELLY TSAI, EVA BEKESI, MAY CHANG, HIROMU SUGINO, AND SUSUMU HONDA 204
30. Production and Purification of Bovine Interferon β
REINHARD AHL AND M. GOTTSCHALK 211
31. Production, Purification, and Characterization of Rat Interferon from Transformed Rat Cells
PETER H. VAN DER MEIDE, JACQUELINE WUBBEN, KITTY VIJVERBERG, AND HUUB SCHELLEKENS 220

Section IV. Modification and Analysis of Interferons

32. Chemical Modifications of Human and Murine Interferons KARL H. FANTES 233
33. Analysis of Different Forms of Recombinant Human Interferons by High-Performance Liquid Chromatography
ARTHUR M. FELIX, EDGAR P. HEIMER, AND S. JOSEPH TARNOWSKI 242
34. Procedure for Reduction and Reoxidation of Human Leukocyte Interferon
JEROME A. LANGER AND SIDNEY PESTKA 248
35. Radiation Inactivation and Target Size Analysis of Interferons
ELLIS S. KEMPNER AND SIDNEY PESTKA 255

Section V. Radiolabeling of Interferons

36. Radiolabeling of Human Leukocyte and Immune Interferons with ^{125}I and Lactoperoxidase FAZLUL H. SARKAR AND SOHAN L. GUPTA 263
37. Radioiodination of Human Alpha Interferons by the Chloramine T Method
KNUD ERIK MOGENSEN AND GILLES UZÉ 267
38. Radiolabeling of Human Interferon Alphas with ^{125}I-Labeled Bolton–Hunter Reagent DOROTHY L. ZUR NEDDEN AND KATHRYN C. ZOON 276
39. Radiolabeling of Human Immune Interferon with ^{125}I-Labeled Bolton–Hunter Reagent PAUL ANDERSON 281
40. *In Vivo* Radiolabeling of Human IFN-β
ERNEST KNIGHT, JR. AND DIANA FAHEY 284
41. Labeling of Recombinant Interferon with [^{35}S]Methionine *in Vivo*
ABBAS RASHIDBAIGI AND SIDNEY PESTKA 286
42. Labeling of Interferons with [^{35}S]Methionine in a Cell-Free DNA-Dependent System HSIANG-FU KUNG 292
43. Phosphorylation of Human Immune Interferon (IFN-γ)
HSIANG-FU KUNG AND EVA BEKESI 296

Section VI. Procedures for Studying the Interferon Receptor and Uptake of Interferon

44. Procedures for Studying Binding of Interferon to Human Cells in Suspension Cultures JEROME A. LANGER AND SIDNEY PESTKA 305
45. Procedures for Studying the Binding of Interferon to Human and Bovine Cells in Monolayer Culture
KATHRYN C. ZOON, DOROTHY ZUR NEDDEN, AND HEINZ ARNHEITER 312
46. Binding of ^{32}P-Labeled Human Recombinant Immune Interferon to U937 Cells
ABBAS RASHIDBAIGI, HSIANG-FU KUNG, AND SIDNEY PESTKA 315
47. Procedures for Studying Binding of Interferon to Mouse Cells
MICHEL AGUET 321
48. Procedures for Binding an Antibody to Receptor-Bound Interferon
HEINZ ARNHEITER AND KATHRYN C. ZOON 326
49. Procedures for Measuring Receptor-Mediated Binding and Internalization of Human Interferon
KATHRYN C. ZOON, HEINZ ARNHEITER, AND DAVID FITZGERALD 332
50. Identification of Interferon Receptors by Chemical Cross-Linking
SOHAN L. GUPTA AND ARATI RAZIUDDIN 340

51. Extraction of Alpha Interferon–Receptor Complexes with Digitonin
PIERRE EID AND KNUD ERIK MOGENSEN 347
52. Measurement of a Receptor for (2′-5′)-Oligoadenylate(trimer) on Macrophages
XIN-YUAN LIU, BO-LIANG LI, AND SHI WU LI 351

Section VII. Procedures for Isolation of Genes and Expression of Interferons in Bacterial and Heterologous Cells

53. A Procedure for Isolation of Alpha Interferon Genes with Short Oligonucleotide Probes
ARTHUR P. BOLLON, MOTOHIRO FUKE, AND RICHARD M. TORCZYNSKI 359
54. Use of the Phage Lambda P_L Promoter for High-Level Expression of Human Interferons in *Escherichia coli*
ERIK REMAUT, PATRICK STANSSENS, GUUS SIMONS, AND WALTER FIERS 366
55. Expression of Human Interferon Genes in *E. coli* with the Lambda P_L Promoter
ROBERT CROWL 376
56. Use of Rous Sarcoma Viral Genome to Express Human Fibroblast Interferon
LINDA MULCAHY, MARIA KAHN, BRUCE KELDER, EDWARD REHBERG, SIDNEY PESTKA, AND DENNIS W. STACEY 383
57. Procedures for Expression, Modification, and Analysis of Human Fibroblast Interferon (IFN-β) Genes in Heterologous Cells
MICHAEL A. INNIS AND FRANK MCCORMICK 397
58. Procedures for *in Vitro* DNA Mutagenesis of Human Leukocyte Interferon Sequences
THOMAS M. DECHIARA, FRAN ERLITZ, AND S. JOSEPH TARNOWSKI 403
59. Yeast Vectors for Production of Interferon
MICHAEL D. SCHABER, THOMAS M. DECHIARA, AND RICHARD A. KRAMER 416
60. Construction of Expression Vectors for Secretion of Human Interferons by Yeast
RONALD A. HITZEMAN, CHUNG NAN CHANG, MARK MATTEUCCI, L. JEANNE PERRY, WILLIAM J. KOHR, JOHN J. WULF, JAMES R. SWARTZ, CHRISTINA Y. CHEN, AND ARJUN SINGH 424
61. Procedures for Isolation of Murine Alpha Interferon Genes and Expression of a Murine Leukocyte Interferon in E. *coli*
BRUCE L. DAUGHERTY AND SIDNEY PESTKA 434
62. Cloning, Expression, and Purification of Rat IFN-α1
PETER H. VAN DER MEIDE, REIN DIJKEMA, MARTIN CASPERS, KITTY VIJVERBERG, AND HUUB SCHELLEKENS 441
63. Cloning, Expression, and Purification of Rat IFN-γ
REIN DIJKEMA, PETER H. VAN DER MEIDE, MARTIN DUBBELD, MARTIN CASPERS, JACQUELINE WUBBEN, AND HUUB SCHELLEKENS 453
64. Isolation of Bovine IFN-α Genes and Their Expression in Bacteria
BARUCH VELAN, SARA COHEN, HAIM GROSFELD, AND AVIGDOR SHAFFERMAN 464
65. The Detection of Individual Cells Containing Interferon mRNA by *in Situ* Hybridization with Specific Interferon DNA Probes
RAINER ZAWATZKY, JAQUELINE DE MAEYER-GUIGNARD, AND EDWARD DE MAEYER 474

66. Detection of a Single Base Substitution between Human Leukocyte IFN-αA and -α2 Genes with Octadecyl Deoxyoligonucleotide Probes
KUNIMOTO HOTTA, KENNETH J. COLLIER, AND SIDNEY PESTKA 481

Section VIII. Enzymes in Interferon Action

67. RNase L, a (2′-5′)-Oligoadenylate-Dependent Endoribonuclease: Assays and Purification of the Enzyme; Cross-Linking to a (2′-5′)-Oligoadenylate Derivative
GEORGIA FLOYD-SMITH AND PETER LENGYEL 489
68. Purification of Double-Stranded RNA-Dependent Protein Kinase from Mouse Fibroblasts
CHARLES E. SAMUEL, GRACE S. KNUTSON, MARLA J. BERRY, JONATHAN A. ATWATER, AND STEPHEN R. LASKY 499
69. Enzymatic Synthesis of (2′-5′)-Linked Oligoadenylates on a Preparative Scale
HIMADRI SAMANTA AND PETER LENGYEL 516
70. Methods for the Synthesis of Analogs of (2′-5′)-Oligoadenylic Acid
PAUL F. TORRENCE, JIRO IMAI, JEAN-CLAUDE JAMOULLE, AND KRYSTYNA LESIAK 522

Section IX. Assay of Interferons

A. Antiviral Assays

71. Objective Antiviral Assay of the Interferons by Computer Assisted Data Collection and Analysis
ROBERT L. FORTI, SHIRLEY S. SCHUFFMAN, HUGH A. DAVIES, AND WILLIAM M. MITCHELL 533
72. Measurement of Interferon in Human Amniotic Fluid and Placental Blood Extract PAULETTE DUC-GOIRAN, PIERRE LEBON, AND CHARLES CHANY 541
73. Use of Ovine and Caprine Cells to Measure Antiviral Effects of Interferon
TILAHUN YILMA 551
74. Quantitation of Neutralization of Interferon by Antibody YOSHIMI KAWADE 558

B. Antiproliferative Assays

75. A Convenient Microassay for Cytolysis and Cytostasis
STEPHEN TYRING, W. ROBERT FLEISCHMANN, JR., AND SAMUEL BARON 574
76. A Simplified Antigrowth Assay Based on Color Change of the Medium
PAUL AEBERSOLD AND SALLY SAMPLE 579

C. Immunoassays

77. A Sandwich Radioimmunoassay for Human IFN-γ
BRUCE KELDER, ABBAS RASHIDBAIGI, AND SIDNEY PESTKA 582
78. Procedures for Measurement of Interferon Dimers and Higher Oligomers by Radioimmunoassay
SIDNEY PESTKA, BRUCE KELDER, AND S. JOSEPH TARNOWSKI 588

Section X. Biology of Interferon Action

79. Selection and Screening of Transformed NIH3T3 Cells for Enhanced Sensitivity to Human Interferons α and β VINCENT JUNG AND SIDNEY PESTKA 597
80. Measurement of the Effect of Interferons on Cellular Differentiation in Murine and Human Melanoma Cells
PAUL B. FISHER, ARMAND F. MIRANDA, AND LEE E. BABISS 611
81. Measurement of the Effect of Interferons on Cellular Differentiation of Human Skeletal Muscle Cells
ARMAND F. MIRANDA, LEE E. BABISS, AND PAUL B. FISHER 619
82. Measurement of the Effect of Interferons on the Proliferative Capacity and Cloning Efficiency of Normal and Leukemic Human Myeloid Progenitor Cells in Culture
PAUL B. FISHER, STEVEN GRANT, JOHN W. GREINER, AND JEFFREY SCHLOM 629
83. Measurement of Effect of Interferons on Cloning Efficiency of Primary Tumor Cells in Culture SYDNEY E. SALMON 635
84. Measurement of the Antiproliferative Effect of Interferon: Influence of Growth Factors
NICOLETTE EBSWORTH, ENRIQUE ROZENGURT, AND JOYCE TAYLOR-PAPADIMITRIOU 643
85. Animal Models for Investigating Antitumor Effects of Interferon
FRANCES R. BALKWILL 649
86. Measurement of Antagonistic Effects of Growth Factors and Interferons
ANNA D. INGLOT AND ELŻBIETA PAJTASZ 657
87. Measurement of Effect of (2′-5′)-Oligoadenylates and Analogs on Protein Synthesis and Growth of Cells
ROBERT J. SUHADOLNIK, CHOONGEUN LEE, AND DAVID H. WILLIS, JR. 667
88. Assay of Effect of (2′-5′)-Oligoadenylate on Macrophages
XIN-YUAN LIU, HONG DA ZHENG, NING WANG, WEN HUA REN, AND T. P. WANG 676
89. Radioimmunoassay for Detection of Changes in Cell Surface Tumor Antigen Expression Induced by Interferon
J. W. GREINER, P. HORAN HAND, D. WUNDERLICH, AND D. COLCHER 682
90. Measurements of Changes in Histocompatibility Antigens Induced by Interferons
MARIANNE HOKLAND, IVER HERON, PETER HOKLAND, PER BASSE, AND KURT BERG 688
91. Purification, Assay, and Characterization of the Interferon Antagonist: Sarcolectin
FRANÇOISE CHANY-FOURNIER, PAN HONG JIANG, AND CHARLES CHANY 694
92. Assay of Effect of Interferon on Virus-Induced Cell Fusion
YOSHIMI TOMITA AND TSUGUO KUWATA 702
93. Measurement of Hyporesponsiveness to Interferon and Interferon Induction with Prostaglandins DALE A. STRINGFELLOW 707

Section XI. Measurement of Effect of Interferons on Drug Metabolism

94. Measurement of Effect of Interferon on Metabolism of Diphenylhydantoin
GERALD SONNENFELD AND DONALD E. NERLAND 715

95. Measurement of Effect of Interferon on Drug Metabolism
GILBERT J. MANNERING 718

Section XII. Interferon and Plant Cells

96. Production, Preparation, and Assay of an Antiviral Substance from Plant Cells
ABDULLAH GERA, SARA SPIEGEL, AND GAD LOEBENSTEIN 729
97. Preparation and Measurement of an Antiviral Protein Found in Tobacco Cells after Infection with Tobacco Mosaic Virus ILAN SELA 734
98. Assay of Effect of Human Interferons on Tobacco Protoplasts ILAN SELA 744
99. Enzymatic Synthesis of Plant Oligoadenylates *in Vitro*
YAIR DEVASH, ILAN SELA, AND ROBERT J. SUHADOLNIK 752
100. Measurement of Effect of (2′-5′)-Oligoadenylates and Analogs on Tobacco Mosaic Virus Replication
YAIR DEVASH, ROBERT J. SUHADOLNIK, AND ILAN SELA 759

VOLUME 121
IMMUNOCHEMICAL TECHNIQUES (PART I: HYBRIDOMA TECHNOLOGY AND MONOCLONAL ANTIBODIES)

Section I. Production of Hybridomas

A. Immunization and Cell Fusion

1. Factors Affecting Production of Monoclonal Antibodies
REGINE J. WESTERWOUDT 3
2. *In Vitro* Immunization for the Production of Antigen-Specific Lymphocyte Hybridomas CHRISTOPHER L. READING 18
3. An Improved *in Vitro* Immunization Procedure for the Production of Monoclonal Antibodies BARBARA D. BOSS 27
4. "Single-Shot" Intrasplenic Immunization for the Production of Monoclonal Antibodies M. SPITZ 33
5. Method for Generating a High Frequency of Hybridomas Producing Monoclonal IgA Antibodies
DAWN E. COLWELL, SUZANNE M. MICHALEK, AND JERRY R. MCGHEE 42
6. Production of IgA-Secreting Rat × Rat Hybridomas
C. J. DEAN, L. A. GYURE, J. G. HALL, AND J. M. STYLES 52
7. The Use of Conventional Antisera in the Production of Specific Monoclonal Antibodies KENDRA B. EAGER AND ROGER H. KENNETT 59
8. Preparation of Monoclonal Antibodies to Preselected Protein Regions
M. ZOUHAIR ATASSI 69
9. The Auto-Anti-Idiotypic Strategy for Preparing Monoclonal Antibodies to Receptor Combining Sites
W. LOUIS CLEVELAND AND BERNARD F. ERLANGER 95

10. Optimal Strategies for Developing Human–Human Monoclonal Antibodies
PAUL G. ABRAMS, JEFFREY L. ROSSIO, H. C. STEVENSON, AND K. A. FOON 107
11. Comparative Phenotypic Analysis of Available Human Hybridoma Fusion Partners
DANUTA KOZBOR, JOHN C. RODER, MICHAEL E. SIERZEGA, SUSAN P. C. COLE, AND CARLO M. CROCE 120
12. The EBV–Hybridoma Technique
J. C. RODER, S. P. C. COLE, AND D. KOZBOR 140
13. Generation of Human Monoclonal Antibodies by Fusion of EBV-Activated B Cells to a Human–Mouse Hybridoma
STEVEN K. H. FOUNG, EDGAR G. ENGLEMAN, AND F. CARL GRUMET 168
14. Improved Hybridoma Technology: Spleen Cell Separation and Soluble Growth Factors P. VAN MOURIK AND W. P. ZEIJLEMAKER 174
15. High Efficiency Fusion Procedure for Producing Monoclonal Antibodies against Weak Immunogens
RICHARD D. LANE, ROBERT S. CRISSMAN, AND STEPHEN GINN 183
16. Methods for High Frequency Production of Soluble Antigen-Specific Hybridomas; Specificities and Affinities of the Monoclonal Antibodies Obtained
MAURIZIO CIANFRIGLIA, MASSIMO MARIANI, DORETTA ARMELLINI, ANNALISA MASSONE, MARGHERITA LAFATA, RIVO PRESENTINI, AND GUIDO ANTONI 193
17. Bispecific Monoclonal Antibodies from Hybrid Hybridomas
M. R. SURESH, A. C. CUELLO, AND C. MILSTEIN 210
18. Human × (Mouse × Human) Hybridomas LARS ÖSTBERG 228
19. Generation of Rat–Rat Hybridomas with the Use of the LOU IR983F Nonsecreting Fusion Cell Line
LIEVE DE CLERCQ, FRANÇOISE CORMONT, AND HERVÉ BAZIN 234
20. Production of Syrian and Armenian Hamster Monoclonal Antibodies of Defined Specificity FRANCISCO SANCHEZ-MADRID AND TIMOTHY A. SPRINGER 239
21. Production and Characterization of Bovine Immunoglobulins from Bovine × Murine Hybridomas
ALBERT J. GUIDRY, S. SRIKUMARAN, AND RICHARD A. GOLDSBY 244

B. Growth and Cloning of Hybridomas

22. Serum-Free Medium for the Growth of NS-1 Mouse Myeloma Cells and the Isolation of NS-1 Hybridomas
TOMOYUKI KAWAMOTO, J. DENRY SATO, DON B. MCCLURE, AND GORDON H. SATO 266
23. Serum-Free Medium for Hybridoma and Parental Myeloma Cell Cultivation
JAN KOVÁŘ AND FRANTIŠEK FRANĚK 277
24. Endothelial Cell Growth Supplements for Promoting the Growth of Monoclonal Antibody-Producing Hybridoma Cells JANET H. RANSOM 293
25. Feeder Effect of Enriched Human Monocytes on Human Hybridoma, Myeloma, and Lymphoma Cell Lines
THOMAS BRODIN, LENNART OLSSON, AND HANS OLOV SJÖGREN 295
26. Removal of Bovine Immunoglobulin from Serum in Hybridoma Culture Media Using Protein A P. A. UNDERWOOD 301
27. A Single-Step Technique for Selecting and Cloning Hybridomas for Monoclonal Antibody Production JOHN M. DAVIS 307

28. Cloning of Murine Hybridoma Cells in Ultra-Low Gelation Temperature Agarose
CURT I. CIVIN AND M. L. BANQUERIGO 322
29. A Simple Method for Cloning Hybridomas in 20-μl Hanging Drops
MARVIN B. RITTENBERG, ABIGAIL BUENAFE, AND MCKAY BROWN 327
30. Clonal Isolation of Hybridomas by Manual Single-Cell Isolation
ALBERT BOEYÉ 332
31. Semiautomated Plaque Method for the Detection of Antibody-Forming Cell Clones BEVERLEY L. PIKE 341
32. Production of Monoclonal Antibodies by Agarose-Entrapped Hybridoma Cells
K. NILSSON, W. SCHEIRER, H. W. D. KATINGER, AND K. MOSBACH 352
33. Automated Production of Monoclonal Antibodies in a Cytostat
S. FAZEKAS DE ST.GROTH 360
34. The Effect of Pristane on Ascites Tumor Formation and Monoclonal Antibody Production NICHOLAS J. HOOGENRAAD AND CHRISTOPHER J. WRAIGHT 375
35. Production of Ascites and Rescue of Hybridomas by Intrasplenic Inoculation
PAMELA L. WITTE AND ROSALIE BER 381
36. Growth of Rat–Mouse Hybridomas in Nude Mice and Nude Rats
DHIRENDRA N. MISRA, SAAD A. NOEMAN, HEINZ W. KUNZ, AND THOMAS J. GILL, III 386
37. Poisson Statistical Analysis of Repetitive Subcloning by the Limiting Dilution Technique as a Way of Assessing Hybridoma Monoclonality
HILARY A. COLLER AND BARRY S. COLLER 412
38. A Method for Freezing Hybridoma Clones in 96-Well Microculture Plates
DAVID E. WELLS AND WILLIAM F. BIBB 417

Section II. Monoclonal Antibodies

A. Screening Assays for Monoclonal Antibodies

39. Multivalent Antibody Radioimmunoassay (MARIA) for Screening Specific Antibody Secretion by Lymphocyte Hybridoma Cultures
J. GHEUENS AND D. E. MCFARLIN 425
40. A Rapid Technique Using Radiolabeled Soluble Antigen to Screen Hybridoma Culture Supernatants K. F. MILLER, D. J. BOLT, AND R. A. GOLDSBY 433
41. Culture-Well Solid-Phase ^{3}H-RIA for the Isotype-Specific Measurement of Antibody from Mouse and Human Antibody-Secreting Cells
PATRICIA K. A. MONGINI AND CLARE BLESSINGER 438
42. The Spaghetti Overlay Procedure W. STEVEN ADAIR 453
43. Solution-Phase RIA and Solid-Phase EIA Using Avidin–Biotin Systems for Analysis of Monoclonal Antibody Epitopes and Affinity Constants
JOHN E. SHIVELY, CHRISTOPH WAGENER, AND BRIAN R. CLARK 459
44. A Simple Method for Ranking the Affinities of Monoclonal Antibodies
VERONICA VAN HEYNINGEN 472
45. The Hazards of Mycoplasma Contamination in the Screening of Hybridoma Supernatants for Effects on [^{3}H]Thymidine Incorporation JAMES P. JAKWAY 481
46. The Dot Immunobinding Assay RICHARD HAWKES 484
47. Detection of Antibody-Secreting Hybridomas with Diazobenzyloxymethyl Paper
GENEVIÈVE MOTTA AND DANIEL LOCKER 491

48. The Use of Alkaline Phosphatase-Conjugated Anti-Immunoglobulin with Immunoblots for Determining the Specificity of Monoclonal Antibodies to Protein Mixtures PETER L. EY AND LEONIE K. ASHMAN 497
49. Slide Immunoenzymatic Assay (SIA) in Hybridoma Technology EVERLY CONWAY DE MACARIO, ALBERTO J. L. MACARIO, AND ROBERT J. JOVELL 509
50. Use of Mouse and Human Monoclonal Antibodies in Enzyme Immunofiltration MARK C. GLASSY AND PATRICK H. CLEVELAND 525
51. Measurement of Monoclonal Immunoglobulin Concentrations in Hybridoma Cultures by Competitive Inhibition Enzyme Immunoassay KENNETH W. HUNTER, JR. AND JOHN M. BOSWORTH, JR. 541
52. The Detection and Characterization of Antigen-Specific Monoclonal Antibodies Using Anti-Immunoglobulin Isotype Antibodies Coupled to Red Blood Cells M. R. CLARK 548
53. Agglutination Assay Using Protein A-Sensitized Erythrocytes for Detection of Immunoglobulin in Tissue Culture Supernatants DAVID MALE AND GARETH PRYCE 556
54. Immunohistochemical Techniques Using Monoclonal Antibodies DANIEL R. CIOCCA AND ROBERT J. BJERCKE 562
55. Method for Rapid Detection of Membrane Antigens by Immunofluorescence and Its Application to Screening Monoclonal Antibodies CLAUDE BOUCHEIX, PATRICIA KRIEF, JEAN-YVES PERROT, MASSOUD MIRSHAHI, AND CLAUDE ROSENFELD 580

B. Purification of Monoclonal Antibodies and Preparation of Antibody Fragments

56. Purification of Mouse Monoclonal Antibodies from Ascitic Fluid by DEAE Affi-Gel Blue Chromatograhy CLAUDINE BRUCK, JEFFREY A. DREBIN, CORINNE GLINEUR, AND DANIEL PORTETELLE 587
57. Purification and Analysis of Monoclonal Antibodies by High-Performance Liquid Chromatography SCOTT W. BURCHIEL 596
58. Separation of IgG Idiotypes by High-Performance Hydroxylapatite Chromatography HECTOR JUAREZ-SALINAS, GARY S. OTT, JENG-CHYH CHEN, TIMOTHY L. BROOKS, AND LARRY H. STANKER 615
59. The Use of Rat Monoclonal Antibodies to Characterize, Quantify, and Purify Polyclonal or Monoclonal Mouse IgM FRANÇOISE CORMONT, PATRICK MANOUVRIEZ, LIEVE DE CLERCQ, AND HERVÉ BAZIN 622
60. Purification of IgG Monoclonal Antibodies from Ascitic Fluid Based on Rivanol Precipitation FRANTIŠEK FRANĚK 631
61. Purification of Rat Monoclonal Antibodies HERVÉ BAZIN, FRANÇOISE CORMONT, AND LIEVE DE CLERCQ 638
62. Preparation of $F(ab')_2$ Fragments from Mouse IgG of Various Subclasses EDMUNDO LAMOYI 652
63. Optimal Conditions for the Preparation of Proteolytic Fragments from Monoclonal IgG of Different Rat IgG Subclasses J. ROUSSEAUX, R. ROUSSEAUX-PREVOST, AND H. BAZIN 663

64. Tritium Radiolabeling of Antibodies to High Specific Activity with *N*-Succinimidyl [2,3-^{3}H]Propionate: Use in Detecting and Analyzing Monoclonal Antibodies
UDO KUMMER 670
65. Determination of the True Immunoreactive Fraction of Monoclonal Antibodies after Radiolabeling TORE LINDMO AND PAUL A. BUNN, JR. 678

Section III. Selected Applications of Monoclonal Antibodies

66. Ultrasensitive Cooperative Immunoassays with Mixed Monoclonal Antibodies
PAUL H. EHRLICH AND WILLIAM R. MOYLE 695
67. Use of Monoclonal Antibody to Increase the Sensitivity and Specificity of Precipitating Immunoassays and Cell Surface Binding Immunoassays
ROBERT R. MONTGOMERY, THOMAS J. KUNICKI, AND L. MICHAEL GLODE 702
68. Immunoaffinity Isolation of Membrane Antigens with Biotinylated Monoclonal Antibodies and Streptavidin–Agarose
TIMOTHY V. UPDYKE AND GARTH L. NICOLSON 717
69. An Indirect Rosette Technique for the Identification and Separation of Lymphocyte Populations by Monoclonal Antibodies KINGSTON MILLS 726
70. Separation of T Cell Subpopulations by Monoclonal Antibodies and Affinity Chromatography RÜDIGER W. BRAUN AND GUNTHER KÜMEL 737
71. A Rapid Micromethod for the Enumeration of T Cell Subsets in Blood by Epifluorescence Microscopy
IAN H. FRAZER, IAN R. MACKAY, AND FRANK BATTYE 748
72. Cytotoxicity Assays Based on the Use of Antibody Secretion as a Measure of Hybridoma Viability J. MARBROOK 759
73. Use of Iron- or Selenium-Coupled Monoclonal Antibodies to Cell Surface Antigens as a Positive Selection System for Cells
MARK BOTHWELL AND TIMOTHY BLOCK 771
74. Screening for Monoclonal Antibodies to Human Cellular and Soluble Antigens
DANIEL P. ESKINAZI, WILLIAM C. EBY, AND GIUSEPPE A. MOLINARO 783
75. Epitope Ratio Analysis for the Simultaneous Study of Several Cell Surface Antigens BENNETT KAUFMAN AND RICHARD GOLDSBY 797
76. Use of Monoclonal Antibodies as Radiopharmaceuticals for the Localization of Human Carcinoma Xenografts in Athymic Mice
DAVID COLCHER, JOSE ESTEBAN, AND FRANÇOISE MORNEX 802
77. Coating of Liposomes with Subunits of Monoclonal IgM Antibody and Targeting of the Liposomes
YOSHIYUKI HASHIMOTO, MINORU SUGAWARA, TOSHIO KAMIYA, AND SHINYA SUZUKI 817
78. Immunohistological Studies with Monoclonal Antibodies
WAYNE W. HANCOCK AND ROBERT C. ATKINS 828
79. Use of Alkaline Phosphatase-Conjugated Antibodies for Detection of Protein Antigens on Nitrocellulose Filters BRYAN M. TURNER 848
80. Stepwise Amplified Immunoenzyme Staining Techniques for the Detection of Monoclonal Antibodies and Antigens
P. M. LANSDORP, A. W. WOGNUM, AND W. P. ZEIJLEMAKER 855

Section IV. Summary

81. Other Articles from Methods in Enzymology Related to Immunochemical Techniques HELEN VAN VUNAKIS AND JOHN J. LANGONE 871

VOLUME 122
VITAMINS AND COENZYMES (PART G)

Section I. Ascorbic Acid

1. High-Performance Liquid Chromatographic Separation of Ascorbic Acid, Erythorbic Acid, Dehydroascorbic Acid, Dehydroerythorbic Acid, Diketogulonic Acid, and Diketogluconic Acid LANDIS W. DONER AND KEVIN B. HICKS 3
2. Glutathione Dehydrogenase (Ascorbate) ROBERT L. STAHL, LEONARD F. LIEBES, AND ROBERT SILBER 10

Section II. Thiamin: Phosphates and Analogs

3. Determination of Thiamin and Its Phosphate Esters by High-Performance Liquid Chromatography TAKASHI KAWASAKI 15
4. Separation and Determination of Thiamin and Its Phosphate Esters by SP-Sephadex Chromatography JACK R. COOPER AND TOSHIO MATSUDA 20
5. Enzymatic Synthesis of Thiamin Triphosphate JACK R. COOPER AND KOHSUKE NISHINO 24

Section III. Pantothenic Acid, Coenzyme A, and Derivatives

6. Pantothenase-Based Assay of Pantothenic Acid R. KALERVO AIRAS 33
7. Enzymatic Hydrolysis of Pantetheine CARL T. WITTWER, BONITA W. WYSE, AND R. GAURTH HANSEN 36
8. Micromethod for the Measurement of Acetyl Phosphate and Acetyl Coenzyme A ARTHUR G. HUNT 43
9. Preparation of Radioactive Acyl Coenzyme A AMIYA K. HAJRA AND JAMES E. BISHOP 50
10. Purification of Plant Acetyl-CoA:Acyl Carrier Protein Transacylase TAKASHI SHIMAKATA AND PAUL K. STUMPF 53

Section IV. Biotin and Derivatives

11. Separation of Biotin and Analogs by High-Performance Liquid Chromatography DELORES M. BOWERS-KOMRO, JANE L. CHASTAIN, AND DONALD B. MCCORMICK 63
12. Fluorometric Assays for Avidin and Biotin R. D. NARGESSI AND D. S. SMITH 67

13. Solid-Phase Assay for *d*-Biotin and Avidin on Cellulose Disks
L. GOLDSTEIN, S. A. YANKOFSKY, AND G. COHEN 72
14. Purification of Avidin and Its Derivatives on 2-Iminobiotin-6-aminohexyl-Sepharose 4B GEORGE A. ORR, GAYLE C. HENEY, AND RON ZEHEB 83
15. Use of Avidin–Iminobiotin Complexes for Purifying Plasma Membrane Proteins
RON ZEHEB AND GEORGE A. ORR 87

Section V. Pyridoxine, Pyridoxamine, Pyridoxal: Analogs and Derivatives

16. Preparation of [^{14}C]Pyridoxal 5′-Phosphate
ALEC LUI, LAWRENCE LUMENG, AND TING-KAI LI 97
17. Cation-Exchange High-Performance Liquid Chromatographic Analysis of Vitamin B_6 STEPHEN P. COBURN AND J. DENNIS MAHUREN 102
18. Highly Sensitive Methods for Assaying the Enzymes of Vitamin B_6 Metabolism
ALFRED H. MERRILL, JR. AND ELAINE WANG 110
19. Modified Purification of Pyridoxamine (Pyridoxine) 5′-Phosphate Oxidase from Rabbit Liver by 5′-Phosphopyridoxyl Affinity Chromatography
DELORES M. BOWERS-KOMRO, TORY M. HAGEN, AND DONALD B. MCCORMICK 116
20. Monoclonal Antibodies to Vitamin B_6
JASON M. KITTLER, DACE VICEPS-MADORE, JOHN A. CIDLOWSKI, NATALIE T. MEISLER, AND JOHN W. THANASSI 120
21. Pyruvoyl-Dependent Histidine Decarboxylase from *Lactobacillus* 30a: Purification and Properties ESMOND E. SNELL 128
22. Prohistidine Decarboxylase from *Lactobacillus* 30a
ESMOND E. SNELL AND QUANG KHAI HUYNH 135
23. Pyridoxal Phosphate-Dependent Histidine Decarboxylase from *Morganella* AM-15 ESMOND E. SNELL AND BEVERLY M. GUIRARD 139

Section VI. Nicotinic Acid: Analogs and Coenzymes

24. Simultaneous Extraction and Combined Bioluminescent Assay of Oxidized and Reduced Nicotinamide Adenine Dinucleotide
MATTI T. KARP AND PAULI I. VUORINEN 147
25. Purification of Reduced Nicotinamide Adenine Dinucleotide by Ion-Exchange and High-Performance Liquid Chromatography
DEXTER B. NORTHROP, COLIN J. NEWTON, AND STEVEN M. FAYNOR 152
26. Desalting of Nucleotides by Reverse-Phase High-Performance Liquid Chromatography DEXTER B. NORTHROP AND COLIN J. NEWTON 155
27. Preparation of 3-Aminopyridine 1,N^6-Ethenoadenine Dinucleotide and 3-Aminopyridine 1,N^6-Ethenoadenine Dinucleotide Phosphate
BRUCE M. ANDERSON AND CONSTANCE D. ANDERSON 156
28. Simultaneous Photometric Determination of Oxygen Consumption and NAD(P)H Formation or Disappearance Using a Continuous Flow Cuvette
MANASE M. DÂNŞOREANU AND MARIUS I. TELIA 161
29. Spectrophotometric Assay of NADase-Catalyzed Reactions
BRUCE M. ANDERSON AND DAVID A. YOST 169

30. Snake Venom NAD Glycohydrolase: Purification, Immobilization, and Transglycosidation
BRUCE M. ANDERSON, DAVID A. YOST, AND CONSTANCE D. ANDERSON 173

Section VII. Flavins and Derivatives

31. Luminometric Determination of Flavin Adenine Dinucleotide
KARL DECKER AND ARI HINKKANEN 185
32. Heavy Riboflavin Synthase from *Bacillus subtilis* ADELBERT BACHER 192
33. Separation of Flavins and Flavin Analogs by High-Performance Liquid Chromatography
ROBERT P. HAUSINGER, JOHN F. HONEK, AND CHRISTOPHER WALSH 199
34. Preparation, Properties, and Separation by High-Performance Liquid Chromatography of Riboflavin Phosphates
PETER NIELSEN, PETER RAUSCHENBACH, AND ADELBERT BACHER 209
35. Competitive Binding Assays for Riboflavin and Riboflavin-Binding Protein
HAROLD B. WHITE, III 221
36. Isolation of Avian Riboflavin-Binding Protein
MARK S. MILLER AND HAROLD B. WHITE, III 227
37. Fluorometric Titration of Urinary Riboflavin with an Apoflavoprotein
JERRY ANN TILLOTSON AND MARK M. BASHOR 234
38. Sensitive Radiosubstrate/High-Performance Liquid Chromatography Assays for Flavocoenzyme-Forming Enzymes
SANG-SUN LEE AND DONALD B. MCCORMICK 237
39. ^{1}H NMR Spectral Analysis of the Ribityl Side Chain of Riboflavin and Its Ring-Substituted Analogs GARY WILLIAMSON AND DALE E. EDMONDSON 240

Section VIII. Pteridines, Analogs, and Pterin Coenzymes (Folate)

40. Purification of Rat Liver Folate-Binding Protein: Cytosol I
ROBERT J. COOK AND CONRAD WAGNER 251
41. Dimethylglycine Dehydrogenase and Sarcosine Dehydrogenase: Mitochondrial Folate-Binding Proteins from Rat Liver
ROBERT J. COOK AND CONRAD WAGNER 255
42. Membrane-Associated Folate Transport Proteins
G. B. HENDERSON AND F. M. HUENNEKENS 260
43. High-Performance Liquid Chromatographic Separation of the Naturally Occurring Folic Acid Derivatives SUSAN D. WILSON AND DONALD W. HORNE 269
44. Separation of Pteridines from Blood Cells and Plasma by Reverse-Phase High-Performance Liquid Chromatography
HANS-JÖRG ZEITLER AND BERTA ANDONDONSKAJA-RENZ 273
45. Separation of Methotrexate Analogs Containing Terminal Lysine or Ornithine and Their Dansyl Derivatives by High-Performance Liquid Chromatography
JAMES H. FREISHEIM AND A. ASHOK KUMAR 293
46. Determination of Pterins by Liquid Chromatography/Electrochemistry
CRAIG E. LUNTE AND PETER T. KISSINGER 300
47. Preparation of (6*S*)-5-Formyltetrahydrofolate Labeled at High Specific Activity with ^{14}C and ^{3}H
RICHARD G. MORAN, KHANDAN KEYOMARSI, AND PAUL D. COLMAN 309

48. Tissue Folate Polyglutamate Chain Length Determination by Electrophoresis as Thymidylate Synthase–Fluorodeoxyuridylate Ternary Complexes
D. G. PRIEST AND M. T. DOIG 313
49. Preparative Electrochemical Reduction of Pteroylpentaglutamate
PETER B. ROWE AND GARRY P. LEWIS 319
50. Identification of Folylpoly(γ-glutamate) Chain Length by Cleavage to and Separation of *p*-Aminobenzoylpoly(γ-glutamates) BARRY SHANE 323
51. Detection of *p*-Aminobenzoylpoly(γ-glutamates) Using Fluorescamine
PETER C. LOEWEN 330
52. Preparation and Analysis of Pteroylpolyglutamate Substrates and Inhibitors
ROWENA G. MATTHEWS 333
53. Analysis of Methotrexate Polyglutamate Derivatives *in Vivo*
B. A. KAMEN AND N. WINICK 339
54. Radioenzymatic Assay for Dihydrofolate Reductase Using [^{3}H]Dihydrofolate
M. PERWAIZ IQBAL AND SHELDON P. ROTHENBERG 346
55. Bacterial Folylpoly(γ-glutamate) Synthase–Dihydrofolate Synthase
ANDREW L. BOGNAR AND BARRY SHANE 349
56. Dihydrofolate Reductase: A Coupled Radiometric Assay
PHILIP REYES AND PRADIPSINH K. RATHOD 360
57. Purification of Folate Oligoglutamate:Amino Acid Transpeptidase
TOM BRODY AND E. L. R. STOKSTAD 367
58. Methylenetetrahydrofolate Reductase from Pig Liver ROWENA G. MATTHEWS 372
59. Enzymatic Assay of 5-Methyl-L-tetrahydrofolate
PHILIP A. KATTCHEE AND ROBERT W. GUYNN 381
60. Purification of 5,10-Methenyltetrahydrofolate Cyclohydrolase from *Clostridium formicoaceticum* LARS G. LJUNGDAHL AND JOAN E. CLARK 385
61. Purification and Properties of 5,10-Methylenetetrahydrofolate Reductase from *Clostridium formicoaceticum* JOAN E. CLARK AND LARS G. LJUNGDAHL 392
62. Assay and Detection of the Molybdenum Cofactor
R. V. HAGEMAN AND K. V. RAJAGOPALAN 399
63. Methanopterin and Tetrahydromethanopterin Derivatives: Isolation, Synthesis, and Identification by High-Performance Liquid Chromatography
JAN T. KELTJENS, GERDA C. CAERTELING, AND GODFRIED D. VOGELS 412

VOLUME 123
VITAMINS AND COENZYMES (PART H)

Section I. Cobalamins and Cobamides (B_{12})

1. Determination of Cobalamins in Biological Material
PETER GIMSING AND WILLIAM S. BECK 3
2. Analysis of Cobalamin Coenzymes and Other Corrinoids by High-Performance Liquid Chromatography
DONALD W. JACOBSEN, RALPH GREEN, AND KENNETH L. BROWN 14
3. Isolation of Native and Proteolytically Derived Ileal Receptor for Intrinsic Factor–Cobalamin BELLUR SEETHARAM AND DAVID H. ALPERS 23

4. Purification of B_{12}-Binding Proteins Using a Photodissociative Affinity Matrix D. W. JACOBSEN AND F. M. HUENNEKENS 28
5. Solid-Phase Immunoassay for Human Transcobalamin II and Detection of the Secretory Protein in Cultured Human Cells MARIJKE FRÀTER-SCHRÖDER 36

Section II. Vitamin A Group

6. Quantitative Analysis of Retinal and 3-Dehydroretinal by High-Pressure Liquid Chromatography MOMOYO MAKINO-TASAKA AND TATSUO SUZUKI 53
7. Enzymatic Synthesis and Separation of Retinyl Phosphate Mannose and Dolichyl Phosphate Mannose by Anion-Exchange High-Performance Liquid Chromatography KIM E. CREEK, DONATA RIMOLDI, MICHELE BRUGH-COLLINS, AND LUIGI M. DE LUCA 61
8. Separation and Quantitation of Retinyl Esters and Retinol by High-Performance Liquid Chromatography A. CATHARINE ROSS 68
9. Separation of Geometric Isomers of Retinol and Retinoic Acid in Nonaqueous High-Performance Liquid Chromatography PANGALA V. BHAT AND ANDRE LACROIX 75
10. Use of Enzyme-Linked Immunosorbent Assay Technique for Quantitation of Serum Retinol-Binding Protein NOBUO MONJI AND EJAL BOSIN 85
11. Fourier Transform Infrared Spectroscopy of Retinoids NATALIE L. ROCKLEY, MARK G. ROCKLEY, BRUCE A. HALLEY, AND ELDON C. NELSON 92
12. Purification of Interstitial Retinol-Binding Protein from the Eye S.-L. FONG, G. I. LIOU, AND C. D. B. BRIDGES 102
13. Quantification of Physiological Levels of Retinoic Acid JOSEPH L. NAPOLI 112

Section III. Vitamin D Group

14. Isolation and Identification of Vitamin D Metabolites JOSEPH L. NAPOLI, NICK J. KOSZEWSKI, AND RONALD L. HORST 127
15. A New Pathway of 25-Hydroxyvitamin D_3 Metabolism GLENVILLE JONES 141
16. Enzyme-Linked Immunoabsorption Assay for Vitamin D-Induced Calcium-Binding Protein BARBARA E. MILLER AND ANTHONY W. NORMAN 154
17. Measurement of 25-Hydroxyvitamin D 1α-Hydroxylase Activity in Mammalian Kidney BRUCE LOBAUGH, JANICE R. ALMOND, AND MARC K. DREZNER 159
18. Quantitation of Vitamin D_2, Vitamin D_3, 25-Hydroxyvitamin D_2, and 25-Hydroxyvitamin D_3 in Human Milk BRUCE W. HOLLIS AND NANCY E. FRANK 167
19. 1,25-Dihydroxyvitamin D Microassay Employing Radioreceptor Techniques TIMOTHY A. REINHARDT AND BRUCE W. HOLLIS 176
20. Assay for 1,25-Dihydroxyvitamin D Using Rabbit Intestinal Cytosol-Binding Protein W. E. DUNCAN, T. C. AW, AND J. G. HADDAD 185
21. Cytoreceptor Assay for 1,25-Dihydroxyvitamin D STAVROS C. MANOLAGAS 190
22. Monoclonal Antibodies as Probes in the Characterization of 1,25-Dihydroxyvitamin D_3 Receptors J. WESLEY PIKE AND MARK R. HAUSSLER 199

Section IV. Vitamin E Group, Tocopherols

23. An HPLC Method for the Simultaneous Determination of Retinol and α-Tocopherol in Plasma or Serum G. L. CATIGNANI 215

Section V. Vitamin K Group

24. Assay of Coumarin Antagonists of Vitamin K in Blood by High-Performance Liquid Chromatography MARTIN J. SHEARER 223
25. Assay of K Vitamins in Tissues by High-Performance Liquid Chromatography with Special Reference to Ultraviolet Detection MARTIN J. SHEARER 235
26. Analysis of Bacterial Menaquinone Mixtures by Reverse-Phase High-Performance Liquid Chromatography JIN TAMAOKA 251

Section VI. Carnitine

27. Radioisotopic Assay of Acetylcarnitine and Acetyl-CoA SHRI V. PANDE 259
28. Short-Chain Acylcarnitines: Identification and Quantitation L. L. BIEBER AND J. KERNER 264
29. Purification and Assay of Carnitine Acyltransferases L. L. BIEBER AND C. FIOL 276
30. Determination of the Specific Activity of Long-Chain Acylcarnitine Esters RICHARD ODESSEY 284
31. Synthesis of Carnitine Precursors and Related Compounds CHARLES J. REBOUCHE 290
32. Synthesis of Radioactive (−)-Carnitine from γ-Aminobutyrate SHRI V. PANDE 297
33. *S*-Adenosylmethionine: ε-*N*-L-Lysine Methyltransferase HARRY P. BROQUIST 303

Section VII. Heme Porphyrins and Derivatives

34. Reflectance Spectrophotometric and Surface Fluorometric Methods for Measuring the Redox State of Nicotinamide Nucleotides and Flavins in Intact Tissues ILMO E. HASSINEN 311
35. Spectrophotometric Determination of Iron in Heme Proteins SERGE N. VINOGRADOV 320
36. Affinity Chromatography of Heme-Binding Proteins: Synthesis and Characterization of Hematin- and Hematoporphyrin-Agarose KENNETH W. OLSEN 324
37. Affinity Chromatography of Heme-Binding Proteins: Synthesis of Hemin-Agarose KEN TSUTSUI 331
38. Porphobilinogen Synthase: A Specific and Sensitive Coupled-Enzyme Assay DAVID F. BISHOP AND ROBERT J. DESNICK 339
39. High-Performance Liquid Chromatographic Analysis of Porphyrins and Their Isomers with Radial Compression Columns ANDREAS SEUBERT AND SIGRID SEUBERT 346
40. High-Performance Liquid Chromatography of Porphyrin Methyl Esters JAMES G. STRAKA 352
41. Synthesis of [^{14}C]Coproporphyrin III by Yeast Cell-Free Extracts PIERRE LABBE 363

42. Isotachophoretic Analysis of Porphyrin Precursors: 5-Aminolevulinic Acid Derivatives — YUZO SHIOI AND TSUTOMU SASA 370
43. Preparation of Stereospecifically Labeled Porphobilinogens — M. AKHTAR AND C. JONES 375
44. High-Performance Liquid Chromatography of Uroporphyrin and Coproporphyrin Isomers — C. K. LIM AND T. J. PETERS 383
45. High-Performance Liquid Chromatographic Separation of the Bilirubin Isomers — C. K. LIM AND T. J. PETERS 389
46. Purification of Hepatic Mitochondrial 5-Aminolevulinate Synthase — IAIN A. BORTHWICK, GOPESH SRIVASTAVA, BYRON A. PIROLA, BRIAN K. MAY, AND WILLIAM H. ELLIOTT 395
47. Purification and Characterization of Mammalian and Chicken Ferrochelatase — HARRY A. DAILEY, JENNIE E. FLEMING, AND BERTILLE M. HARBIN 401
48. Purification and Characterization of Bacterial Ferrochelatase — HARRY A. DAILEY 408
49. Uroporphyrinogen Decarboxylase Purification from Chicken Erythrocytes — YOSHIHIKO SEKI, SHOSUKE KAWANISHI, AND SEIYO SANO 415
50. Purification of Solubilized Chlorophyllase from *Chlorella protothecoides* — YUZO SHIOI AND TSUTOMU SASA 421
51. Purification of Porphobilinogen Synthase from Bovine Liver — PETER M. JORDAN AND JASBIR S. SEEHRA 427
52. Purification of 5-Aminolevulinate Synthase — PETER M. JORDAN AND AGHDAS LAGHAI-NEWTON 435

VOLUME 124
HORMONE ACTION (PART J: NEUROENDOCRINE PEPTIDES)

Section I. Preparation of Chemical Probes

1. Computationally Directed Biorational Drug Design of Peptides — FRANK A. MOMANY AND HIROSHI CHUMAN 3
2. Comparative Methods for the Radiolabeling of Peptides — A. E. BOLTON 18
3. Luciferin Derivitization of Ligands for Development of Nonisotopic Binding Assays — JAMES M. SCHAEFFER 29
4. Synthesis and Use of Colloidal Gold-Coupled Receptor Ligands — LOTHAR JENNES, P. MICHAEL CONN, AND WALTER E. STUMPF 36
5. Preparation and Use of Biotinylated Neuroendocrine Peptides — ELI HAZUM 47
6. Synthesis and Use of Diacylglycerols as Activators of Protein Kinase C in Neuroendocrine Tissue — P. MICHAEL CONN, BARRY R. GANONG, JAMES EBELING, DAPHNE STALEY, JAMES E. NIEDEL, AND ROBERT M. BELL 57

Section II. Equipment and Technology

7. Computer-Controlled Perifusion System for Neuroendocrine Tissues: Development and Applications — ANDRES NEGRO-VILAR AND MICHAEL D. CULLER 67

8. Perifusion of Anterior Pituitary Cells: Release of Gonadotropins and Somatotropins — THOMAS M. BADGER 79
9. Measurement of Intracellular Ionized Calcium with Aequorin — ANDRÉ B. BORLE AND KENNETH W. SNOWDOWNE 90
10. Use of Intracellular Calcium and Membrane Potential Fluorescent Indicators in Neuroendocrine Cells — ALESSANDRO M. CAPPONI, P. DANIEL LEW, WERNER SCHLEGEL, AND TULLIO POZZAN 116
11. The Design and Synthesis of Detergents for Membrane Biochemistry — LEONARD M. HJELMELAND 135
12. Solubilization of Neuropeptide Receptors — MARILYN H. PERRIN 164
13. Solubilization and Characterization of Opiate Receptors — R. SUZANNE ZUKIN AND RHODA MANECKJEE 172
14. Temperature Measurement and Control of Small Volumes: Applications for Single Channel Recording — ANTHONY AUERBACH, FREDERICK SACHS, JAMES NEIL, AND RICHARD MCGARRIGLE 190
15. Techniques for Studying the Role of Electrical Activity in Control of Secretion by Normal Anterior Pituitary Cells — W. T. MASON AND C. D. INGRAM 207

Section III. Preparation and Maintenance of Biological Materials

16. Culture and Characterization of Follicular Cells of the Bovine Anterior Pituitary and Pars Tuberalis — NAPOLEONE FERRARA, PAUL GOLDSMITH, DENNIS FUJII, AND RICHARD WEINER 245
17. Peptidergic Secretory Granules Isolated from the Brain: A Model System for Study of Intracellular Aspects of the Process of Peptide Secretion — AYALLA BARNEA 254

Section IV. Quantitation of Neuroendocrine Substances

18. RNA Dot and Blot Hybridization: Selected Procedures for Endocrine and Neuroendocrine Studies — BRUCE A. WHITE, THOMAS LUFKIN, GREGORY M. PRESTON, AND CARTER BANCROFT 269
19. Gene Transfer Methods for Studying the Regulation and Expression of Neuropeptide Genes — AUDREY SEASHOLTZ, MICHAEL COMB, MITCHEL MARTIN, AND EDWARD HERBERT 278
20. Use of Vaccinia Virus as a Neuropeptide Expression Vector — DENNIS E. HRUBY, GARY THOMAS, EDWARD HERBERT, AND CHRISTINE A. FRANKE 295
21. Detection of Neuroendocrine Peptide Precursor cDNA Clones Using Synthetic Oligonucleotides — JAMES DOUGLASS AND EDWARD HERBERT 309
22. Detection of a High-Molecular-Weight LHRH Precursor by Cell-Free Translation of mRNA from Human, Rat, and Mouse Hypothalamus — ANN CURTIS, MICHAEL SZELKE, AND GEORGE FINK 318

23. Methods for the Identification of Neuropeptide Processing Products: Somatostatin and the Tachykinins ANTHONY J. HARMAR AND PETER M. KEEN 335
24. Assay and Purification of Protein Kinase C
TATSURO KITANO, MASAYOSHI GO, USHIO KIKKAWA, AND YASUTOMI NISHIZUKA 349
25. Mixed Micelle Assay of Protein Kinase C
ROBERT M. BELL, YUSUF HANNUN, AND CARSON LOOMIS 353
26. A Tissue Culture Model for the Study of Peptide Synthesis and Secretion from Microdissected Hypothalamic Explants
MARIANA MORRIS, ROBERT L. ESKAY, AND DAVID K. SUNDBERG 359
27. Measurement of Growth Hormone-Releasing Factor
LAWRENCE A. FROHMAN AND THOMAS R. DOWNS 371
28. Assay of Growth Hormone-Releasing Factor
WYLIE VALE, JOAN VAUGHAN, DIANE JOLLEY, GAYLE YAMAMOTO, THOMAS BRUHN, HANS SEIFERT, MARILYN PERRIN, MICHAEL THORNER, AND JEAN RIVIER 389
29. Microbore HPLC of Biogenic Amines in Small Biological Samples
IVAN N. MEFFORD, EDWARD J. CALIGURI, RICHARD K. GRADY, JR., PETER CAPELLA, TRACY A. DURKIN, AND PAMELA CHEVALIER 402
30. Somatostatin Release from Dissociated Cerebral Cortical Cell Cultures
ROXANNE M. LANDON AND RICHARD J. ROBBINS 412
31. Measurement of Phospholipid Turnover in Cultured Hormone Responsive Pituitary Cells THOMAS F. J. MARTIN 424
32. Detection and Measurement of Secretion from Individual Neuroendocrine Cells Using a Reverse Hemolytic Plaque Assay
PHILIP F. SMITH, ENRIQUE H. LUQUE, AND JIMMY D. NEILL 443
33. Measurement of Neuropeptide Release: *In Vitro* and *in Vivo* Procedures
J. E. LEVINE AND V. D. RAMIREZ 466

Section V. Localization of Neuroendocrine Substances

34. *In Situ* Hybridization for the Study of Gene Expression in the Brain
BRENDA D. SHIVERS, BETH S. SCHACHTER, AND DONALD W. PFAFF 497
35. *In Situ* cDNA:mRNA Hybridization: Development of a Technique to Measure mRNA Levels in Individual Cells
JOSIAH N. WILCOX, CONNIE E. GEE, AND JAMES L. ROBERTS 510
36. Location of Gene Expression in CNS Using Hybridization Histochemistry
J. D. PENSCHOW, J. HARALAMBIDIS, P. ALDRED, G. W. TREGEAR, AND J. P. COGHLAN 534
37. Measurement of Neuroendocrine Peptide mRNA in Discrete Brain Regions
JEFFREY D. WHITE, KIM D. STEWART, AND JEFFREY F. MCKELVY 548
38. Corticotropin-Releasing Factor Receptors in the Pituitary Gland and Central Nervous System: Methods and Overview
ERROL B. DE SOUZA AND MICHAEL J. KUHAR 560
39. Autoradiographic Localization of Brain Receptors for Peptide Hormones: Angiotensin II, Corticotropin-Releasing Factor, and Gonadotropin-Releasing Hormone
MONICA MILLAN, GRETI AGUILERA, PETER C. WYNN, FREDERICK A. O. MENDELSOHN, AND KEVIN J. CATT 590

40. Localization of Somatostatin Receptors
G. PELLETIER, P. LEROUX, AND G. MOREL 607
41. Methods for Estimating the Contribution of Proenkephalin A and Proenkephalin B Input to Neuronal Areas
DAVID K. SUNDBERG AND CLAUD E. DUNLAP III 617

Section VI. Summary

42. Previously Published Articles from Methods in Enzymology Related to Neuroendocrine Peptides 629

VOLUME 125

BIOMEMBRANES (PART M: TRANSPORT IN BACTERIA, MITOCHONDRIA, AND CHLOROPLASTS: GENERAL APPROACHES AND TRANSPORT SYSTEMS)

Section I. General Approaches

1. Energetics of Proton Transport and Secondary Transport HAGAI ROTTENBERG 3

A. Membrane Characteristics and Chemistry

2. Isolation of Highly Coupled Heart Mitochondria in High Yield Using a Bacterial Collagenase
PETER P. TOTH, SHELAGH M. FERGUSON-MILLER, AND CLARENCE H. SUELTER 16
3. Alkyl Glycoside Detergents: Synthesis and Applications to the Study of Membrane Proteins
T. VANAKEN, S. FOXALL-VANAKEN, S. CASTLEMAN, AND S. FERGUSON-MILLER 27
4. Lipid Enrichment and Fusion of Mitochondrial Inner Membranes
CHARLES R. HACKENBROCK AND BRAD CHAZOTTE 35
5. Changes of Surface and Membrane Potentials in Biomembranes
NAOKI KAMO AND YONOSUKE KOBATAKE 46
6. Rate of Redox Reactions Related to Surface Potential and Other Surface-Related Parameters in Biological Membranes
SHIGERU ITOH AND MITSUO NISHIMURA 58
7. Use of *N*,*N*′-Dicyclohexylcarbodiimide to Study Membrane-Bound Enzymes
MACIEJ J. NAŁĘCZ, ROBERT P. CASEY, AND ANGELO AZZI 86
8. Voltammetric Measurements of Quinones
ROGER C. PRINCE, PAUL LLOYD-WILLIAMS, J. MALCOLM BRUCE, AND P. LESLIE DUTTON 109
9. Determination of Partition Coefficients of Quinones by High-Performance Liquid Chromatography
BARRY S. BRAUN, ULI BENBOW, PAUL LLOYD-WILLIAMS, J. MALCOLM BRUCE, AND P. LESLIE DUTTON 119

B. Membrane Assembly, Mutations, and Cloning Strategy

10. Function of Protonmotive Force in Translocation of Protein across Membranes
LINDA L. RANDALL 129
11. Use of Secretion Cloning Vectors for Guiding the Localization of Proteins in *Escherichia coli*
CHARLES A. LUNN, MASAYASU TAKAHARA, AND MASAYORI INOUYE 138
12. Use of *lac* Gene Fusions to Study Transport Proteins HOWARD A. SHUMAN 150
13. Use of Transposons to Isolate and Characterize Mutants Lacking Membrane Proteins, Illustrated by the Sugar Transport Systems of *Escherichia coli*
MAURICE C. JONES-MORTIMER AND PETER J. F. HENDERSON 157
14. ECF Locus in *Escherichia coli*: Defect in Energization for ATP Synthesis and Active Transport JEN-SHIANG HONG 180
15. Mutations in the *eup* Locus of *Escherichia coli*, Energy Uncoupled Phenotype
CHARLES A. PLATE 187
16. Methods for Mutagenesis of the Bacterioopsin Gene
MARIE A. GILLES-GONZALEZ, NEIL R. HACKETT, SIMON J. JONES, H. GOBIND KHORANA, DAE-SIL LEE, KIN-MING LO, AND JOHN M. MCCOY 190
17. Oligonucleotide-Directed Site-Specific Mutagenesis of the *lac* Permease of *Escherichia coli*
HEMANTA K. SARKAR, PAUL V. VIITANEN, ETANA PADAN, WILLIAM R. TRUMBLE, MOHINDAR S. POONIAN, WARREN MCCOMAS, AND H. RONALD KABACK 214
18. DNA Sequence and Transcription of Genes for ATP Synthase Subunits from Photosynthetic Bacteria
VICTOR L. J. TYBULEWICZ, GUNNAR FALK, AND JOHN E. WALKER 230

Section II. Bacterial Transport

19. Membrane Transport in Rickettsiae HERBERT H. WINKLER 253
20. Transport in Mycoplasmas SHLOMO ROTTEM AND VINCENT P. CIRILLO 259
21. Transport through the Outer Membrane of Bacteria HIROSHI NIKAIDO 265
22. Binding Protein-Dependent Active Transport in *Escherichia coli* and *Salmonella typhimurium* CLEMENT E. FURLONG 279
23. Calcium-Induced Permeabilization of the Outer Membrane: A Method for Reconstitution of Periplasmic Binding Protein-Dependent Transport Systems in *Escherichia coli* and *Salmonella typhimurium*
JOHANN M. BRASS 289
24. Reconstitution of Periplasmic Binding Protein-Dependent Glutamine Transport in Vesicles ARTHUR G. HUNT AND JEN-SHIANG HONG 302
25. Isolation and Crystallization of Bacterial Porin
R. MICHAEL GARAVITO AND JÜRG P. ROSENBUSCH 309
26. Ion Extrusion Systems in *Escherichia coli* BARRY P. ROSEN 328
27. Intracellular pH Regulation in Bacterial Cells
ETANA PADAN AND SHIMON SCHULDINER 337
28. Regulation of Internal pH in Acidophilic and Alkalophilic Bacteria
T. A. KRULWICH AND A. A. GUFFANTI 352
29. Peptide Transport in Bacteria
CHRISTOPHER F. HIGGINS AND MARGARET M. GIBSON 365

30. Methods for the Study of the Melibiose Carrier of *Escherichia coli*
DOROTHY M. WILSON, TOMOFUSA TSUCHIYA, AND T. HASTINGS WILSON 377
31. Assay, Genetics, Proteins, and Reconstitution of Proton-Linked Galactose, Arabinose, and Xylose Transport Systems of *Escherichia coli*
PETER J. F. HENDERSON AND ANDREW J. S. MACPHERSON 387
32. Purification, Reconstitution, and Characterization of the *lac* Permease of *Escherichia coli*
PAUL VIITANEN, MICHAEL J. NEWMAN, DAVID L. FOSTER, T. HASTINGS WILSON, AND H. RONALD KABACK 429
33. Preparation of Monoclonal Antibodies and Site-Directed Polyclonal Antibodies against the *lac* Permease of *Escherichia coli*
NANCY CARRASCO, DORIS HERZLINGER, WALEED DANHO, AND H. RONALD KABACK 453
34. Glycerol Facilitator in *Escherichia coli* E. C. C. LIN 467
35. Polyhydric Alcohol Transport by Bacteria JOSEPH W. LENGELER 473
36. Alanine Carrier from Thermophilic Bacteria H. HIRATA 485
37. Generation of a Protonmotive Force in Anaerobic Bacteria by End-Product Efflux BART TEN BRINK AND WIL N. KONINGS 492
38. Light-Driven Chloride Transport in Halorhodopsin-Containing Cell Envelope Vesicles BRIGITTE SCHOBERT AND JANOS K. LANYI 510
39. Sodium Translocation by NADH Oxidase of *Vibrio alginolyticus*: Isolation and Characterization of the Sodium Pump-Defective Mutants HAJIME TOKUDA 520
40. Preparation, Characterization, and Reconstitution of Oxaloacetate Decarboxylase from *Klebsiella aerogenes*, a Sodium Pump PETER DIMROTH 530
41. Sodium Pump Methylmalonyl-CoA Decarboxylase from *Veillonella alcalescens*
WILHELM HILPERT AND PETER DIMROTH 540
42. Biotin-Dependent Decarboxylases as Bacterial Sodium Pumps: Purification and Reconstitution of Glutaconyl-CoA Decarboxylase from *Acidaminococcus fermentans* WOLFGANG BUCKEL 547
43. Anion Exchange in Bacteria: Reconstitution of Phosphate:Hexose 6-Phosphate Antiport from *Streptococcus lactis*
SURESH V. AMBUDKAR AND PETER C. MALONEY 558
44. Proton-Driven Bacterial Flagellar Motor ROBERT M. MACNAB 563
45. Sodium-Driven Flagellar Motors of Alkalophilic *Bacillus*
YASUO IMAE, HIROSHI MATSUKURA, AND SYOYU KOBAYASHI 582

Section III. Transport in Mitochondria and Chloroplasts

A. Mitochondria

46. Mitochondrial Outer Membrane Channel (VDAC, Porin) Two-Dimensional Crystals from *Neurospora* CARMEN A. MANNELLA 595
47. Reconstitution of ADP/ATP Translocase in Phospholipid Vesicles
REINHARD KRÄMER 610
48. Fluorescent Nucleotide Analogs as Active Site Probes for the ADP/ATP Carrier and the Uncoupling Protein MARTIN KLINGENBERG 618
49. Techniques for NMR and ESR Studies on the ADP/ATP Carrier
KLAUS BEYER AND ANTON MUNDING 631

50. Fluorescent Probes of the Mitochondrial ADP/ATP Carrier Protein
MARC R. BLOCK, FRANÇOIS BOULAY, GÉRARD BRANDOLIN, YVES DUPONT, GUY J. M. LAUQUIN, AND PIERRE V. VIGNAIS 639
51. ADP/ATP Carrier: Analysis of Transmembrane Folding Using Pyridoxal Phosphate WERNER BOGNER AND HEINRICH AQUILA 650
52. Chemical Modifications and Active Site Labeling of the Mitochondrial ADP/ATP Carrier
MARC R. BLOCK, FRANÇOIS BOULAY, GÉRARD BRANDOLIN, GUY J. M. LAUQUIN, AND PIERRE V. VIGNAIS 658
53. Measurement of Citrate Transport in Tumor Mitochondria
RONALD S. KAPLAN, RISA A. PARLO, AND PETER S. COLEMAN 671
54. Partial Purification and Reconstitution of the Tricarboxylate Carrier from Rat Liver Mitochondria
FERDINANDO PALMIERI, ITALO STIPANI, GIROLAMO PREZIOSO, AND REINHARD KRÄMER 692
55. Isolation and Reconstitution of the Phosphate Transport Protein from Mitochondria HARTMUT WOHLRAB, HANNO V. J. KOLBE, AND ANNE COLLINS 697

B. Chloroplasts

56. Transport of Metabolites across the Chloroplast Envelope
H. W. HELDT AND U. I. FLÜGGE 705
57. Chloroplast Phosphate–Triose Phosphate–Phosphoglycerate Translocator: Its Identification, Isolation, and Reconstitution
U. I. FLÜGGE AND H. W. HELDT 716

VOLUME 126

BIOMEMBRANES (PART N: TRANSPORT IN BACTERIA, MITOCHONDRIA, AND CHLOROPLASTS: PROTONMOTIVE FORCE)

1. Protonmotive Force and Secondary Transport: Historical Perspectives and Unifying Principles YASUO KAGAWA 1

Section I. Electron Transfer

A. Cytochrome Oxidase

2. Measurement of the H^+-Pumping Activity of Reconstituted Cytochrome Oxidase
ROBERT P. CASEY 13
3. Structure of Beef Heart Cytochrome-*c* Oxidase Obtained by Combining Studies of Two-Dimensional Crystals with Biochemical Experiments
RODERICK A. CAPALDI AND YU-ZHONG ZHANG 22
4. Isozymes of Cytochrome-*c* Oxidase: Characterization and Isolation from Different Tissues
BERNHARD KADENBACH, ANNEMARIE STROH, MARGIT UNGIBAUER, LUCIA KUHN-NENTWIG, URSULA BÜGE, AND JOCHEN JARAUSCH 32

5. Techniques for the Study of Bovine Cytochrome-*c* Oxidase Monomer–Dimer Association
KATARZYNA A. NAŁĘCZ, REINHARD BOLLI, AND ANGELO AZZI 45
6. Affinity Chromatography Purification of Cytochrome-*c* Oxidase from Bovine Heart Mitochondria and Other Sources
CLEMENS, BROGER, KURT BILL, AND ANGELO AZZI 64
7. Resolution of Cytochrome-*c* Oxidase
ANGELO AZZI, CLEMENS BROGER, KURT BILL, AND MICHAEL J. CORBLEY 72
8. Functional Reconstitution of Proton-Pumping Cytochrome-*c* Oxidase in Phospholipid Vesicles
MICHELE MÜLLER, MARCUS THELEN, PAUL O'SHEA, AND ANGELO AZZI 78
9. Purification and Reconstitution of the Cytochrome *d* Terminal Oxidase Complex from *Escherichia coli* MICHAEL J. MILLER AND ROBERT B. GENNIS 87
10. Purification and Properties of Two Terminal Oxidase Complexes of *Escherichia coli* Aerobic Respiratory Chain
KIYOSHI KITA, KIYOSHI KONISHI, AND YASUHIRO ANRAKU 94
11. Purification and Reconstitution of the Cytochrome *o*-Type Oxidase from *Escherichia coli*
KAZUNOBU MATSUSHITA, LEKHA PATEL, AND H. RONALD KABACK 113
12. Functional Reconstitution of Bacterial Cytochrome Oxidases in Planar Lipid Bilayers TOSHIRO HAMAMOTO AND MAURICIO MONTAL 123
13. Purification of the aa_3-Type Cytochrome-*c* Oxidase from *Rhodopseudomonas sphaeroides* ANGELO AZZI AND ROBERT B. GENNIS 138
14. Cytochrome Oxidase from Thermophilic Bacterium PS3 NOBUHITO SONE 145
15. Cytochrome *c* Oxidase from *Paracoccus denitrificans* BERND LUDWIG 153
16. Purification, Enzymatic Properties, and Reconstitution of Cytochrome-*c* Oxidase from *Bacillus subtilis*
WIM DE VRIJ, BERT POOLMAN, WIL N. KONINGS, AND ANGELO AZZI 159

B. Cytochrome bc_1 Complex

17. Preparation of Complex III from Yeast Mitochondria and Related Methodology
LIVIU CLEJAN AND DIANA S. BEATTIE 173
18. Purification of Cytochrome bc_1 Complexes from Phylogenically Diverse Species by a Single Method
PER O. LJUNGDAHL, JEFFREY D. PENNOYER, AND BERNARD L. TRUMPOWER 181
19. Preparation of Membrane Crystals of Ubiquinol–Cytochrome-*c* Reductase from *Neurospora* Mitochondria and Structure Analysis by Electron Microscopy
HANNS WEISS, SVEN HOVMÖLLER, AND KEVIN LEONARD 191
20. Reconstitution of Ubiquinol–Cytochrome-*c* Reductase from *Neurospora* Mitochondria with Regard to Subunits I and II PETRA LINKE AND HANNS WEISS 201
21. Resolution and Reconstitution of the Iron–Sulfur Protein of the Cytochrome bc_1 Segment of the Mitochondrial Respiratory Chain
CAROL A. EDWARDS AND BERNARD L. TRUMPOWER 211
22. Isolation of the Eleven Protein Subunits of the bc_1 Complex from Beef Heart
H. SCHÄGGER, TH. A. LINK, W. D. ENGEL, AND G. VON JAGOW 224
23. Preparation of Hinge Protein and Its Requirement for Interaction of Cytochrome *c* with Cytochrome c_1 TSOO E. KING AND CHONG H. KIM 238
24. Use of Specific Inhibitors on the Mitochondrial bc_1 Complex
G. VON JAGOW AND TH. A. LINK 253

25. Preparations of Electrogenic, Proton-Transporting Cytochrome Complexes of the b_6f-Type (Chloroplasts and Cyanobacteria) and bc_1-Type (*Rhodopseudomonas sphaeroides*) GÜNTER HAUSKA 271
26. Reconstitution of H^+ Translocation and Photophosphorylation with Photosystem I Reaction Centers, PMS, and CF_1CF_0 GÜNTER HAUSKA AND NATHAN NELSON 285
27. Construction of the Photosynthetic Reaction Center–Mitochondrial Ubiquinol–Cytochrome-*c* Oxidoreductase Hybrid System CHRISTOPHER C. MOSER, KATHLEEN M. GIANGIACOMO, KATSUMI MATSUURA, SIMON DEVRIES, AND P. LESLIE DUTTON 293
28. Isolation of Ubiquinol Oxidase from *Paracoccus denitrificans* EDWARD A. BERRY AND BERNARD L. TRUMPOWER 305

C. Other Electron Transfer

29. Isolation of a Three-Subunit Cytochrome bc_1 Complex from *Paracoccus denitrificans* XIAOHANG YANG AND BERNARD L. TRUMPOWER 316
30. A Simple, One-Step Purification for Cytochrome *b* from the bc_1 Complexes of Bacteria WILLIAM E. PAYNE AND BERNARD L. TRUMPOWER 325
31. Cooperative Proton-Transfer Reactions in the Respiratory Chain: Redox Bohr Effects S. PAPA, F. GUERRIERI, AND G. IZZO 331
32. Preparation of Two-Dimensional Crystals of Complex I and Image Analysis EGBERT J. BOEKEMA, MARIN G. VAN HEEL, AND ERNST F. J. VAN BRUGGEN 344
33. Purification and Reconstitution of Bovine Heart Mitochondrial Transhydrogenase LICIA N. Y. WU, JULIA A. ALBERTA, AND RONALD R. FISHER 353
34. Isolation of the Iron–Sulfur-Containing Polypeptides of NADH:Ubiquinone Oxidoreductase C. IAN RAGAN AND YOUSSEF HATEFI 360
35. Monoclonal Affinity Purification of D-Lactate Dehydrogenase from *Escherichia coli* EUGENIO SANTOS AND H. RONALD KABACK 370
36. Fumarate Reductase of *Escherichia coli* BERNARD D. LEMIRE AND JOEL H. WEINER 377
37. Reconstitution of a Functional Electron-Transfer Chain from Purified Formate Dehydrogenase and Fumarate Reductase Complexes GOTTFRIED UNDEN AND ACHIM KRÖGER 387
38. Molecular Properties, Genetics, and Biosynthesis of *Bacillus subtilis* Succinate Dehydrogenase Complex LARS HEDERSTEDT 399

Section II. Reversible ATP Synthase (F_0F_1-ATPase)

A. Preparation and Reconstitution

39. Preparation of a Highly Coupled H^+-Transporting ATP Synthase from Pig Heart Mitochondria DANIÈLE C. GAUTHERON, FRANÇOIS PENIN, GILBERT DELÉAGE, AND CATHERINE GODINOT 417
40. Resolution and Reconstitution of F_0F_1-ATPase in Beef Heart Submitochondrial Particles L. ERNSTER, T. HUNDAL, AND G. SANDRI 428
41. Electron Microscopy of Single Molecules and Crystals of F_1-ATPases CHRISTOPHER W. AKEY, STANLEY D. DUNN, VITALY SPITSBERG, AND STUART J. EDELSTEIN 434

42. Isolation and Reconstitution of Membrane-Bound Pyrophosphatase from Beef Heart Mitochondria
I. S. Kulaev, S. E. Mansurova, and Yu. A. Shakhov 447

43. Use of Monoclonal Antibodies to Purify Oligomycin Sensitivity-Conferring Protein and to Study Its Interactions with F_0 and F_1
Philippe Archinard, François Penin, Catherine Godinot, and Danièle C. Gautheron 455

44. Purification and Properties of the ATPase Inhibitor from Bovine Heart Mitochondria Maynard E. Pullman 460

45. Purification of the Proton-Translocating ATPase from Rat Liver Mitochondria Using the Detergent 3-[(3-Cholamidopropyl)dimethylammonio]-1-propane Sulfonate Maureen W. McEnery and Peter L. Pedersen 470

46. Rapid Purification of F_1-ATPase from Rat Liver Mitochondria Using a Modified Chloroform Extraction Procedure Coupled to High-Performance Liquid Chromatography Noreen Williams and Peter L. Pedersen 477

47. Purification of α and β Subunits and Subunit Pairs from Rat Liver Mitochondrial F_1-ATPase Noreen Williams and Peter L. Pedersen 484

48. Isolation and Hydrodynamic Characterization of the Uncoupling Protein from Brown Adipose Tissue Martin Klingenberg and Chi-shui Lin 490

49. Nucleotide Binding Assay for Uncoupling Protein from Brown Fat Mitochondria
Martin Klingenberg, Maria Herlt, and Edith Winkler 498

50. Isolation and Characterization of an Inactivated Complex of F_1F_0-ATPase and Its Inhibitory Factors from Yeast
Kunio Tagawa, Tadao Hashimoto, and Yukuo Yoshida 504

51. Isolation and Reconstitution of CF_0-F_1 from Chloroplasts Uri Pick 512

52. Extraction, Purification, and Reconstruction of the Chloroplast *N,N'*-Dicyclohexylcarbodiimide-Binding Proteolipid
Angelo Azzi, Kristine Sigrist-Nelson, and Nathan Nelson 520

53. Selective Extraction and Reconstitution of F_1 Subunits from *Rhodospirillum rubrum* Chromatophores
Zippora Gromet-Elhanan and Daniel Khananshvili 528

54. Preparation and Reconstitution of the Protein-Pumping Membrane-Bound Inorganic Pyrophosphatase from *Rhodospirillum rubrum*
Margareta Baltscheffsky and Pål Nyrén 538

55. Purification of F_1F_0 H^+-ATPase from *Escherichia coli*
Robert H. Fillingame and David L. Foster 545

56. Use of λ-*unc* Transducing Phages in Genetic Analysis of H^+-ATPase Mutants of *Escherichia coli* Robert H. Fillingame and Mary E. Mosher 558

57. Proton-Conducting Portion (F_0) from *Escherichia coli* ATP Synthase: Preparation, Dissociation into Subunits, and Reconstitution of an Active Complex
Erwin Schneider and Karlheinz Altendorf 569

58. Preparation and Reconstitution of F_1F_0 and F_0 from *Escherichia coli*
Peter Friedl and Hans Ulrich Schairer 579

59. Use of Isolated Subunits of F_1 from *Escherichia coli* for Genetic and Biochemical Studies Masamitsu Futai and Hiroshi Kanazawa 588

60. Analysis of *Escherichia coli* Mutants of the H^+-Transporting ATPase: Determination of Altered Site of the Structural Genes
Hiroshi Kanazawa, Takato Noumi, and Masamitsu Futai 595

61. Proton Permeability of Membrane Sector (F_0) of H^+-Transporting ATP Synthase (F_0F_1) from a Thermophilic Bacterium
Nobuhito Sone and Yasuo Kagawa 604

B. Kinetics, Modification, and Other Characterization

62. Rate Constants and Equilibrium Constants for the Elementary Steps of ATP Hydrolysis by Beef Heart Mitochondrial ATPase HARVEY S. PENEFSKY 608
63. Refinements in Oxygen-18 Methodology for the Study of Phosphorylation Mechanisms KERSTIN E. STEMPEL AND PAUL D. BOYER 618
64. ATP Formation in Mitochondria, Submitochondrial Particles, and F_0F_1 Liposomes Driven by Electric Pulses YASUO KAGAWA AND TOSHIRO HAMAMOTO 640
65. Synthesis of Enzyme-Bound Adenosine Triphosphate by Chloroplast Coupling Factor CF_1 RICHARD I. FELDMAN 643
66. Divalent Azido-ATP Analog for Photoaffinity Cross-Linking of F_1 Subunits HANS-JOCHEN SCHÄFER 649
67. Synthesis and Use of an Azido-Labeled Form of the ATPase Inhibitor Peptide of Rat Liver Mitochondria KLAUS SCHWERZMANN AND PETER L. PEDERSEN 660
68. Benzophenone-ATP: A Photoaffinity Label for the Active Site of ATPase NOREEN WILLIAMS, SHARON H. ACKERMAN, AND PETER S. COLEMAN 667
69. Use of ADP Analogs for Functional and Structural Analysis of F_1-ATPase GÜNTER SCHÄFER, UWE LÜCKEN, AND MATHIAS LÜBBEN 682
70. Modifiers of F_1-ATPases and Associated Peptides MICHEL SATRE, JOËL LUNARDI, ANNE-CHRISTINE DIANOUX, ALAIN DUPUIS, JEAN PAUL ISSARTEL, GÉRARD KLEIN, RICHARD POUGEOIS, AND PIERRE V. VIGNAIS 712
71. A Nonlinear Approach for the Analysis of Different Models of Protein–Ligand Interaction: Nucleotide Binding to F_1-ATPase FRENS PETERS AND UWE LÜCKEN 733
72. Identification of Essential Residues in the F_1-ATPases by Chemical Modification WILLIAM S. ALLISON, DAVID A. BULLOUGH, AND WILLIAM W. ANDREWS 741
73. Monoclonal Antibodies to F_1-ATPase Subunits as Probes of Structure, Conformation, and Functions of Isolated or Membrane-Bound F_1 CATHERINE GODINOT, MAHNAZ MORADI-AMELI, AND DANIÈLE C. GAUTHERON 761
74. Determination of the Stoichiometry and Arrangement of α and β Subunits in F_1-ATPase Using Monoclonal Antibodies in Immunoelectron Microscopy KARIN EHRIG, HEINRICH LÜNSDORF, PETER FRIEDL, AND HANS ULRICH SCHAIRER 770

VOLUME 127

BIOMEMBRANES (PART O: PROTONS AND WATER: STRUCTURE AND TRANSLOCATION)

Section I. Interactions between Water, Ions, and Biomolecules

1. Methods to Determine Structure in Water and Aqueous Solutions HAROLD L. FRIEDMAN 3
2. Structural Chemistry of Biomolecular Hydration via Computer Simulation: The Proximity Criterion MIHALY MEZEI AND DAVID L. BEVERIDGE 21

3. Theoretical and Computational Studies of Hydrophobic Interactions
LAWRENCE R. PRATT AND DAVID CHANDLER 48
4. Theoretical Methods for Obtaining Free Energies of Biomolecular Equilibria in Aqueous Solutions ANDREW POHORILLE AND LAWRENCE R. PRATT 64
5. Proton Tunneling AKINORI SARAI AND DON DEVAULT 79
6. Low-Frequency Raman Scattering from Water and Aqueous Solutions: A Direct Measure of Hydrogen Bonding G. E. WALRAFEN AND M. R. FISHER 91
7. Vibrational Methods for Water Structure in Nonpolar Media
HERBERT L. STRAUSS 106
8. Simulations of Complex Chemical Systems ENRICO CLEMENTI 114
9. Calorimetric Methods for Measurement of Interactions of Biomolecules with Water GORDON C. KRESHECK 141
10. Nuclear Magnetic Resonance Study of Water Interactions with Proteins, Biomolecules, Membranes, and Tissues B. M. FUNG 151
11. Determination of Water Structure around Biomolecules Using X-Ray and Neutron Diffraction Methods HUGH SAVAGE AND ALEXANDER WLODAWER 162
12. Orientation of Amino Acid Side Chains: Intraprotein and Solvent Interactions
GEORGE NÉMETHY 183
13. Correlation of Protein Dynamics with Water Mobility: Mössbauer Spectroscopy and Microwave Absorption Methods FRITZ PARAK 196
14. Protein Dynamics and Hydration
HANS FRAUENFELDER AND ENRICO GRATTON 207
15. Quantitation of Water in Membranes by Neutron Diffraction and X-Ray Techniques R. B. KNOTT AND B. P. SCHOENBORN 217
16. *Artemia* Cysts as a Model for the Study of Water in Biological Systems
JAMES S. CLEGG 230
17. Determining Intracellular Viscosity from the Rotational Motion of Spin Labels
PHILIP D. MORSE, II 239
18. Computation of Energy Profiles in the Gramicidin A Channel
ALBERTE PULLMAN 250
19. Proton Movements in Inorganic Materials J. B. GOODENOUGH 263
20. Solid-Phase Protein Hydration Studies
PHILIP L. POOLE AND JOHN L. FINNEY 284
21. Antifreeze Glycopeptides and Peptides: Interactions with Ice and Water
ARTHUR L. DEVRIES 293
22. Methods for the Study of Water in Ice Phases W. F. KUHS 303
23. Diffraction Techniques for the Study of Pure Water and Aqueous Solutions
J. E. ENDERBY AND G. W. NEILSON 319
24. A Method of Solvent Structure Analysis for Proteins Using D_2O–H_2O Neutron Difference Maps JOSEPH SHPUNGIN AND ANTHONY A. KOSSIAKOFF 329
25. Fourier Transform Infrared Studies of an Active Proton Transport Pump
KENNETH J. ROTHSCHILD 343
26. Direct Methods for Measuring Conformational Water Forces (Hydration Forces) between Membrane and Other Surfaces
JACOB ISRAELACHVILI AND JOHAN MARRA 353
27. Cell Water Viscosity ALEC D. KEITH AND ANDREA M. MASTRO 360
28. Theoretical Approaches to Solvation of Biopolymers
CHARLES L. BROOKS, III AND MARTIN KARPLUS 369
29. Osmotic Stress for the Direct Measurement of Intermolecular Forces
V. A. PARSEGIAN, R. P. RAND, N. L. FULLER, AND D. C. RAU 400

Section II. Protons and Membrane Functions

A. Theoretical/Model Membrane Methods

30. Proton Conduction through Proteins: An Overview of Theoretical Principles and Applications Z. SCHULTEN AND K. SCHULTEN 419
31. Proton Polarizability of Hydrogen Bonds: Infrared Methods, Relevance to Electrochemical and Biological Systems GEORG ZUNDEL 439
32. Theoretical Calculation of Energetics of Proton Translocation through Membranes STEVE SCHEINER 456
33. Barrier Models for the Description of Proton Transport across Membranes P. LÄUGER 465
34. Proton Permeation through Model Membranes DAVID W. DEAMER AND JOHN GUTKNECHT 471
35. Lipid Phase Transitions: Water and Proton Permeability ALFRED BLUME 480
36. Planar Bilayers: A Powerful Tool to Study Membrane Proteins Involved in Ion Transport ALBERTO DARSZON 486
37. Electron Paramagnetic Resonance Methods for Measuring H^+/OH^- Fluxes across Phospholipid Vesicles DAVID S. CAFISO 502
38. Depth of Water Penetration into Lipid Bilayers S. A. SIMON AND T. J. MCINTOSH 511
39. Application of the Laser-Induced Proton Pulse for Measuring the Protonation Rate Constants of Specific Sites on Proteins and Membranes MENACHEM GUTMAN 522

B. Natural Membrane Methods

40. Localized Protonic Coupling: Overview and Critical Evaluation of Techniques DOUGLAS B. KELL 538
41. Methods for the Determination of Membrane Potential in Bioenergetic Systems J. BAZ. JACKSON AND DAVID G. NICHOLLS 557
42. Correlation between the Structure and Efficiency of Light-Induced Proton Pumps A. WARSHEL 578
43. Resonance Raman Methods for Proton Translocation in Bacteriorhodopsin MARK S. BRAIMAN 587
44. Water Channels ROBERT I. MACEY AND TERESA MOURA 598
45. Displacement Current in Proteins as a Measure of Proton Movements: Bacteriorhodopsin LAJOS KESZTHELYI 610
46. Measurement of Density and Location of Solvent Associated with Biomolecules by Small-Angle Neutron Scattering GIUSEPPE ZACCAI 619
47. Use of Hydrogen Exchange Kinetics in the Study of the Dynamic Properties of Biological Membranes ANDREAS ROSENBERG 630
48. Bacteriorhodopsin: Fourier Transform Infrared Methods for Studies of Protonation of Carboxyl Groups GAVIN DOLLINGER, LAURA EISENSTEIN, SHUO-LIANG LIN, KOJI NAKANISHI, KAZUNORI ODASHIMA, AND JOHN TERMINI 649
49. Deuteration as a Tool in Investigating the Role of Water in the Structure and Function of Excitable Membranes V. VASILESCU AND EVA KATONA 662
50. Spectroscopic Methods for the Determination of Membrane Surface Charge Density BENJAMIN EHRENBERG 678

51. Water and Carbohydrate Interactions with Membranes: Studies with Infrared Spectroscopy and Differential Scanning Calorimetry Methods
JOHN H. CROWE AND LOIS M. CROWE 696
52. Freeze-Fracture Methods: Preparation of Complementary Replicas for Evaluating Intracellular Ice Damage in Ultrarapidly Cooled Specimens
M. JOSEPH COSTELLO AND RICHARD D. FETTER 704
53. Electron Microscopy of Frozen Hydrated Specimens: Preparation and Characteristics JEAN LEPAULT AND JACQUES DUBOCHET 719
54. Bacterial Nucleation of Ice in Plant Leaves
SUSAN S. HIRANO AND CHRISTEN D. UPPER 730
55. Measurement of Transmembrane Proton Movements with Nitroxide Spin Probes
ROLF J. MEHLHORN, LESTER PACKER, ROBERT MACEY, ALEXANDRU T. BALABAN, AND ILEANA DRAGUTAN 738
56. Determination of the Net Proton–Hydroxide Ion Permeability across Vesicular Lipid Bilayers and Membrane Proteins by Optical Probes
NORBERT A. DENCHER, PETRA A. BURGHAUS, AND STEPHAN GRZESIEK 746
57. Methods to Study the Freezing Process in Plants
C. B. RAJASHEKAR AND M. J. BURKE 761
58. Reconstitution of an H^+ Translocator, the "Uncoupling Protein" from Brown Adipose Tissue Mitochondria, in Phospholipid Vesicles
MARTIN KLINGENBERG AND EDITH WINKLER 772

VOLUME 128
PLASMA LIPOPROTEINS (PART A: PREPARATION, STRUCTURE, AND MOLECULAR BIOLOGY)

Section I. Overview

1. Introduction to the Plasma Lipoproteins
ANTONIO M. GOTTO, JR., HENRY J. POWNALL, AND RICHARD J. HAVEL 3
2. Molecular and Cell Biology of Lipoprotein Biosynthesis
DONNA M. DRISCOLL AND GODFREY S. GETZ 41
3. Comparative Analysis of Mammalian Plasma Lipoproteins
M. JOHN CHAPMAN 70
4. Impact of Technology on the Plasma Lipoprotein Field ANGELO M. SCANU 144

Section II. Preparation of Plasma Lipoproteins

5. Precautionary Measures for Collecting Blood Destined for Lipoprotein Isolation
CELINA EDELSTEIN AND ANGELO M. SCANU 151
6. Sequential Flotation Ultracentrifugation
VERNE N. SCHUMAKER AND DONALD L. PUPPIONE 155
7. Density Gradient Ultracentrifugation of Serum Lipoproteins in a Swinging Bucket Rotor JIM L. KELLEY AND ARTHUR W. KRUSKI 170

8. Single Vertical Spin Density Gradient Ultracentrifugation
BYUNG H. CHUNG, JERE P. SEGREST, MARJORIE J. RAY, JOHN D. BRUNZELL, JOHN E. HOKANSON, RONALD M. KRAUS, KEN BEAUDRIE, AND JOHN T. CONE 181

Section III. Isolation and Physical–Chemical Characterization of Plasma Apolipoproteins

9. Delipidation of Plasma Lipoproteins JAMES C. OSBORNE, JR. 213
10. Isolation and Characterization of Apolipoproteins A-I, A-II, and A-IV
H. BRYAN BREWER, JR., ROSEMARY RONAN, MARTHA MENG, AND CHERI BISHOP 223
11. Isolation and Characterization of Apolipoprotein B-100
WALDO R. FISHER AND VERNE N. SCHUMAKER 247
12. Isolation and Characterization of Apolipoprotein B-48
DAVID A. HARDMAN AND JOHN P. KANE 262
13. Isolation and Characterization of Apolipoprotein E
STANLEY C. RALL, JR., KARL H. WEISGRABER, AND ROBERT W. MAHLEY 273
14. Isolation and Properties of Human Apolipoproteins C-I, C-II, and C-III
RICHARD L. JACKSON AND GEORGE HOLDSWORTH 288
15. Isolation and Characterization of Other Apolipoproteins
W. J. MCCONATHY AND P. ALAUPOVIC 297
16. Serum Amyloid A (ApoSAA) and Lipoproteins
NILS ERIKSEN AND EARL P. BENDITT 311
17. Structure and Lipid Binding Properties of Serum Albumin
ARTHUR A. SPECTOR 320
18. High-Performance Liquid Chromatography of Apolipoproteins
CELINA EDELSTEIN AND ANGELO M. SCANU 339
19. Immunochemical Isolation and Identification of Glucosylated Apolipoproteins
LINDA K. CURTISS, MICHAEL G. PEPE, AND JOSEPH L. WITZTUM 354
20. Isothermal Calorimetry of Apolipoproteins M. ROSSENEU 365
21. Solution Properties of Apolipoproteins
JAMES C. OSBORNE, JR., NANCY S. LEE, AND GRACE M. POWELL 375
22. Studies of Apolipoproteins at the Air–Water interface
MICHAEL C. PHILLIPS AND KEITH E. KREBS 387
23. Thermodynamics of Apolipoprotein–Phospholipid Association
JOHN B. MASSEY AND HENRY J. POWNALL 403

Section IV. Structure of Intact and Reconstituted Plasma Lipoproteins

24. Nondenaturing Polyacrylamide Gradient Gel Electrophoresis
ALEX V. NICHOLS, RONALD M. KRAUS, AND THOMAS A. MUSLINER 417
25. Isoelectric Focusing of Plasma Lipoproteins
YVES L. MARCEL AND PHILIP K. WEECH 432
26. Electron Microscopy of Negatively Stained Lipoproteins
TRUDY M. FORTE AND ROBERT W. NORDHAUSEN 442
27. Studies of Lipoproteins by Freeze-Fracture and Etching Electron Microscopy
LAWRENCE P. AGGERBECK AND TADEUSZ GULIK-KRZYWICKI 457

28. Nuclear Magnetic Resonance Studies of Lipoproteins
JAMES A. HAMILTON AND JOEL D. MORRISETT 472
29. Spectroscopic Studies of Lipoproteins H. J. POWNALL AND J. B. MASSEY 515
30. Circular Dichroism of Lipoprotein Lipids G. CHI CHEN AND JOHN P. KANE 519
31. Immunochemical Methods for Studying Lipoprotein Structure
ELAINE S. KRUL AND GUSTAV SCHONFELD 527
32. Reconstitution of High-Density Lipoproteins ANA JONAS 553
33. Reassembly of Low-Density Lipoproteins
MARY T. WALSH AND DAVID ATKINSON 582
34. Reconstitution of Hydrophobic Core of Low-Density Lipoprotein
MONTY KRIEGER 608
35. Use of Cross-Linking Reagents to Study Lipoprotein Structure
JOHN B. SWANEY 613
36. Synthetic Peptide Analogs of Apolipoproteins G. M. ANANTHARAMAIAH 627
37. Lipoprotein–Liposome Interactions
ALAN R. TALL, IRA TABAS, AND KEVIN J. WILLIAMS 647
38. Carboxyfluorescein Leakage Assay for Lipoprotein–Liposome Interaction
JOHN N. WEINSTEIN, ROBERT BLUMENTHAL, AND RICHARD D. KLAUSNER 657

Section V. Molecular Biology of Plasma Lipoproteins

39. Measurement of Apolipoprotein mRNA by DNA-Excess Solution Hybridization with Single-Stranded Probes
DAVID L. WILLIAMS, THOMAS C. NEWMAN, GREGORY S. SHELNESS, AND DAVID A. GORDON 671
40. Intra- and Extracellular Modifications of Apolipoproteins
VASSILIS I. ZANNIS, SOTIRIOS K. KARATHANASIS, GAYLE M. FORBES, AND JAN L. BRESLOW 690
41. Characterization of the Apolipoprotein A-I–C-III Gene Complex
SOTIRIOS K. KARATHANASIS, VASSILIS I. ZANNIS, AND JAN L. BRESLOW 712
42. Genetic Polymorphism in the ApoA-I–C-III Complex
CAROL C. SHOULDERS AND F. E. BARALLE 727
43. Molecular Cloning and Sequence Analysis of Human Apolipoprotein A-II cDNA
LAWRENCE CHAN, MARSHA N. MOORE, AND YUAN-KAI TSAO 745
44. Rat Apolipoprotein A-IV: Application of Computational Methods for Studying the Structure, Function, and Evolution of a Protein
MARK S. BOGUSKI, NABIL A. ELSHOURBAGY, JOHN M. TAYLOR, AND JEFFREY I. GORDON 753
45. Detecting Internally Repeated Sequences and Inferring the History of Duplication WALTER M. FITCH, TEMPLE SMITH, AND JAN L. BRESLOW 773
46. Isolation of cDNA and Genomic Clones for Apolipoprotein C-II
CYNTHIA L. JACKSON, GAIL A. P. BRUNS, AND JAN L. BRESLOW 788
47. Cloning of the cDNA for Rat and Human Apolipoprotein E mRNA
JOHN M. TAYLOR, ROBERT W. MAHLEY, AND CHIKAFUSA FUKAZAWA 801
48. Cloning and Expression of the Human Apolipoprotein E Gene
CATHERINE A. REARDON, YOUNG-KI PAIK, DAVID J. CHANG, GLENN E. DAVIES, ROBERT W. MAHLEY, YUN-FAI LAU, AND JOHN M. TAYLOR 811
49. Genetic Polymorphism in Human Apolipoprotein E VASSILIS I. ZANNIS 823
50. Genetic Mapping of Apolipoprotein Genes in the Human Genome by Somatic Cell Hybrids FA-TEN KAO AND LAWRENCE CHAN 851

51. Chromosomal Fine Mapping of Apolipoprotein Genes by *in Situ* Nucleic Acid Hybridization to Metaphase Chromosomes
MARY E. HARPER AND LAWRENCE CHAN 863
52. Genetic Control of Plasma Lipid Transport: Mouse Model
ALDONS J. LUSIS AND RENEE C. LEBOEUF 877
53. Molecular Cloning of Bovine LDL Receptor cDNAs
DAVID W. RUSSELL AND TOKUO YAMAMOTO 895

VOLUME 129

PLASMA LIPOPROTEINS (PART B: CHARACTERIZATION, CELL BIOLOGY, AND METABOLISM)

Section I. Characterization of Plasma Lipoproteins

1. Zonal Ultracentrifugation JOSEF R. PATSCH AND WOLFGANG PATSCH 3
2. Analytic Ultracentrifugation of Plasma Lipoproteins
TALWINDER S. KAHLON, LAURA A. GLINES, AND FRANK T. LINDGREN 26
3. Separation and Analysis of Lipoproteins by Gel Filtration
LAWRENCE L. RUDEL, CAROL A. MARZETTA, AND FRED L. JOHNSON 45
4. High-Performance Liquid Chromatography of Serum Lipoproteins
ICHIRO HARA AND MITSUYO OKAZAKI 57
5. Precipitation Methods for Quantification of Lipoproteins
PAUL S. BACHORIK AND JOHN J. ALBERS 78
6. Enzymatic Methods for Quantification of Lipoprotein Lipids
G. RUSSELL WARNICK 101
7. Characterization of Apolipoprotein B-Containing Lipoproteins JOHN P. KANE 123
8. Characterization of Apolipoprotein A-Containing Lipoproteins
MARIAN C. CHEUNG 130
9. Characterization of Apolipoprotein E-Containing Lipoproteins
KARL H. WEISGRABER AND ROBERT W. MAHLEY 145
10. Quantitation, Isolation, and Characterization of Human Lipoprotein (a)
JOHN W. GAUBATZ, GARY L. CUSHING, AND JOEL D. MORRISETT 167
11. Isolation of Apolipoprotein E-Containing Lipoproteins by Immunoaffinity Chromatography
JOYCE COREY GIBSON, ARDON RUBINSTEIN, HENRY N. GINSBERG, AND W. VIRGIL BROWN 186

Section II. Cell Biology of Plasma Lipoproteins

12. Methods for Visualization of the LDL Pathway in Cultured Human Fibroblasts
RICHARD G. W. ANDERSON 201
13. Isolation and Assay of the Ac-LDL Receptor
DAVID P. VIA, HANS A. DRESEL, AND ANTONIO M. GOTTO, JR. 216
14. Isolation of Somatic Cell Mutants with Defects in the Endocytosis of Low-Density Lipoprotein MONTY KRIEGER 227
15. Detection of Animal Cell LDL Mutants by Replica Plating JEFFREY D. ESKO 237
16. Immunochemical Measurement of Apolipoprotein Synthesis in Cell and Organ Culture DAVID L. WILLIAMS AND PAUL A. DAWSON 254

17. Lipoprotein Synthesis and Secretion by Cultured Hepatocytes
ROGER A. DAVIS 272
18. Studies of Nascent Lipoproteins in Isolated Hepatic Microsomes and Golgi Cell Fractions DEBENDRANATH BANERJEE 283
19. Morphological Localization of Apolipoproteins and Their mRNA by Immunocytochemistry and *in Situ* Nucleic Acid Hybridization
CHIN-TARNG LIN AND LAWRENCE CHAN 297
20. The Role of Apolipoprotein Processing in Receptor Recognition of VLDL
SANDRA H. GIANTURCO AND WILLIAM A. BRADLEY 319

Section III. Metabolism of Plasma Lipoproteins

A. *In Vivo* Metabolism of Lipoproteins

21. Plasma Lipoprotein Conversions SHLOMO EISENBERG 347
22. The Methodology of Compartmental Modeling as Applied to the Investigation of Lipoprotein Metabolism
LOREN A. ZECH, RAYMOND C. BOSTON, AND DAVID M. FOSTER 366
23. Kinetic Analysis Using Specific Radioactivity Data
NGOC-ANH LE, RAJASEKHAR RAMAKRISHNAN, RALPH B. DELL, HENRY N. GINSBERG, AND W. VIRGIL BROWN 384
24. Metabolism of the Apolipoprotein B-Containing Lipoproteins
GEORGE STEINER, MARY E. POAPST, STEVEN L. SHUMAK, AND DAVID M. FOSTER 395
25. Metabolism of Apolipoproteins A-I, A-II, and A-IV
ERNST J. SCHAEFER AND JOSE M. ORDOVAS 420
26. Metabolism of Apolipoprotein C NOEL H. FIDGE AND PAUL J. NESTEL 443
27. Direct Determination of Apolipoprotein C-III Specific Activity Using Immunoaffinity Chromatography
NGOC-ANH LE, PHILLIP R. BUKBERG, HENRY N. GINSBERG, JOYCE C. GIBSON, AND W. VIRGIL BROWN 457
28. Metabolism of Postprandial Lipoproteins ALAN R. TALL 469
29. *In Vivo* Metabolism of Apolipoprotein E in Humans
RICHARD E. GREGG AND H. BRYAN BREWER, JR. 482

B. Cellular and Tissue Metabolism of Plasma Lipoproteins

30. Hepatic Lipoprotein Biosynthesis JULIAN B. MARSH 498
31. High-Density Lipoprotein Formation by the Intestine
ROBERT M. GLICKMAN AND ARTHUR M. MAGUN 519
32. Phosphorylation–Dephosphorylation of Apolipoprotein B
ROGER A. DAVIS AND ROY A. BORCHARDT 536
33. Lipoprotein–Receptor Interactions
THOMAS L. INNERARITY, ROBERT E. PITAS, AND ROBERT W. MAHLEY 542
34. Receptor-Independent Low-Density Lipoprotein Catabolism
J. SHEPHERD AND C. J. PACKARD 566
35. Role of the Liver in Lipoprotein Catabolism RICHARD J. HAVEL 591
36. Methods for Assessment of Tissue Sites of Lipoprotein Degradation
RAY C. PITTMAN AND CLINTON A. TAYLOR, JR. 612

37. Reverse Cholesterol Transport
GEORGE H. ROTHBLAT, MARK BAMBERGER, AND MICHAEL C. PHILLIPS 628
38. Receptor-Mediated Transport of Cholesterol between Cultured Cells and High-Density Lipoproteins JOHN F. ORAM 645
39. Interstitial Fluid (Peripheral Lymph) Lipoproteins
LADISLAV DORY, CHARLES H. SLOOP, AND PAUL S. ROHEIM 660
40. Lipoproteins and Steroid Hormone-Producing Tissues
JOHN T. GWYNNE AND DARIEN D. MAHAFFEE 679

C. Enzymes of Lipoprotein Metabolism

41. Preparation, Characterization, and Measurement of Lipoprotein Lipase
PER-HENRIK IVERIUS AND ANN-MARGRET ÖSTLUND-LINDQVIST 691
42. Assays of the *in Vitro* Metabolism of Very-Low-Density Lipoproteins in Whole Plasma by Purified Lipoprotein Lipase
BYUNG HONG CHUNG AND JERE P. SEGREST 704
43. Preparation, Characterization, and Measurement of Hepatic Lipase
CHRISTIAN EHNHOLM AND TIMO KUUSI 716
44. Mechanism of Action of Lipoprotein Lipase
LARRY R. MCLEAN, RUDY A. DEMEL, LILIAN SOCORRO, MASAKI SHINOMIYA, AND RICHARD L. JACKSON 738
45. Isolation, Characterization, and Assay of Lecithin–Cholesterol Acyltransferase
JOHN J. ALBERS, CHING-HONG CHEN, AND ANDRAS G. LACKO 763
46. Mechanisms of Action of Lecithin–Cholesterol Acyltransferase
CHRISTOPHER J. FIELDING 783
47. Lysolecithin Acyltransferase of Human Plasma: Assay and Characterization of Enzyme Activity P. V. SUBBAIAH 790
48. Isolation, Characterization, and Assay of Plasma Lipid Transfer Proteins
JOHN H. TOLLEFSON AND JOHN J. ALBERS 797

D. Methods

49. Synthesis of Ether Analogs of Lipoprotein Lipids and Their Biological Applications GIDEON HALPERIN, OLGA STEIN, AND YECHEZKIEL STEIN 816
50. Fluorescent Labeling of Lipoproteins DAVID P. VIA AND LOUIS C. SMITH 848
51. Digital Imaging Fluorescence Microscopy
LOUIS C. SMITH, DOUGLAS M. BENSON, ANTONIO M. GOTTO, JR., AND JOSEPH BRYAN 857

VOLUME 130
ENZYME STRUCTURE (PART K)

Section I. Macromolecular Self-Associations and Structural Assemblies

1. Criteria for Distinguishing Self-Associations in Velocity Sedimentation
V. PRAKASH AND SERGE N. TIMASHEFF 3

2. Calculation of Asymptotic Boundary Shapes from Experimental Mass Migration Patterns LAWRENCE W. NICHOL AND DONALD J. WINZOR 6
3. Effects of Microheterogeneity on Sedimentation Patterns of Interacting Proteins and the Sedimentation Behavior of Systems Involving Two Ligands JOHN R. CANN 19
4. Measurement of Linear Polymer Formation by Small-Angle Light Scattering ROBLEY C. WILLIAMS, JR. 35
5. The Measurement of Cooperative Protein Self-Assembly by Turbidity and Other Techniques JOSE MANUEL ANDREU AND SERGE N. TIMASHEFF 47
6. Thermodynamic Parameters for Protein–Protein and Protein–Ligand Interaction by Differential Scanning Microcalorimetry HANS-JÜGEN HINZ 59
7. Structure and Interactions of Proteins in Solution Studied by Small-Angle Neutron Scattering SOW-HSIN CHEN AND DALILA BENDEDOUCH 79
8. Use of Gel Electrophoresis to Characterize Multimolecular Aggregates PHILIP SERWER 116
9. Quantitative DNase Footprint Titration: A Method for Studying Protein–DNA Interactions MICHAEL BRENOWITZ, DONALD F. SENEAR, MADELINE A. SHEA, AND GARY K. ACKERS 132

Section II. Macromolecular Conformation: Spectroscopy

10. Prediction of Protein Structure PATRICK ARGOS AND J. K. MOHANARAO 185
11. Calculation of Protein Conformation from Circular Dichroism JEN TSI YANG, CHUEN-SHANG C. WU, AND HUGO M. MARTINEZ 208
12. Magnetic Circular Dichroism BARTON HOLMQUIST 270
13. Resolution-Enhanced Fourier Transform Infrared Spectroscopy of Enzymes HEINO SUSI AND D. MICHAEL BYLER 290
14. Protein Secondary Structure Analysis Using Raman Amide I and Amide III Spectra ROBERT W. WILLIAMS 311
15. Ultraviolet Resonance Raman Spectroscopy of Biopolymers BRUCE HUDSON AND LELAND MAYNE 331
16. Resonance Raman Studies of Ligand Binding NAI-TENG YU 350

Section III. Macromolecular Conformational Stability and Transitions

17. Calculation of Electrostatic Interactions in Proteins JAMES B. MATTHEW AND FRANK R. N. GURD 413
18. Stabilization and Destabilization of Protein Structure by Charge Interactions JAMES B. MATTHEW AND FRANK R N. GURD 437
19. Measurements of Diffusion and Chemical Kinetics by Fluorescence Photobleaching Recovery and Fluorescence Correlation Spectroscopy N. O. PETERSEN AND E. L. ELSON 454
20. Fluorescence Lifetimes with a Synchrotron Source STEVEN G. BOXER AND RODNEY R. BUCKS 484
21. Fluorescence Polarization at High Pressure ALEJANDRO A. PALADINI, JR. 493
22. Stopped-Flow Circular Dichroism MOLLIE PFLUMM, JEREMY LUCHINS, AND SHERMAN BEYCHOK 519

23. The Measurement of the Kinetics of Lipid Phase Transitions: A Volume-Perturbation Kinetic Calorimeter
MICHAEL L. JOHNSON, WILLIAM W. VAN OSDOL, AND RODNEY L. BILTONEN 534

VOLUME 131
ENZYME STRUCTURE (PART L)

Section I. Unfolding and Refolding of Proteins

1. Protein Folding: Introductory Comments ROBERT L. BALDWIN 3
2. Scanning Microcalorimetry in Studying Temperature-Induced Changes in Proteins PETER L. PRIVALOV AND SERGEY A. POTEKHIN 4
3. Kinetic Mechanisms of Protein Folding
HIROYASU UTIYAMA AND ROBERT L. BALDWIN 51
4. Fast-Folding and Slow-Folding Forms of Unfolded Proteins
FRANZ X. SCHMID 70
5. Disulfide Bonds as Probes of Protein Folding Pathways
THOMAS E. CREIGHTON 83
6. Proline Isomerization Studied with Proteolytic Enzymes
JOHN F. BRANDTS AND LUNG-NAN LIN 107
7. Folding Intermediates Studied by Circular Dichroism A. M. LABHARDT 126
8. Amide Proton Exchange as a Probe of Protein Folding Pathways
PETER S. KIM 136
9. Detection of Folding Intermediates Using Urea-Gradient Electrophoresis
THOMAS E. CREIGHTON 156
10. Protein Folding in Cryosolvents and at Subzero Temperatures
ANTHONY L. FINK 173
11. Complementation in Folding and Fragment Exchange
HIROSHI TANIUCHI, GARY R. PARR, AND MARCEL A. JUILLERAT 185
12. Refolding and Association of Oligomeric Proteins
RAINER JAENICKE AND RAINER RUDOLPH 218
13. Mutational Analysis of Protein Folding Pathways: The P22 Tailspike Endorhamnosidase JONATHAN KING AND MYEONG-HEE YU 250
14. Determination and Analysis of Urea and Guanidine Hydrochloride Denaturation Curves C. N. PACE 266

Section II. Structural Dynamics and Mobility of Proteins

15. Internal Dynamics of Proteins MARTIN KARPLUS 283
16. Observation of Internal Motility of Proteins by Nuclear Magnetic Resonance in Solution GERHARD WAGNER AND KURT WÜTHRICH 307
17. Protein Dynamics by Solid State Nuclear Magnetic Resonance
STANLEY J. OPELLA 327
18. Internal Motion of Proteins: Nuclear Magnetic Resonance Measurements and Dynamic Simulations C. M. DOBSON AND M. KARPLUS 362

19. Study of Protein Dynamics by X-Ray Diffraction
DAGMAR RINGE AND GREGORY A. PETSKO 389
20. Protein Dynamics Investigated by Neutron Diffraction A. A. KOSSIAKOFF 433
21. Protein Conformational Dynamics Measured by Hydrogen Isotope Exchange Techniques ROGER B. GREGORY AND ANDREAS ROSENBERG 448
22. Individual Breathing Reactions Measured by Functional Labeling and Hydrogen Exchange Methods
JOSE R. ROGERO, JOAN J. ENGLANDER, AND S. W. ENGLANDER 508
23. Fluorescence Studies of Structural Fluctuations in Macromolecules as Observed by Fluorescence Spectroscopy in the Time, Lifetime, and Frequency Domains
JOSEPH R. LAKOWICZ 518
24. Mössbauer Effect in the Study of Structure Dynamics
FRITZ PARAK AND LOU REINISCH 568

VOLUME 132
IMMUNOCHEMICAL TECHNIQUES (PART J: PHAGOCYTOSIS AND CELL-MEDIATED CYTOTOXICITY)

Section I. Phagocytosis: General Methodology

1. Phagocytosis: An Overview CAREL J. VAN OSS 3
2. Measurement of Surface Properties of Phagocytes, Bacteria, and Other Particles
DARRYL R. ABSOLOM 16
3. Basic Methods for the Study of Phagocytosis DARRYL R. ABSOLOM 95

Section II. Special Methods for Measuring Phagocytosis

4. *In Vitro* and *in Vivo* Measurement of Phagocytosis by Flow Cytometry
CARLETON C. STEWART, BRUCE E. LEHNERT, AND JOHN A. STEINKAMP 183
5. Oil-Droplet Method for Measuring Phagocytosis THOMAS P. STOSSEL 192
6. Methods for Distinguishing Ingested from Adhering Particles JAN HED 198
7. Methods for the Study of Receptor-Mediated Phagocytosis
SAMUEL D. WRIGHT 204

Section III. Specific Methods for the Isolation of Cells and Cellular Components

8. Purification and Cryopreservation of Phagocytes from Human Blood
DIRK ROOS AND MARTIN DE BOER 225
9. Suspension Culture of Human Monocytes
PAUL J. MILLER AND HENRY C. STEVENSON 243
10. Preparation and Cryopreservation of Cytoplasts from Human Phagocytes
DIRK ROOS AND ALWIN A. VOETMAN 250
11. Assay of Phagosome–Lysosome Fusion MARGARET KIELIAN 257

12. Methods for Assessing Exocytosis by Neutrophil Leukocytes
BEATRICE DEWALD AND MARCO BAGGIOLINI 267

Section IV. Opsonization and Nonspecific Promoters of Phagocytosis

13. Opsonins and Dysopsonins: An Overview DARRYL R. ABSOLOM 281
14. Tuftsin VICTOR A. NAJJAR, DANUTA KONOPINSKA, AND JAMES LEE 318
15. Substance P and Neurotensin ZVI BAR-SHAVIT AND RACHEL GOLDMAN 326
16. Opsonic Activity of Fibronectin
FRANK A. BLUMENSTOCK, THOMAS M. SABA, PINA CARDARELLI, AND CLAIRE H. DAYTON 334

Section V. Enzymes and Metabolic Activity of Phagocytes

17. Respiratory Burst during Phagocytosis: An Overview
MANFRED L. KARNOVSKY AND JOHN A. BADWEY 353
18. NADPH Oxidase from Polymorphonuclear Cells
THEODORE G. GABIG AND BRUCE A. LEFKER 355
19. NADH Oxidase from Guinea Pig Polymorphonuclear Leukocytes
JOHN A. BADWEY AND MANFRED L. KARNOVSKY 365
20. Human Myeloperoxidase and Hemi-myeloperoxidase
PATRICIA C. ANDREWS AND NORMAN I. KRINSKY 369
21. Cytochrome *b*-245
ANTHONY W. SEGAL, ANGELA M. HARPER, ANDREW R. CROSS, AND OWEN T. G. JONES 378
22. Measurement of Hydrogen Peroxide Production by Phagocytes Using Homovanillic Acid and Horseradish Peroxidase
MARCO BAGGIOLINI, WALTER RUCH, AND PHILIP H. COOPER 395
23. Assay of the Extracellular Hydrogen Peroxide Pool Generated by Phagocytes
SAMUEL T. TEST AND STEPHEN J. WEISS 401
24. Microassays for Superoxide and Hydrogen Peroxide Production and Nitroblue Tetrazolium Reduction Using an Enzyme Immunoassay Microplate Reader
EDGAR PICK 407
25. Determination of Adenylate Cyclase and Guanylate Cyclase Activities in Cells of the Immune System CHANDRA K. MITTAL 422
26. Measurements of Cyclic Adenosine Monophosphate and Cyclic Guanosine Monophosphate Levels in Polymorphonuclear Leukocytes, Macrophages, and Lymphocytes CHANDRA K. MITTAL 428

Section VI. Cell-Mediated Cytotoxicity

27. Cytolytic Activity Mediated by T Lymphocytes
HOWARD D. ENGERS, K. THEODOR BRUNNER, AND JEAN-CHARLES CEROTTINI 437
28. Cytotoxicity by Natural Killer Cells: Analysis of Large Granular Lymphocytes
JOHN R. ORTALDO 445

29. Cytotoxicity by Macrophages and Monocytes
ROEL A. DE WEGER, BERT A. RUNHAAR, AND WILLEM DEN OTTER 458
30. Cytolytic T-Cell Clones and Hybridomas
HARALD VON BOEHMER AND WERNER HAAS 467

Section VII. Assays for Cytotoxic Activity

31. Quantitative Cytotoxic Assay Using a Coulter Counter
YOSHITAKA NAGAI, MACHIKO IWAMOTO, AND TOSHIHARU MATSUMURA 481
32. Continuous Assay for Cytolytic Activity
STEVE J. MCFAUL AND JOHANNES EVERSE 491
33. Measurement of Phagocytosis and Cell-Mediated Cytotoxicity by Chemiluminescence
KNOX VAN DYKE, MICHAEL R. VAN SCOTT, AND VINCENT CASTRANOVA 498
34. Activated Macrophage-Mediated Cytotoxicity: Use of the *in Vitro* Cytotoxicity Assay for Study of Bioenergetic and Biochemical Changes That Develop in Tumor Target Cells JOHN B. HIBBS, JR. AND READ R. TAINTOR 508
35. Intracellular Killing of Bacteria and Fungi
LINDA L. MOORE AND JAMES R. HUMBERT 520

Section VIII. Cytotoxic Activity of Macrophages

36. Induction of Specific Macrophage Cytotoxicity
ROEL A. DE WEGER AND WILLEM DEN OTTER 531
37. Activation of Macrophages with Oxidative Enzymes
DORIS L. LEFKOWITZ, STANLEY S. LEFKOWITZ, RU-QI WEI, AND JOHANNES EVERSE 537
38. Assay of a Cytocidal Protein Excreted by Activated Macrophages
BETH-ELLEN DRYSDALE, CHARLES M. ZACHARCHUK, MASAKO OKAJIMA, AND HYUN S. SHIN 549
39. Assays Detecting the Antibody-Dependent and Independent Binding and Cytolysis of Tumor Cells by Murine Macrophages
WILLIAM J. JOHNSON AND DOLPH O. ADAMS 555
40. Preparation and Characterization of Chloramines
EDWIN L. THOMAS, MATTHEW B. GRISHAM, AND M. MARGARET JEFFERSON 569
41. Cytotoxicity of Chloramines
EDWIN L. THOMAS, MATTHEW B. GRISHAM, AND M. MARGARET JEFFERSON 585
42. Liposome Encapsulation of Muramyl Peptides for Activation of Macrophage Cytotoxic Properties
RAJIV NAYAR, ALAN J. SCHROIT, AND ISAIAH J. FIDLER 594
43. Intracellular Parasitism of Macrophages in Leishmaniasis: *In Vitro* Systems and Their Applications
K.-P. CHANG, CAROL ANN NACY, AND RICHARD D. PEARSON 603
44. Parasiticidal Activity of Macrophages against *Toxoplasma*
SOMESH D. SHARMA, JAMES R. CATTERALL, AND JACK S. REMINGTON 626
45. Parasiticidal Activity of Macrophages against *Plasmodium*
HANNAH LUSTIG SHEAR AND CHRISTIAN F. OCKENHOUSE 637
46. Parasiticidal Activity of Macrophages against Schistosomes
STEPHANIE L. JAMES 649

VOLUME 133

BIOLUMINESCENCE AND CHEMILUMINESCENCE (PART B)

Section I. Bioluminescence

1. Cloning Firefly Luciferase
JEFFREY R. DE WET, KEITH V. WOOD, DONALD R. HELINSKI, AND MARLENE DELUCA 3
2. Extraction of Adenosine Triphosphate from Microbial and Somatic Cells
PHILIP E. STANLEY 14
3. Detection of Bacteriurea by Bioluminescence BRUCE A. HANNA 22
4. Estimation of Biomass in Growing Cell Lines by Adenosine Triphosphate Assay
ARNE LUNDIN, MICHAEL HASENSON, JÖRGEN PERSSON, AND ÅKE POUSETTE 27
5. Sensitive Bioluminescent Assay for α-Bungarotoxin Binding Sites
JAMES M. SCHAEFFER 43
6. Commercially Available Firefly Luciferase Reagents
FRANKLIN R. LEACH AND JOANN J. WEBSTER 51
7. Cloning and Expression of the Genes from the Bioluminescent System of Marine Bacteria
CAROL MIYAMOTO, MICHAEL BOYLAN, ANGUS GRAHAM, AND EDWARD MEIGHEN 70
8. Techniques for Cloning and Analyzing Bioluminescence Genes from Marine Bacteria JOANNE ENGEBRECHT AND MICHAEL SILVERMAN 83
9. Purification of Bacterial Luciferase by Affinity Methods
THOMAS O. BALDWIN, THOMAS F. HOLZMAN, RITA B. HOLZMAN, AND VICKI A. RIDDLE 98
10. Purification of Bacterial Luciferase by High-Performance Liquid Chromatography
DENNIS J. O'KANE, MUSHTAQ AHMAD, IAIN B. C. MATHESON, AND JOHN LEE 109
11. Bacterial Luciferase 4a-Hydroperoxyflavin Intermediates: Stabilization, Isolation, and Properties SHIAO-CHUN TU 128
12. Bacterial Luciferase Intermediates: The Neutral Flavin Semiquinone, Its Reaction with Superoxide, and the Flavin 4a-Hydroxide
M. KURFÜRST, S. GHISLA, AND J. W. HASTINGS 140
13. Purification and Properties of Lumazine Proteins from *Photobacterium* Strains
DENNIS J. O'KANE AND JOHN LEE 149
14. Fatty Acid Reductase from *Photobacterium phosphoreum*
ANGEL RODRIGUEZ, LEE WALL, DENIS RIENDEAU, AND EDWARD MEIGHEN 172
15. Bioluminescence-Related Acyltransferases from *Photobacterium phosphoreum* and *Vibrio harveyi*
DAVID BYERS, ANGEL RODRIGUEZ, LUC CAREY, AND EDWARD MEIGHEN 183
16. Bioluminescent Analysis of Insect Pheromones
DAVID MORSE, ROSE SZITTNER, GARY GRANT, AND EDWARD MEIGHEN 189
17. Bioluminescent Assays Using Coimmobilized Enzymes
G. WIENHAUSEN AND M. DELUCA 198
18. Bioluminescent Assays of Estrogens
JEAN-CLAUDE NICOLAS, ANNE MARIE BOUSSIOUX, AND ANDRÉ CRASTES DE PAULET 209

19. Immobilized Enzymes for Assaying Bile Acids
J. SCHOELMERICH, A. RODA, AND M. DELUCA 215
20. Continuous-Flow Bioluminescent Assays Employing Sepharose-Immobilized Enzymes DANIEL C. VELLOM AND LARRY J. KRICKA 229
21. Flow Systems Utilizing Nylon-Immobilized Enzymes
ALDO RODA, STEFANO GIROTTI, AND GIACOMO CARREA 238
22. Active Center-Based Immunoassay Approach Using Bacterial Luciferase
THOMAS O. BALDWIN, THOMAS F. HOLZMAN, AND RITA B. HOLZMAN 248
23. Bioluminescence Test for Genotoxic Agents S. ULITZUR 264
24. Determination of Antibiotic Activities with the Aid of Luminous Bacteria
S. ULITZUR 275
25. Amplified Bioluminescence Assay Using Avidin–Biotin Technology
G. BARNARD, E. A. BAYER, M. WILCHEK, Y. AMIR-ZALTSMAN, AND F. KOHEN 284
26. Isolation and Expression of a cDNA Coding for Aequorin, the Ca^{2+}-Activated Photoprotein from *Aequorea victoria*
DOUGLAS C. PRASHER, RICHARD O. MCCANN, AND MILTON J. CORMIER 288
27. Molecular Cloning of Apoaequorin cDNA
MASATO NOGUCHI, FREDERICK I. TSUJI, AND YOSHIYUKI SAKAKI 298
28. Cell-Free Components in Dinoflagellate Bioluminescence. The Particulate Activity: Scintillons; The Soluble Components: Luciferase, Luciferin, and Luciferin-Binding Protein J. WOODLAND HASTINGS AND JAY C. DUNLAP 307

Section II. Chemiluminescence

29. Enhanced Chemiluminescent Reactions Catalyzed by Horseradish Peroxidase
GARY H. G. THORPE AND LARRY J. KRICKA 331
30. Luminescence Immunoassays for Haptens and Proteins
W. GRAHAM WOOD, JÖRG BRAUN, AND UWE HANTKE 354
31. Immunoassays Using Acridinium Esters
IAN WEEKS, MARIA STURGESS, RICHARD C. BROWN, AND J. STUART WOODHEAD 366
32. Surface Chemiluminescent Immunoassays of Steroids
F. KOHEN, J. DE BOEVER, AND J. B. KIM 387
33. Photographic Detection of Chemiluminescent and Bioluminescent Reactions
LARRY J. KRICKA AND GARY H. G. THORPE 404
34. On-Line Computer Analysis of the Kinetics of Chemiluminescent Reactions: Application to Luminescent Immunoassays
MARIO PAZZAGLI AND MARIO SERIO 420
35. Chemiluminescence Detection System for High-Performance Liquid Chromatography KAZUHIRO IMAI 435
36. Phagocytic Leukocyte Oxygenation Activities and Chemiluminescence: A Kinetic Approach to Analysis ROBERT C. ALLEN 449
37. Cellular Chemiluminescence Associated with Disease States
KNOX VAN DYKE AND CHRIS VAN DYKE 493
38. Luminescent Bioassays Based on Macrophage Cell Lines
PATRICK DE BAETSELIER AND ERIC SCHRAM 507

39. Stable 1,2-Dioxetanes as Labels for Thermochemiluminescent Immunoassay
JAN C. HUMMELEN, THEO M. LUIDER, AND HANS WYNBERG 531
40. Synthesis of Chemiluminogenic Protein Conjugates with Dimethyl Adipimidate for Sensitive Immunoassays
G. MESSERI, H. R. SCHROEDER, A. L. CALDINI, AND C. ORLANDO 557
41. Chemiluminescent Probes for Singlet Oxygen in Biological Reactions
AMBLER THOMPSON, HOWARD H. SELIGER, AND GARY H. POSNER 569

Section III. Instrumentation

42. Characteristics of Commercial Radiometers PHILIP E. STANLEY 587

VOLUME 134

STRUCTURAL AND CONTRACTILE PROTEINS (PART C: THE CONTRACTILE APPARATUS AND THE CYTOSKELETON)

Section I. Actin-Associated Proteins

1. Preparation and Assay of Cytoplasmic and Secreted Gelsolin H. L. YIN 3
2. Purification of Severin SUSAN S. BROWN 9
3. Isolation of Fascin, an Actin-Bundling Protein, and SU45, an Actin-Severing/Capping Protein from Sea Urchin Eggs JOSEPH BRYAN 13
4. Purification of the Intestinal Microvillus Cytoskeletal Proteins Villin, Fimbrin, and Ezrin ANTHONY BRETSCHER 24
5. Purification of Caldesmon WILLIAM LYNCH AND ANTHONY BRETSCHER 37
6. Purification of Fodrin from Mammalian Brain
RICHARD CHENEY, JOEL LEVINE, AND MARK WILLARD 42
7. Purification of Brain Analogs of Red Blood Cell Membrane Skeletal Proteins: Ankyrin, Protein 4.1 (Synapsin), and Spectrin Subunits
VANN BENNETT, ANTHONY J. BAINES, AND JONATHAN DAVIS 55
8. Purification and Assay of Vinculin, Metavinculin, and Talin
T. O'HALLORAN, L. MOLONY, AND K. BURRIDGE 69
9. Isolation of Metavinculin from Chicken Smooth Muscle
JANET D'ANGELO SILICIANO AND SUSAN W. CRAIG 78

Section II. Microtubules

A. Tubulin and Microtubule-Associated Proteins

10. Reversible Assembly Purification of Microtubules without Assembly-Promoting Agents and Further Purification of Tubulin, Microtubule-Associated Proteins, and MAP Fragments RICHARD B. VALLEE 89
11. Purification of Brain Microtubules and Microtubule-Associated Protein 1 Using Taxol RICHARD B. VALLEE 104

12. Purification of Microtubules and Microtubule-Associated Proteins from Sea Urchin Eggs and Cultured Mammalian Cells Using Taxol, and Use of Exogenous Taxol-Stabilized Brain Microtubules for Purifying Microtubule-Associated Proteins RICHARD B. VALLEE AND CHRISTINE A. COLLINS 116
13. Isolation of Sea Urchin Egg Tubulin H. WILLIAM DETRICH, III 128
14. Direct Identification of Microtubule-Associated Proteins by Selective Extraction of Cultured Cells FRANK SOLOMON 139
15. Purification and Assay of Primate 210K Microtubule-Associated Protein JEANNETTE CHLOË BULINSKI 147
16. Purification of Tau Protein from Brain DAVID DRUBIN AND MARC KIRSCHNER 156
17. Purification and Assay of Cold-Stable Microtubules and STOP Protein ROBERT L. MARGOLIS, CHARLES T. RAUCH, AND DIDIER JOB 160
18. Isolation and Purification of Tubulin Tyrosine Ligase JÜRGEN WEHLAND, HEINZ C. SCHRÖDER, AND KLAUS WEBER 170
19. Proteolytic Modification of Tubulin LUIS SERRANO, FRANCISCO WANDOSELL, JAVIER DE LA TORRE, AND JESÚS AVILA 179

B. Microtubule-Containing and -Organizing Organelles

20. Isolation of Sea Urchin Spindles and Cytasters RYOKO KURIYAMA 190
21. Isolation of Native, Membrane-Containing Mitotic Apparatus from Sea Urchin Embryos ROBERT B. SILVER 200
22. Isolation of Mitotic Spindles from Mammalian Cells R. C. BRADY, M. J. SCHIBLER, AND F. CABRAL 217
23. Isolation of Mitotic Spindles and Spindle Pole Bodies from Yeast JEREMY S. HYAMS AND STEPHEN M. KING 225
24. Isolation of the Erythrocyte Marginal Band WILLIAM D. COHEN AND MARY F. GINSBURG 232
25. Isolation of Basal Bodies from *Chlamydomonas* WILLIAM J. SNELL 252
26. Isolation of Mammalian Centrosomes T. J. MITCHISON AND M. W. KIRSCHNER 261
27. Isolation of a Kinetochore–Centromere Fraction from HeLa Metaphase Chromosomes MANUEL M. VALDIVIA AND B. R. BRINKLEY 268
28. Isolation of *Chlamydomonas* Flagella and Flagellar Axonemes GEORGE B. WITMAN 280

C. Dynein

29. Purification and Characterization of *Chlamydomonas* Flagellar Dyneins STEPHEN M. KING, TIM OTTER, AND GEORGE B. WITMAN 291
30. Preparation and Properties of Dynein from *Tetrahymena* Cilia KENNETH A. JOHNSON 306
31. Extraction and Properties of Dyneins from Bull Spermatozoa CLAUDE GAGNON 318
32. Purification of Cytoplasmic Dynein from *Strongylocentrotus* Sea Urchin Eggs M. M. PRATT 325
33. Purification of Cytoplasmic Dynein from Sea Urchin Eggs SHIN-ICHI HISANAGA AND HIKOICHI SAKAI 337

Section III. Intermediate Filaments and Other Cytoskeletal Elements

34. Separation of Cytokeratin Polypeptides by Gel Electrophoretic and Chromatographic Techniques and Their Identification by Immunoblotting
THOMAS ACHTSTAETTER, MECHTHILD HATZFELD, ROY A. QUINLAN, DAVID C. PARMELEE, AND WERNER W. FRANKE 355

35. Isolation of Intermediate Filaments from Neuronal Tissues by Cycles of Assembly/Disassembly ROBERT V. ZACKROFF 371

36. Purification of Neurofilaments and Their Constituent Polypeptides
RONALD K. H. LIEM 380

37. Isolation of Paired Helical Filaments and Amyloid Fibers from Human Brain
DENNIS J. SELKOE AND CARMELA R. ABRAHAM 388

38. Methods for Studying Spasmin W. BRAD AMOS 404

39. Isolation of the Calcium-Modulated Contractile Protein of Striated Flagellar Roots J. L. SALISBURY, K. W. AEBIG, AND D. E. COLING 408

40. Isolation of Nematode Major Sperm Proteins
SAMUEL WARD AND MICHAEL R. KLASS 414

Section IV. Immunological, Pharmacological, and Fluorescence Probes

41. Antibody Inhibitors of Nonmuscle Myosin Function and Assembly
DANIEL P. KIEHART, DONALD A. KAISER, AND THOMAS D. POLLARD 423

42. Preparation of Antibodies Reactive with Specific Regions of Cytoskeletal Proteins JEANNETTE CHLOË BULINSKI AND GREGG G. GUNDERSEN 453

43. Purification of Isoform-Selective Actin Antibody from Polyclonal Antiserum
SUSAN W. CRAIG, SHEILA K. SANDERS, AND JOSÉ V. PARDO 460

44. Analysis of Cytoskeletal Structures Using Blot-Purified Monospecific Antibodies
J. B. OLMSTED 467

45. Preparation of *N*-Ethylmaleimide-Modified Heavy Meromyosin and Its Use as a Functional Probe of Actomyosin-Based Motility W. ZACHEUS CANDE 473

46. Erythro-9-[3-(2-hydroxynonyl)]adenine and Vanadate as Probes for Microtubule-Based Cytoskeletal Mechanochemistry STEPHEN M. PENNINGROTH 477

47. Preparation of a Fluorescent Analog: Acetamidofluoresceinyl-Labeled *Dictyostelium discoideum* α-Actinin JOHN R. SIMON AND D. LANSING TAYLOR 487

48. Preparation, Assay, and Microinjection of Fluorescently Labeled Cytoskeletal Proteins: Actin, α-Actinin, and Vinculin THOMAS E. KREIS 507

49. Preparation and Characterization of Fluorescent Analogs of Tubulin
PATRICIA WADSWORTH AND E. D. SALMON 519

Section V. Analysis of Function, Distribution, and Structure of Cytoskeletal Proteins

50. Myosin Movement *in Vitro*: A Quantitative Assay Using Oriented Actin Cables from *Nitella*
MICHAEL P. SHEETZ, STEVEN M. BLOCK, AND JAMES A. SPUDICH 531

51. Methods to Study the Microtubule–Neurofilament Network *in Vitro*
ERIC J. AAMODT AND ROBLEY C. WILLIAMS, JR. 544

52. Gel Overlay Methods for Detecting Specific Protein–Protein Interactions
JOANN J. OTTO 555

53. Viewing Single Microtubules by Video Light Microscopy BRUCE J. SCHNAPP 561
54. Agar Overlay Method: High-Resolution Immunofluorescence for the Study of the Contractile Apparatus
YOSHIO FUKUI, SHIGEHIKO YUMURA, TOSHIKO K. YUMURA, AND HIROSHI MORI 573
55. Polyethylene Glycol Embedding Protocol for Immunofluorescence Microscopy
JOHN J. WOLOSEWICK 580
56. Preembedding for Localization by Electron Microscopy of Cytoskeletal Antigens in Cultured Cell Monolayers Using Gold-Labeled Antibodies
J. DE MEY, G. LANGANGER, G. GEUENS, R. NUYDENS, AND M. DE BRABANDER 592
57. Quick-Freeze, Deep Etch of the Cytoskeleton NOBUTAKA HIROKAWA 598
58. Cryo-Electron Microscopy of Unstained Frozen-Hydrated Microtubules
EVA-MARIA MANDELKOW AND ECKHARD MANDELKOW 612
59. X-Ray Diffraction of Cytoskeletal Fibers ECKHARD MANDELKOW 633
60. Time-Resolved X-Ray Scattering of Microtubule Assembly Using Synchrotron Radiation ECKHARD MANDELKOW AND JOAN BORDAS 657
61. Rapid Kinetic Analysis of Mechanochemical Adenosinetriphosphatases
KENNETH A. JOHNSON 677

METHODS IN ENZYMOLOGY

EDITED BY

Sidney P. Colowick and Nathan O. Kaplan

VANDERBILT UNIVERSITY
SCHOOL OF MEDICINE
NASHVILLE, TENNESSEE

DEPARTMENT OF CHEMISTRY
UNIVERSITY OF CALIFORNIA
AT SAN DIEGO
LA JOLLA, CALIFORNIA

I. Preparation and Assay of Enzymes
II. Preparation and Assay of Enzymes
III. Preparation and Assay of Substrates
IV. Special Techniques for the Enzymologist
V. Preparation and Assay of Enzymes
VI. Preparation and Assay of Enzymes (*Continued*)
Preparation and Assay of Substrates
Special Techniques
VII. Cumulative Subject Index

METHODS IN ENZYMOLOGY

EDITORS-IN-CHIEF

Sidney P. Colowick and Nathan O. Kaplan

VOLUME VIII. Complex Carbohydrates
Edited by ELIZABETH F. NEUFELD AND VICTOR GINSBURG

VOLUME IX. Carbohydrate Metabolism
Edited by WILLIS A. WOOD

VOLUME X. Oxidation and Phosphorylation
Edited by RONALD W. ESTABROOK AND MAYNARD E. PULLMAN

VOLUME XI. Enzyme Structure
Edited by C. H. W. HIRS

VOLUME XII. Nucleic Acids (Parts A and B)
Edited by LAWRENCE GROSSMAN AND KIVIE MOLDAVE

VOLUME XIII. Citric Acid Cycle
Edited by J. M. LOWENSTEIN

VOLUME XIV. Lipids
Edited by J. M. LOWENSTEIN

VOLUME XV. Steroids and Terpenoids
Edited by RAYMOND B. CLAYTON

VOLUME XVI. Fast Reactions
Edited by KENNETH KUSTIN

VOLUME XVII. Metabolism of Amino Acids and Amines (Parts A and B)
Edited by HERBERT TABOR AND CELIA WHITE TABOR

VOLUME XVIII. Vitamins and Coenzymes (Parts A, B, and C)
Edited by DONALD B. MCCORMICK AND LEMUEL D. WRIGHT

VOLUME XIX. Proteolytic Enzymes
Edited by GERTRUDE E. PERLMANN AND LASZLO LORAND

VOLUME XX. Nucleic Acids and Protein Synthesis (Part C)
Edited by KIVIE MOLDAVE AND LAWRENCE GROSSMAN

VOLUME XXI. Nucleic Acids (Part D)
Edited by LAWRENCE GROSSMAN AND KIVIE MOLDAVE

VOLUME XXII. Enzyme Purification and Related Techniques
Edited by WILLIAM B. JAKOBY

VOLUME XXIII. Photosynthesis (Part A)
Edited by ANTHONY SAN PIETRO

VOLUME XXIV. Photosynthesis and Nitrogen Fixation (Part B)
Edited by ANTHONY SAN PIETRO

VOLUME XXV. Enzyme Structure (Part B)
Edited by C. H. W. HIRS AND SERGE N. TIMASHEFF

VOLUME XXVI. Enzyme Structure (Part C)
Edited by C. H. W. HIRS AND SERGE N. TIMASHEFF

VOLUME XXVII. Enzyme Structure (Part D)
Edited by C. H. W. HIRS AND SERGE N. TIMASHEFF

VOLUME XXVIII. Complex Carbohydrates (Part B)
Edited by VICTOR GINSBURG

VOLUME XXIX. Nucleic Acids and Protein Synthesis (Part E)
Edited by LAWRENCE GROSSMAN AND KIVIE MOLDAVE

VOLUME XXX. Nucleic Acids and Protein Synthesis (Part F)
Edited by KIVIE MOLDAVE AND LAWRENCE GROSSMAN

VOLUME XXXI. Biomembranes (Part A)
Edited by SIDNEY FLEISCHER AND LESTER PACKER

VOLUME XXXII. Biomembranes (Part B)
Edited by SIDNEY FLEISCHER AND LESTER PACKER

VOLUME XXXIII. Cumulative Subject Index Volumes I–XXX
Edited by MARTHA G. DENNIS AND EDWARD A. DENNIS

VOLUME XXXIV. Affinity Techniques (Enzyme Purification: Part B)
Edited by WILLIAM B. JAKOBY AND MEIR WILCHEK

VOLUME XXXV. Lipids (Part B)
Edited by JOHN M. LOWENSTEIN

VOLUME XXXVI. Hormone Action (Part A: Steroid Hormones)
Edited by BERT W. O'MALLEY AND JOEL G. HARDMAN

VOLUME XXXVII. Hormone Action (Part B: Peptide Hormones)
Edited by BERT W. O'MALLEY AND JOEL G. HARDMAN

VOLUME XXXVIII. Hormone Action (Part C: Cyclic Nucleotides)
Edited by JOEL G. HARDMAN AND BERT W. O'MALLEY

VOLUME XXXIX. Hormone Action (Part D: Isolated Cells, Tissues, and Organ Systems)
Edited by JOEL G. HARDMAN AND BERT W. O'MALLEY

VOLUME XL. Hormone Action (Part E: Nuclear Structure and Function)
Edited by BERT W. O'MALLEY AND JOEL G. HARDMAN

VOLUME XLI. Carbohydrate Metabolism (Part B)
Edited by W. A. WOOD

VOLUME XLII. Carbohydrate Metabolism (Part C)
Edited by W. A. WOOD

VOLUME XLIII. Antibiotics
Edited by JOHN H. HASH

VOLUME XLIV. Immobilized Enzymes
Edited by KLAUS MOSBACH

VOLUME XLV. Proteolytic Enzymes (Part B)
Edited by LASZLO LORAND

Volume XLVI. Affinity Labeling
Edited by William B. Jakoby and Meir Wilchek

Volume XLVII. Enzyme Structure (Part E)
Edited by C. H. W. Hirs and Serge N. Timasheff

Volume XLVIII. Enzyme Structure (Part F)
Edited by C. H. W. Hirs and Serge N. Timasheff

Volume XLIX. Enzyme Structure (Part G)
Edited by C. H. W. Hirs and Serge N. Timasheff

Volume L. Complex Carbohydrates (Part C)
Edited by Victor Ginsburg

Volume LI. Purine and Pyrimidine Nucleotide Metabolism
Edited by Patricia A. Hoffee and Mary Ellen Jones

Volume LII. Biomembranes (Part C: Biological Oxidations)
Edited by Sidney Fleischer and Lester Packer

Volume LIII. Biomembranes (Part D: Biological Oxidations)
Edited by Sidney Fleischer and Lester Packer

Volume LIV. Biomembranes (Part E: Biological Oxidations)
Edited by Sidney Fleischer and Lester Packer

Volume LV. Biomembranes (Part F: Bioenergetics)
Edited by Sidney Fleischer and Lester Packer

Volume LVI. Biomembranes (Part G: Bioenergetics)
Edited by Sidney Fleischer and Lester Packer

Volume LVII. Bioluminescence and Chemiluminescence
Edited by Marlene A. DeLuca

Volume LVIII. Cell Culture
Edited by William B. Jakoby and Ira Pastan

Volume LIX. Nucleic Acids and Protein Synthesis (Part G)
Edited by Kivie Moldave and Lawrence Grossman

VOLUME LX. Nucleic Acids and Protein Synthesis (Part H)
Edited by KIVIE MOLDAVE AND LAWRENCE GROSSMAN

VOLUME 61. Enzyme Structure (Part H)
Edited by C. H. W. HIRS AND SERGE N. TIMASHEFF

VOLUME 62. Vitamins and Coenzymes (Part D)
Edited by DONALD B. MCCORMICK AND LEMUEL D. WRIGHT

VOLUME 63. Enzyme Kinetics and Mechanism (Part A: Initial Rate and Inhibitor Methods)
Edited by DANIEL L. PURICH

VOLUME 64. Enzyme Kinetics and Mechanism (Part B: Isotopic Probes and Complex Enzyme Systems)
Edited by DANIEL L. PURICH

VOLUME 65. Nucleic Acids (Part I)
Edited by LAWRENCE GROSSMAN AND KIVIE MOLDAVE

VOLUME 66. Vitamins and Coenzymes (Part E)
Edited by DONALD B. MCCORMICK AND LEMUEL D. WRIGHT

VOLUME 67. Vitamins and Coenzymes (Part F)
Edited by DONALD B. MCCORMICK AND LEMUEL D. WRIGHT

VOLUME 68. Recombinant DNA
Edited by RAY WU

VOLUME 69. Photosynthesis and Nitrogen Fixation (Part C)
Edited by ANTHONY SAN PIETRO

VOLUME 70. Immunochemical Techniques (Part A)
Edited by HELEN VAN VUNAKIS AND JOHN J. LANGONE

VOLUME 71. Lipids (Part C)
Edited by JOHN M. LOWENSTEIN

VOLUME 72. Lipids (Part D)
Edited by JOHN M. LOWENSTEIN

VOLUME 73. Immunochemical Techniques (Part B)
Edited by JOHN J. LANGONE AND HELEN VAN VUNAKIS

VOLUME 74. Immunochemical Techniques (Part C)
Edited by JOHN J. LANGONE AND HELEN VAN VUNAKIS

VOLUME 75. Cumulative Subject Index Volumes XXXI, XXXII, XXXIV–LX
Edited by EDWARD A. DENNIS AND MARTHA G. DENNIS

VOLUME 76. Hemoglobins
Edited by ERALDO ANTONINI, LUIGI ROSSI-BERNARDI, AND EMILIA CHIANCONE

VOLUME 77. Detoxication and Drug Metabolism
Edited by WILLIAM B. JAKOBY

VOLUME 78. Interferons (Part A)
Edited by SIDNEY PESTKA

VOLUME 79. Interferons (Part B)
Edited by SIDNEY PESTKA

VOLUME 80. Proteolytic Enzymes (Part C)
Edited by LASZLO LORAND

VOLUME 81. Biomembranes (Part H: Visual Pigments and Purple Membranes, I)
Edited by LESTER PACKER

VOLUME 82. Structural and Contractile Proteins (Part A: Extracellular Matrix)
Edited by LEON W. CUNNINGHAM AND DIXIE W. FREDERIKSEN

VOLUME 83. Complex Carbohydrates (Part D)
Edited by VICTOR GINSBURG

VOLUME 84. Immunochemical Techniques (Part D: Selected Immunoassays)
Edited by JOHN J. LANGONE AND HELEN VAN VUNAKIS

VOLUME 85. Structural and Contractile Proteins (Part B: The Contractile Apparatus and the Cytoskeleton)
Edited by DIXIE W. FREDERIKSEN AND LEON W. CUNNINGHAM

VOLUME 86. Prostaglandins and Arachidonate Metabolites
Edited by WILLIAM E. M. LANDS AND WILLIAM L. SMITH

VOLUME 87. Enzyme Kinetics and Mechanism (Part C: Intermediates, Stereochemistry, and Rate Studies)
Edited by DANIEL L. PURICH

VOLUME 88. Biomembranes (Part I: Visual Pigments and Purple Membranes, II)
Edited by LESTER PACKER

VOLUME 89. Carbohydrate Metabolism (Part D)
Edited by WILLIS A. WOOD

VOLUME 90. Carbohydrate Metabolism (Part E)
Edited by WILLIS A. WOOD

VOLUME 91. Enzyme Structure (Part I)
Edited by C. H. W. HIRS AND SERGE N. TIMASHEFF

VOLUME 92. Immunochemical Techniques (Part E: Monoclonal Antibodies and General Immunoassay Methods)
Edited by JOHN J. LANGONE AND HELEN VAN VUNAKIS

VOLUME 93. Immunochemical Techniques (Part F: Conventional Antibodies, Fc Receptors, and Cytotoxicity)
Edited by JOHN J. LANGONE AND HELEN VAN VUNAKIS

VOLUME 94. Polyamines
Edited by HERBERT TABOR AND CELIA WHITE TABOR

VOLUME 95. Cumulative Subject Index Volumes 61–74, 76–80
Edited by EDWARD A. DENNIS AND MARTHA G. DENNIS

VOLUME 96. Biomembranes [Part J: Membrane Biogenesis: Assembly and Targeting (General Methods; Eukaryotes)]
Edited by SIDNEY FLEISCHER AND BECCA FLEISCHER

VOLUME 97. Biomembranes [Part K: Membrane Biogenesis: Assembly and Targeting (Prokaryotes, Mitochondria, and Chloroplasts)]
Edited by SIDNEY FLEISCHER AND BECCA FLEISCHER

VOLUME 98. Biomembranes (Part L: Membrane Biogenesis: Processing and Recycling)
Edited by SIDNEY FLEISCHER AND BECCA FLEISCHER

VOLUME 99. Hormone Action (Part F: Protein Kinases)
Edited by JACKIE D. CORBIN AND JOEL G. HARDMAN

VOLUME 100. Recombinant DNA (Part B)
Edited by RAY WU, LAWRENCE GROSSMAN, AND KIVIE MOLDAVE

VOLUME 101. Recombinant DNA (Part C)
Edited by RAY WU, LAWRENCE GROSSMAN, AND KIVIE MOLDAVE

VOLUME 102. Hormone Action (Part G: Calmodulin and Calcium-Binding Proteins)
Edited by ANTHONY R. MEANS AND BERT W. O'MALLEY

VOLUME 103. Hormone Action (Part H: Neuroendocrine Peptides)
Edited by P. MICHAEL CONN

VOLUME 104. Enzyme Purification and Related Techniques (Part C)
Edited by WILLIAM B. JAKOBY

VOLUME 105. Oxygen Radicals in Biological Systems
Edited by LESTER PACKER

VOLUME 106. Posttranslational Modifications (Part A)
Edited by FINN WOLD AND KIVIE MOLDAVE

VOLUME 107. Posttranslational Modifications (Part B)
Edited by FINN WOLD AND KIVIE MOLDAVE

VOLUME 108. Immunochemical Techniques (Part G: Separation and Characterization of Lymphoid Cells)
Edited by GIOVANNI DI SABATO, JOHN J. LANGONE, AND HELEN VAN VUNAKIS

VOLUME 109. Hormone Action (Part I: Peptide Hormones)
Edited by LUTZ BIRNBAUMER AND BERT W. O'MALLEY

VOLUME 110. Steroids and Isoprenoids (Part A)
Edited by JOHN H. LAW AND HANS C. RILLING

VOLUME 111. Steroids and Isoprenoids (Part B)
Edited by JOHN H. LAW AND HANS C. RILLING

VOLUME 112. Drug and Enzyme Targeting (Part A)
Edited by KENNETH J. WIDDER AND RALPH GREEN

VOLUME 113. Glutamate, Glutamine, Glutathione, and Related Compounds
Edited by ALTON MEISTER

VOLUME 114. Diffraction Methods for Biological Macromolecules (Part A)
Edited by HAROLD W. WYCKOFF, C. H. W. HIRS, AND SERGE N. TIMASHEFF

VOLUME 115. Diffraction Methods for Biological Macromolecules (Part B)
Edited by HAROLD W. WYCKOFF, C. H. W. HIRS, AND SERGE N. TIMASHEFF

VOLUME 116. Immunochemical Techniques (Part H: Effectors and Mediators of Lymphoid Cell Functions)
Edited by GIOVANNI DI SABATO, JOHN J. LANGONE, AND HELEN VAN VUNAKIS

VOLUME 117. Enzyme Structure (Part J)
Edited by C. H. W. HIRS AND SERGE N. TIMASHEFF

VOLUME 118. Plant Molecular Biology
Edited by ARTHUR WEISSBACH AND HERBERT WEISSBACH

VOLUME 119. Interferons (Part C)
Edited by SIDNEY PESTKA

VOLUME 120. Cumulative Subject Index Volumes 81–94, 96–101

VOLUME 121. Immunochemical Techniques (Part I: Hybridoma Technology and Monoclonal Antibodies)
Edited by JOHN J. LANGONE AND HELEN VAN VUNAKIS

VOLUME 122. Vitamins and Coenzymes (Part G)
Edited by FRANK CHYTIL AND DONALD B. McCORMICK

VOLUME 123. Vitamins and Coenzymes (Part H)
Edited by FRANK CHYTIL AND DONALD B. McCORMICK

VOLUME 124. Hormone Action (Part J: Neuroendocrine Peptides)
Edited by P. MICHAEL CONN

VOLUME 125. Biomembranes (Part M: Transport in Bacteria, Mitochondria, and Chloroplasts: General Approaches and Transport Systems)
Edited by SIDNEY FLEISCHER AND BECCA FLEISCHER

VOLUME 126. Biomembranes (Part N: Transport in Bacteria, Mitochondria, and Chloroplasts: Protonmotive Force)
Edited by SIDNEY FLEISCHER AND BECCA FLEISCHER

VOLUME 127. Biomembranes (Part O: Protons and Water: Structure and Translocation)
Edited by LESTER PACKER

VOLUME 128. Plasma Lipoproteins (Part A: Preparation, Structure, and Molecular Biology)
Edited by JERE P. SEGREST AND JOHN J. ALBERS

VOLUME 129. Plasma Lipoproteins (Part B: Characterization, Cell Biology, and Metabolism)
Edited by JOHN J. ALBERS AND JERE P. SEGREST

VOLUME 130. Enzyme Structure (Part K)
Edited by C. H. W. HIRS AND SERGE N. TIMASHEFF

VOLUME 131. Enzyme Structure (Part L)
Edited by C. H. W. HIRS AND SERGE N. TIMASHEFF

VOLUME 132. Immunochemical Techniques (Part J: Phagocytosis and Cell-Mediated Cytotoxicity)
Edited by GIOVANNI DI SABATO AND JOHANNES EVERSE

VOLUME 133. Bioluminescence and Chemiluminescence (Part B)
Edited by MARLENE DELUCA AND WILLIAM D. McELROY

VOLUME 134. Structural and Contractile Proteins (Part C: The Contractile Apparatus and the Cytoskeleton)
Edited by RICHARD B. VALLEE

VOLUME 135. Immobilized Enzymes and Cells (Part B)
Edited by KLAUS MOSBACH

VOLUME 136. Immobilized Enzymes and Cells (Part C)
Edited by KLAUS MOSBACH

VOLUME 137. Immobilized Enzymes and Cells (Part D)
Edited by KLAUS MOSBACH

VOLUME 138. Complex Carbohydrates (Part E)
Edited by VICTOR GINSBURG

VOLUME 139. Cellular Regulators (Part A: Calcium- and Calmodulin-Binding Proteins)
Edited by ANTHONY R. MEANS AND P. MICHAEL CONN

VOLUME 140. Cumulative Subject Index Volumes 102–119, 121–134

VOLUME 141. Cellular Regulators (Part B: Calcium and Lipids)
Edited by P. MICHAEL CONN AND ANTHONY R. MEANS

VOLUME 142. Metabolism of Aromatic Amino Acids and Amines
Edited by SEYMOUR KAUFMAN

VOLUME 143. Sulfur and Sulfur Amino Acids
Edited by WILLIAM B. JAKOBY AND OWEN GRIFFITH

VOLUME 144. Structural and Contractile Proteins (Part D: Extracellular Matrix)
Edited by LEON W. CUNNINGHAM

VOLUME 145. Structural and Contractile Proteins (Part E: Extracellular Matrix)
Edited by LEON W. CUNNINGHAM

VOLUME 146. Peptide Growth Factors (Part A)
Edited by DAVID BARNES AND DAVID A. SIRBASKU

VOLUME 147. Peptide Growth Factors (Part B)
Edited by DAVID BARNES AND DAVID A. SIRBASKU

VOLUME 148. Plant Cell Membranes
Edited by LESTER PACKER AND ROLAND DOUCE

VOLUME 149. Drug and Enzyme Targeting (Part B)
Edited by RALPH GREEN AND KENNETH J. WIDDER

VOLUME 150. Immunochemical Techniques (Part K: *In Vitro* Models of B and T Cell Functions and Lymphoid Cell Receptors)
Edited by GIOVANNI DI SABATO

VOLUME 151. Molecular Genetics of Mammalian Cells
Edited by MICHAEL M. GOTTESMAN

VOLUME 152. Guide to Molecular Cloning Techniques
Edited by SHELBY L. BERGER AND ALAN R. KIMMEL

VOLUME 153. Recombinant DNA (Part D)
Edited by RAY WU AND LAWRENCE GROSSMAN

VOLUME 154. Recombinant DNA (Part E)
Edited by RAY WU AND LAWRENCE GROSSMAN

VOLUME 155. Recombinant DNA (Part F)
Edited by RAY WU

VOLUME 156. Biomembranes (Part P: ATP-Driven Pumps and Related Transport: The Na,K-Pump)
Edited by SIDNEY FLEISCHER AND BECCA FLEISCHER

VOLUME 157. Biomembranes (Part Q: ATP-Driven Pumps and Related Transport: Calcium, Proton, and Potassium Pumps)
Edited by SIDNEY FLEISCHER AND BECCA FLEISCHER

VOLUME 158. Metalloproteins (Part A)
Edited by JAMES F. RIORDAN AND BERT L. VALLEE

VOLUME 159. Initiation and Termination of Cyclic Nucleotide Action
Edited by JACKIE D. CORBIN AND ROGER A. JOHNSON

VOLUME 160. Biomass (Part A: Cellulose and Hemicellulose)
Edited by WILLIS A. WOOD AND SCOTT T. KELLOGG

VOLUME 161. Biomass (Part B: Lignin, Pectin, and Chitin) (in preparation)
Edited by WILLIS A. WOOD AND SCOTT T. KELLOGG

VOLUME 162. Immunochemical Techniques (Part L: Chemotaxis and Inflammation) (in preparation)
Edited by GIOVANNI DI SABATO

VOLUME 163. Immunochemical Techniques (Part M: Chemotaxis and Inflammation) (in preparation)
Edited by GIOVANNI DI SABATO

VOLUME 164. Ribosomes (in preparation)
Edited by HARRY F. NOLLER, JR., AND KIVIE MOLDAVE

VOLUME 165. Microbial Toxins: Tools in Enzymology (in preparation)
Edited by SIDNEY HARSHMAN

VOLUME 166. Branched-Chain Amino Acids (in preparation)
Edited by ROBERT A. HARRIS AND JOHN R. SOKATCH

VOLUME 167. Cyanobacteria (in preparation)
Edited by LESTER PACKER AND ALEXANDER N. GLAZER

VOLUME 168. Hormone Action (Part K: Neuroendocrine Peptides) (in preparation)
Edited by P. MICHAEL CONN

VOLUME 169. Platelets: Receptors, Adhesion, Secretion (Part A) (in preparation)
Edited by JACK HAWIGER

Subject Index

Boldface numerals indicate volume number.

A

Abrin
A chain
–antibody conjugates, cytotoxicity, **112**, 222
coupling to antibodies, **112**, 218
isolation, **112**, 210
radiolabeling, **112**, 213
–antibody conjugates
biological evaluation, **112**, 221
cytotoxicity, **112**, 222
physicochemical characterization, **112**, 220
preparation, **112**, 213
purification, **112**, 219
storage, **112**, 219
coupling to antibodies
with chlorambucil *N*-hydroxysuccinimidyl ester, **112**, 214
with *N*-hydroxysuccinimidyl ester of iodoacetic acid, **112**, 217
with *N*-succinimidyl-3-(2-pyridyldithio)propionate, **112**, 215
handling, safety considerations, **112**, 224
purification, **112**, 209
radiolabeling, **112**, 213
Absorption spectroscopy
apolipoproteins, **128**, 516
calmodulin, **102**, 86
carotenoids, **111**, 135
chloramines, **132**, 577
ecdysteroid conjugates, **111**, 414
vitamin D and metabolites, **123**, 132
Acedapsone
loading in poly(lactic/glycolic acid) polymers, **112**, 441
Acer pseudoplatanus, *see* Sycamore
Acetamidofluorescein
–α-actinin conjugates
characterization, **134**, 501
preparation, **134**, 496
Acetaminophen
prodrug structure, effect on metabolism, **112**, 340
Acetate esters
pheromone, bioluminescent analysis, **133**, 194
Acetate kinase
in acetyl phosphate and acetyl-CoA determination, **122**, 44
Acetoacetate–CoA ligase
rat liver
assays, **110**, 3
properties, **110**, 9
purification, **110**, 6
Acetoacetyl-CoA synthetase, *see* Acetoacetate–CoA ligase
Acetoacetyl-CoA thiolase, *see* Acetyl-CoA acetyltransferase
Acetone
in precipitation of thymosin fraction 5 and 5A, **116**, 221
S-Acetonyl-CoA
as inhibitor of protein acetylation, **106**, 182
synthesis, **106**, 181
Acetylation
actin amino-terminal
assays, **106**, 187, 189
inhibition, **106**, 181
in reticulocyte lysate system, **106**, 183
cyclic GMP, in radioimmunoassay, **109**, 831
protein amino-terminal, overview, **106**, 165
protein side-chain, analytical techniques, **107**, 224
Acetylcarnitine
radioisotopic assay, **123**, 259
Acetylcholine receptors
monoclonal antibodies to
assay in hybridoma supernatants, **121**, 103
preparation, **121**, 97
specificity assay, **121**, 104
reconstitution in lipid vesicles, **104**, 342
Acetyl-CoA
bacterial, determination, **122**, 43
radioisotopic assay, **123**, 259
Acetyl-CoA acetyltransferase
Clostridium kluyveri, selenium incorporation, **107**, 621
Acetyl-CoA:acyl carrier protein transacylase, *see* [Acyl-carrier-protein] acetyltransferase

Acetyl-CoA carboxylase synthetase
  assays, **107**, 265
  properties, **107**, 262, 274
  purification, **107**, 271
  reaction mechanism, **107**, 275
  sequence homology with other synthetases, **107**, 264
O^{β}-(*N*-Acetyl-α-glucosamine-1-phosphoryl)serine
  in proteinase I from *Dictyostelium discoideum*
    chemical characterization, **107**, 173
    spectroscopic characterization, **107**, 176
N-Acetyl-β-glucosaminidase
  human neutrophil, assay, **132**, 274
N-Acetylglucosaminylphosphotransferase, *see* UDP-*N*-acetylglucosamine–lysosomal enzyme *N*-acetylglucosamine-phosphotransferase
N-Acetylglutamate synthase, *see* Amino-acid acetyltransferase
β-*N*-Acetylhexosaminidase
  as endosome marker, determination, **109**, 267
N-Acetyllactosaminide α-1,3-galactosyltransferase
  as endosome marker, determination, **109**, 267
ε-*N*-Acetyllysine
  deacetylation, **107**, 236
  detection and quantitation, **107**, 226
  formation *in vivo* and *in vitro*, **107**, 226
Acetyl phosphate
  bacterial, determination, **122**, 43
N-Acetyltransferase
  assays, **107**, 231
  Drosophila melanogaster, purification, **103**, 489
  in radioenzymatic assay for serotonin, **103**, 491
  rat liver, purification, **103**, 489
  wheat germ
    assay, **106**, 194
    properties, **106**, 196
N^{α}-Acetyltransferase
  rat, distribution in various organs, **106**, 176
  rat pituitary
    activity in anterior and posterior lobes, **106**, 177
    kinetic constants, **106**, 178
    properties, **106**, 173
    subcellular localization, **106**, 175
Achromobacter superficialis
  purification of taurine-glutamate transaminase, **113**, 104
Acidaminococcus fermentans
  purification of glutaconyl-CoA decarboxylase, **125**, 552
Acid hydrolysis
  phosphoproteins, **107**, 10
  sulfated proteins from polyacrylamide gels, **107**, 213
Acid phosphatase
  assay
    for assessment of phagocytic function, **132**, 123, 163
    in hepatic Golgi subfractions, **109**, 224
  in hydrolysis of polyprenol pyrophosphates, **110**, 153
  in phagocyte staining, **132**, 133, 174
Acid stability
  O-phosphates in proteins, **107**, 4
Acinar cells
  dispersed pancreatic, preparation, **109**, 289
Acinetobacter glutaminsificans
  purification of glutaminase, **113**, 259
Acini
  dispersed pancreatic, preparation, **109**, 291
Aconitine
  interactions with ion channels in neuroendocrine tissue, **103**, 401
Acridine orange
  in assay of microbial killing by phagocytes, **132**, 523
  –ethidium bromide, in neutrophil viability assessment, **132**, 96, 138
  in phagosome–lysosome fusion assay, **132**, 260
Acridinium esters
  chemiluminescence, mechanism, **133**, 366
  immunoassays with, **133**, 379
  as labels
    reactions, **133**, 374
    synthesis, **133**, 371
Acrylamide
  copolymers with unsaturated ligand derivatives, preparation, **104**, 279
ACTH, *see* Adrenocorticotropin

Actin
amino-terminal acetylation
assays, **106**, 187, 189
inhibition, **106**, 181
in reticulocyte lysate system, **106**, 183
antibodies
isoform-specific
production, **134**, 461
purification, **134**, 462
region-specific
production, **134**, 453
specificity, **134**, 458
fluorescently labeled
assay, **134**, 511
microinjection, **134**, 513
preparation, **134**, 508
gel overlays, assay for protein–protein interactions, **134**, 558
gels, extraction of fascin, **134**, 17
nonacetylated, preparation, **106**, 182
oriented filaments, myosin movement
assays, **134**, 531, 538
inhibition by antibodies, **134**, 452
purification, **134**, 20, 508
pyrene labeling, **134**, 20
–Sepharose affinity resins, preparation, **134**, 463
time-resolved X-ray scattering, **134**, 675
α-Actinin
fluorescently labeled
assay, **134**, 511
microinjection, **134**, 513
preparation, **134**, 501, 510
properties, **134**, 501
isolation, **134**, 496, 510
Actinomycin D, *see* Dactinomycin
Active enzyme centrifugation
in measurement of hydrodynamic properties
applications, **117**, 101
data analysis, **117**, 107
principles, **117**, 99
simulations, **117**, 114
Actomyosin
-based motility, analysis with *N*-ethylmaleimide-modified heavy meromyosin, **134**, 474
Actomyosin ATPase, *see* Myosin ATPase
Acyclic isoprenoid hydrocarbons
chromatographic analysis, **111**, 186
N-Acylation
amines, in amide prodrug formation, **112**, 357
imides, in amide prodrug formation, **112**, 357
proteins, mass spectrometric identification, **106**, 37, 40
Acylcarnitines
long-chain
specific activity, assay, **123**, 288
TLC, **123**, 286
short-chain
identification, **123**, 267
isolation, **123**, 265, 271
quantitation, **123**, 268
radioactive, separation, **123**, 273
[Acyl-carrier-protein] acetyltransferase
spinach leaves
assay, **122**, 53
function, **122**, 58
properties, **122**, 56
purification, **122**, 54
Acyl-CoA
radioactive, preparation, **122**, 50
Acyl:CoA-*O*-acyltransferase
metabolic function, **128**, 17
1-Acylglycerophosphocholine acyltransferase
human plasma
assays, **129**, 790
properties, **129**, 794, 795
N-Acyloxymethylation
amides, in prodrug formation, **112**, 355
carbamates, in prodrug formation, **112**, 355
hydantoins, in prodrug formation, **112**, 355
imides, in prodrug formation, **112**, 355
Acyl phosphates
diagnostic tests, **107**, 29
Acyl-protein synthetase
Photobacterium phosphoreum, assay, **133**, 174
Acyltransferase
bioluminescence-related bacterial
assay, **133**, 183
properties, **133**, 187
purification, **133**, 185
N-(Adamantylideneadamantane 1,2-dioxetane-4-eq-yl)succinic acid monoamide *N*-hydroxysuccinimide ester
isomers, synthesis, **133**, 538

N-(Adamantylideneadamant-4-eq-yl)succinic acid monoamide
synthesis, **133**, 537
N-(Adamantylideneadamant-4-eq-yl)succinic acid monoamide *N*-hydroxysuccinimide ester
synthesis, **133**, 537
Adenosine 5′-diphosphate, *see* ADP (adenosine 5′-diphosphate)
Adenosine 5′-monophosphate, *see* AMP (adenosine 5′-monophosphate)
Adenosinetriphosphatase, *see* ATPase
Adenosine 5′-triphosphate, *see* ATP (adenosine 5′-triphosphate)
S-Adenosylmethionine:calmodulin (lysine) *N*-methyltransferase
partial purification, **102**, 161
radiometric assay, **102**, 160
S-Adenosylmethionine-cycloartenol methyltransferase
plant, active cell-free preparations, isolation, **111**, 324
S-Adenosylmethionine:ε-*N*-L-lysine methyltransferase
Neurospora crassa
assay, **123**, 304
properties, **123**, 308
purification, **123**, 306
S-Adenosylmethionine:protein (arginine) *N*-methyltransferase, *see* Protein-arginine methyltransferase
S-Adenosylmethionine:protein-carboxyl *O*-methyltransferase, *see* Protein-glutamate methyltransferase
S-Adenosylmethionine:protein (lysine) *N*-methyltransferase, *see* Histone-lysine methyltransferase
S-Adenosylmethionine:γ-tocopherol methyltransferase
Capsicum chromoplasts, catalyzed reaction, characterization, **111**, 544
Adenylate cyclase
activation by iodinated glucagon, in glucagon receptor assay, **109**, 11
associated coupling proteins N_s and N_i
ADP-ribosylation by pertussis and cholera toxins, **109**, 566
analytical procedures and assays, **109**, 450
purification, **109**, 446, 454
effects of toxin ADP-ribosyltransferases, **106**, 411
as endosome marker, determination, **109**, 267
functional analysis with pertussis toxin as probe, **109**, 563
fusion to hepatic membranes, in analysis of glucagon receptor function, **109**, 346
hormonal regulation, assay, **109**, 333
leukocyte, assays, **132**, 422
related mutations in Y1 adrenal tumor cells, characterization, **109**, 355
Adenylosuccinate synthetase
Escherichia coli, purification, **104**, 109
Adenyl-3′,5′-uridine
hydrated, X-ray and neutron diffraction analysis, **127**, 169
Adenylylation
glutamine synthetase, detection methods, **107**, 187
Adenylyl cyclase, *see* Adenylate cyclase
Adenylyltransferase
assay, **107**, 197
Adipocytes
derived fibroblast-like cells with redifferentiating ability
adipose conversion, **109**, 383
characterization, **109**, 384, 385
establishment and handling, **109**, 378
insulin receptors, photolabeling, **109**, 175
lipogenesis, effect of photoreactive insulin covalent binding, **103**, 64
Adipose tissue
brown, isolation of uncoupling protein, **126**, 491
Adoptive transfer
in rodents
overview, **108**, 35
representative examples, **108**, 38
ADP (adenosine 5′-diphosphate)
–ATP carrier protein, mitochondrial
activity, influencing factors, **125**, 617
atractyloside-binding site, photolabeling, **125**, 665
chemical modification, **125**, 659
detection with fluorescent nucleotide analogs, **125**, 620
ESR studies, **125**, 638
extrinsic fluorescent probes, **125**, 640

intrinsic fluorescent probes, **125**, 648
isolation, **125**, 610
^{31}P NMR studies in detergent micelles, **125**, 634
reconstitution in phospholipid vesicles, **125**, 613, 633
spin labeling, **125**, 635
transmembrane folding, analysis with pyridoxal phosphate, **125**, 650
transport assay, **125**, 615
cardiac mitochondrial, assay, **125**, 19
cellular
assay, **133**, 36
extraction, **133**, 32
ester analogs
application in analysis of F_1-ATPase, **126**, 699
synthesis, **126**, 688
formation, in assay for
carbamyl phosphate synthetase, **113**, 308
γ-glutamylcysteine synthetase, **113**, 380
radiolabeled azido derivatives, synthesis, **125**, 665
reversed affinity chromatography with silica-bound alcohol dehydrogenase, **104**, 220
transport in chloroplast envelope, **125**, 715
ADP-ribose
monomeric
levels *in vivo*, fluorescence-based measurement, **106**, 490
protein-bound, determination, **106**, 473
polymeric, *see* Poly(ADP-ribose)
–protein conjugates
characterization, **106**, 466
purification, **106**, 462
reversed affinity chromatography with silica-bound alcohol dehydrogenase, **104**, 220
ADP-ribosylation
diphthamide in elongation factor 2, **106**, 379, 392
histones, **106**, 450
membrane components by pertussis and cholera toxins, **109**, 564, 566
overview, **106**, 438
RNA polymerase, coliphage-induced, **106**, 418
ADP-ribosylguanidines
enzymatic synthesis, **106**, 402
ADP-ribosyltransferase
arginine-dependent
assays, **106**, 430
chemical and enzymatic properties, **106**, 399
general properties, **106**, 433, 437
purification from turkey erythrocytes, **106**, 433, 437
coliphage T4 *alt* and *mod* gene-encoded
assay conditions and kinetic properties, **106**, 426
mechanisms, **106**, 424, 425
physiological function, **106**, 428
properties, **106**, 418, 419
purification, **106**, 427
substrate specificities, **106**, 419
guanidine-specific
chromatographic assay, **106**, 405
spectrophotometric assay, **106**, 407
toxin
assays, **106**, 414, 416
demonstration, **106**, 413
effects on adenylate cyclase systems, **106**, 411
Adrenal cortex
bovine
preparation of glomerulosa cells, **124**, 126
purification of
low-density lipoprotein receptors, **109**, 410
plasma membranes, **109**, 120
rat, preparation of glomerulosa cells, **109**, 121
Adrenal medulla
bovine, chromaffin cells
culture stability, **103**, 311
isolation, **103**, 306
plating and maintenance, **103**, 310
β-Adrenergic receptors
reconstitution into lipid vesicles, **104**, 344
transfer during cell fusion, **109**, 371
Adrenochrome
formation from epinephrine, in assay for superoxide production, **105**, 395, 430

Adrenocortical cells
lipoprotein metabolism, analytical techniques, **129**, 687
steroidogenesis, activation by photoaffinity labeling, **103**, 65
Adrenocorticotropic hormone, *see* Adrenocorticotropin
Adrenocorticotropin
biotinylated, preparation, **109**, 432
immunoassay, **103**, 428
with sample purification, **103**, 431
without sample purification, **103**, 430
photoreactive
characterization, **103**, 60
nonradioactive, applications, **103**, 63
photolysis reactions, **103**, 62
radiolabeled, applications, **103**, 66
synthesis, **103**, 60
regulation of cAMP levels in primary glial cultures, analytical techniques, **109**, 344
resistant Y1 adrenal tumor cells
characteristics, **109**, 355
preparation, **109**, 353
selection, **109**, 354
Adrenocorticotropin receptors
photoaffinity labeling, **103**, 66
Adriamycin, *see* Doxorubicin
Adsorption
on controlled-pore glass, in purification of interferon γ, **119**, 459
hydrophobic, in reconstitution of cytochrome-*c* oxidase, **126**, 79
mevalonate kinase on calcium phosphate gel, **110**, 74
monoclonal antibodies on microspheres, **112**, 70, 71
Aequorin
cDNA
construction, **133**, 291
expression in *Escherichia coli* extracts, **133**, 294
identification, **133**, 293
in cytosolic Ca^{2+} measurement, **124**, 95
intracellular incorporation, **124**, 93
storage, **124**, 92
Affi-Gel 102
cytochrome *c*-linked, preparation, **126**. 71
Affinity chromatography
actin isoform-specific antibodies, **134**, 463
$(ADP\text{-ribosyl})_n$ proteins, **106**, 465
adsorbent preparation, **104**, 212
adsorption and elution strategies, **104**, 18
antiisotypic antibodies, **116**, 149
apolipoprotein E-containing lipoproteins, **129**, 160
applications, **104**, 218
avidin on 2-iminobiotin–6-aminohexyl-Sepharose 4B, **122**, 83
bacterial luciferase, **133**, 106
with boronate gels, in detection of nonenzymatic protein glycosylation, **106**, 82
calcineurin, **102**, 215, 235
calmodulin, **102**, 15, 17, 107, 187
calmodulin-dependent protein phosphatase, **102**, 246
class I MHC antigens, **108**, 510
concanavalin A–trypsin conjugates, **112**, 252
cytochrome-*c* oxidase, **126**, 64, 158, 161
cytoplasmic dynein, **134**, 349
enzymes, **104**, 21, 22
farnesylpyrophosphate synthetase, **110**, 171
fibronectin, **132**, 335
glutaconyl-CoA decarboxylase from *Acidaminococcus fermentans*, **125**, 553
glutamine synthetase, **107**, 192; **113**, 222, 224
glutathione transferase, **113**, 508
heme-binding proteins
with hematin- and hematoporphyrin-agarose columns, **123**, 324
with hemin-agarose, **123**, 331
hepatic lipase, **129**, 721, 733
HLA antigens, **108**, 624
Ia antigens, **108**, 538
IgA, **116**, 138
IgA secretory component, **116**, 71, 74
IgG
comparison with DEAE Affi-Gel Blue chromatography, **121**, 595
on staphylococcal protein A, **116**, 11
IgG-binding factor, **116**, 408
IgG subclasses, murine, **116**, 131
Ig isotypes from bovine × murine hybridomas, **121**, 251

IgM
on protein A–Sepharose, **116**, 30
on Sepharose immunoadsorbents, **116**, 31
2-iminobiotinylated plasma membrane proteins, **122**, 92
with immobilized dyes, **104**, 97, 109
dye selection, considerations, **104**, 105
operational parameters, **104**, 107
insulin receptors, **109**, 399, 401, 603
interferon α, **119**, 138
interferon β, **119**, 215
interferon γ, **119**, 90, 91, 146
interferons from human placenta, **119**, 545
interstitial retinol-binding protein, **123**, 110
lectins, **104**, 54
lipoprotein lipase, **129**, 698
low-density lipoprotein receptors, **109**, 411
lymphocytes
on immobilized peanut agglutinin, **108**, 177
on lectin–Sepharose gels, **108**, 153
on protein A gels, **108**, 132
L-lysine 6-aminotransferase, **113**, 96
membrane proteins, **104**, 337
methylmalonyl-CoA decarboxylase, **125**, 543
monoclonal antibodies to insulin, **109**, 712
monoclonal immunoglobulins, **116**, 80, 128
neuropeptides, **103**, 73
oxaloacetate decarboxylase, **125**, 532
peptidergic secretion granules on immobilized lectins, **124**, 259
phosphodiesterase on calmodulin–Sepharose 4B, **102**, 215
phosphorylated insulin receptors, **109**, 614
plant antiviral protein, **119**, 742
prolactin receptors, **109**, 158
protein A-binding Ig in fetal bovine serum, **121**, 304
protein methylase II, **106**, 301
proteins with immobilized antibodies, **104**, 21, 47
pyridoxamine (pyridoxine) 5′-phosphate oxidase, **122**, 118
receptors and other binding proteins, **104**, 50, 54
^{35}S-labeled interferon β, **119**, 285
with sulfonated supports, **104**, 65
transhydrogenase, **126**, 354
tubulin-tyrosine ligase, **106**, 227
ubiquinol–cytochrome-*c* reductase, **126**, 202
vitamin B_{12}-binding proteins, **123**, 29
Affinity constants
antibody
analysis by solution-phase radioimmunoassay, **121**, 468
determination by dilution analysis, **121**, 472
Affinity crosslinking
receptors for insulin and insulin-like growth factors I and II, **109**, 179
Affinity electrophoresis
applicability and limitations, **104**, 278
experimental modifications, **104**, 276
macromolecular carriers for, preparation, **104**, 279
methodology, **104**, 280
Affinity partitioning
proteins, **104**, 356
determining factors, **104**, 358
system selection strategy, **104**, 360
Affinity precipitation
glutamate dehydrogenase, **104**, 367
lactate dehydrogenase, **104**, 366, 368
Agarose
activated
coupling capacity, prediction, **104**, 11
determination, **104**, 8, 14
ligand coupling, **104**, 17
monoclonal antibody coupling, **121**, 742
activation with
carbamate derivatives, **104**, 11
1-cyano-4-dimethylaminopyridinium tetrafluoroborate, **104**, 7
cyanogen bromide, **104**, 4
N-cyanotriethylammonium, **104**, 7
p-nitrophenyl cyanate, **104**, 7
sulfonyl chlorides, **104**, 58
in affinity precipitation of dehydrogenases, **104**, 368

alkylated, *see* Alkylagarose
entrapment of hybridoma cells
methodology, **121**, 354
preparation for, **121**, 353
in gel chromatographic separation of lipoproteins, **129**, 47
hematin-coupled
in affinity chromatography of heme-binding proteins, **123**, 327
preparation, **123**, 324
hematoporphyrin-coupled
in affinity chromatography of heme-binding proteins, **123**, 327
preparation, **123**, 324
hemin-coupled
in affinity chromatography of heme-binding proteins, **123**, 336
preparation, **123**, 333
properties, **123**, 334
for high-resolution preparative isoelectric focusing
description, **104**, 263
load capacity, **104**, 265
protein detection techniques, **104**, 269
horizontal slab gels, for multimolecular aggregate characterization
description, **130**, 117
preparation, **130**, 118
immobilization of triazine dyes, **104**, 102
immobilized triazine dyes, concentration determination, **104**, 105
isoelectric focusing of lipoproteins, **128**, 435, 438
ligand-substituted meltable derivatives, preparation, **104**, 280
in microdroplet assay for migration inhibitory factor, **116**, 383
optimal pH conditions, determination by electrophoretic titration curves, **104**, 225, 229
retention mapping, **104**, 227, 229
overlay method, for study of contractile apparatus, **134**, 573
–polyacrolein microspheres
applications, **112**, 171
synthesis and properties, **112**, 169
radioactivity in, fluorographic detection, **104**, 462
–streptavidin
biotin-binding activity, **121**, 722
in membrane antigen immunoaffinity isolation, **121**, 724
preparation, **121**, 721
stability, **121**, 723
sulfonated
in affinity chromatography, **104**, 65
coupling of ligands and proteins, **104**, 60
in high-performance liquid affinity chromatography, **104**, 66
immobilized enzymes, stability, **104**, 67
ultra-low gelation temperature, hybridoma cloning in, **121**, 323
Age
effect on hypothalamic dopamine secretion into hypophysial portal blood, **103**, 616
Neurospora crassa conidia, assay, **105**, 263
role in excitotoxin lesion distribution, **103**, 383
Agglutination
selective, lymphocytes
with peanut agglutinin, **108**, 172, 174
with soybean agglutinin, **108**, 175
Agglutinins, *see specific agglutinins*
Agrobacterium tumefaciens
Ti plasmid
introduction of foreign genes, **118**, 628
transformation of plants, **118**, 581, 634
Air
exhaled, ethane collection and concentration, **105**, 306
negative ions
concentration measurements, **105**, 244
laboratory generation, **105**, 240
natural generation, **105**, 238
Alafosfalin
in selection of bacterial mutants deficient in tripeptide permease, **125**, 368
Alanine
carrier protein, in *Escherichia coli*
purification, **125**, 486
reconstitution into proteoliposomes, **125**, 490
transport assay, **125**, 490
Alanine aminotransferase
assays, **113**, 70
natural occurrence, **113**, 69

pyridoxal phosphate-binding site, isolation and sequence determination, **106**, 123
Alanine dipeptide
hydration
conformational effects, **127**, 382
Monte Carlo simulation with proximity criterion analysis, **127**, 37
Albumin
serum, *see* Serum albumin
Albumin microspheres
chemically crosslinked
labeling with ^{99m}Tc, **112**, 44
preparation, **112**, 12
as drug delivery system, **112**, 3
heat-crosslinked
diagnostic applications, **112**, 6
drug entrapment, **112**, 7
drug release, **112**, 8, 9
preparation, **112**, 4
heat-denatured, labeling with ^{99m}Tc, **112**, 44
hydrophilic
drug incorporation, **112**, 40
assessment, **112**, 32
fractionation, **112**, 31
with incorporated drug
physical characteristics, **112**, 35
physicochemical considerations, **112**, 37
lysophilization and storage, **112**, 31
minimicrosphere production, **112**, 32
preparation, **112**, 19, 27
release of adriamycin *in vitro*, **112**, 22
uptake of adriamycin, **112**, 21
magnetically responsive
behavior in blood flow, **112**, 62
physical properties, analytical techniques, **112**, 60
preparation, **112**, 58
Alcaligenes denitrificans
uptake of α-aminoadipate, **113**, 644
Alcohol consumption
effect on postprandial lipoprotein metabolism, **129**, 480
Alcohol dehydrogenase
equine liver
affinity HPLC, **104**, 218
configuration, stability after coupling to sulfonated supports, **104**, 67
coupling to sulfonated support, **104**, 61
silica-bound, in reversed affinity chromatography, **104**, 220
Alcohols
polyhydric, *see* Polyhydric alcohols
Aldehydes
in biological samples
continuous-flow assay, **133**, 240
immobilization on nylon, **133**, 239
properties, **133**, 247
spectrophotometry, **133**, 239
dithionite assay, **133**, 190
luciferase response to, **133**, 191
Alkaline hydrolysis
effect on halogenated tyrosine derivatives, **107**, 431
Alkaline phosphatase
assay for assessment of phagocytic function, **132**, 123, 163
conjugation to anti-Ig antibodies for Ig detection on blots, **121**, 499
delivery systems, polymeric and copolymeric
kinetic analysis, **112**, 414
preparation, **112**, 400
gastrointestinal, targeting for prodrug absorption, **112**, 367
linked antibodies, in protein detection on nitrocellulose, **121**, 848
in phagocyte staining, **132**, 133, 173
^{31}P NMR studies, **107**, 73
stepwise amplified immunoenzymatic staining, **121**, 855
Alkanal monooxygenase (FMN-linked)
in active center-based enzyme immunoassays, **133**, 248
assays, **133**, 99, 113, 200
binding to 2,2-diphenylpropylamine–Sepharose, **133**, 102
in bioluminescent assay for
aldehydes, **133**, 247
bile acids, **133**, 215, 246
ethanol, **133**, 246
glycerol, **133**, 247
NADH, **133**, 235
NAD(P)H, **133**, 244
6-phosphogluconic acid, **133**, 235
bioluminescent response to aldehydes, **133**, 191
–diaphorase, assay, **133**, 200

in dithionite assay for aldehydes, **133**, 190
flavin intermediates
formation and properties, **133**, 141, 146
reaction with superoxide, **133**, 144
4a-hydroperoxyflavin intermediates
isolation, **133**, 131, 133
properties, **133**, 135
reaction products, **133**, 138
stabilization, **133**, 130
immobilization, **133**, 201, 218, 233
–oxidoreductase system, stability, **133**, 203
purification, **133**, 106, 119
Alkanes
measurement, in detection of lipid peroxidation products, **105**, 292
Alkylagarose
homologous series
action mechanism, **104**, 92
in hydrophobic chromatography, **104**, 70, 90
preparation, **104**, 73
properties and characterization, **104**, 75
selection for optimal purification, **104**, 80
Alkylamines
coupling to activated agarose, **104**, 18
Alkylation
IgA, for monomer preparation, **116**, 139
IgG, for chain separation, **116**, 20
IgM
for chain separation, **116**, 35
for monomer preparation, **116**, 139
Alkyl glycoside detergents
applications, **125**, 33
properties, **125**, 31
synthesis, **125**, 28
ε-Alkyllysinase, *see* N^6-Methyl-lysine oxidase
Allergic reactions
modeling, applications of nude mouse, **108**, 356
Allogeneic effect factor
bioassays, **116**, 429
biochemical properties, **116**, 435
overview, **116**, 364
physicochemical properties, **116**, 435
production, **116**, 428
cellular and genetic requirements, **116**, 433
Alloxan
cytotoxic effects, assessment, **105**, 514
Amacrine cells
isolated from dispersed retinal cells, identification, **103**, 363
α-Amanitin
inhibition of glucagon receptor induction in transformed MDCK cells, **109**, 359
Amatoxins
–protein conjugates, prepared by mixed anhydride coupling, properties, **112**, 233
protein coupling
with carbodiimides, **112**, 227
with mixed anhydrides, **112**, 231
ω-Amidase
rat liver
in assay for α-ketoglutaramate, **113**, 356
assays, **113**, 352, 353
properties, **113**, 355
purification, **113**, 353, 357
Amidation
protein C-terminal, mass spectrometric identification, **106**, 44
Amides
N-acyloxymethylation, in prodrug formation, **112**, 355
N-Mannich bases
application as prodrugs, **112**, 351
hydrolysis, **112**, 348
preparation, **112**, 348
peptide carboxy-terminal
biosynthesis, **106**, 220
detection, **106**, 219
occurrence, **106**, 218
prodrugs, formation by N-acylation, **112**, 357
proton exchange, for detection of protein folding pathways, **131**, 137
Raman spectroscopy, **130**, 314, 322, 326
Amidophosphoribosyltransferase
Escherichia coli
assays, **113**, 265
properties, **113**, 272
purification, **113**, 270

Amines
N-acylation, in amide prodrug formation, **112**, 357
biogenic, in small biological samples, microbore HPLC, **124**, 402
enamine derivatives, as prodrugs, **112**, 359
N-Mannich bases
application as prodrugs, **112**, 351
hydrolysis, **112**, 348
preparation, **112**, 348
Amino-acid acetyltransferase
rat liver
assays, **113**, 28
properties, **113**, 32
purification, **113**, 29
Amino acid analyzers
in γ-carboxyglutamic acid assay, **107**, 503
Amino acid composition
apolipoprotein B-48, **128**, 267
apolipoprotein C, **128**, 295
apolipoprotein E, **128**, 283
apolipoproteins A-I, A-II, and A-IV, **128**, 236
apolipoproteins D, F, G, and H, **128**, 301
bacterial chemotaxis-related carboxylmethylesterase, **106**, 326, 328
calmodulin, **102**, 300
glutamate-aspartate transaminase, **113**, 88
glutaredoxins, **113**, 540
3-hydroxy-3-methylglutaryl-CoA synthase, **110**, 25
insulin-like growth factor I, **109**, 779, 791
insulin-like growth factor II, **109**, 792
lecithin–cholesterol acyltransferase, **129**, 769
lymphotoxin, **116**, 446
L-lysine 6-aminotransferase and subunits, **113**, 100
methylenetetrahydrofolate reductase, **122**, 379
osteocalcin, **107**, 524
platelet-derived growth factor, **109**, 771
prenyltransferase, **110**, 163
recombinant interferons, **119**, 11, 190
sterol carrier protein, **111**, 300
tumor necrosis factor, **116**, 454
ubiquitin–histone 2A conjugate, **106**, 241
wool and morphological components, **106**, 66
Amino acids
abbreviations for, **119**, 21
characteristics, comparison for protein structure prediction, **130**, 200
concentrated solutions, for protein partial specific volume determinations, **117**, 60
coupling to activated agarose, **104**, 18
derivatives, ion-exchange chromatography, **106**, 17
excitotoxic, *see* Excitotoxins
incorporation, measurement in isolated chloroplast translation reactions, **118**, 292
inhibition of γ-glutamyl transpeptidase, **113**, 425
in mammalian proteins, *in vivo* racemization
analysis, **106**, 107, 110
applications and implications, **106**, 113
detection, analytical methods, **106**, 99
kinetics and mechanism, **106**, 101
neuropeptide, analytical techniques, **103**, 24
phosphorylated basic
determination, **107**, 24
diagnostic tests, **107**, 28
posttranslationally modified
HPLC probes, **106**, 22
mass spectrometric identification, **106**, 29
radiolabeled, in identification of H^+–sugar transport proteins in *Escherichia coli*, **125**, 417
reactivities with heavy atoms, **114**, 149
side chains
conformational preferences, **127**, 189
distributions in proteins, **127**, 192
intraprotein and solvent interactions
computational methods, **127**, 187
effect on stability, **127**, 194
influencing factors, **127**, 185
as substrate for γ-glutamyl transpeptidase, **113**, 426
tracer, for analysis of hypothalamic peptide precursors, central delivery, **103**, 527

transport in chloroplast envelope, **125**, 714
α-Amino acids
analogs and modified forms, symbolism, **106**, 6
modifications, symbolism for, **106**, 10, 14, 17
nomenclature, **106**, 3
substitutions in, symbolism for, **106**, 7
symbolism, **106**, 4
Amino acid sequence
amino-terminal
cutinase, **106**, 217
cytochrome-*c* oxidase isozymes, **126**, 44
ubiquitin–histone 2A conjugate, **106**, 245
vitamin K-dependent plasma proteins, **107**, 511
apoaequorin, determination, **133**, 303
apolipoprotein A-I, **128**, 225
apolipoprotein A-II, **128**, 237
apolipoprotein A-IV, **128**, 756
apolipoprotein C-I, **128**, 294
apolipoprotein C-II, **128**, 295
apolipoprotein C-III, **128**, 296
apolipoprotein E, **128**, 283
apolipoprotein signal peptides, **128**, 697
ATPase inhibitor and stabilizing factors from yeast, **126**, 508
carboxy-terminal, ubiquitin–histone 2A conjugate, **106**, 245
class I MHC antigens, determination, **108**, 513
flavin peptides from various enzymes, **106**, 374
hinge protein, **126**, 245
histidine decarboxylase from *Lactobacillus* 30a, **122**, 133
homology, computational methods, **128**, 758
homology tests for protein structure predictions, **130**, 197
insulin from various sources, **109**, 706
interferon α, **119**, 6, 13
interferon αWA, **119**, 365
interferon β, **119**, 8
interferon γ, **119**, 10
interleukin 3, **116**, 550
lymphotoxin, **116**, 447
mitochondrial ATPase inhibitor, **126**, 468
partial, recombinant interferon β, **119**, 91
polypeptide β_1, **116**, 244
prothymosin α, **116**, 262
pyridoxal phosphate-binding site in enzymes, **106**, 127, 135
Qa and TL antigens, determination, **108**, 555
recombinant interferon γ, determination, **119**, 206
Thy-1 antigens, determination, **108**, 662
thymopoietin, **116**, 290
thymosin α_1, **116**, 244
thymosin β_4, **116**, 253
thymosin β_4-related peptides, **116**, 267
thymulin, **116**, 274
tumor necrosis factor, **116**, 455
ubiquitin, **106**, 238; **116**, 290
ubiquitin–histone 2A conjugate, **106**, 252
9-Aminoacridine
in measurement of pH changes across lipid bilayers and membrane proteins, **127**, 754
Aminoacylation
chloroplast and cytoplasmic tRNAs, **118**, 219
Aminoacyl-tRNA synthetase
Escherichia coli, purification, **118**, 218
plant, purification, **118**, 218
4-Aminoadamantylideneadamantane
synthesis, **133**, 535
α-Aminoadipate
cellular uptake, **113**, 644
metabolism, **113**, 645
properties, **113**, 640
related enzymes, **113**, 664
role in β-lactam antibiotics, **113**, 657
separation and identification, **113**, 642
synthesis and resolution, **113**, 643
L-α-Aminoadipate aminotransferase
rat kidney
assay, **113**, 666
properties, **113**, 672
purification, **113**, 669
β-Amino alcohols
oxazolidines as prodrugs, **112**, 359
p-Aminobenzoate synthase
Bacillus subtilis
assay, **113**, 293

properties, **113**, 296
subunit G, purification, **113**, 294
Escherichia coli
assay, **113**, 293
properties, **113**, 297
p-Aminobenzoylpoly(γ-glutamates)
detection with fluorescamine, **122**, 332
in identification of folylpoly(γ-glutamate) chain length, **122**, 323
preparation and purification, **122**, 331
γ-Aminobutyrate
synthesis of radioactive (−)-carnitine from, **123**, 297
9-Amino-6-chloro-2-methoxyacridine
fluorescence, in assay for ATP synthase subunits, **126**, 571, 581
Aminoethylcarbazole
as chromagen in immunohistological staining, **121**, 837
N^6-(6-Aminohexyl)-5′-adenosine monophosphate
coupling to sulfonated supports, **104**, 60, 61
6-Aminohexyl-Sepharose 4B
in affinity purification of avidin, **122**, 85
N^ε-(4-Amino-2-hydroxybutyl)lysine, *see* Hypusine
5-Aminolevulinate synthase
chicken embryo hepatic mitochondria
assay, **123**, 395
purification, **123**, 397
rat hepatic mitochondria
assay, **123**, 395
purification, **123**, 401
Rhodopseudomonas sphaeroides
applications, **123**, 439
assay, **123**, 435
properties, **123**, 441
purification, **123**, 437
5-Aminolevulinic acid
derivatives, isotachophoresis, **123**, 370
stereospecifically labeled, biosynthesis, **123**, 439
Aminomalonic acid
synthesis, **113**, 145
6-Aminonicotinamide
–polylysine conjugates
preparation, **112**, 272
properties, **112**, 282
L-2-Amino-4-oxo-5-chloropentanoate
inhibition of glutamate synthase, **113**, 331
Aminopeptidase
mucosal, targeting for gastrointestinal absorption of prodrugs, **112**, 368
Aminopeptidase M, *see* Microsomal aminopeptidase
3-Aminopyridine 1,N^6-ethenoadenine dinucleotide
preparation, **122**, 157
properties, **122**, 159
3-Aminopyridine 1,N^6-ethenoadenine dinucleotide phosphate
preparation, **122**, 158
properties, **122**, 159
Aminopyrine *N*-demethylase
assay for analysis of interferon effects on hepatic drug metabolism, **119**, 722
Ammonium acetate
concentration, effect on HPLC separation of bilirubin isomers, **123**, 394
in HPLC separation of uro- and coproporphyrin isomers, **123**, 387
Ammonium sulfate
in fractionation of bacterial cell lysate, **133**, 104
–polyethylene glycol, in concentration of human immune interferon, **119**, 58
as precipitant for protein partial specific volume determinations, **117**, 64
in precipitation of
calmodulin, **102**, 16
cytochrome-*c* oxidase, **126**, 164
IgG, **116**, 6
IgM, **116**, 29
interferon δ, **119**, 99
mevalonate kinase, **110**, 74
monoclonal IgE, **116**, 79
myeloperoxidase, **132**, 373
polyclonal IgE, **116**, 81
RNase L, **119**, 494
thymosin fraction 5 and 5A, **116**, 221
Amniotic fluid
human, detection of interferon, **119**, 542
AMP (adenosine 5′-monophosphate)
binding to fructose-1,6-bisphosphatase, assay, **117**, 345

cellular
assay, **133**, 36
extraction, **133**, 32
reversed affinity chromatography with silica-bound alcohol dehydrogenase, **104**, 220
Amphidicolin
in cultured cells, measurement, **118**, 91
synchrony induction in cultured cells and meristems, **118**, 93
assay, **118**, 89
effectiveness, **118**, 96
Amphiuma tridactylum, *see* Salamander
Ampholytes
in displacement chromatography of proteins, **104**, 115
Amyloid A
serum, *see* Serum amyloid A
Anabaena variabilis
cytochrome b_6f complex, isolation, **126**, 279
Anaphylaxis
passive cutaneous, for IgE antibody assay, **116**, 93
Angiotensin II
effects on
Ca^{2+} fluxes
in isolated hepatocytes, **109**, 554, 555
in perfused liver, **109**, 553
phosphorylase in hepatocytes, **109**, 557
monoiodinated
preparation and storage, **109**, 112
receptor binding studies, **109**, 118
specific activity and maximum binding activity determinations, **109**, 117
regulation of cGMP levels in hybrid cells, testing procedures, **109**, 337
Angiotensin II receptors
autoradiographic localization in brain, **124**, 597
binding assay with [^{125}I]angiotensin II, **109**, 118
properties, **109**, 113, 124
Anhydrides
in drug release from poly(ortho ester) polymer disks, **112**, 432
mixed, in protein coupling
to amatoxins, **112**, 231
to phallotoxins, **112**, 236
Aniline hydroxylase
assay for analysis of interferon effects on hepatic drug metabolism, **119**, 722
Anion exchange
in bacteria, **125**, 558
Ankyrin
bovine and porcine brain, purification, **134**, 62
Anostracans
carotenoid–lipovitellin complexes, isolation, **111**, 247
Antamanide
hydrated, X-ray and neutron diffraction analysis, **127**, 169
Anterior pituitary
cell enrichment, **124**, 211
cell isolation, **103**, 292
cell lines, **124**, 208
cell separation
affinity techniques, **103**, 287
cell-sorting methods, **103**, 287
by density gradient centrifugation, **103**, 272
by electrophoresis, **103**, 277
velocity sedimentation techniques, **103**, 260
cell types, identification, **124**, 215, 220
channel activation, **124**, 237
corticotropin-releasing factor, cell culture assay, **103**, 565
derived follicular cells
culture technique, **124**, 246
ultrastructural analysis, **124**, 248
dispersed cells
continuous perifusion, **103**, 298
comparison with static incubation techniques, **103**, 296
culture on extracellular matrix, **103**, 287
and fragments, experimental limitations, **103**, 294
dispersion of tissue, **103**, 289
dopamine receptors, direct radioligand measurement
on dispersed cells, **103**, 585
in membranes, **103**, 579
in solubilized preparations of membranes, **103**, 587
electrical activity, recording, **124**, 217

growth hormone-releasing factor, measurement, **124**, 373, 396
isolated cells for secretion studies
cultures and incubations, **109**, 296
preparation, **109**, 295
membrane binding of luciferin-derivatized GnRH analogs, **124**, 32
peptide secretion, reverse hemolytic plaque assay, **124**, 447
perifused cells, hormone secretion, **124**, 79
primary cultures, preparation, **124**, 209
regional functional specialization, **124**, 214
serum-free medium for, **103**, 293
Anthracene
substituted derivatives, in singlet oxygen trapping, **105**, 40
Anthranilate synthase
Serratia marcescens
assay, **113**, 288
component ASI, purification, **113**, 290
properties, **113**, 292
purification, **113**, 289
9,10-Anthraquinone
voltametric measurements, **125**, 112
Anthrone
in colorimetric assay for hexoses, **118**, 25
Antibiotics
activity assays with luminous bacteria
bacteriophage coupled test, **133**, 283
bioluminescence test, **133**, 277
induced test, **133**, 278
lysis test, **133**, 276
in serum, **133**, 281
dysopsonic function, overview, **132**, 307
β-lactam, role of α-aminoadipate, **113**, 657
Antibodies, *see also* Antiserum
actin
production, **134**, 453, 461
purification, **134**, 462
antiallotypic
applications, **116**, 163
assays, **116**, 162
preparation, **116**, 158
anti-Fab
in chromatographic separation of lymphocytes, **108**, 150
preparation and purification, **108**, 149
–antigen reactions, sedimentation patterns, microheterogeneity effects, **130**, 25
antiidiotypic
immunoassays with, **116**, 178
preparation, **116**, 174
secreting hybridoma clones, screening, **121**, 101
anti-Ig
conjugation to alkaline phosphatase for Ig detection on blots, **121**, 499
coupling to erythrocytes, **121**, 550
in isolation of surface immunoglobulins, **108**, 432
in localization of intracellular immunoglobulins, **108**, 397
in monoclonal antibody assays, **121**, 552, 553
in multivalent antibody radioimmunoassay, **121**, 431
–peroxidase conjugates, preparation, **108**, 394
purification, **121**, 550
radiolabeled
in culture-well radioimmunoassay, **121**, 442
preparation, **108**, 432
antiisotypic
applications, **116**, 149
characterization, **116**, 148
isolation, **116**, 148
preparation, **116**, 146
apolipoprotein A-containing lipoprotein
applications, **129**, 136
preparation, **129**, 134
specificity analysis, **129**, 135
apolipoprotein B, preparation, **129**, 537
to ATP synthase subunits, preparation, **118**, 358
binding to
polyacrolein-activated hybrid beads, **112**, 158
polyacrolein microspheres, **112**, 158
bound monolayers
lymphocyte separation on, **108**, 130
preparation, **108**, 128, 130
to cell-surface receptors
binding-inhibition assay, **109**, 657
immunoprecipitation assay, **109**, 660
insulin-like bioactivity, **109**, 663

to colony-stimulating factor 1
assays, **116**, 578
preparation, **116**, 577
and complement, for purification of lymphocyte subsets, **108**, 249
coupled to
culture plates, separation of lymphocytes, **108**, 118, 124
immunoadsorbent matrices, separation of lymphocytes, **108**, 139
latex spheres, detection of cell surface markers, **108**, 56
coupling to intact toxins
with chlorambucil *N*-hydroxysuccinimidyl ester, **112**, 214
with *N*-hydroxysuccinimidyl ester of iodoacetic acid, **112**, 217
with *N*-succinimidyl-3-(2-pyridyldithio)propionate, **112**, 215
coupling to toxin A chains, **112**, 218
cytokinin-specific
affinity purification, **110**, 351
binding to cellulose, **110**, 352
preparation, **110**, 349
cytoskeletal protein, gold-labeled, production, **134**, 596
β-endorphin, production, **103**, 548
enkephalin, production, **103**, 548
enzyme-labeled, preparation, **103**, 413
fluorescein isothiocyanate-labeled, in staining of glycoproteins in gels, **104**, 449
fluorescent
in detection of surface immunoglobulins, **108**, 421
in patching and capping of surface immunoglobulins, **108**, 371
preparation, **108**, 384, 415
glucitollysine-specific, preparation, **128**, 357
γ-glutamyl transpeptidase, preparation, **113**, 407
–hapten reactions, sedimentation patterns, microheterogeneity effects, **130**, 29
hemagglutinin, multivalent antibody radioimmunoassay, **121**, 428
immobilized, in affinity chromatography of proteins, **104**, 21, 47
interleukin 1
applications, **116**, 489
preparation, **116**, 488
interleukin 2
assay, **116**, 507
isotyping, **116**, 509
production, **116**, 506
low-density lipoprotein receptor, preparation, **129**, 205
mediated cytotoxicity, assay with Coulter counter, **132**, 481
to microtubule-associated 210K protein, preparation, **134**, 150
mixtures produced by hybridomas, analytical techniques, **121**, 337
monoclonal, *see* Monoclonal antibodies
monospecific, production on immunoblots, **134**, 468
for neuropeptide immunohistochemistry
application, **103**, 458
preparation, **103**, 448
neutralization of interferon
analysis of mixed-type interferon, **119**, 571
specificity of determination, **119**, 569
titer definition and computation, **119**, 562
titration methods, **119**, 559, 568
OKT6, in isolation of epidermal Langerhans cells, **108**, 360
to photosystem II reaction center subunits, preparation, **118**, 358
polyclonal, *see* Polyclonal antibodies
production, for use in enzyme immunoassays, **103**, 413
radioiodination for spaghetti overlay technique, **121**, 455
radiolabeling with tritium
applications, **121**, 674, 675
methodology, **121**, 670
to S-100 protein, preparation, **102**, 256
secretion by hybridomas in KSLM medium, **121**, 275
to striated flagellar root protein, immunofluorescence screening, **134**, 412
thymosin α_1, preparation, **116**, 244
thyroid-stimulating, assays
receptor modulation procedures, **109**, 677
technique comparison, **109**, 690

thyroid stimulation procedures, **109**, 686
–toxin conjugates
biological evaluation, **112**, 221
cytotoxicity, comparison with A chain conjugates, **112**, 222
physicochemical characterization, **112**, 220
purification and storage, **112**, 219
transferrin receptor
coupling to ricin A chain, **112**, 241
modification for toxin conjugation, **112**, 240
tubulin, production, **134**, 458
Antibody-forming cells
detection by semiautomated plaque method, **121**, 341
Antibody responses
anti-dinitrophenyl, suppression by IgG-binding factor, **116**, 412
anti-sheep erythrocyte
enhancement by allogeneic effect factor, **116**, 429
suppression by IgG-binding factor, **116**, 411
to free myoglobin peptides
adjuvant effects, **121**, 78
associated antibody classes, **121**, 78
associated antibody titers, **121**, 78
effect of peptide size, **121**, 77
genetic control, **121**, 83
optimum peptide dose, **121**, 76
peptide antigenicity, analysis, **121**, 79
Antibody-secreting cells
screening with culture-well solid-phase radioimmunoassay, **121**, 438
Antigenic determinants, *see* Epitopes
Antigens
allotypic, cellular expression, detection methods, **116**, 169
–antibody reactions, sedimentation patterns, microheterogeneity effects, **130**, 25
bacterial, slide immunoenzymatic assay, **121**, 516, 522
cell surface
analysis with antiisotypic antibodies, **116**, 155
epitope ratio analysis, **121**, 797
immunofluorescence assay, application to monoclonal antibodies, **121**, 580
cellular and soluble, antibodies to, assays, **121**, 783
coating of microtiter plates, **121**, 551
cytoskeletal, localization with gold-labeled antibodies, **134**, 592
fodrin, preparation, **134**, 51
GAT polypeptide, specific suppressor factors
assays, **116**, 328
characterization, **116**, 337
production, **116**, 327
purification, **116**, 329
H-2
biosynthetic labeling, **108**, 453
immunoprecipitation, **108**, 454
isolation, **108**, 506
separation, **108**, 456
structural analysis, **108**, 513, 516
haptenic and macromolecular, in preparation of anti-receptor antibodies, **121**, 96
HLA
detection in serum, **108**, 616
isolation and purification, **108**, 622
serologic and chemical properties, **108**, 614
HLA-A and HLA-B
detection assays, **108**, 595
gel electrophoresis, **108**, 598
purification, **108**, 589, 593
radiochemical purification, **108**, 599
solubilization, **108**, 585
HLA-DR
in generation of cytotoxic T cells, **108**, 611
purification, **108**, 602, 605, 608
reconstitution, **108**, 610
separation of α and β chains, **108**, 603
human MHC, measurement of interferon-induced changes, **119**, 688
human T-lymphocyte
characterization, **108**, 637
distribution and immunological function, **108**, 627
identification, **108**, 632
Ia
biosynthetic labeling, **108**, 453

component separation, **108**, 540
detection on epidermal Langerhans cells, **108**, 688
immunoprecipitation, **108**, 454
molecular cloning, **108**, 567
monoclonal antibodies, production, **108**, 559, 565
nonpolymorphic components, **108**, 525
polymorphic components, **108**, 524
purification, **108**, 527, 536, 560
separation, **108**, 456
kinetochore, identification, **134**, 271, 277
lipoprotein (a)-specific
discovery, **129**, 167
identification, **129**, 169
quantitation, **129**, 174
Ly, listing, **108**, 677
lymphocyte, isolation by binding immune complexes with protein A, **108**, 467
Lyt-1, Lyt-2, and Lyt-3
alloantisera and monoclonal antibodies, **108**, 667
gene mapping, **108**, 667
structural analysis, **108**, 670
macrophage, monoclonal antibody-defined, characterization, **108**, 314
membrane
chemiluminescent assay, **133**, 523
immunoaffinity isolation, **121**, 724
radiolabeling and solubilization, **121**, 723
murine MHC
characterization techniques, **108**, 486, 490
class I molecules, overview, **108**, 439
class II molecules, overview, **108**, 445
class III molecules, overview, **108**, 450
detection systems, **108**, 482, 490
homology, **108**, 449
neuroendocrine cell expression, immunofluorescence analysis, **103**, 238
nuclear, EBV-induced, in detection of lymphoblastoid cells, **121**, 125
Qa and TL
purification, **108**, 549
structural analysis, **108**, 555
radioiodination for solution-phase radioimmunoassay, **121**, 434
radiolabeled, in lymphocyte suicide assay, **108**, 255
stepwise amplified immunoenzymatic staining, **121**, 855
TGAL, specific helper factor
assays, **116**, 347
biochemical characterization, **116**, 349
isolation, **116**, 349
sources, **116**, 341
Thy-1
amino acid sequencing, **108**, 662
assay, **108**, 655
carbohydrate structure, determination, **108**, 664
molecular weight determination, **108**, 660
purification, **108**, 656, 659
Thy-1.1 and Thy-1.2
biochemical nature, **108**, 649
discovery and gene mapping, **108**, 642
functional studies, **108**, 652
monoclonal antibodies, production and applications, **108**, 645, 652
tissue distribution, **108**, 643
tumor cell surface
interferon-induced, radioimmunoassay, **119**, 686
monoclonal antibodies to, preparation, **119**, 683
Antimycin, *see* Citrinin
Antiserum, *see also* Antibodies
antiallotypic
preparation, **116**, 161
screening, **116**, 162
avidin, preparation, **122**, 90
to calcium-binding protein, preparation, **102**, 292
to calmodulin-dependent protein phosphatase, production, **102**, 251
enkephalin peptide
characterization, **124**, 620
preparation, **124**, 619
growth hormone-releasing factor, preparation, **124**, 380, 391
heterologous, isolation of IgG, **116**, 148
to immunoglobulin light chains
in analysis of light chain conformation, **116**, 120
immunodiffusion analysis, **116**, 108
preparation, **116**, 106

recognition of hidden determinants on C and V light chain fragments, **116**, 120
intact immunoglobulins, **116**, 117
specificity, **116**, 110
immunoglobulin purification from, **118**, 748
to interferons, as reference reagents, **119**, 15
LHRH precursor, preparation, **124**, 324
to Lyt-1, Lyt-2, and Lyt-3 antigens, description, **108**, 667
α_2-macroglobulin, in monoclonal antibody production, **121**, 59
polyclonal, to lipoproteins
applications, **128**, 542
production, **128**, 528
prolactin receptor
assay, **109**, 668
inhibition of prolactin binding, **109**, 669
mono- and bivalent fragments, effects, **109**, 673
production, **109**, 668
somatostatin, preparation, **124**, 337
substance P, preparation, **124**, 337
testing for antibody idiotypes, **116**, 184
transcobalamin II, preparation, **123**, 40
tuftsin, preparation, **132**, 323
Antiviral assay
cross-species, bovine leukocyte interferon, **119**, 140
heterologous interferons, with ovine and caprine cells, **119**, 551
interferon from human amniotic fluid, **119**, 542
interferon γ with murine L-929 and P388 cells, **119**, 575
interferons, computer-assisted
assay procedure, **119**, 536
cell lines in, **119**, 534
clinical blood samples, **119**, 540
cytopathic indicator virus, **119**, 534
data reduction, **119**, 536
reference standards, **119**, 535
solutions and media, **119**, 534
relative, interferon αA on various cell lines, **119**, 19
in screening of cells for sensitivity to interferon, **119**, 605
Antiviral factor
from virus-infected tobacco plants
antiviral activity, assay, **119**, 735
properties, **119**, 744
purification, **119**, 739
Apoaequorin
cDNA
cloning, **133**, 298
construction, **133**, 301
expression in *Escherichia coli*, **133**, 294
identification, **133**, 306
screening, **133**, 303
Apo-D-amino-acid oxidase
porcine kidney, in luminometric FAD assay
inactivation during assay, **122**, 190
preparation, **122**, 187
Apoflavoproteins
in fluorometric titration of urinary riboflavin, **122**, 234
Apolipoprotein A
-containing lipoproteins
characterization, **129**, 138, 143
isolation, **129**, 132
types of, **129**, 132
Apolipoprotein A-I
–C-III gene complex, restriction fragment length polymorphisms
analysis, **128**, 740
detection, **128**, 730
as genetic markers, **128**, 735
cDNA clones, isolation and characterization, **128**, 713
characterization, **128**, 232
degradation sites in tissues, assessment methods, **129**, 626
encoding genes
DNA insertion, familial segregation, **128**, 725
isolation, **128**, 717
linkage to C-III genes, **128**, 720
role in premature atherosclerosis, **128**, 721
isolation, **128**, 224; **129**, 422
metabolism
analytical techniques, **129**, 428
in dyslipidemic subjects, **129**, 438
kinetics, comparison with apolipoprotein A-II, **129**, 433

in normal subjects, **129**, 435
radioiodination, **129**, 425
Apolipoprotein A-II
cDNA clones
preparation and sequence analysis, **128**, 749
screening, **128**, 747
characterization, **128**, 240
isolation, **128**, 235; **129**, 422
metabolism
analytical techniques, **129**, 428
in dyslipidemic subjects, **129**, 438
kinetics, comparison with apolipoprotein A-I, **129**, 433
in normal subjects, **129**, 435
radioiodination, **129**, 425
Apolipoprotein A-IV
characterization, **128**, 243
functional domains, localization, **128**, 766
isolation, **128**, 240; **129**, 422
metabolism, analytical techniques, **129**, 428
radioiodination, **129**, 425
secondary structure, prediction methods, **128**, 765
sequence analysis, computer searching methods, **128**, 754
sequence comparisons, analytical techniques, **128**, 758
Apolipoprotein B
antibody preparation, **129**, 537
characterization, **129**, 340
chemically modified, properties, **129**, 406
-containing lipoproteins
characterization, **129**, 126
isolation, **129**, 412
metabolism, analytical techniques, **129**, 410
modification mechanisms, **129**, 124
–lipid complexes
biological characteristics, **128**, 606
physical characteristics, **128**, 600
preparation, **128**, 590
metabolic conversions, analysis with specific radioactivity data, **129**, 385
^{32}P-labeled
dephosphorylation, **129**, 541
isolation, **129**, 538, 540
phosphoamino acid analysis, **129**, 540
phosphorylation, **129**, 539
protease processing, **129**, 339
radiolabeling, **129**, 401
solubilization, **128**, 585
Apolipoprotein B-48
amino acid analysis, **128**, 267
molecular weight determination, **128**, 269
purification, **128**, 263
Apolipoprotein B-100
characterization, **128**, 257
isolation, **128**, 250
molecular weight determination, **128**, 260
purity assessment, **128**, 251
quantitation, **128**, 256
Apolipoprotein C
exchange and transfer reactions, **129**, 445
metabolism, analytical techniques, **129**, 449
plasma levels, determination, **129**, 448
synthesis, overview, **129**, 446
Apolipoprotein C-I
properties, **128**, 294
purification, **128**, 288
Apolipoprotein C-II
binding sites on lipoprotein lipase, identification, **129**, 742
cDNA clones, isolation, **128**, 789
effect on lipoprotein lipase binding to lipid interfaces, **129**, 743
encoding gene, chromosomal localization, **128**, 796
genomic clones, isolation, **128**, 793
properties, **128**, 296
purification, **128**, 288
Apolipoprotein C-III
–A-I gene complex, restriction fragment length polymorphisms
analysis, **128**, 740
detection, **128**, 730
as genetic markers, **128**, 735
cDNA clones, isolation and characterization, **128**, 713
gene linkage to A-I genes, **128**, 720
metabolism, analysis with specific radioactivity data, **129**, 392
properties, **128**, 296
purification, **128**, 288
role in premature atherosclerosis, **128**, 721

specific activity, determination by immunoaffinity chromatography, **129**, 457
Apolipoprotein D
characterization, **128**, 300
isolation, **128**, 297
Apolipoprotein E
amino acid analysis, **128**, 283
characterization, **129**, 340
-containing lipoproteins
characterization, **129**, 162
isolation, **129**, 146, 186
properties, **129**, 156
separation, **129**, 159
functional tests, **128**, 287
genetic polymorphisms
analytical techniques, **128**, 825
genetic basis, **128**, 844
phenotype determination, **128**, 833
internally repeated sequences
detection, **128**, 777
duplication history, inference, **128**, 781
isoform purification, **128**, 276
isolation, **128**, 273
metabolism, analytical techniques, **129**, 482
mRNA, cDNA cloning and characterization, **128**, 801
phenotyping, **128**, 279
protease processing, **129**, 339
radiolabeled
characterization, **129**, 485
physicochemical and lipid-binding properties, **129**, 486
preparation, **129**, 484
storage and handling, **128**, 279
transfected
in bacteriophage λ, screening, **128**, 813
in cosmid vector
characterization, **128**, 818
expression analysis, **128**, 820
product characterization, **128**, 821
screening, **128**, 816
Apolipoprotein F
characterization, **128**, 305
isolation, **128**, 304
Apolipoprotein G
isolation and characterization, **128**, 309
Apolipoprotein H
characterization, **128**, 308
isolation, **128**, 306
Apolipoproteins
adsorbed monolayers at air–water interface
surface concentration measurement, **128**, 395
surface potential measurement, **128**, 397
surface pressure measurement, **128**, 392
adsorption spectroscopy, **128**, 516
analytical techniques, impact of technology, **128**, 144
assembly and secretion, **128**, 19, 66
characteristics, **128**, 9
cholesterol-induced, **128**, 142
components, **128**, 5
definition, **128**, 375
effect on hepatic lipase activity, **129**, 730
encoding genes
chromosomal location, **128**, 49
expression regulation, **128**, 53
organization and expression, **128**, 42
enthalpy measurements
from crystalline structure, **128**, 370
for ionization, **128**, 372
during lipid binding, **128**, 368
for phospholipid associations, **128**, 367
epitope assignment, **128**, 535, 542
extracellular modifications, analytical techniques, **128**, 705
fluorescence properties, **128**, 517
function, linkage to epitopes, **128**, 551
gene mapping
by hybridization to metaphase chromosomes, **128**, 863
with somatic cell hybrids, **128**, 851
glucosylated, immunochemical isolation and identification, **128**, 354
high-density lipoprotein-associated, preparation, **128**, 554
immunocytochemical detection, **129**, 298
interactions with liposomes
carboxyfluorescein leakage assay, **128**, 657
mechanisms, **128**, 668
intracellular modifications, analytical techniques, **128**, 691

isolation, **128**, 10
lipid-associating regions
helical amphipathic moment, **128**, 406
thermodynamics, measurement, **128**, 407
–lipid complexes
characterization, **128**, 569
chemical modification, **129**, 548
preparation, **128**, 558; **129**, 546
metabolic labeling, **129**, 255
metabolic roles, **128**, 13
metabolic theory, **129**, 375
molecular conformations, **128**, 378
molecular species in solution, **128**, 376
monoclonal antibodies
applications, **128**, 535
assays, **128**, 530
production, **128**, 529
purification, **128**, 534
mRNA
cDNA probe synthesis, **128**, 672
hybridization assays, **128**, 680; **129**, 315
nascent
intracellular localization, **129**, 294
isolation, **129**, 290
radiolabeling, **129**, 285
nomenclature and classification, **128**, 75
peripheral lymph, synthesis, analytical techniques, **129**, 674
–phospholipid complexes
application in cholesterol efflux studies, **129**, 637
preparation, **129**, 634
physicochemical characteristics, **128**, 126
polyclonal antisera
applications, **128**, 542
production, **128**, 528
processing, role in receptor recognition, **129**, 319
properties, **128**, 341
role in hepatic lipoprotein catabolism, **129**, 607
separation, **128**, 244, 340
signal peptide sequences, **128**, 691, 696
solution properties
analytical techniques, **128**, 381
relevance to lipoprotein structure and metabolism, **128**, 386
spread monolayers at air–water interface
spreading methods, **128**, 392
surface pressure measurement, **128**, 390
synthesis
analysis in cultured hepatocytes, **129**, 281
immunochemical measurement, **129**, 254
and intracellular targeting, **128**, 56
synthetic peptide analogs
characterization, **128**, 637, 642, 644
design, **128**, 629
interactions with lipids, **128**, 636
synthesis, **128**, 632
Aprotinin
hydrogel delivery system, structure, function, and release rates, **112**, 519
Arabinose
transport system in *Escherichia coli*
assay
in intact cells, **125**, 388
in right-side out subcellular vesicles, **125**, 399
associated membrane proteins, identification
by gel electrophoresis, **125**, 411, 419
after radiolabeling, **125**, 404, 417
genetics, **125**, 403
properties, **125**, 389
reconstitution in liposomes, **125**, 425
Arachidonic acid
^{3}H-labeled, in analysis of phospholipid turnover in cultured cells, **124**, 429
Area detectors
multiwire, diffractometers with
crystal examination and alignment programs, **114**, 456
crystal orientation, refinement program, **114**, 457
data acquisition programs, **114**, 458
data collection procedure and results, **114**, 467
data output and reduction, **114**, 465
Mark I system, description and operation, **114**, 418
Mark II system, description and operation, **114**, 440
operation software, program organization, **114**, 454

parallax in, **114**, 435
position-sensitive
data processing, automated peak fitting procedure, **114**, 530
in neutron diffraction data collection, **114**, 518, 551
television, diffractometers with
camera system, **114**, 488
data collection strategy, **114**, 484
description, **114**, 474
detector availability, **114**, 509
detector performance, **114**, 499
memory system, **114**, 496
operation, **114**, 483
phosphor efficiencies and characteristics, **114**, 498
scan control and image processing electronics, **114**, 493
system calibration, **114**, 497
system configuration, **114**, 504
Argentation chromatography
sterols, **111**, 8
Arginine
in F_1-ATPase, chemical modification, **126**, 728
reactivity with heavy atoms, **114**, 150
Arginine deiminase
assays, **107**, 628
properties, **107**, 631
Arginine vasopressin
in neuroendocrine cells, radioimmunoassay, **103**, 144
Argininosuccinate lyase
localization in gels by simultaneous capture technique, **104**, 436
Arginyltransferase
assay, **106**, 198
deficient mutants, characterization, **106**, 204
purification from rabbit liver, **106**, 201
substrate identification, **106**, 202
L-Arginyl-tRNA:protein transferase, *see* Arginyltransferase
Aromatic-amino-acid aminotransferase
assay, **113**, 74
natural occurrence, **113**, 73
Artemia
cysts, as model for cell water analysis
decapsulation, **127**, 233
hydration, **127**, 234
physical properties of cell water, **127**, 236
preparation, **127**, 231
sources, **127**, 230
Artificial cells
containing multienzyme systems
applications, **112**, 196
cofactor recycling in, **112**, 200
preparation, **112**, 196
Aryl hydrocarbon hydroxylase, *see* Unspecific monooxygenase
Asbestos
in depletion of macrophages, **108**, 330, 333
Ascaris lumbricoides
sperm, isolation, **134**, 419
Ascites fluid
murine
antibody production by hybridomas grown in, **121**, 163
production by
intraperitoneal inoculation, pristane effects, **116**, 125; **121**, 375
intrasplenic inoculation, **121**, 381
purification of
IgG, **121**, 588, 600, 631, 803
IgM, **116**, 33
monoclonal Igs, **116**, 128
rat, purification of monoclonal antibodies, **121**, 646
Ascorbate
in preparation of oxidatively modified proteins, **107**, 372
Ascorbic acid
in detection of soluble immune response suppressor, **116**, 397
HPLC, **122**, 3
interference in dichlorofluorescein assay for lipid hydroperoxides, **105**, 355
as supplement in preparation of SFH medium, **121**, 284
Asialofetuin
–amatoxin conjugates
biological properties, **112**, 230
preparation by carbodiimide coupling, **112**, 227
–trifluorothymidine conjugates, preparation and applications, **112**, 302

Asialoglycoprotein receptors
mammalian hepatocyte
internalization with ligand, **109**, 236
intracellular, role in recycling, **109**, 237
number on cell surface, measurement, **109**, 234
postinternalization events, **109**, 242
recycling in absence of ligand, **109**, 241
survival half-life, calculation, **109**, 233
Asialoglycoproteins
in mammalian hepatocytes
postinternalization events in endocytosis, **109**, 242
role in
receptor internalization, **109**, 236
receptor recycling, **109**, 241
survival half-life, calculation, **109**, 233
Asolectin
vesicles, reconstitution of mitochondrial F_0F_1-ATPase, **126**, 474
Asparaginase
bacterial
assay, **113**, 609
modifications, **113**, 616
properties, **113**, 615
purification, **113**, 610
mammalian, occurrence and properties, **113**, 618
Asparagine
reactivity with heavy atoms, **114**, 152
Asparagine–oxo-acid aminotransferase
rat liver
assay, **113**, 603
purification, **113**, 605
substrate specificity, **113**, 606
Asparagine transaminase, *see* Asparagine–oxo-acid aminotransferase
Aspartate aminotransferase
assays, **113**, 67
3-chloro-L-alanine-labeled active site
effect on catalytic mechanism, **106**, 133
isolation, **106**, 131
structural analysis, **106**, 133
Escherichia coli
assay, **113**, 83
properties, **113**, 87
purification, **113**, 84, 86
natural occurrence, **113**, 66
Aspartate ammonia-lyase
Escherichia coli
assay, **113**, 619
properties, **113**, 622
purification, **113**, 620
Pseudomonas fluorescens
assay, **113**, 624
properties, **113**, 626
purification, **113**, 624
Aspartate carbamoyltransferase
Bacillus subtilis
assay, **113**, 628
properties, **113**, 632
purification, **113**, 629
localization in gels, simultaneous capture and postincubation capture techniques, **104**, 437
time-resolved X-ray scattering, **134**, 674
L-Aspartate α-decarboxylase
Escherichia coli
assay, **113**, 589
properties, **113**, 594
purification, **113**, 591
Aspartate kinase
Escherichia coli
assay, **113**, 596
properties, **113**, 599
purification, **113**, 597
Aspartate-semialdehyde dehydrogenase
Escherichia coli
assay, **113**, 600
properties, **113**, 601
purification, **113**, 600
Aspartate transcarbamylase, *see* Aspartate carbamoyltransferase
Aspartic acid
radiolabeled
in analysis of bacteriorhodopsin carboxyl groups by FTIR, **127**, 651
synthesis, **127**, 655
reactivity with heavy atoms, **114**, 151
D-Aspartic acid β-methyl esters
from erythrocyte carboxyl methylated proteins
characterization, **106**, 336
configuration, determination, **106**, 338
isolation, **106**, 332, 341
occurrence, **106**, 343
N^{ε}-(β-Aspartyl)lysine
identification in colostrum, **107**, 259

Aspergillus terreus
aspulvinone dimethylallyltransferase
purification, **110**, 322
Aspirin
prodrug structure, effect on metabolism, **112**, 343
Aspulvinone
radiolabeled, preparation, **110**, 321
Aspulvinone dimethylallyltransferase
Aspergillus terreus
assay, **110**, 320
properties, **110**, 323
purification, **110**, 322
Asteriarubin
properties, **111**, 241
purification, **111**, 241
Asters
sea urchin egg
induction by taxol, **134**, 199
isolation, **134**, 199
Atherosclerosis
premature, role of apolipoprotein A-I–C-III gene complex, **128**, 721, 740
Atomic absorption spectroscopy
hepatic Ca^{2+}, in assay of hormone effects, **109**, 553, 554
Atoms
free, in extending and refining phases, **115**, 112
ATP (adenosine 5′-triphosphate)
–ADP carrier protein, mitochondrial
activity, influencing factors, **125**, 617
atractyloside-binding site, photolabeling, **125**, 665
chemical modification, **125**, 659
detection with fluorescent nucleotide analogs, **125**, 620
ESR studies, **125**, 638
extrinsic fluorescent probes, **125**, 640
intrinsic fluorescent probes, **125**, 648
isolation, **125**, 610
^{31}P NMR studies in detergent micelles, **125**, 634
reconstitution in phospholipid vesicles, **125**, 613, 633
spin labeling, **125**, 635
transmembrane folding, analysis with pyridoxal phosphate, **125**, 650
transport assay, **125**, 615
binding sites in receptor-associated kinases, analysis, **109**, 816
cellular
assay, **133**, 36
extraction, **133**, 14, 32
electric pulse-driven formation, **126**, 640
enzyme-bound, synthesis by CF_1-ATPase
characteristics, **126**, 647
methodology, **126**, 644
extractants, properties, **133**, 15, 21
hydrolysis by mitochondrial ATPase, measurement of equilibrium and rate constants, **126**, 608
inhibitory effect on 3-hydroxy-3-methylglutaryl-CoA reductase, **110**, 48
luminometric assay for cell biomass, **134**, 27
comparison with other methods, **133**, 39
^{18}O-labeled, synthesis, **126**, 633
–$^{32}P_i$ exchange, assay, **126**, 580
reversed affinity chromatography with silica-bound alcohol dehydrogenase, **104**, 220
synthesis defects in *Escherichia coli ecf* mutants, **125**, 184
transport in chloroplast envelope, **125**, 715
ATPase
Ca^{2+}-
–phospholamban kinase–phospholamban complex, purification, **102**, 269
sarcoplasmic reticulum, assay with *N,N′*-dicyclohexylcarbodiimide, **125**, 96
dynein-associated, *see* Dynein ATPase
epidermal Langerhans cell, detection, **108**, 686
F_0F_1, *see* ATP synthase (H^+-transporting)
H^+-transporting, *see* ATP synthase (H^+-transporting)
mechanochemical, kinetic analysis, **134**, 677
Na^+,K^+-, as endosome marker, determination, **109**, 267
proteolipids, proton conduction, **127**, 424
ATPase inhibitor
bovine cardiac mitochondria
assays, **126**, 461

properties, **126**, 468
purification, **126**, 464
rat hepatic mitochondria
assays, **126**, 661
azido-labeled form
applications, **126**, 664
synthesis, **126**, 663
yeast mitochondria
properties, **126**, 508
purification, **126**, 506
ATP synthase (H^+-transporting)
bacterial, F_1 portion
inactivating modifications, **126**, 713
noninactivating modifications, **126**, 729
chloroplast
assays, **126**, 518
F_1 portion
inactivating modifications, **126**, 713
noninactivating modifications, **126**, 729
synthesis of enzyme-bound ATP, **126**, 643
individual subunits
antibody preparation, **118**, 358
isolation, **118**, 358
quantitative estimation, **118**, 360
structure and function, **118**, 355
purification, **126**, 513
reconstitution, **126**, 516
synthesis during development, analytical techniques, **118**, 367
Escherichia coli
assays, **126**, 548
F_0 portion
assays, **126**, 569, 580
dissociation into subunits, **126**, 574
properties, **126**, 574, 587
purification, **126**, 572, 583, 588
reconstitution, **126**, 577, 578, 579
F_1 portion
assays, **126**, 569
defective, identification, **126**, 593
dissociation, **126**, 574
immunoelectron microscopy, **126**, 774
monoclonal antibodies, preparation, **126**, 771
properties, **126**, 574, 594
purification, **126**, 572, 590
reconstitution, **126**, 577, 592
F_1F_0 portion
assays, **126**, 580
properties, **126**, 587
purification, **126**, 583, 588
reconstitution, **126**, 579
mutants
altered nucleotide sequence, determination, **126**, 602
analysis with λ-*unc* transducing phages, **126**, 558
anomalous migration of ssDNA fragments, detection, **126**, 602
DNA cloning, **126**, 599
mapping by genetic recombination, **126**, 597
purification, **126**, 550, 553
reconstitution, **126**, 555
Micrococcus luteus, F_1 portion, photoaffinity labeling, **126**, 655
mitochondrial
assays, **125**, 94; **126**, 419, 472
F_1 portion
analysis with monoclonal antibodies, **126**, 766
assays, **126**, 478, 484
crystals, electron microscopy, **126**, 441
essential residues, identification, **126**, 741
functional and structural analysis with ADP analogs, **126**, 699
inactivating modifications, **126**, 713
monoclonal antibodies, preparation, **126**, 763
noninactivating modifications, **126**, 729
nucleotide binding, nonlinear analysis, **126**, 733
photolabeling, **126**, 678
properties, **126**, 482
purification, **126**, 479
single molecules, electron microscopy, **126**, 435
α and β subunits and subunit pairs, purification, **126**, 485
inactivated complex
characteristics, **126**, 511
purification of inhibitor and stabilizing factors, **126**, 505

mechanism of ATP hydrolysis, **126**, 608
oligomycin sensitivity-conferring protein
interactions with F_0 and F_1, **126**, 458
purification, **126**, 455
properties, **126**, 423, 476
purification, **126**, 417, 473
reconstitution, **126**, 432, 474
subunit resolution, **126**, 431
Rhodopseudomonas blastica, encoding gene
cloning and DNA sequencing, **125**, 231
transcriptional mapping, **125**, 245
Rhodospirillum rubrum
encoding gene
cloning and DNA sequencing, **125**, 231
transcriptional mapping, **125**, 245
F_1 subunits
properties, **126**, 535
purification, **126**, 531, 533
reconstitutive activity, **126**, 529
thermophilic bacteria PS3, F_0 portion
H^+ translocation in vesicles
assays, **126**, 605
influencing factors, **126**, 607
purification, **126**, 604
reconstitution, **126**, 605
Atractyloside
binding site on mitochondrial ADP/ATP carrier protein, photolabeling, **125**, 665
fluorescent derivatives, as probes of mitochondrial ADP/ATP carrier protein, **125**, 644
radiolabeled azido derivatives, synthesis, **125**, 665
Autoimmunity
modeling, applications of nude mouse, **108**, 356
Autophosphorylation
polyamine-dependent protein kinase, assay, **107**, 158
Autoradiography
in assay for calcium channels, **109**, 520, 546
chloroplast 32-kDa membrane protein, **118**, 391
CRF receptors in pituitary and central nervous system, **124**, 566
in gel screening, spaghetti overlay technique, **121**, 453
hepatocyte subcellular peptide hormone receptors, **109**, 229
^{3}H-labeled metaphase chromosomes hybridized to apolipoprotein-specific probes, **128**, 872
with immunocytochemistry
steroid hormone-receiving cells in CNS, **103**, 639
steroid hormones and neuropeptides in brain, **103**, 631
insulin receptor subunits, **109**, 608
lipoprotein subpopulations, **128**, 419
low-density lipoprotein mutants, **129**, 245
macrophages, applications of monoclonal antibodies, **108**, 321
nascent lipoproteins, **129**, 295
neuropeptide mRNA–DNA hybrids, **124**, 543
neuropeptide receptors in brain, **124**, 590
photolabeled insulin receptors, **109**, 178
somatostatin receptors in brain and pituitary, **124**, 608
viral RNAs, **118**, 721
Autoxidation
6-hydroxy-benzo[*a*]pyrene, generated products, **105**, 543
Avena sativa, *see* Oats
AVF, *see* Antiviral factor
Avidin
affinity purification on 2-iminobiotin–6-aminohexyl-Sepharose 4B, **122**, 85
antiserum preparation, **122**, 90
–biotin conjugates
in immunoassay for human chorionic gonadotropin, **133**, 286
immunohistochemical techniques with monoclonal antibodies, **121**, 576, 839
in monoclonal antibody attachment to microspheres, **112**, 74
preparation, **133**, 285
bound ligands, displacement by biotin, **109**, 443
fluorescein-labeled, preparation, **122**, 68
fluorometric assay, **122**, 67

–iminobiotin conjugates, in purification of plasma membrane proteins, **122**, 87
in modification of allotype-specific ELISA, **116**, 168
modification and radiolabeling, **109**, 441
solid-phase assay on cellulose disks, **122**, 81
Axial ratios
protein, estimation from covolume measurements, **117**, 182
Axonemes
Chlamydomonas flagella, isolation, **134**, 289
Axon-sparing lesions
in hypothalamus, induction by excitotoxins, **103**, 381, 393
Azide
binding to hemoproteins, resonance Raman spectroscopy, **130**, 396
effect on chemiluminescence of probes for phagocyte oxygenation activity, **133**, 469
Azido-ATP
divalent analog
interactions with F_1-ATPase, **126**, 655
properties, **126**, 653
synthesis, **126**, 652
3-Azido-1-butanol
preparation, **110**, 127
o-Azidophenylethanol
^{3}H-labeled, preparation, **110**, 126
preparation, **110**, 125
p-Azidophenylethanol
preparation, **110**, 126

B

Bacillus
alkalophilic
bioenergetic parameters, **125**, 591
chemotaxis, **125**, 592
cultivation, **125**, 583
flagella and flagellar basal bodies, isolation, **125**, 591
Na^+-driven flagellar motors, motility measurements, **125**, 584, 586
tethered cells, preparation, **125**, 591
B. acidocaldarius, intracellular pH regulation, **125**, 354
B. firmus, intracellular pH regulation, **125**, 361
B. licheniformis, purification of
L-glutamic acid-activating enzyme, **113**, 150, 168
poly-γ-D-glutamate synthetase, **113**, 150
B. sphaericus, purification of D-glutamate-D-amino acid transaminase, **113**, 110
B. subtilis
cultivation, **110**, 199; **126**, 400
purification of
p-aminobenzoate synthase subunit G, **113**, 294
aspartate transcarbamylase, **113**, 629
cytochrome-*c* oxidase, **126**, 160
glutamate-$tRNA^{Gln}$ amidotransferase, **113**, 304
glutamyl-tRNA synthetase, **113**, 51
heavy riboflavin synthase, **122**, 194
heptaprenylpyrophosphate synthetase, **110**, 200
succinate dehydrogenase, **126**, 403
succinate dehydrogenase-deficient mutants
complementation analysis, **126**, 411
isolation, **126**, 409
Bacilysin
in selection of bacterial mutants deficient in dipeptide permease, **125**, 370
Bacteria, *see also specific bacteria*
anaerobic
generation of protonmotive force
by end-product translocation, **125**, 497
energy recycling model, **125**, 495
solute transport mechanisms, **125**, 493
bioluminescent, selection and growth, **133**, 156
dysopsonic activity, overview, **132**, 310
flagellar motor, proton-driven
characterization, **125**, 572
energetic analysis, **125**, 579
reconstitution, **125**, 571
gram-negative
broad host range vectors, application of *Escherichia coli* plasmids, **118**, 657

site-specific insertion of selective markers, **118**, 649, 656
transposon mutagenesis, **118**, 644
vector plasmid mobilization in, **118**, 641, 654
killing by
neutrophils, in assessment of phagocytic function, **132**, 124, 168
phagocytes, assays, **132**, 521, 523, 525
marine, bioluminescent system, gene transfer to *Escherichia coli*, **133**, 70
radiolabeling for assessment of phagocytic function, **132**, 141
surface tension determination
overview of techniques, **132**, 22
procedures, **132**, 70
thermophilic PS3, purification of cytochrome oxidase, **126**, 146
Bacterial luciferase, *see* Alkanal monooxygenase (FMN-linked)
Bacterio-opsin
encoding gene
amber mutations, introduction by generalized mutagenesis, **125**, 193
deletion mutagenesis, **125**, 198
oligonucleotide-directed point mutagenesis, **125**, 202
site-specific mutagenesis, **125**, 206
Bacteriophages
Escherichia coli, active sites, determination by selective oxidative free radical probes, **105**, 497
lambda
human genomic library, screening for apolipoprotein E, **128**, 813
P_L promoter, in interferon synthesis, **119**, 366, 376
ribulose-bisphosphate carboxylase small subunit genes, cloning, **118**, 397
lambda-*uncX*
construction, **126**, 560
genetic complementation of *Escherichia coli* ATPase mutants, **126**, 562
and luminescent bacteria, in determination of antibiotic activities, **133**, 283
M13, in site-specific mutagenesis of *Escherichia coli lac* permease, **125**, 217
serotyping by gel electrophoresis, **130**, 129
T4, *alt* and *mod* genes
encoded ADP-ribosyltransferases, purification and properties, **106**, 419
mapping, **106**, 418, 419
Bacteriorhodopsin
carboxyl groups, protonation, FTIR studies, **127**, 649
ion flux, measurement with optical probes, **127**, 746
proton translocation
blue light effect, **127**, 433
detection by FTIR, **127**, 344
measurement by displacement currents, **127**, 610
resonance Raman spectroscopy, **127**, 587
response to oscillating electrical fields, **127**, 436
role of carboxyl groups, **127**, 347
at Schiff base, **127**, 349
purple membrane surface charge density, spectroscopic analysis, **127**, 693
structural analysis with mutant bacterio-opsins, **125**, 190
Bacteriurea
bioluminescence screening, **133**, 24
detection methods, **133**, 23
significant, definition, **133**, 23
sources of infection, **133**, 22
Barium citrate
in isolation of vitamin K-dependent plasma proteins, **107**, 507
Barium sulfate
in isolation of vitamin K-dependent plasma proteins, **107**, 507
Barley
chlorina *f*2 mutant, isolation of chlorophyll-binding proteins, **118**, 348
infection with *in vitro* viral transcripts, **118**, 715
propagation of brome mosaic virus, **118**, 688
purification of brome mosaic virus, **118**, 688

Basal bodies
Chlamydomonas
fractionation, **134**, 259
negative staining, **134**, 260
Base lability
O-phosphates in proteins, **107**, 5
Batrachotoxin
interactions with ion channels in neuroendocrine tissue, **103**, 402
B-cell differentiation factors
assay, **116**, 378
purification, **116**, 378
B-cell growth factors
overview, **116**, 364
type I, *see* B-cell stimulatory factor 1
type II
assay, **116**, 375
purification, **116**, 376
B-cell maturation factors
assay, **116**, 378
purification, **116**, 378
B cells
antigen-specific, selection for viral immortalization, **121**, 146
fusion with human × murine hybridoma, **121**, 171
human, *in vitro* immunization with, **121**, 26
hybridomas, growth promoters, **121**, 9
immortalization with Epstein–Barr virus
application to human hybridoma production, **121**, 154
overview, **121**, 141
procedure, **121**, 145
subsets, separation by indirect rosette technique, **121**, 734
transformation with Epstein–Barr virus, **121**, 170
B-cell stimulatory factor 1
assay, **116**, 373
purification, **116**, 375
Bean golden mosaic virus
propagation in plants, **118**, 703
Benzene
hydration, Monte Carlo simulation with proximity criterion analysis, **127**, 37
Benzidine
bisdiazotized, neuropeptide conjugation, **103**, 451
Benzo[*a*]pyrene-1,6-dione
preparation, **105**, 541
redox cycles, **105**, 545
reduction, **105**, 542
Benzo[*a*]pyrene-3,6-dione
preparation, **105**, 541
redox cycles, **105**, 545
reduction, **105**, 542
Benzo[*a*]pyrene-6,12-dione
preparation, **105**, 541
redox cycles, **105**, 545
reduction, **105**, 542
Benzo[*a*]pyrene hydroxylase
assay for analysis of interferon effects on hepatic drug metabolism, **119**, 723
3′-*O*-(4-Benzoyl)benzoyl-ATP
characterization, **126**, 676
photolabeling of F_1-ATPase, **126**, 678
purification, **126**, 674
synthesis, **126**, 673
Berberine
delivery to brain by dihydroberberine administration, **112**, 381
Bile acids
bioluminescent assay with immobilized enzymes, **133**, 215, 240, 246
biosynthesis-related cytochrome *P*-450
assays, **111**, 365
properties, **111**, 375
purification, **111**, 368, 370, 372, 374
conjugated
mass spectra, **111**, 96
reverse-phase liquid chromatography, **111**, 55
free
mass spectra, **111**, 78
reverse-phase liquid chromatography, **111**, 59
HPLC mobility studies, **111**, 52
labeling with stable isotopes in kinetic studies, **111**, 111
mass spectrometry
ionization methods, **111**, 64
quantitative analysis, **111**, 98, 104
sample introduction methods, **111**, 71
normal-phase chromatography, **111**, 63
Bile salts
micelles, protein transfer by gel filtration, **104**, 325

S^{β}-(Bilin)cysteine derivatives
quantitation, **106**, 364
spectroscopic properties, **106**, 361
structure, **106**, 359
Bilirubin
isomers, HPLC separation, **123**, 389
Bioautography
cobalamins, **123**, 8
Bioluminescence
bacterial luciferase 4a-hydroperoxyflavin intermediates in response to aldehydes, **133**, 138
bacterial luciferase in response to aldehydes, **133**, 191
-based reagents
applications, **133**, 68
commercial suppliers, **133**, 53
determination, apparatus for, **133**, 266
dinoflagellate, cell-free components, isolation and purification, **133**, 307
encoding *lux* genes from marine bacteria, cloning, **133**, 83
lumazine proteins
excitation mechanism, **133**, 170
properties, **133**, 168
photographic detection
applications, **133**, 413, 418
assays, **133**, 412
film types, **133**, 404
instrumentation, **133**, 405
related acyltransferases from bacteria
assay, **133**, 183
properties, **133**, 187
purification, **133**, 185
Bioluminescence test
genotoxic agents
applications, **133**, 274
assay modifications, **133**, 269
data analysis, **133**, 268, 271
experimental apparatus, **133**, 266
experimental methods, **133**, 267
polymyxin B in milk, **133**, 277
Bioluminescent assay
amplified, with avidin–biotin system, **133**, 284
bile acids, with immobilized enzymes, **133**, 215
with coimmobilized enzymes, **133**, 198
continuous flow, with immobilized enzymes, **133**, 229, 240
estrogens, **133**, 209
oxidized and reduced NAD, **122**, 147
in screening for bacteriurea, **133**, 24
Biotin
and analogs, HPLC separation, **122**, 63
attachment to peptides and proteins, **109**, 418
–avidin conjugates
in immunoassay for human chorionic gonadotropin, **133**, 286
immunohistochemical techniques with monoclonal antibodies, **121**, 576, 839
in monoclonal antibody attachment to microspheres, **112**, 74
preparation, **133**, 285
binding by streptavidin–agarose, assay, **121**, 722
displacement of ligands from avidin, succinoylavidin, and streptavidin, **109**, 443
in epitope analysis by solid-phase enzyme immunoassay, **121**, 463, 466
fluorometric assay, **122**, 67
in modification of allotype-specific ELISA, **116**, 168
–neuroendocrine peptide conjugates
applications, **124**, 51
preparation, **124**, 48
solid-phase assay on cellulose disks, **122**, 78
Biotinylation
ACTH, **109**, 432
insulin, **109**, 429
monoclonal antibodies, **121**, 719
nucleotides, **109**, 439
N^{ε}-(Biotinyl)lysine
formation in biotin enzymes, **107**, 261
Birefringence
transient electric, *see* Transient electric birefringence
window, in high-pressure bomb, correction, **130**, 507
N,*N′*-Bisgluconamidopropylcholanamides
physical properties, **124**, 162
synthesis, **124**, 159
Bisoxonol
as indicator of membrane potential
applications, **124**, 128
general methodology, **124**, 124

3-(N-{β-[3,4-Bis(pivalyloxy)phenyl]ethyl}-
carbamoyl)pyridine
synthesis, **112**, 391
Bis(sulfosuccinimidyl)suberate
in structural analysis of lipoproteins,
128, 613
Bleomycin
in assay for iron salts *in vivo*, **105**, 56
mutagenicity, testing with *Salmonella
typhimurium* TA102, **105**, 250
rate-controlled delivery by osmotic
system, **112**, 474
Blood, *see also* Plasma; Serum
catalase measurement, **105**, 123
chemiluminescence, assay, **132**, 501
human, isolation of
granulocytes, **108**, 91
leukocytes, **108**, 91
lipoproteins, precautionary measures,
128, 151
lymphocytes, **108**, 51, 95, 242
monocytes, **108**, 96, 242
mononuclear cells, **108**, 91
neutrophils, **132**, 380, 381
pteridines, **122**, 273
somatomedin C, **109**, 814
hypophysial portal, hypothalamic dopamine secretion into, **103**, 607
interferon antiviral assays with clinical
samples, **119**, 539, 540
osteocalcin levels, variation, **107**, 533
porcine, isolation of neutrophils, **132**,
381
somatomedins in, physical state, **109**,
800
unseparated, T-cell subset enumeration
by epifluorescence microscopy, **121**,
754
vitamin E levels, analytical techniques,
105, 138
whole, phagocyte activity assay, **133**,
488
Blood–brain barrier
role in excitotoxin lesion distribution,
103, 383
Blood flow
effect on magnetically responsive albumin microspheres, **112**, 62
Blotting
(ADP-ribosyl)$_n$ proteins on gels, **106**, 468

electro-, *see* Electroblotting
flagellar dynein proteins on nitrocellulose, **134**, 303
immuno-, *see* Immunoblotting
mRNA on nitrocellulose, in analysis of
hormonal treatment effects, **109**, 579
Western, *see* Western blotting
Blowfly
isolation of
ecdysone oxidase, **111**, 422
ecdysteroid carrier proteins, **111**, 435
Bluetongue virus
induction of bovine interferon β, **119**,
212
Bohr effects
redox, measurement in cytochrome
system of mitochondria, **126**, 331
Bolton–Hunter reagent
^{125}I-labeled, in radiolabeling of
interferon α, **119**, 277, 278
interferon αA, **119**, 306
interferon γ, **119**, 281
peptides, **124**, 26
Bone
chicken embryo, parathyroid hormone
receptor assay, **109**, 55
isolation of osteocalcin, **107**, 518
normalization parameters, **107**, 538
rat, calcitonin receptors in
binding assays, **109**, 43
regulation, **109**, 47
Bone marrow
human
preparation of single cell suspensions,
116, 567
progenitor cells, interferon effects on
growth, **119**, 631
quantification of secretory transcobalamin II, **123**, 48
murine
preparation of single cell suspensions,
116, 566
sedimentation profile of cells, **108**,
80
Borate
in colorimetric assay for dopa and dopa
proteins, **107**, 399
Bordetella pertussis
cultivation, **109**, 559
purification of pertussis toxin, **109**, 561

toxin ADP-ribosyltransferase
  assays, **106,** 416
  effect on adenylate cyclase system,
    106, 411
Borohydride
  in detection of nonenzymatic protein
    glycosylation, **106,** 83
Boundary shapes
  asymptotic
    calculations, **130,** 8
    for defining self-association equilibria,
      130, 10
    experimental determinations, **130,** 12
Bradford assay
  for cell wall proteins, **118,** 26
Bradykinin
  regulation of cGMP levels in hybrid
    cells, testing procedures, **109,**
    337
  stimulation of prostaglandin production
    by dermal fibroblasts
    analytical chromatographic tech-
      niques, **109,** 493
    effects of inhibitors of arachidonate
      release and metabolism, **109,** 486
    radioimmunoassay, **109,** 485
Bragg reflections
  from oscillation photographs, intensity
    determination
    data collection strategy, **114,** 277
    film processing techniques, **114,** 241
    postrefinement methods, **114,** 263
Brain
  bovine, purification of
    Ca^{2+}-dependent phosphodiesterase,
      102, 40
    calmodulin, **102,** 41
    calmodulin-dependent enzymes, **102,**
      215, 245
    carboxypeptidase-tubulin, **106,** 235
    dolichyl-pyrophosphate phosphatase,
      111, 550
    erythrocyte membrane skeletal protein
      analogs, **134,** 56
    microtubules, **134,** 106
    protein methylase II, **106,** 303, 305
    S-100 protein, **102,** 256
    tau protein, **134,** 157
    thiamin-diphosphate kinase, **122,** 26
  drug targeting methods, **112,** 381

fetal rat
  explant cultures, preparation, **103,** 356
  tissue collection and dissociation, **103,**
    352
guinea pig
  calcium channels, localization, **109,**
    513
  ubiquinones, determination, **105,** 147
human
  amyloid fibers
    assays, **134,** 391
    purification, **134,** 399
    structure, **134,** 389
  paired helical filaments
    assays, **134,** 391
    purification, **134,** 393
    structure, **134,** 389
mammalian
  catecholamine neurotransmitters
    fluorescence-based histochemistry,
      103, 620
    immunohistochemistry, **103,** 623
  excitotoxin direct injection, in analysis
    of neuroendocrine function, **103,**
    391
  extraction of β-endorphin and
    enkephalins, **103,** 552
  neuropeptide-specific pathways, trac-
    ing techniques, **103,** 663
  nuclei, micropunch sampling, **103,**
    368
  purification of
    fodrin, **134,** 44, 51
    microtubules and microtubule-
      associated proteins, **134,** 124
  sectioning for micropunch technique,
    103, 370
  steroid hormones and neuropeptides,
    simultaneous localization, **103,**
    631
murine perinatal, primary glial cultures
  establishment, **109,** 343
  hormonal regulation of cAMP levels,
    analytical techniques, **109,** 344
porcine, purification of
  cytochrome-*c* oxidase, **126,** 34
  erythrocyte membrane skeletal protein
    analogs, **134,** 56
  glutaminase, **113,** 249
  tubulin-tyrosine ligase, **106,** 227

primate, distribution of CRF receptors, **124**, 587
rat
CRF receptors, distribution, **124**, 576
dissociated cortical cell cultures
preparation, **124**, 413
somatostatin content, measurement, **124**, 419
somatostatin release, measurement, **124**, 417
LHRH secretion granules, isolation, **124**, 257
neuropeptide gene expression, analysis by *in situ* hybridization, **124**, 497, 510
neuropeptide mRNA in discrete regions, measurement, **124**, 548
neuropeptide receptors, localization, **124**, 597, 600, 604, 612
purification of
S-adenosylmethionine:calmodulin (lysine) *N*-methyltransferase, **102**, 161
S-adenosylmethionine:protein carboxylmethyltransferase, **102**, 165
L-glutamate decarboxylase, **113**, 4
phosphodiesterase, **102**, 171
protein kinase C, **124**, 351
thiamin and phosphates, **122**, 17
Thy-1 antigens, **108**, 659
tissue fixation techniques, **103**, 671
tissue sectioning methods, **103**, 675
uptake of α-aminoadipate, **113**, 645
Branched-chain amino-acid aminotransferase
assays, **113**, 72
natural occurrence, **113**, 71
Branched-chain α-ketoacid dehydrogenase kinase
properties and function, **107**, 94
Brassica rapa, *see* Turnip
Brome mosaic virus
infectious cDNA clones
production, **118**, 707
properties, **118**, 705
in vitro transcription, **118**, 712
infectious RNA from cloned cDNA
plant infection with, **118**, 715
production and analysis, **118**, 712
propagation in barley, **118**, 688
properties, **118**, 687
purification techniques, **118**, 688, 693
RNA, translation in wheat germ system, **118**, 154
Bromination
enzymatic, proteins, **107**, 441
4-Bromoadamantylideneadamantane
synthesis, **133**, 535
5-Bromo-4-chloro-3-indolyl phosphate
in protein immunostaining on nitrocellulose, **121**, 852
5-Bromodeoxyuridine
and light, in depletion of immunoreactive T cells, **108**, 263, 268
Bromoperoxidase
assay, **107**, 442
protein bromination, **107**, 441, 444
purification from *Penicillus capitatus*, **107**, 442
6-Bromotryptophan
separation from molluscan periostracum, **107**, 429
3-Bromotyrosine
synthesis, **107**, 421
Buffers
broad-range, preparation, **104**, 410
deuterated, preparation and applications, **104**, 412
effect on glutamate carboxylase activity, **107**, 556
for enzyme immunofiltration assays, **121**, 528
at low temperature, **104**, 412
metal ion content, control, **140**, 409
in NMR studies, **104**, 414
organic solvent effects, **104**, 411
preparation factors, **104**, 408
selection, **104**, 404
for solubilization of membrane proteins, choice, **104**, 310
α-Bungarotoxin
luciferin-conjugated
fluorescent properties, **133**, 45
membrane binding properties, **133**, 47
preparation, **133**, 43
Bursectomy
chicken
chemical techniques, **108**, 8

hormonal techniques, **108**, 5
surgical techniques, **108**, 3, 6
2,3-Butadione
in hydrophilic albumin microsphere preparation, **112**, 31
n-Butane
aqueous, conformational structure, **127**, 57, 59
4-(3-Butoxy-4-methoxybenzyl)-2-imidazolidinone
induction of glucagon responsiveness in transformed MDCK cells, **109**, 357
t-Butyl hydroperoxide
mutagenicity, testing with *Salmonella typhimurium* TA102, **105**, 250
t-Butylperoxymaleic acid
mutagenicity, testing with *Salmonella typhimurium* TA102, **105**, 250
γ-Butyrobetaine
chromatographic properties, **123**, 294
^{14}C-labeled, synthesis, **123**, 295

C

Caenorhabditis elegans
growth in culture, **134**, 416
isolation of males, **134**, 417
major sperm protein purification, **134**, 418
sperm isolation, **134**, 418
Calcineurin
assay, **102**, 229
properties, **102**, 239
purification from bovine brain, **102**, 215, 234
tissue distribution and localization, **102**, 243
Calcitonin
radiolabeled
metabolism in bone and kidney, **109**, 46
preparation, **109**, 42
in receptor binding assays, **109**, 43
regulation of cAMP levels in primary glial cultures, analytical techniques, **109**, 344
Calcitonin receptors
in bone and kidney, binding assays, **109**, 41
regulation, **109**, 47
Calcium
binding to calmodulin
analytical techniques, **102**, 135
in skeletal muscle, assessment, **102**, 74
effect on phagocytosis, **132**, 97
extrusion from *Escherichia coli*, assay, **125**, 335
hepatic, hormone effects, assessment, **109**, 550
ionized intracellular
fluorescent probes, **124**, 116
measurement, **124**, 90
mitochondrial retention, hydroperoxide effects, **105**, 438
in reconstitution of bacterial periplasmic binding protein-dependent transport
applicability for whole cells, **125**, 300
assays, **125**, 295
in spheroplasts and treated cells, **125**, 291
removal methods, **124**, 91
role in neutrophil activation, **105**, 389
in solution, coordination number, **127**, 325
transport in human erythrocytes
assay in inside-out vesicles, **102**, 54
calmodulin effects, **102**, 59
-treated *malE* mutants, *Escherichia coli*
maltose-binding protein entry and function, influencing factors, **125**, 296
maltose transport characteristics, **125**, 297
Calcium-binding proteins
antiserum preparation, **102**, 292
ELISA, **102**, 294
purification from chick intestine, **102**, 291
radioimmunoassay, **102**, 292
vitamin D-induced, ELISA, **123**, 154
Calcium channels
autoradiographic localization, **109**, 520, 546
basic ligand-binding assay, **109**, 518
differential labeling in brain, skeletal muscle, and heart, **109**, 536
direct ligand-binding studies, **109**, 521
labeling with
(−)-[^{3}H]desmethoxyverapamil or (±)-[^{3}H]verapamil, **109**, 543

[^{3}H]*d-cis*-diltiazem, **109**, 542
[^{125}I]iodipine, **109**, 539
solubilization, **109**, 518, 546
Calcium lactate
release from poly(ortho ester) polymer disks, **112**, 429
Calcium phosphate gels
mevalonate kinase adsorption, **110**, 74
Caldesmon
purification from chicken gizzard smooth muscle, **134**, 38
Calli
plant
establishment, **118**, 545
growth curves, establishment, **118**, 549
Calliphora vicina, see Blowfly
Calmodulin
amino acid composition, **102**, 300
antagonistic drugs
comparative studies, **102**, 193
synthesis and properties, **102**, 195
antibodies, preparation, **102**, 105
assay
with Ca^{2+}-dependent phosphodiesterase, **102**, 42
and localization in neutrophils, **105**, 390
binding of Ca^{2+}
analytical techniques, **102**, 135
in skeletal muscle, assessment, **102**, 74
binding polypeptides, detection by gel overlay assay, **102**, 205
binding sites, immunofluorescent localization, **102**, 117
carboxylmethylation
electrophoretic assay, **102**, 163
radiometric assay, **102**, 164, 166
chemically modified, preparation and application, **102**, 308
crystallization, **102**, 143
dependent phosphodiesterase, purification from bovine brain, **102**, 215
dependent protein phosphatase
antiserum production, **102**, 251
assay, **102**, 231
purification from bovine brain, **102**, 245
radioimmunoassay, **102**, 253
radioiodination, **102**, 251
effect on Ca^{2+} transport by human erythrocytes, **102**, 59
fluorescent derivatives
applications, **102**, 149, 153
biological activity, **102**, 152
preparation, **102**, 6, 150
properties, **102**, 156
function, myosin light chain phosphorylation as probe, **102**, 62
gel electrophoresis, **102**, 298
gel overlays, in assay for protein–protein interactions, **134**, 559
immunofluorescent localization
in fixed tissue sections and cultured cells, **102**, 111
in unfixed tissue sections, **102**, 122
interactions with binding proteins, quantitative fluorescence methods, **102**, 149
N-methylation
electrophoretic assay, **102**, 159
radiometric assay, **102**, 160
modification by converting enzyme, **102**, 167
–naphthalenesulfonamide interactions, analytical techniques, **102**, 185
–phenothiazine interactions, analytical techniques, **102**, 171
phosphorylase kinase-bound, quantitation, **102**, 220
purification
by affinity chromatography, **102**, 17, 107
from animal tissues, **102**, 2
from bovine brain, **102**, 41
from *Dictyostelium discoideum*, **102**, 160
by immunoprecipitation, **102**, 107
by naphthalenesulfonamide–Sepharose chromatography, **102**, 187
from plant tissues, **102**, 11
role in neutrophil activation, **105**, 391, 392
–Sepharose 4B conjugate, preparation, **102**, 211
spectroscopic analyses, **102**, 82
structural analysis, **102**, 303
subunits, interaction with phosphorylase kinase, **102**, 223
Calmodulin carboxylmethylesterase
radiometric assay, **102**, 166

Calmodulin-converting enzyme
rat anterior pituitary
assay, **102**, 169
properties, **102**, 170
Calorimetry
differential scanning, *see* Differential scanning calorimetry
isothermal, apolipoproteins
applications, **128**, 367
instrumentation, **128**, 365
for measurement of water–biomolecule interactions, **127**, 141
volume-perturbation, kinetics of lipid phase transitions, **130**, 534
Cameras
in photographic detection of luminescence, **133**, 406, 408, 409
rotation, with synchrotron radiation source
applications in crystallography, **114**, 281
description and operation, **114**, 286
performance, **114**, 292
Cancer
detection with cellular chemiluminescence, **133**, 499
Candida albicans
ingestion by macrophages, effect of (2′-5′)-oligoadenylates, **119**, 678
killing by neutrophils, in assessment of phagocytic function, **132**, 125, 170
Cannula
push-pull, for measurement of neuropeptide release, **124**, 480
Cannulation
thoracic duct, for lymphocyte collection, **108**, 270
Capillary tubes
in assay for migration inhibitory factor, **116**, 385
crystallization in, **114**, 128
Capping
surface immunoglobulins, **108**, 371
Caprylic acid
in precipitation of IgG, **116**, 13, 136
Capsicum fruit
chromoplasts
S-adenosylmethionine:γ-tocopherol methyltransferase reaction, **111**, 544
carotene synthesis, **110**, 244
plastids, prenylation of chlorophyllide *a*, **110**, 274
Carbamates
N-acyloxymethylation, in prodrug formation, **112**, 355
Carbamyl-phosphate synthetase (glutamine-utilizing)
Escherichia coli
assays, **113**, 306
heavy and light subunits
interactions, **113**, 318
separation, **113**, 314, 315
inhibition by oxidation, **113**, 320
mechanism of action, **113**, 322
properties, **113**, 316, 317
purification, **113**, 309
N-Carbobenzyloxy-γ-aminobutyric acid
in preparation of haptens for small oligopeptide radioimmunoassay, **103**, 442
N-Carbobenzyloxy-ω-amino-*n*-caproic acid
in preparation of haptens for small oligopeptide radioimmunoassay, **103**, 442
Carbocyclase
Citrus limonum
assay, **110**, 407
product identification, **110**, 409
properties, **110**, 415
purification, **110**, 413
radioactive substrates, synthesis, **110**, 410
Carbodiimides
in coupling monoclonal antibodies to microspheres, **112**, 76
–*p*-diazonium phenylacetic acid, neuropeptide conjugation methods, **103**, 452
modification of F_1-ATPase carboxylic groups, **126**, 714
neuropeptide conjugation methods, **103**, 451
in protein coupling
to amatoxins, **112**, 227
to phallotoxins, **112**, 236
Carbohydrates
attachment site on interferon β, gene sequence for, oligonucleotide-directed mutagenesis, **119**, 399
interaction with membranes
differential scanning calorimetry, **127**, 698

infrared spectroscopy, **127**, 698
Thy-1 antigen, measurement, **108**, 664
Carbon
activated, in preparation of high-purity water, **104**, 394
removal capabilities and limitations, **104**, 400
Carbon dioxide
respiratory, formation from radiolabeled substrates
apparatus, **113**, 461
precautions, **113**, 466
procedures, **113**, 464
Carbon monoxide
binding to hemoproteins, resonance Raman spectroscopy, **130**, 366
reaction with cytochrome *b*-245, spectrophotometric assay, **132**, 391
Carbon tetrachloride
aqueous solutions, water structure, infrared spectroscopy, **127**, 109
Carbonyldiimidazole
activation of affinity resins, **104**, 12
in spin-labeling of ADP–ATP carrier protein, **125**, 636
synthesis, **125**, 635
Carbonyl iron
in macrophage depletion from homogeneous cell populations
magnetic column method, **108**, 310
plastic tissue culture dish method, **108**, 309
β-Carboxyaspartic acid
chemical properties, **107**, 574
derivatization, **107**, 571
racemic, synthesis, **107**, 564; **113**, 145
Carboxyfluorescein
in assay for liposome–lipoprotein interactions, **128**, 662
characteristics, **128**, 658
purification, **128**, 659
4-Carboxyglutamate
excretion in urine, **107**, 537
quantitation, **107**, 533
γ-Carboxyglutamic acid
assay methods, **107**, 503
in proteins
assays, **113**, 138
biosynthesis *in vitro*, **113**, 136
decarboxylation, **107**, 549
mass spectrometric identification, **106**, 45
protection from decarboxylation, **107**, 550
quantitation, **113**, 140
tritium labeling, **107**, 545
Carboxylates
poly(ADP-ribosyl) derivatives in histone, evidence for, **106**, 450
Carboxyl groups
F_1-ATPase
chemical modifications, **126**, 714
identification, **126**, 755
Carboxylmethylation
calmodulin
electrophoretic assay, **102**, 163
radiometric assays, **102**, 164, 166
Carboxylmethylesterase
Salmonella typhimurium chemotaxis-related
assay, **106**, 322
properties, **106**, 325
purification, **106**, 323
Carboxylmethyltransferase
Escherichia coli and *Salmonella typhimurium* chemotaxis-related
assays, **106**, 312
differences, **106**, 318
effects of attractants, **106**, 318
general properties, **106**, 310
partial purification, **106**, 318
product identification, **106**, 316
Carboxyltransferase
oxaloacetate decarboxylase-associated
assay, **125**, 535
isolation, **125**, 536
Carboxymethyldextrans
characterization, **104**, 122
in displacement chromatography of proteins, **104**, 118, 124
preparation, **104**, 120
N-3-Carboxymethylriboflavin
ribityl side chain, ^{1}H NMR, **122**, 241
Carboxypeptidase
mucosal, targeting for gastrointestinal absorption of prodrugs, **112**, 368
pancreatic, types A and B, molecular replacement study, **115**, 70
tubulin-specific
assay, **106**, 234

properties, **106**, 236
purification from bovine brain, **106**, 235
substrate preparation, **106**, 234
Carcinogenesis
role of free radicals, *in vitro* analysis, **105**, 464
Carnitine
radioactive
characterization, **123**, 301
synthesis from γ-aminobutyrate, **123**, 297
L-Carnitine
chromatographic properties, **123**, 294
^{3}H-labeled, synthesis, **123**, 296
Carnitine acetyltransferase
bovine cardiac mitochondria
assays, **123**, 281
purification, **123**, 280
murine hepatic peroxisomes
assays, **123**, 281
purification, **123**, 279
Carnitine octanoyltransferase
murine hepatic peroxisomes
assays, **123**, 281
purification, **123**, 278
Carnitine palmitoyltransferase
bovine cardiac mitochondria
assays, **123**, 281
purification, **123**, 280
Carnitine precursors
chromatographic properties, **123**, 294
synthesis, **123**, 290
Carotene
in *Capsicum* chromoplasts
assay, **110**, 250
composition, **110**, 246
synthesis, **110**, 248
in daffodil chromoplasts, synthesis, **110**, 272
mutants, *Phycomyces blakesleeanus*
applications, **110**, 235
conservation, **110**, 234
description, **110**, 223
detection, **110**, 220
genetic analyses, **110**, 229
genetic nomenclature and strain list, **110**, 237
preparation, **110**, 226
in spinach chloroplasts, synthesis, **110**, 268
β-Carotene
as spectroscopic probe of order in lipoprotein lipids, **128**, 524
Carotenogenic enzymes
Neurospora crassa
localization, **110**, 262
membrane-bound, properties, **110**, 264
solubilization, **110**, 266
Carotenoids
extraction from tissue samples, **105**, 156; **111**, 120
handling precautions, **111**, 117
HPLC, **105**, 157; **111**, 189
normal-phase techniques, **111**, 200
reverse-phase techniques, **111**, 193
identification, **105**, 160
in *Neurospora crassa*
content, **110**, 257
isolation and identification, **110**, 260
photoregulation, **110**, 254
synthesis, **110**, 256, 259
nomenclature, **111**, 114
quantitative determination, **105**, 161; **111**, 147
saponification, **111**, 122
separation and purification, **111**, 123
spectroscopic analysis, **111**, 135
TLC, **105**, 157
Carotenoproteins
Echinodermata
properties, **111**, 241
purification, **111**, 241
Hydrocorallina
extraction, **111**, 242
properties, **111**, 242
lobster carapace
apoprotein preparation, **111**, 224
apoprotein subunit isolation, **111**, 227
crystallization, **111**, 238
extraction, **111**, 219, 239
polyacrylamide gel electrophoresis, **111**, 233
properties, **111**, 217, 231
reconstitution, **111**, 225
separation, **111**, 221, 229

type B
extraction and purification, **111**, 244, 247
properties, **111**, 243, 246
Carrot
calli and suspension cultures, establishment, **118**, 545
protoplasts, isolation, **118**, 576
seeds, aseptic germination, **118**, 540
in suspension cultures, synchronization by amphidicolin, **118**, 93
appropriate conditions for, definition, **118**, 91
assay, **118**, 89
effectiveness of method, **118**, 96
Casein
mRNA, phosphoserine codon usage, **106**, 158
Casein kinase
properties and function, **107**, 92
Casomorphine
computationally directed biorational drug design, **124**, 7
Catabolism, *see also* Metabolism
hepatic, lipoproteins
analytical techniques, **129**, 596, 599
pathway evaluation, **129**, 609
role of
apolipoproteins, **129**, 607
parenchymal and nonparenchymal cells, **129**, 605
low-density lipoproteins
analytical techniques, **129**, 567
receptor-independent
characteristics, **129**, 586
measurement, **129**, 570, 576
receptor-mediated, measurement, **129**, 570
in tissues, lipoproteins, assessment methods, **129**, 612
Catalase
assays, **105**, 122, 126
delivery systems, polymeric and copolymeric
kinetic analysis, **112**, 414
preparation, **112**, 400
in detection of
drug-induced H_2O_2 formation, **105**, 508
soluble immune response suppressor, **116**, 397
kinetic properties *in vitro*, **105**, 121
release of O_2, in measurement of H_2O_2, **105**, 397
Catecholamines
in biological samples
HPLC with electrochemical detection, **103**, 470
microbore HPLC, **124**, 404
radioenzymatic assays, **103**, 475, 484
in situ voltammetry, **103**, 479
in hypophysial portal blood, **103**, 609
hypothalamic
extraction from tissue and incubation media, **103**, 498
HPLC with electrochemical detection, **103**, 500
measurement in tissue culture, **124**, 362
steady-state and non-steady-state turnover, comparison, **103**, 506
neuron systems
fluorescence histochemistry, **103**, 620
immunohistochemistry, **103**, 623
Catechol *O*-methyltransferase
purification from rat liver, **103**, 485
in radioenzymatic assay for catecholamines, **103**, 486
Catheterization
intraatrial, with push–pull perfusion, in freely behaving unanesthetized rat, **103**, 182
Cauliflower
purification of elongation factor 1, **118**, 147
Cauliflower mosaic virus
cloning vectors
construction, rationale for, **118**, 661
introduction and expression in plants, **118**, 660, 663
preparation, **118**, 659
propagation in turnip, **118**, 697
properties, **118**, 696
CDPdiacylglycerol:inositol transferase
rat liver, assay, **109**, 474
Celite
in preparation of fatty acid–albumin solutions, **128**, 324
Cell adhesion
in cell surface tension determination, **132**, 34, 74

Cell counting
visual, in assay of monocyte/macrophage cytotoxicity, **132**, 461
Cell cultures
adrenal medullary chromaffin cells, preparation and maintenance, **103**, 305
anterior pituitary cells
corticotropin-releasing factor assay, **103**, 565
dispersed, on extracellular matrix, **103**, 287
preparation for secretion studies, **109**, 296
apolipoprotein synthesis, immunochemical measurement, **129**, 254
chicken embryo
interferon induction, **119**, 120
preparation, **119**, 115
dermal cells
induction of fibroblast interferon in, **119**, 132
preparation, **119**, 130
dissociated cerebral cortical
preparation, **124**, 413
somatostatin content, measurement, **124**, 419
somatostatin release, measurement, **124**, 417
dissociated hypothalamic cells
in serum-free media, **103**, 329
in serum-supplemented media, **103**, 322, 328
fibroblasts
interferon induction in, **119**, 127
preparation, **119**, 126
follicular cells, establishment, **124**, 246
hepatocytes
applications, **129**, 278, 281
preparation, **129**, 278
hypothalamic, synthesis of luteinizing hormone-releasing hormone, **103**, 521
kidney cells
interferon induction, **119**, 212
preparation, **119**, 212
mixed lymphocyte, preparation for generation of cytotoxic activity, **132**, 438
monolayer, bovine and human, interferon binding analysis, **119**, 312
mycoplasma, methodology, **125**, 261
neuroblastoma × glioma hybrid, preparation, **109**, 321
(neuroblastoma × glioma) × neuroblastoma hybrid–hybrid, preparation, **109**, 321
pituitary, short-term
application for simultaneous hormone determination, **103**, 257
cell adhesion and aggregation in, **103**, 252
cell density, effect on function, **103**, 253
cell dispersion and culturing conditions, **103**, 249
effects of incubation length, **103**, 255
media for, choice, **103**, 251
plant callus
establishment, **118**, 545
growth curves, establishment, **118**, 549
primary dispersed, peptidergic neurons, **103**, 348
primary glial, perinatal murine brain
establishment, **109**, 343
hormonal regulation of cAMP levels, analytical techniques, **109**, 344
protoplast
influencing factors, **118**, 562
preparation, **118**, 570
as tool in study of free radicals and radical modifiers in carcinogenesis, **105**, 464
Cell elution
in cell surface tension determination, **132**, 41
Cell lines, *see also* Tumor cells
A375 (human melanoma)
cultivation, **119**, 683
IgG-based localization in athymic mice, **121**, 808, 811, 812, 814
interferon-induced changes in surface antigen expression, radioimmunoassay, **119**, 684
A875 (human melanoma), growth stimulation by ferritin-coupled IgG, **121**, 776

CHO (Chinese hamster ovary), *see* Chinese hamster ovary cells
clonal hypothalamic, establishment, **103**, 331
Daudi (human lymphoblastoid cells)
extraction of interferon–receptor complexes with digitonin, **119**, 348
growth inhibition by interferon, assay, **119**, 579
interferon receptors, identification by chemical crosslinking, **119**, 342
fibroblast-like, with capacity to redifferentiate into adipocyte-like cells
adipose conversion, **109**, 383
characterization, **109**, 384, 385
establishment and handling, **109**, 378
HeLa (human cervical carcinoma)
cultivation, **134**, 148
purification of
metaphase chromosomes, **134**, 269
microtubule protein, **134**, 149
HL-60 (human promyelocytic leukemia)
coculturing with, in production of interferon γ, **119**, 65
histocompatibility antigen expression, interferon-induced changes, **119**, 688
production of tumor necrosis factor, **116**, 449
KB (human nasopharyngeal carcinoma), isolation of receptosomes, **109**, 257
K_c cells (*Drosophila melanogaster*)
assay of juvenile hormone binding proteins, **111**, 499
culture conditions, **111**, 496
photoaffinity labeling of juvenile hormone-binding proteins, **111**, 521
preparation and characterization of ecdysteroid receptors, **111**, 462
LS-174T (human colon carcinoma), IgG-based localization in athymic mice, **121**, 812, 814
mammalian, SV40 *tsA* mutant-transformed, general properties, **109**, 387
MCF-7 (human breast carcinoma)
cultivation, **119**, 683
interferon-induced changes in surface antigen expression, radioimmunoassay, **119**, 684
MDBK (Madin–Darby bovine kidney), *see* Madin–Darby bovine kidney cells
MDCK (Madin–Darby canine kidney), *see* Madin–Darby canine kidney cells
murine C-243
binding of interferon
data analysis, **119**, 325
in monolayer cultures, **119**, 325
in suspension cultures, **119**, 323
detection of interferon mRNA by *in situ* hybridization, **119**, 478
murine macrophage, isolation of acetylated low-density lipoprotein receptors, **129**, 219
Namalwa (human lymphoblastoid cells), large-scale production of interferon, **119**, 35
PC12 (rat pheochromocytoma)
cytosolic Ca^{2+} changes, monitoring, **124**, 132
growth stimulation by coupled IgG, **121**, 776
membrane potential changes, monitoring, **124**, 132
Ratec (transformed rat embryo)
interferon production, **119**, 220
isolation of interferon α mRNA, **119**, 442
stably transfected, for analysis of neuropeptide gene expression, **124**, 282
3T12, cytolytic activity of peroxidase-activated macrophages, **132**, 539
U937 (human histiocytic lymphoma)
binding of ^{32}P-labeled interferon γ, **119**, 317
preparation for binding assay, **119**, 316
WiDr (human colon carcinoma)
cultivation, **119**, 683
interferon-induced changes in surface antigen expression, radioimmunoassay, **119**, 684
WISH (human cells)
interferon receptors, identification by chemical crosslinking, **119**, 340
in monolayer culture, binding of interferon, **119**, 313
Y1 (murine adrenal tumor)
ACTH-resistant mutants
characteristics, **109**, 355

selection, **109**, 354
cultivation, **109**, 351
mutagenesis, **109**, 353
properties, **109**, 351
Cellobiose
–tyramine, in assay for tissue sites of lipoprotein degradation, **129**, 622
Cell sorters
fluorescence-activated, *see* Fluorescence-activated cell sorter
Cellulose disks
avidin–biotin-linked
in biotin assay, **122**, 78
preparation, **122**, 77
avidin-linked
in biotin assay, **122**, 78
preparation, **122**, 75
biotin-linked
in avidin assay, **122**, 81
preparation, **122**, 76
isonitrile-linked, preparation, **122**, 74
Cell walls
plant
component extraction and purification, **118**, 8
isolation, **118**, 6
polysaccharides in, analytical methods, **118**, 25
sources, **118**, 4
Central nervous system
CRF receptors, analytical techniques, **124**, 575
gene expression, analysis by hybridization histochemistry, **124**, 534
Centrifugal elutriation
countercurrent, *see* Countercurrent centrifugal elutriation
hepatic macrophages, **108**, 286
monocytes, **132**, 238
Centrifugation
active enzyme, *see* Active enzyme centrifugation
bacterial luciferase 4a-hydroperoxyflavin intermediates on Sephadex, **133**, 133
continuous gradient, intact Golgi complex for analysis of peptide hormone receptors, **109**, 225
cytochrome-*c* oxidase, in determination of sedimentation coefficients, **126**, 53
density gradient, *see* Density gradient centrifugation
differential, Golgi fractions for analysis of peptide hormone receptors, **109**, 219
isopycnic, *see* Isopycnic centrifugation
monocytes, **116**, 383
in purification of
chloroplast DNA, **118**, 167
plant mitochondria, **118**, 439, 491
plant mitochondrial DNA, **118**, 439
RNA plant viruses, **118**, 687
in separation of chloroplast inner and outer envelope membranes, **118**, 320
spleen cells on Percoll solution, **121**, 178
in subfractionation of thylakoid membranes, **118**, 335
Centromere
–kinetochore fraction from metaphase chromosomes, preparation, **134**, 270
Centrosome
mammalian
assay, **134**, 263
isolation, **134**, 264, 267
Cerebrospinal fluid
extraction of
catecholamines, **124**, 404
β-endorphin and enkephalins, **103**, 555
Ceroid
fluorescent, measurement, **105**, 486
identification, **105**, 484
quantification, **105**, 485
Cesium
gramicidin A channel, energy profiles, computational methods, **127**, 261
Cetrimonium bromide
in leukocyte solubilization for myeloperoxidase purification, **132**, 372
Cetyltrimethylammonium bromide, *see* Cetrimonium bromide
Cevadine
interactions with ion channels in neuroendocrine tissue, **103**, 403
CHAPS, *see* 3-[(3-Cholamidopropyl)dimethylammonio]-1-propane sulfonate
Chemical shifts
phosphoproteins, in ^{31}P NMR studies, **107**, 53

Chemiluminescence
acridinium esters
applications, **133**, 370
measurement, **133**, 269
mechanism, **133**, 366
cellular
associated with disease states, **133**, 493
measurement, **132**, 501, 505
assay systems, **133**, 501
data analysis, **133**, 506
instrumentation for, **133**, 504
detection system for HPLC
applications, **133**, 447
description, **133**, 439
experimental methods, **133**, 441
instrumentation, **133**, 445
horseradish peroxidase-catalyzed reactions
enhancement, **133**, 332
enhancers
concentration effects, **133**, 344
identification, **133**, 336
pH effects, **133**, 343
specificity, **133**, 341
synergistic action, **133**, 340
mechanisms, **133**, 332
as indicator of singlet oxygen in biological systems, **105**, 222
photographic detection
applications, **133**, 413, 418
assays, **133**, 412
film types, **133**, 404
instrumentation, **133**, 405
reaction kinetics
computer-aided shape analysis, **133**, 422
measurement, **133**, 421
Chemiluminescent assay
FAD, **122**, 186
horseradish peroxidase, **133**, 331
lipid peroxidation products, **105**, 289
lipoxygenase, **105**, 129
macrophage activation, **132**, 544
with macrophage cell lines, **133**, 507
oxygen, **105**, 129
oxygen radical production by neutrophils, **105**, 398
1,4-pentadiene fatty acids, **105**, 130
Chemiluminescent immunoassay
with acridinium esters, **133**, 379
computer-aided
applications, **133**, 428
instrumentation, **133**, 424
α_1-fetoprotein, **133**, 381
free thyroxine, **133**, 384
photographic enhanced
chromium, **133**, 418
commercial systems, **133**, 420
cytomegalovirus-specific antibody, **133**, 413
steroids, **133**, 387
thyrotropin, **133**, 382
transferrin in seminal plasma, **133**, 561
Chemiluminescent probes
for phagocyte oxygenation activity analysis
applications, **133**, 461
azide effects, **133**, 469
mechanism, **133**, 456
preparation, **133**, 457
superoxide dismutase effects, **133**, 473
temperature effects, **133**, 485
for singlet oxygen in biological reactions, **133**, 569
Chinese hamster ovary cells
expression of
human interferon β, **119**, 398
rat interferon γ, **119**, 458
lysis, **134**, 219
mitotic spindle
isolation, **134**, 219
morphology, **134**, 220
purification, **134**, 221
mutants defective in
low-density lipoprotein endocytosis, selection, **129**, 227
low-density lipoprotein metabolism
immobilization, **129**, 241
induction, **129**, 240
screening, **129**, 245
synchronization, **134**, 218
Chlamydomonas
basal bodies, fractionation, **134**, 259
cell wall removal, **134**, 257
cultivation, **134**, 282
flagella
fractionation, **134**, 289
isolation, **134**, 287
removal, **134**, 259
flagellar dyneins
analytical techniques, **134**, 297

purification, **134**, 293
harvesting, **134**, 286
negative staining, **134**, 260
Chlorambucil
N-hydroxysuccinimidyl ester, in antibody coupling to toxins, **112**, 214
Chloramines
identification and quantitation, **132**, 577
neutrophil-produced, cytotoxic effects on
bacteria, **132**, 592
erythrocytes, **132**, 590
neutrophils, **132**, 587
tumor cells, **132**, 592
preparation, **132**, 571
Chloramine-T
in radioiodination of
interferon α, **119**, 267
peptides, **124**, 22
thyroid hormones, **107**, 483
Chlorella protothecoides
purification of chlorophyllase, **123**, 421
Chloride
chloramine-derived, assay, **132**, 578
transport in halorhodopsin-containing envelope vesicles
assays, **125**, 515, 518
effect on halorhodopsin absorption spectrum, **125**, 518
light-generated gradients, **125**, 516
related changes in vesicle properties, measurement, **125**, 514, 515
Chlorination
dyes, in assay of chloramines, **132**, 582
Chlorine
in solution, coordination number, **127**, 327
Chloris striate mosaic virus
propagation in plants, **118**, 703
3-Chloro-L-alanine
labeled active site of aspartate aminotransferase
effect on catalytic mechanism, **106**, 133
isolation, **106**, 131
structural analysis, **106**, 133
3-Chloro-5-bromotyrosine
synthesis, **107**, 421
Chloroform
in purification of mitochondrial F_1-ATPase, **126**, 480
Chloroformates
activation of affinity resins, **104**, 12
p-Chloromercuribenzene sulfonate
in identification of membrane phosphate transport protein, **125**, 718
4-Chloronaphthol
as chromagen in immunohistological staining, **121**, 837
7-Chloro-4-nitrobenzofurazan
^{14}C-labeled, identification of F_1-ATPase essential residues, **126**, 748
modification of F_1-ATPase tyrosine residues, **126**, 724
Chlorophyll
binding proteins, isolation from
barley chlorina *f*2 mutant, **118**, 348
pea leaves, **118**, 342
spinach, **118**, 351
Capsicum plastid, composition, **110**, 274
chromatographic analysis, **110**, 305
spectrophotometric analysis, **110**, 306
Chlorophyllase
Chlorella protothecoides
assays, **123**, 422
properties, **123**, 426
purification, **123**, 424
Chlorophyllide *a*
prenylation in *Capsicum annuum* plastids, **110**, 274
Chloroplasts
Capsicum annuum
chlorophyll composition, **110**, 274
chlorophyllide prenylation, **110**, 276
preparation, **110**, 274
Euglena gracilis
isolation, **118**, 256
purification of elongation factor Tu, **118**, 296, 305
transcriptionally active extracts
identification of promoter sequences for RNA polymerase, **118**, 267
preparation, **118**, 257
properties, **118**, 265
in vitro transcription reactions, **118**, 259
lettuce
isolation, **110**, 308
purification of
membrane proteolipid, **126**, 520
violaxanthin deepoxidase, **110**, 311
violaxanthin cycle in, **110**, 309

maize
gene translation initiation site, determination, **118**, 309
transcriptionally active extracts
preparation, **118**, 234
properties and applications, **118**, 250
stability, **118**, 245
temperature optimization, **118**, 245
in vitro transcription reactions, **118**, 239
pea
envelope membranes
characterization, **118**, 323
preparation, **118**, 319
separation, **118**, 320
gene translation initiation site, determination, **118**, 309
isolated, translation reactions, **118**, 286
isolation, **118**, 283
transcriptionally active extracts
preparation, **118**, 234
properties and applications, **118**, 250
stability, **118**, 245
temperature optimization, **118**, 245
in vitro transcription reactions, **118**, 239
spinach
carotene synthesis *in vitro*, **110**, 268
isolation, **110**, 268; **118**, 215, 255, 299
photogeneration of
hydrogen peroxide, **105**, 425
superoxide anion, **105**, 423
purification of
aminoacyl-tRNA synthetase, **118**, 218
CF_0-F_1, **126**, 512
cytochrome b_6f, **126**, 277
elongation factors, **118**, 296, 298
tRNAs, **118**, 215
scavenging of
hydrogen peroxide, **105**, 426
superoxide, **105**, 426
transcriptionally active extracts
identification of promoter sequences for RNA polymerase, **118**, 267
preparation, **118**, 234, 257
properties and applications, **118**, 250, 265
stability, **118**, 245
temperature optimization, **118**, 245
in vitro transcription reactions, **118**, 239, 259
tobacco
coded mutations, induction and selection, **118**, 611
controlled traits
identification in cybrids, **118**, 603
transfer by protoplast fusion, **118**, 595
gene translation initiation site, determination, **118**, 309
isolation, **118**, 272
lysate coupled transcription–translation system
applications, **118**, 280
preparation, **118**, 272
in vitro reactions, **118**, 274
purification of
ribosomal proteins and rRNA, **118**, 207
ribosomes, **118**, 203
from various plants
coupling factor complex
assays, **126**, 288
isolation, **126**, 286
reconstitution, **126**, 288
envelope membrane
metabolite transport processes, **125**, 705
phosphate transport proteins, identification and isolation, **125**, 716, 723
proton transport, stoichiometry, **125**, 6
isolation, general methodology, **118**, 188
NaBr-treated, preparation, **118**, 353
plastoquinone biosynthesis, **110**, 316, 318
polypeptide synthesis during development, analytical techniques, **118**, 367
psbA gene product, identification and characterization, **118**, 384
purification of
DNA, **118**, 167
DNA polymerase, **118**, 188
photosystem II reaction center, **118**, 354

quantitative estimation of protein subunits, **118**, 360
3-[(3-Cholamidopropyl)dimethylammonio]-1-propane sulfonate
in mitochondrial F_0F_1-ATPase purification, **126**, 473
in opioid receptor solubilization, **124**, 180
Cholanamidopropylsulfobetaines
physical properties, **124**, 160
synthesis, **124**, 156
Cholanylsulfobetaines
synthesis, **124**, 154
Cholate
dialysis, in reconstitution of cytochrome-*c* oxidase, **126**, 81
Cholecystokinin
radiolabeled
binding studies with dispersed pancreatic acini, **109**, 66
preparation, **109**, 65
Cholecystokinin receptors
on pancreatic acini, binding assay, **109**, 64
Cholera toxin
ADP-ribosylation of membrane N_s and N_i components, **109**, 566
ADP-ribosyltransferase
assays, **106**, 414
demonstration, **106**, 413
effect on adenylate cyclase system, **106**, 411
Cholesterol
association with coronary artery disease, **129**, 101
cellular
efflux, quantitation, **129**, 637
purification, **129**, 636
radiolabeling, **129**, 630, 653
enzymatic quantitation, **129**, 104
esterification rate, in assay for lecithin–cholesterol acyltransferase, **129**, 778
induced lipoproteins and apolipoproteins in mammals, **128**, 142
lipoprotein, mass determination, **129**, 557
metabolic theory, **129**, 379
plasma, continuous-flow analysis, **128**, 190
receptor-mediated transport, analytical techniques, **129**, 645
serum, HPLC elution pattern, **129**, 66
side-chain cleavage
enzymatic reaction, **111**, 354
gas chromatography–mass spectrometry in selected ion monitoring mode, **111**, 355
product determination, **111**, 354
in singlet oxygen trapping, **105**, 40, 42
transport and lecithin–cholesterol acyltransferase reaction, **111**, 267
Cholesterol acyltransferase
rat liver
activation, **111**, 293
assay, **111**, 286
inhibitors, **111**, 291
substrate specificity, **111**, 290
Cholesterol epoxide
in assay of epoxide hydrolases, **111**, 309
Cholesteryl alkyl ether
biological applications, **129**, 824
labeled lipoproteins, analysis, **129**, 826
labeled liposomes
analysis, **129**, 832
preparation, **129**, 831
properties, **129**, 822
synthesis, **129**, 817
Cholesteryl esters
cellular clearance, quantitation, **129**, 642
lipoprotein
formation assay, **129**, 556
mass determination, **129**, 557
transfer assay for plasma lipid transfer proteins I and II, **129**, 811
transfer protein
assays, **111**, 275
purification from human plasma, **111**, 280
Chorion
from fertilized eggs, occurrence of Glu–Lys crosslink, **107**, 256
Choroid plexus
ovine
cultivation, **119**, 552
sensitivity to heterologous interferons, **119**, 554
Chromaffin cells
adrenal medullary
culture stability, **103**, 311

isolation, **103**, 306
plating and maintenance, **103**, 310
Chromatin
nucleosomal, poly(ADP-ribosyl)ation, **106**, 513
analytical immunofractionation methods, **106**, 518
electrophoretic analysis, **106**, 517
time-resolved X-ray scattering, **134**, 675
ubiquitin–histone 2A conjugate, isolation, **106**, 241
Chromatofocusing
calcineurin, **102**, 237
IgG-binding factor, **116**, 414
insulin-like growth factors I and II, **109**, 786, 794
interleukin 2, **116**, 515
Chromatographic columns
affinity, packing, **104**, 218
anti-Fab
preparation, **108**, 150
in separation of lymphocytes, **108**, 151
for cytoplasmic dynein, preparation, **134**, 347
gelatin–Sepharose, preparation, **132**, 335
HPLC
installation, **104**, 134
for malonaldehyde determination, selection and life performance, **105**, 327
operation, **104**, 136
repair, **104**, 150
storage, **104**, 143
testing, **104**, 135
troubleshooting, **104**, 143
hydroxyapatite
guidelines for use, **117**, 375
protein adsorption, **117**, 371
protein elution, **117**, 374
for IgG, overview, **121**, 599
immunoadsorbent, in T-suppressor factor assay, **116**, 321
immunoaffinity
for cytokinin
applications, **110**, 352
preparation, **110**, 349
properties, **110**, 357
for interleukin 2, testing, **116**, 509
for separation of T-cell subsets, preparation, **121**, 741
ion-exchange
cleaning, **104**, 188
elution, **104**, 179
loading capacity, **104**, 178
resolution, optimization, **104**, 183
retention, **104**, 170
selection, **104**, 173
selectivity, **104**, 185
switching, **104**, 189
for lipoprotein assay, selection, **129**, 58
nickel chelate, preparation, **119**, 201
nylon wool
preparation, **108**, 299
in separation of lymphoid cells, **108**, 300
reversed-phase
mobile- and stationary phase parameters, selection, **104**, 199
resolution optimization, **104**, 192
retention mechanisms and behavior, **104**, 191, 194
Sephadex G-10
preparation, **108**, 304
in separation of macrophages and lymphocytes, **108**, 305
size-exclusion
calibration with globular proteins, **117**, 147
selection and maintenance, **104**, 163
universal calibration curve, **117**, 144
Chromatography, *see also specific chromatographic techniques*
N-acetylglutamate synthase, **113**, 31
allogeneic effect factor, **116**, 435
α-aminoadipate, **113**, 642
p-aminobenzoylpoly(γ-glutamates), **122**, 331
apolipoprotein A-containing lipoproteins, **129**, 143
apolipoprotein C, **128**, 290
apolipoprotein E, **128**, 278
apolipoprotein E-containing lipoproteins, **129**, 159
bacterial luciferase 4a-hydroperoxyflavin intermediates on Sephadex, **133**, 131
bile acids, **111**, 63
Ca^{2+}-protease I, **102**, 285
calcineurin on DE-23 cellulose, **102**, 234
calmodulin
on DEAE-cellulose, **102**, 3

on phenyl-Sepharose, **102**, 6
in γ-carboxyglutamic acid assay, **107**, 504
carotenoids, **111**, 124
carotenoproteins, **111**, 221
chloramines, **132**, 584
chlorophylls, **110**, 305
colony-stimulating factor 1, **116**, 570
cytochrome *b*-245, **132**, 385
deglycosylated glycoprotein hormones, **109**, 729
double-stranded mitochondrial RNAs, **118**, 502
double-stranded RNA-dependent protein kinase, **119**, 504
double-stranded viral RNAs, **118**, 719
Glu–Lys isodipeptide, **107**, 246
L-glutamate decarboxylase, **113**, 5, 6
granulocyte colony-stimulating factor, **116**, 606
granulocyte–macrophage colony-stimulating factor, **116**, 592
halogenated tyrosine derivatives, **107**, 418, 423
Ia antigens on hydroxylapatite, **108**, 547
IgG
on BioGel HPHT, **116**, 135
on DEAE Affi-Gel Blue, **121**, 588
IgG_1
on DE-52, **116**, 133
on DEAE Affi-Gel Blue, **116**, 134
IgG-binding factor, **116**, 416
IgG idiotypes on hydroxylapatite, **121**, 615
IgG proteolytic fragments, **121**, 666
IgG subclasses, **121**, 650
IgM, **116**, 136; **121**, 650
inositol phospholipids, **124**, 436
interferon β, **119**, 218, 401
interferon δ, **119**, 99
interferon γ, **119**, 86, 194, 199
interleukin 2, **116**, 533
mevalonate kinase, **110**, 74
neuropeptides, **103**, 22, 74
nitrogenase from *Klebsiella pneumoniae*, **118**, 516
osteocalcin, **107**, 521
phosphorylated basic amino acids, **107**, 28
plant antiviral protein, **119**, 741
plant oligoadenylates, **119**, 756
polyacrolein nanospheres, **112**, 156
proteins on hydroxyapatite columns, **117**, 370
radioactive sterol metabolites, **111**, 326
recombinant interferon α2, **119**, 170
recombinant interferon β, **119**, 179, 185
retinoic acid isomers for nonaqueous HPLC, **123**, 78
retinol isomers for nonaqueous HPLC, **123**, 76
RNase L, **119**, 495, 496
sterols, **111**, 5
striated flagellar root protein on phenyl–Sepharose, **134**, 412
thiamin and phosphate esters on SP-Sephadex column, **122**, 20
thymopoietin, **116**, 288
ubiquitin, **116**, 288
xanthophylls, **110**, 305
Chromatophores
bacterial
in measurement of membrane potential, **127**, 557
preparation, **127**, 560
Rhodospirillum rubrum, properties after F_1-ATPase subunit depletion, **126**, 535
Chromic chloride
as crosslinking reagent in Ig coupling to erythrocytes, **121**, 550, 727
Chromium-51
release assay for
monocyte cytotoxicity, **132**, 465
T-cell-mediated cytotoxicity, **132**, 440
tumor cell cytotoxicity, **132**, 559, 565
Chromoplasts
Capsicum annuum
S-adenosylmethionine:γ-tocopherol methyltransferase reaction, **111**, 544
carotene composition, **110**, 246
carotene synthesis, **110**, 248
chlorophyll composition, **110**, 274
chlorophyllide prenylation, **110**, 276
preparation, **110**, 244, 274
purity criteria, **110**, 245
daffodil
carotene synthesis, **110**, 267
isolation, **110**, 272

tomato, purification of
isopentenyl pyrophosphate isomerase, **110**, 217
phytoene synthetase, **110**, 216
prenyltransferase, **110**, 217
Chromosomes
in bovine × murine hybridomas, number determination, **121**, 260
human, localization of apolipoprotein genes, **128**, 49, 796, 851, 863
in hybrid cells, analysis, **121**, 5
metaphase, from HeLa cells
fractionation, **134**, 269
kinetochore antigen identification, **134**, 271, 277
kinetochore–centromere complex
immunofluorescence and electron microscopy, **134**, 271
isolation, **134**, 270
Chronic granulomatous disease
detection with cellular chemiluminescence, **133**, 495
Chymotrypsin
proteolysis of tubulin, **134**, 182
Cilia
Tetrahymena thermophila
axonemal microtubules
complexing with dynein, **134**, 316
repolymerization, **134**, 314
purification of dynein, **134**, 309
Circular dichroism
apolipoproteins in solution, **128**, 383
calmodulin, **102**, 86
lipoprotein lipids, **128**, 519
magnetic
analytical applications, **130**, 276
measurement and instrumentation, **130**, 272
principles of operation, **130**, 271
peptide–lipid complexes, **128**, 639
for protein conformation analysis
comparison of methods, **130**, 239
computer programs for, **130**, 256
data analysis methods, **130**, 221
experimental measurements, **130**, 217
expression of experimental data, **130**, 216
historical development, **130**, 208
reference spectra for, **130**, 222, 225
protein folding intermediates, **131**, 126
reconstituted low-density lipoprotein complexes, **128**, 603
stopped-flow, *see* Stopped-flow circular dichroism
Citrate
transport in tumor mitochondria
kinetics
determination, **125**, 674
temperature effects, **125**, 684
metabolite flux, determination, **125**, 685
Citrinin
inhibition of cytochrome bc_1, **126**, 267
Citrus limonum, *see* Lemon
Citrulline
in proteins
identification and estimation, **107**, 626
occurrence, **107**, 625
Claviceps
growth in culture, **110**, 335
purification of
isopentenyl-diphosphate Δ-isomerase, **110**, 96
tryptophan dimethylallyltransferase, **110**, 336
Clonidine
transdermal application, rate-controlled system, **112**, 456
Cloning
anti-receptor antibody-secreting hybridomas, **121**, 100
apolipoprotein A-II cDNA, **128**, 745
apolipoprotein E cDNA, **128**, 801
ATP synthase subunit genes from photosynthetic bacteria, **125**, 231
cauliflower mosaic virus DNA in bacterial vectors, **118**, 659
cDNA copies of viral RNA, **118**, 707
Epstein–Barr virus-transformed B cells, **121**, 146
F_1-ATPase mutant genes, **126**, 599
hamster × murine hybridomas, **121**, 242
human × human hybridomas, **121**, 115
hybridomas
by limiting dilution, **103**, 466
in methylcellulose, **121**, 315
in SFH medium, **121**, 290
in 20-μl hanging drops, **121**, 327
in ultra-low gelation temperature agarose, **121**, 323

Ia antigen genes, **108**, 567
IgA-secreting rat × rat hybridomas, **121**, 58
interferon α gene, **119**, 442
interferon γ gene, **119**, 453
β-lactamase gene in pIN III-*ompA*, **125**, 143
lipoprotein mRNAs, **128**, 42
low-density lipoprotein receptor cDNAs, **128**, 895
lux genes from marine bacteria, **133**, 83
β_2-microglobulin gene, **108**, 501
myeloid progenitor cells in culture, interferon effects, **119**, 629
plant mitochondrial DNA, **118**, 458, 476
rat × rat hybridomas produced with IR983F cells, **121**, 238
ribulose-bisphosphate carboxylase genes, **118**, 420
ribulose-bisphosphate carboxylase large subunit genes, **118**, 429
ribulose-bisphosphate carboxylase small subunit genes, **118**, 397
staphylococcal nuclease A gene, **125**, 146, 148
tumor cells
comparative feeder cell effects, **121**, 298
in culture, interferon effects, **119**, 635
Clostridium
C. formicoaceticum
growth in culture, **122**, 388, 395
purification of
5,10-methenyltetrahydrofolate cyclohydrolase, **122**, 388
5,10-methylenetetrahydrofolate reductase, **122**, 392
C. tetanomorphum, purification of glutamate mutase components, **113**, 124, 128
Clostripain
proteolysis of tubulin, **134**, 182
Cobalamin, *see* Vitamin B_{12}
Cobalt
binding to tRNA crystals, **114**, 165
Cockroach
corpora allata, radiochemical assay for juvenile hormone III biosynthesis, **111**, 531
photoaffinity labeling of juvenile hormone binding proteins, **111**, 516
Codons
phosphoserine, usage in casein mRNA, **106**, 158
Coenzyme B_{12}
hydrated, X-ray and neutron diffraction analysis, **127**, 169
Collagen
as substrate in tissue culture of sympathetic neurons, **103**, 341
time-resolved X-ray scattering, **134**, 675
Collagenase
in isolation of
cardiac mitochondria, **125**, 17
Leydig cells, **109**, 277
Colletotrichum lindemuthianum
purification of endopolygalacturonase, **118**, 10
Colloidal gold
chemistry, **124**, 37
particles, sizing, **124**, 39
peptide conjugation, **124**, 43
Colony-stimulating factor
granulocyte–macrophage-specific
bioassays, **116**, 592
properties and sources, **116**, 588
purification, **116**, 590
radioiodination, **116**, 598
granulocyte-specific
assays, **116**, 603
binding analysis, **116**, 617
properties, **116**, 601
purification, **116**, 605
radioiodination, **116**, 617
macrophage-specific
antibodies, preparation, **116**, 577
assays, **116**, 565, 579
binding to cellular receptors, analytical techniques, **116**, 582
properties, **116**, 569
purification, **116**, 570
radioiodination, **116**, 576
overview, **116**, 365
Colony-stimulating factor receptors
binding activity, analysis, **116**, 582, 617
Colorimetric assay
5-aminolevulinate synthase, **123**, 395
arginine deiminase, **107**, 628
γ-carboxyglutamic acid, **107**, 506

for cell wall polysaccharides, **118**, 25
citrulline in proteins, **107**, 626
hydroxyl radicals, **105**, 51
lecithin–cholesterol acyltransferase, **129**, 781
prolyl hydroxylase, **107**, 365
proteins in dilute solutions, **104**, 415
Colostrum
human, purification of
IgA, **116**, 49
IgA secretory component, **116**, 71
identification of N^{ε}-(β-aspartyl)lysine, **107**, 259
Compactin, *see* Mevastin
Complement
and antibody, for purification of lymphocyte subsets, **108**, 249
C3b, and IgG, cooperative effect on phagocytosis, **132**, 302
coating of culture surfaces for phagocytosis studies, **132**, 213
deficiencies, effect on opsonization, **132**, 314
derived opsonins, overview, **132**, 296
Complementation
H^+-ATPase mutants with λ-*unc* transducing phages, **126**, 562
Complement receptors
C3, phagocytosis-promoting capacity, activation by fibronectin, **132**, 218
phagocytic, for opsonic complement, overview, **132**, 301
Complex III, *see* Cytochrome bc_1
Computer applications, *see also* Software
biological sequencing, data bank search techniques, **128**, 754
chemiluminescent reaction kinetics, **133**, 422
data reduction for interferon antiviral assays, **119**, 536
in fluorescence correlation spectroscopy, **130**, 465
in fluorescence photobleaching recovery, **130**, 465
perifusion system for neuroendocrine tissues, **124**, 67
in scanning gel chromatography for active enzymes, **117**, 116
simulation of
biomolecular hydration via proximity criterion analysis, **127**, 33
biopolymer solvation, **127**, 371
complex chemical systems, **127**, 114
conformational structure of aqueous *n*-butane, **127**, 58
in transient electric birefringence system, **117**, 205
Computer graphics
interactive, in crystallography, **115**, 157
in molecular biology, **115**, 381
Concanavalin A
–amatoxin conjugates
biological properties, **112**, 233
preparation by mixed anhydride coupling, **112**, 231
bound monolayers
lymphocyte separation on, **108**, 130
preparation, **108**, 128
demetallized, constrained–restrained least-squares structure refinement, **115**, 300
induction of human interferon γ, **119**, 89
–Sepharose conjugates
in affinity chromatographic purification of insulin receptor, **109**, 399, 402
immobilization of NAD glycohydrolase, **122**, 178
–trypsin conjugates
activity assays, **112**, 250
characterization, **112**, 253
preparation, **112**, 251
purification, **112**, 252
Conidia
Neurospora crassa, aging rate, assay, **105**, 263
Contact angle measurements
in cell surface tension determination
methodology, **132**, 70
overview, **132**, 22
Contractile apparatus
ameboid cell, agar overlay immunofluorescence, **134**, 573
Coomassie Blue
protein staining in gels, **104**, 439
Copper sandwich holders
for specimen freezing, **127**, 706
Coproporphyrin
^{14}C-labeled, synthesis in cell-free extracts, **123**, 363
isomers, HPLC separation, **123**, 386
and uroporphyrin isomers, simultaneous separation, **123**, 387

Coral
extraction of carotenoproteins, **111**, 242
Cordycepin
analog of 2,5-p_3A_n, inhibition of protein synthesis and cell growth, **119**, 668
Coronary artery disease
cholesterol association, **129**, 101
Corpora allata
Diploptera punctata, radiochemical assay for juvenile hormone III biosynthesis, **111**, 531
Corrinoids, *see also specific corrinoids*
extraction from cells, **123**, 16
HPLC, **123**, 17
photodissociative matrices
preparation, **123**, 29
in purification of vitamin B_{12}-binding proteins, **123**, 32
synthesis, **123**, 16
Corticotropin-releasing factor
cell culture assay, **103**, 565
extraction techniques, **103**, 575
radioimmunoassay, **103**, 569
Corticotropin-releasing factor receptors
in brain, localization, **124**, 600
in central nervous system
distribution, **124**, 576
general characteristics, **124**, 575
localization, **124**, 563
in pituitary
differential modulation, **124** 572
distribution, **124**, 569
general characteristics, **124**, 567
localization, **124**, 563, 600
Corynebacterium
growth in culture, **122**, 351
purification of folylpoly(γ-glutamate) synthase–dihydrofolate synthase, **122**, 351
Coulter counter
in assay for
antibody-mediated cytotoxicity, **132**, 481
tissue culture cell viability, **132**, 488
in estimation of lymphoreticular cell number, **108**, 46, 76
Countercurrent centrifugal elutriation
in isolation of mononuclear leukocyte subsets, **108**, 242
Countercurrent distribution
in purification of neuropeptides, **103**, 20
in separation of luteinizing hormone subunits, **109**, 742, 744
Counterion electrophoresis
in measurement of ligand–protein interactions, **117**, 392
Coupling factors
chloroplast
assays, **126**, 518
purification, **126**, 286, 513
reconstituted in liposomes
assays, **126**, 288
photophosphorylation in, **126**, 290
reconstitution, **126**, 288, 516
synthesis of enzyme-bound ATP, **126**, 643
Covalent chromatography
glutathione peroxidase, **107**, 611
membrane proteins, **104**, 338
Covolume
in estimation of protein axial ratios, **117**, 182
Crambin
hydrated, X-ray and neutron diffraction analysis, **127**, 169
Crayfish
hypodermis, preparation of ecdysteroid receptors, **111**, 459
Creatine kinase
in measurement of interferon effect on cellular differentiation, **119**, 619
Cresol red
in assay for pantetheine-hydrolyzing enzyme on polyacrylamide gels, **122**, 38
CRF, *see* Corticotropin-releasing factor
Crosslinking
affinity, *see* Affinity crosslinking
(2′-5′)$(A)_3[^{32}P]$pCp to RNase L, **119**, 497
bound radioiodinated glucagon to hepatic membranes, **109**, 209
chemical, in identification of interferon receptors, **119**, 340
human chorionic gonadotropin to luteal receptors, **109**, 203
neuropeptides by ethylene glycol bis-(succinimidyl succinate), **103**, 53
photoaffinity, *see* Photoaffinity labeling
in proteins, mass spectrometric identification, **106**, 52

Crosslinking reagents
for structural analysis of lipoproteins
applications, **128**, 620
performance, **128**, 616
selection, **128**, 614
Crown ether
in generation of superoxide anion in nonaqueous solutions, **105**, 75
Crustacyanins
extraction, **111**, 219, 239
properties, **111**, 217, 231
separation, **111**, 221, 229
Cryo-electron microscopy
applications, **134**, 623, 629
image analysis, **134**, 616
principles, **134**, 613
Cryofixation
lipoproteins, **128**, 459
Cryopreservation
lymphoid cells, **108**, 363
monocytes, **132**, 234, 241
neutrophil cytoplasts, **132**, 255
Cryosolvents
effects on protein folding, **131**, 174
Cryptate
in generation of superoxide anion in nonaqueous solutions, **105**, 75
Crystallization
bacterial porin
microdialysis technique, **125**, 323
vapor diffusion method, **125**, 320
calmodulin
from 2-methyl-2,4-pentanediol, **102**, 144
from polyethylene glycol, **102**, 146
carotenoproteins, **111**, 238
heavy riboflavin synthase from *Bacillus subtilis*, **122**, 195
immunoglobulins and fragments, **116**, 195
macromolecules
general principles, **114**, 112
influencing factors, **114**, 119
heterogeneity, contributing factors, **114**, 120
organic solvents in, **114**, 118
with polyethylene glycol, **114**, 120
precipitants in, **114**, 118
by variation of pH and temperature, **114**, 125
nucleic acids and oligonucleotides, **114**, 168
proteins, **104**, 372, 378
in capillary tubes, **114**, 128
by free interface diffusion, **114**, 140
growth cessation phase, **114**, 108
growth phase, **114**, 101
influencing factors, **104**, 370
nucleation phase, **114**, 80
precipitating agents in, **104**, 373
seeding methods, **104**, 380
Crystallography
applications of
interactive computer graphics program, **115**, 166, 167
rotation camera with synchrotron radiation source, **114**, 281
rotation and translation functions, **115**, 55
crystal treatment and manipulation in, **114**, 136
direct phase determination by anomalous scattering, **115**, 41
least-squares refinement methods, **115**, 23
macromolecular
resolution of phase ambiguity, **115**, 90
structure refinement
definition, **115**, 228
limits, **115**, 229
model accuracy, **115**, 232
monitoring, **115**, 231
number of observations, **115**, 230
proteins
diffraction methods in, **114**, 20
dynamic information from X-ray diffraction data, **115**, 117
experimental procedures in, **114**, 41
historical review, **114**, 3
neutron diffraction in, **114**, 510
structure refinement, **115**, 313
water in ice phases, **127**, 311
Crystals
Bragg reflections from oscillation photographs, intensity determination
data collection strategy, **114**, 277
film processing techniques, **114**, 241
postrefinement methods, **114**, 263
cytochrome-*c* oxidase
gross structure, **126**, 24

preparation, **126**, 22, 23
diffraction data collection
by microdensitometry, **114**, 200
with multiwire area detector diffractometer, **114**, 416, 452
oscillation method with large unit cells, **114**, 211
with rotation camera using synchrotron radiation source, **114**, 281
with television area detector diffractometers, **114**, 472, 486
with X-ray film, **114**, 204
F_1-ATPase, electron microscopy, **126**, 441
macromolecular structure, stereochemically restrained refinement
applications, **115**, 268
computer program design, **115**, 260
observational functions, **115**, 253
strategy, **115**, 264
weighting considerations, **115**, 263
membrane ubiquinol–cytochrome-*c* reductase, preparation, **126**, 191
mitochondrial outer membrane channel
characteristics, **125**, 601
mechanism of formation, **125**, 606
preparation, **125**, 597
structure, **125**, 607
mounting techniques, **114**, 136, 141
NADH dehydrogenase
electron microscopy, **126**, 345
image analysis, **126**, 348
preparation, **126**, 345
nucleic acid
binding to heavy metal ions, **114**, 156
heavy-atom derivatives, preparation, **114**, 173
protein
density measurements
with aqueous Ficoll solutions, **114**, 187
in hydration determination, **114**, 176
methodology, **114**, 179
in molecular weight determination, **114**, 176
under salt-free conditions, **114**, 182
growth cessation, mechanism, **114**, 108
isomorphous heavy-atom derivatives, preparation, **114**, 147
molecular motion, X-ray diffraction study, **115**, 127
motions in, **131**, 394
postnucleation growth, analytical techniques, **114**, 101
structure
automated electron density map analysis, **115**, 206
Fourier refinement, **115**, 235
thermal motion, X-ray diffraction study, **115**, 117, 122
repeated seeding technique, **114**, 134
seeding technique, **114**, 133
seed sources, **114**, 132
slippage, reduction methods, **114**, 139
water networks, computer simulations, **127**, 122
CTP synthetase
Escherichia coli
assay, **113**, 282
properties, **113**, 286
purification, **113**, 283
Cucumber mosaic virus
infectivity, effect of plant IVR, **119**, 733
Culture media
for callus and suspension cultures, preparation, **118**, 543
color change, in assay for cell growth inhibition by interferon, **119**, 579
for monocyte suspension cultures, **132**, 245
serum-free
for dispersed anterior pituitary cells, **103**, 293
for myeloma and hybridoma cells, **121**, 266, 277
purification of IgG, **121**, 602
standard, abbreviations for, **119**, 22
for sympathetic neuron culturing, preparation, **103**, 343
Culture plates
antibody-coated, in lymphocyte separation, **108**, 118, 124
Cumene hydroperoxide
mutagenicity, testing with *Salmonella typhimurium* TA102, **105**, 250
Cutinase
Fusarium solani pisi
amino-terminal glucuronamide linkage occurrence, **106**, 216

detection, **106**, 210
amino-terminal region, structure, **106**, 217
biosynthesis *in vitro*, **106**, 216
Cuvettes
continuous-flow, in simultaneous determination of oxygen consumption and NAD(P)H formation or disappearance, **122**, 162
Cyanide
binding to hemoproteins, resonance Raman spectroscopy, **130**, 391
insensitive respiration, in detection of paraquat-mediated oxyradical formation, **105**, 525
Cyanogen bromide
activation of affinity resins, **104**, 4
in coupling monoclonal antibodies to microspheres, **112**, 77
Cyanuric chloride
–[^{14}C]sucrose, in assay for tissue sites of lipoprotein degradation, **129**, 619
Cyclic AMP
intracellular levels, hormonal regulation, testing procedures, **109**, 327
leukocyte
extraction, **132**, 429
purification, **132**, 430
radioimmunoassay, **132**, 432
in MDCK cells
inhibition of glucagon receptor induction, **109**, 360
radioimmunoassay, **109**, 362
in primary glial cultures, hormonal regulation, analytical techniques, **109**, 344
production after cell fusion, kinetics, **109**, 373
Cyclic GMP
intracellular levels, hormonal regulation in hybrid cells, testing procedures, **109**, 337
leukocyte
extraction, **132**, 429
purification, **132**, 430
radioimmunoassay, **132**, 432
radioimmunoassay with femtomole sensitivity, **109**, 827
Cyclic voltammetry
quinones, **125**, 112
α-Cyclodextrin
hydrated, X-ray and neutron diffraction analysis, **127**, 169
β-Cyclodextrin
hydrated, X-ray and neutron diffraction analysis, **127**, 169
Cycloeucalenol–obtusifoliol isomerase
plant, active cell-free preparations, isolation, **111**, 323
1,2-Cyclohexanedione
modification of low-density lipoproteins for catabolism studies, **129**, 576
Cycloheximide
inhibition of glucagon receptor induction in transformed MDCK cells, **109**, 359
Cyclophosphamide
bursectomy with, **108**, 8
with interferon treatment, antitumor effects on human tumor xenografts, **119**, 655
Cystathionase, *see* Cystathionine γ-lyase
Cystathionine γ-lyase
precipitating immunoassay, **121**, 704
Cysteamine
in apolipoprotein E phenotyping, **128**, 280
Cysteine
binding to hemoproteins, resonance Raman spectroscopy, **130**, 407
–glyceride modification, in membrane penicillinase
identification criteria, **106**, 366
nature of, **106**, 365
sequence required for, **106**, 368
in human and murine interferon, chemical modification, **119**, 237
reactivity with heavy atoms, **114**, 149
Cysteine-*S*-conjugate *N*-acetyltransferase
rat kidney
assay, **113**, 516
properties, **113**, 519
purification, **113**, 517
Cysteine-conjugate β-lyase
rat liver
assay, **113**, 511
properties, **113**, 514
purification, **113**, 512
Cysteine thiosulfonate
characterization, **113**, 563
synthesis, **113**, 558

Cysteinylglycine
hydrolysis, catalyzing microvillus membrane peptidases, **113**, 471
Cystine
in human and murine interferon, chemical modification, **119**, 237
reactivity with heavy atoms, **114**, 149
Cytasters
sea urchin egg, preparation, **134**, 196
Cytochemistry
horseradish peroxidase in neuroendocrine cells, **103**, 187
Cytochrome *b*
bacterial
properties, **126**, 328
purification from cytochrome bc_1 complex, **126**, 326
Cytochrome b_{245}
human neutrophil
electrophoresis, **132**, 394
purification, **132**, 384
spectrophotometric assays, **132**, 385
Cytochrome b_{559}
NADPH oxidase-associated
characterization, **132**, 364
preparation, **132**, 358
Cytochrome bc_1, *see also* Ubiquinol–cytochrome-*c* reductase
bovine cardiac mitochondria
iron–sulfur protein, purification, **126**, 212
protein subunits, isolation, **126**, 227
purification, **126**, 183, 186, 225
reaction with *N,N'*-dicyclohexylcarbodiimide, **125**, 102
redox Bohr effects, measurement, **126**, 338
resolution, **126**, 218, 231
mitochondrial, inhibition by
antimycin, **126**, 267
2,3-dimercaptopropanol, **126**, 270
5,5'-dithiobis(2-nitrobenzoic acid), **126**, 271
funiculosin, **126**, 268
hydroxyquinoline *N*-oxides, **126**, 270
hydroxyquinones, **126**, 263
β-methoxyacrylate, **126**, 257
Neurospora crassa
isolation, **126**, 203
structural properties, **126**, 206
Paracoccus denitrificans
cytochrome *b* purification, **126**, 326
properties, **126**, 323
purification, **126**, 318
Rhodopseudomonas sphaeroides
assays, **126**, 283
cytochrome *b* purification, **126**, 326
purification, **126**, 183, 186, 282
yeast mitochondrial
assays, **126**, 178
properties, **126**, 178
purification, **126**, 174, 183, 185
reconstitution, **126**, 179
Cytochrome b_6f
chloroplast
assays, **126**, 283
isolation, **126**, 277, 279
cyanobacterial
assays, **126**, 279
isolation, **126**, 279
Cytochrome *c*
acetylated
in assay for superoxide anion, **105**, 14, 377
preparation, **105**, 14, 376
in assessment of phagocytic function, **132**, 112, 152
effect on cytochrome-*c* oxidase isozyme carboxyl group reactivity, **126**, 39
equine heart
CNBr-activated Sepharose 4B-linked, preparation, **126**, 65
fragment complementing systems
folding kinetics and intermediates, **131**, 205
fragment exchange in, **131**, 207
preparation, **131**, 197
high-affinity binding site on cytochrome-*c* oxidase, structural analysis, **126**, 30
interaction with cytochrome c_1, role of hinge protein, **126**, 245
reduced, preparation, **134**, 215
yeast
activated thiol-Sepharose 4B-linked, preparation, **126**, 67
Affi-Gel 102-SMPB-linked, preparation, **126**, 71

Cytochrome c'
crystal intensity, multiwire area detector diffractometry, **114**, 470
Cytochrome c_1
interaction with cytochrome *c*, role of hinge protein, **126**, 245
purification of hinge protein, **126**, 238
Cytochrome-*c* oxidase
Bacillus subtilis
properties, **126**, 165
purification, **126**, 160
reconstitution, **126**, 170
bovine heart
activation and dispersion with alkyl glycoside detergents, **125**, 33
crystal preparation, **126**, 22, 23
gross structure, **126**, 24
high-affinity site for cytochrome *c*, structure, **126**, 30
H^+ pumping activity, measurement, **126**, 15, 20
monomeric and dimeric forms
centrifugation, **126**, 53, 61
detergent binding assay, **126**, 51
gel filtration, **126**, 47
sedimentation equilibrium measurements, **126**, 56
sedimentation velocity measurements, **126**, 57
polypeptide components, topography, **126**, 25
prosthetic groups, locus, **126**, 29
purification, **126**, 64
reaction with *N,N'*-dicyclohexylcarbodiimide, **125**, 98
reconstitution, **126**, 14, 78
redox Bohr effects, measurement, **126**, 338
subcomplex splitting, **126**, 73
subunit III depletion, **126**, 74
chicken cardiac mitochondria, assay, **125**, 18
general properties, **105**, 31
mitochondrial, tissue-specific isozymes
amino-terminal sequence analysis, **126**, 44
carboxylic groups, reactivity, **126**, 39
immunological characterization, **126**, 41
isolation, **126**, 34
properties, **126**, 36
subunit isolation, **126**, 43
oxygen intermediates
detection, methodological approach, **105**, 32
general features, **105**, 34
Paracoccus denitrificans, purification, **126**, 156
Rhodopseudomonas sphaeroides aa_3-type
characterization, **126**, 141
purification, **126**, 138
sea urchin embryo, assay, **134**, 215
thermophilic bacterium PS3
assays, **126**, 131, 146
properties, **126**, 149
purification, **126**, 146
reconstitution, **126**, 124, 151
Cytochrome-*c* peroxidase
in assay for hydrogen peroxide production by mitochondria, **105**, 433
crystal intensity, multiwire area detector diffractometry, **114**, 470
Cytochrome *d*
terminal oxidase complex, *Escherichia coli*
assays, **126**, 88, 89
characterization, **126**, 90
oxidase activity, effect of oxygen concentration, **126**, 109
oxidation–reduction potential, **126**, 105
polypeptide composition, **126**, 99
purification, **126**, 89, 97
reconstitution, **126**, 91, 109
spectral properties, **126**, 103
ubiquinol oxidase assay, **126**, 106
Cytochrome *o*
terminal oxidase complex, *Escherichia coli*
oxidase activity, effect of oxygen concentration, **126**, 109
oxidation–reduction potential, **126**, 105
polypeptide composition, **126**, 99
purification, **126**, 95
reconstitution, **126**, 109
spectral properties, **126**, 101
ubiquinol oxidase assay, **126**, 106

Cytochrome-*o* oxidase
-containing proteoliposomes, characterization, **126**, 121
Escherichia coli
assays, **126**, 114, 131
properties, **126**, 119
purification, **126**, 115
reconstitution, **126**, 118, 124
Cytochrome *P*-450
active in bile acid synthesis
assays, **111**, 365
properties, **111**, 375
purification
from rabbit liver, **111**, 372, 374
from rat liver, **111**, 368, 370
assay for analysis of interferon effects on hepatic drug metabolism, **119**, 721
Cytokeratin
identification, **134**, 365
polypeptide separation, **134**, 361, 368
purification, **134**, 367, 368
Cytokines
afferent immunoenhancing, overview, **116**, 360
–cell interactions, sequential immunoenhancing, **116**, 367
efferent, acting on nonlymphoid cells, overview, **116**, 369
immunosuppressive, overview, **116**, 368
Cytokinins
biosynthetic enzyme
assay, **110**, 344
properties, **110**, 346
purification, **110**, 343, 346
HPLC, **110**, 354
immunoaffinity purification, **110**, 352
radioimmunoassay, **110**, 355
Cytolysis
erythrocytes, continuous assay, **132**, 491
interferon-mediated, microassay, **119**, 574
lymphocytes, assay
with 5-bromodeoxyuridine and light, **108**, 262
hot antigen technique, **108**, 255
hot pulse technique, **108**, 259
tumor cells by macrophages, assays, **132**, 559, 563, 565, 567
Cytoplasts
neutrophil
applications, **132**, 256
cryopreservation, **132**, 255
preparation, **132**, 253
properties, **132**, 254
Cytosine
methylation, effect on restriction analysis of nuclear DNA, **118**, 69
Cytoskeleton
antigens in cell monolayers
antibody production and labeling, **134**, 596
embedding, **134**, 596, 597
fixation and permeabilization, **134**, 593, 595
cytokeratin-enriched
polypeptide separation, **134**, 361
preparation, **134**, 358
fibers, X-ray diffraction analysis, **134**, 643
intestinal brush border cells, quick freeze–deep etching, **134**, 606
microtubule-based mechanochemistry
erythro-9-[3-(2-hydroxynonyl)]adenine as probe, **134**, 481
vanadate as probe, **134**, 486
nerve cells, quick freeze–deep etching, **134**, 606
nonmammalian vertebrate erythrocytes, description, **134**, 232
polyethylene glycol embedding protocol for immunofluorescence microscopy, **134**, 580
Cytostasis
interferon-mediated, microassay, **119**, 574
Cytostat
application in antibody production, **121**, 360
Cytotoxicity
antibody-mediated, assay with Coulter counter, **132**, 481
assays
with anti-erythrocyte hybridomas as targets, **121**, 761, 764
by chemiluminescence, **132**, 498
chemoimmunoliposomes, assays, **121**, 825, 827
chloramine-mediated, assays, **132**, 587, 590, 592

large granular lymphocyte-mediated, assay, **132**, 454
macrophage-mediated, assays, **132**, 458, 516, 531, 539, 565, 567, 601
monocyte-mediated, assays, **132**, 458
ricin A chain–antibody conjugates, comparison, **112**, 243
T-cell-mediated, ^{51}Cr release assay, **132**, 440

D

Dactinomycin
-containing immunoliposomes, in tumor therapy, **121**, 827
Daffodil
chromoplasts, carotene synthesis, **110**, 267
N-Dansylaziridine
-labeled calmodulin
applications, **102**, 149, 153
biological activity, **102**, 152
preparation, **102**, 150
properties, **102**, 156
Dansyl chloride
in measurement of peptide uptake by bacteria, **125**, 372
Dansyl hydrazine
and periodic acid, in staining of glycoproteins in gels, **104**, 448
Data banks
biological sequences, search methods, **128**, 754
Daucus carota, *see* Carrot
Daunomycin
–melanotropin conjugates, preparation, **112**, 259
–polylysine conjugates
preparation, **112**, 274
properties, **112**, 277
–thyrotropin conjugates, preparation, **112**, 267
Deacetylase
assay, **107**, 237
purification and properties, **107**, 236
Deacetylation
ε-*N*-acetyllysine in proteins, **107**, 236
Dealkylation
phytosterols in insects, analytical techniques, **111**, 346
5-Deazaflavins
HPLC separation, **122**, 205
5-Deazariboflavin
phosphorylation, in preparation of riboflavin phosphates, **122**, 217
n-Decane
aqueous solutions, water structure, infrared spectroscopy, **127**, 110
Decarboxylation
γ-carboxyglutamic acid in proteins, **107**, 549
protection from, **107**, 550
α-ketoglutarate, in assay for prolyl hydroxylase, **107**, 365
thermal, for radiolabeling γ-carboxyglutamic acid, **113**, 143
Deep etching
and freeze-fracturing, quick frozen lung slices, **134**, 605
Deglycosylation
glycoprotein hormones
with anhydrous HF, **109**, 726
with trifluoromethanesulfonic acid, **109**, 733
Degranulation
assay for assessment of phagocytic function, **132**, 108, 146
Dehydroalanine
identification, **107**, 485
Dehydroascorbate reductase, *see* Glutathione dehydrogenase (ascorbate)
Dehydroascorbic acid
HPLC, **122**, 3
Dehydroerythorbic acid
HPLC, **122**, 3
3-Dehydroretinal
extraction as oxime, **123**, 53
HPLC, **123**, 55
applications, **123**, 58
comparison with bleaching analysis, **123**, 58
12-*cis*-Dehydrosqualene
squalene synthetase-catalyzed formation
assay, **110**, 374
properties, **110**, 375
Deimination
enzymatic, arginine residues, **107**, 624
Delipidation
apolipoprotein E-containing lipoproteins, **128**, 274

high-density lipoproteins, **128**, 343, 348, 350
low-density lipoproteins for apolipoprotein B-100 isolation, **128**, 252
plasma lipoproteins, **128**, 213
serum albumin, **128**, 322
triglyceride-rich lipoproteins, **128**, 289
very-low-density lipoproteins, **128**, 345, 349, 351
Demethylation
enzymatic, protein-bound methyllysine, **106**, 282
4-Demethylsterols
radiolabeled
incubation with marine invertebrates and plant tissues, **111**, 320
preparation, **111**, 317
Denaturation
solvent, FTIR spectroscopy, **130**, 308
Densitometry
lipoproteins on stained gels, **128**, 422
micro-, in X-ray diffraction data collection, **114**, 199
phosphatidylinositol in hepatic plasma membranes, **109**, 512
Density
protein crystals, measurement
with aqueous Ficoll solutions, **114**, 187
in hydration determination, **114**, 176
methodology, **114**, 179
in molecular weight determination, **114**, 176
under salt-free conditions, **114**, 182
Density gradient centrifugation
cytochrome-*c* oxidase, in determination of sedimentation coefficients, **126**, 61
F_1F_0-ATPase, **126**, 551
IgG, **116**, 10
lymphoid cells on albumin and metrizamide, **108**, 102
monocytes, **132**, 207
in separation of
anterior pituitary cells, **103**, 272
dispersed retinal cells, **103**, 363
lymphoreticular cells, **108**, 61
mononuclear cells and granulocytes from blood, **108**, 88
whole bone, in analysis of normalization parameters, **107**, 539
Density gradients
in active enzyme centrifugation, **117**, 101
Deoxycholate
in preparation of cytochrome-*c* oxidase two-dimensional crystals, **126**, 23
Deoxyhemoglobin
human, molecular replacement study, **115**, 72
sickle cell, intermolecular forces, direct measurement by osmotic stress, **127**, 415
Deoxyoligonucleotides
octadecyl, as probes for base substitution in interferon αA and α2 genes
hybridization of plasmid DNAs with, **119**, 483
preparation, **119**, 483
Deoxyribonuclease I
footprint titrations, in analysis of protein–DNA interactions, **130**, 132
Deoxyribonucleic acid, *see* DNA
Deoxyuridine triphosphate
biotinylation, **109**, 439
Dephosphorylation
^{32}P-labeled apolipoprotein B, **129**, 541
Desalting
nucleotides from NADH preparations, **122**, 155
(−)-Desmethoxyverapamil
^{3}H-labeled, labeling of calcium channels, **109**, 543
Des(24–28)-thymosin α_1
isolation, **116**, 256
Detergents, *see also specific detergents*
aqueous micelles, physical properties, **124**, 139
bile acid derivatives, synthesis, **124**, 153
as buffer for determination of protein molecular weight, **117**, 41
classification, **124**, 137
common, properties, **104**, 321
in crystallization of membrane proteins, **104**, 377
in delipidation of lipoproteins, **128**, 215
design criteria, **124**, 148
effect on glutamate carboxylase activity, **107**, 554
hydrophilic group, properties, **124**, 149
hydrophobic group, properties, **124**, 152

micelles
dimensions, **104**, 320
and phospholipid vesicles, separation by gel filtration, **104**, 326
protein-containing, dimensions, **104**, 320, 323
in preparation of apolipoprotein–lipid complexes, **128**, 560
–protein complex, nature of, **124**, 145
–protein ratio, for optimum solubilization, **104**, 313
and proteins, separation by gel filtration of crude membrane extracts, **104**, 323
in purification of hydrophobic proteins, **104**, 319
removal from protein-containing micelles, **104**, 327
in solubilization of
apolipoprotein B, **128**, 585
functional membrane proteins, choice, **104**, 306
membranes, **124**, 146
opioid receptors, **124**, 178
Dethiobiotin
in biotinylation of peptide hormones and other biological materials, **109**, 429
preparation, **109**, 422
Deuterium
in analysis of water role in structure and function of membranes, **127**, 662
–H_2O neutron difference maps, in solvent structure analysis of proteins, **127**, 329
–hydrogen exchange in proteins
analytical techniques, **131**, 453
applications, **131**, 502
mechanisms, **131**, 471, 494
methodology, **131**, 452
multiple-site, data analyses, **131**, 480
ratio determination, **131**, 444
Dextran
activated, coupling to monoclonal antibodies, **121**, 742
in coupling monoclonal antibodies to microspheres, **112**, 81
crystallized microspheres
magnetically responsive, preparation for drug targeting, **112**, 126
for slow release of proteins
biological activity of entrapped proteins, assay, **112**, 121
preparation, **112**, 120
in vitro release studies, **112**, 122
fluorescent, as pinocytic marker in leukocytes
applications, **108**, 339
merits, **108**, 337
ligand-substituted, preparation, **104**, 280
in osmotic stress measurement of membrane intermolecular forces, **127**, 407
–protein conjugates
pharmacological properties, **112**, 292
preparation, **112**, 286
therapeutic applications, **112**, 296
Diabetes
detection with cellular chemiluminescence, **133**, 500
Diacylglycerol kinase
rat liver, assay, **109**, 475
Diacylglycerols
as activators of protein kinase C, **124**, 60
synthesis, **124**, 58
Diafiltration
in ligand binding studies, **117**, 369
1,2-*O*-Dialkyl-*sn*-glycero-3-phosphocholine
biological applications, **129**, 846
synthesis, **129**, 843
Dialkyl glyceryl ether
biological applications, **129**, 842
synthesis, **129**, 835
Dialysis
ascites fluid for IgG purification, **121**, 633
with cholate, in reconstitution of cytochrome-*c* oxidase, **126**, 81
in crystallization of proteins, **104**, 372
in determination of protein phosphoramidates, **107**, 35
equilibrium, *see* Equilibrium dialysis
lipoproteins, **128**, 219, 444
micro-, in crystallization of bacterial porin, **125**, 323
Diaminobenzidine
as chromagen in immunohistological staining, **121**, 837

–photooxidation product, in identification of triple-labeled neuroendocrine cells, **103**, 143
Diaphorase, *see* Dihydrolipoamide dehydrogenase
Diaphragm
porcine, purification of cytochrome-*c* oxidase, **126**, 34
Diazobenzyloxymethyl paper
protein blotting onto, **104**, 456
in screening hybridoma supernatants, **121**, 491
Diazoethane
synthesis, **126**, 622
in triethyl phosphate synthesis, **126**, 625
p-Diazonium phenylacetic acid
–carbodiimide, neuropeptide conjugation methods, **103**, 452
Diazophenylthio paper
protein blotting onto, **104**, 456
2-Diazo-3-trifluoropropionyloxyprenyl analogs
allylic pyrophosphate
absorption spectrum and photolability, **110**, 121
photolysis conditions, **110**, 120
as prenyltransferase substrates and inhibitors, **110**, 121
preparation, **110**, 117
Dicarboxylates
transport in chloroplast envelope, **125**, 709
Dichlorofluorescein
in assay for picomole levels of lipid hydroperoxides, **105**, 352
5-(4,6-Dichlorotriazin-2-yl)amino fluorescein
–tubulin conjugates
characterization, **134**, 524, 525
preparation, **134**, 522
3,5-Dichlorotyrosine
synthesis, **107**, 420
Dictyostelium discoideum
purification of
α-actinin, **134**, 496
calmodulin, **102**, 160
severin, **134**, 10
N,N'-Dicyclohexylcarbodiimide
in analysis of membrane-bound enzymes, **125**, 89, 92
in assay for ATP synthase subunit activity, **126**, 570, 582
binding proteolipid, chloroplast
characterization, **126**, 523
purification, **126**, 522
radiolabeling, **126**, 521
reconstituted, proton conduction measurement, **126**, 527
reconstitution, **126**, 526
chemical properties, **125**, 87
^{14}C-labeled, in identification of F_1-ATPase carboxyl groups, **126**, 755
Dienes
in detection of lipid peroxidation, **105**, 287, 294
lipid-conjugated, spectrophotometric detection, **105**, 331
in measurement of singlet oxygen production by neutrophils, **105**, 396
Diethylaminoethyl-Sephacel
in purification of human interferon γ, **119**, 90
Diethyldithiocarbamate
enhancement of indoleamine 2,3-dioxygenase intracellular activity, **105**, 66
Difference sedimentation velocity
macromolecule hydrodynamic properties
applications, **117**, 38
practical aspects, **117**, 28
sources of variation, **117**, 35
Difference spectroscopy
in measurement of ligand–protein interactions, **117**, 382
Differential scanning calorimetry
instrumentation for, **130**, 60; **131**, 11
peptide–lipid complexes, **128**, 640
polymerization reactions, **130**, 73
protein domain interactions, **130**, 75
protein–ligand stability measurements, **130**, 71
protein–nucleic acid interactions, **130**, 69
protein–protein interactions
apparent specific heat capacity measurements, **130**, 64
enthalpy measurements, **130**, 65
reconstituted low-density lipoprotein complexes, **128**, 601
subunit interactions, **130**, 70
temperature-induced changes in proteins
applications, **131**, 15, 33

data analysis, **131**, 24, 45
experimental methods, **131**, 6
water and carbohydrate interactions with membranes, **127**, 698
Differentiation factors
B-cell, *see* B-cell differentiation factors
Diffractometers
adventitious quasi-simultaneous data collection methods, **114**, 414
five-circle
data collection strategy, **114**, 411
instrumentation, **114**, 409
quasi-simultaneous measurements of reflections, **114**, 403
with position-sensitive detector, for neutron diffraction data collection, **114**, 551
for simultaneous and quasi-simultaneous measurement of reflections, design principles, **114**, 398
with television area detectors
camera system, **114**, 488
data collection strategy, **114**, 484
description, **114**, 474
detector availability, **114**, 509
detector performance, **114**, 499
memory system, **114**, 496
operation, **114**, 483
phosphor efficiencies and characteristics, **114**, 498
scan control and image processing electronics, **114**, 493
system calibration, **114**, 497
system configuration, **114**, 504
X-ray, with multiwire area detector
crystal examination and alignment programs, **114**, 456
crystal orientation, refinement program, **114**, 457
data acquisition programs, **114**, 458
data collection procedure and results, **114**, 467
data output and reduction, **114**, 465
Mark I system, description and operation, **114**, 418
Mark II system, description and operation, **114**, 440
operation software, program organization, **114**, 454
parallax in, **114**, 435
Diffractometry
comparison with other methods, **114**, 330
crystal geometry and orientation task, **114**, 336
data collection strategies, **114**, 358
diffraction geometry, **114**, 347
instrumentation, design and operation, **114**, 374
single-crystal, overview, **114**, 331
X-ray, large proteins, data collection
background determination, **114**, 391
monochromators, **114**, 390
radiation decay corrections, **114**, 386
strategies, **114**, 393
in X-ray tube, **114**, 392
Diffusion
double, in Ig idiotype assay, **116**, 178
drugs in polymer membranes, rates and mechanisms, **112**, 500
free interface, in protein crystallization, **114**, 140
immuno-, *see* Immunodiffusion
lateral, measurement techniques, **130**, 454
norethisterone from poly(lactide-*co*-glycolide) microcapsules, kinetics, **112**, 113, 115
in polymer membranes, principles, **112**, 496
protons in inorganic materials, comparison with tunneling, **127**, 267
vapor, *see* Vapor diffusion
Diffusion coefficients
macromolecules, by active enzyme centrifugation, **117**, 108
proteins in solutions, from small-angle X-ray scattering data, **117**, 170
1,5-Difluoro-2,4-dinitrobenzene
in structural analysis of lipoproteins, **128**, 613
Digital imaging fluorescence microscopy
applications to lipoproteins, **129**, 867
theory of operation, **129**, 857
unique problems
calibration, **129**, 864
out-of-plane fluorescence, **129**, 866
photobleaching, **129**, 860

Digitonin
in extraction of
human interferon α2–receptor complex, **119**, 348
opioid receptors, **124**, 181
in localization of superoxide dismutase in rat liver, **105**, 112
Digoxin
serum, chemiluminescence immunoassay, **133**, 401
Dihydroberberine
in delivery of berberine to brain, **112**, 381
Dihydrofolate
^{3}H-labeled, in radioenzymatic assay for dihydrofolate reductase, **122**, 346
Dihydrofolate reductase
bacterial, crystal intensity, multiwire area detector diffractometry, **114**, 468
gene amplification, FACS analysis, **108**, 237
radiometric assays, **122**, 346, 360
Dihydrofolate synthase
bacterial
assay, **122**, 351
properties, **122**, 356, 359
purification, **122**, 351
Dihydrolipoamide dehydrogenase
immobilized, bioluminescent assays with, **133**, 198
–luciferase luminescent assay, **133**, 200
spectrophotometric assay, **133**, 201
17β-[(1,4-Dihydro-1-methyl-3-pyridinylcarbonyl)oxy]androst-4-en-3-one
oxidation kinetics in biological fluids, determination, **112**, 394
quaternary derivative, enzymatic hydrolytic cleavage and oxidation rate, determination, **112**, 393
synthesis, **112**, 392
Dihydropterins
quinonoid, detection by liquid chromatography/electrochemistry system, **122**, 306
Dihydropteroylpolyglutamate
preparation and analysis, **122**, 335
7,8-Dihydroxy-7,8-dihydrobenzo[*a*]pyrene
epoxidation, detection during lipid peroxidation, **105**, 347
3,4-L-Dihydroxyphenylalanine, *see* Dopa
1,25-Dihydroxyvitamin D_2
extraction methods, **123**, 181
radioreceptor assay, **123**, 176
1,25-Dihydroxyvitamin D_3
assay with rabbit intestinal cytosol-binding protein, **123**, 185
cytoreceptor assay, **123**, 190
extraction methods, **123**, 179, 186, 192
radioreceptor assay, **123**, 176
1,25-Dihydroxyvitamin D_3 receptors
characterization with monoclonal antibodies, **123**, 199
Diimidoesters
in structural analysis of lipoproteins, **128**, 613
3,5-Diiodotyrosine aminotransferase
rat kidney, assay, **113**, 668
Diisodecyl phthalate
in oil-droplet assay of phagocytosis, **132**, 194
Diketogluconic acid
HPLC, **122**, 3
Diketogulonic acid
HPLC, **122**, 3
Dilauroylphosphatidylcholine
bilayer intermolecular forces, direct measurement by osmotic stress, **127**, 412
d-cis-Diltiazem
^{3}H-labeled, labeling of calcium channels with, **109**, 539, 542
Dilution analysis
in antibody affinity determination, **121**, 472
limiting, *see* Limiting dilution analysis
2,3-Dimercaptopropanol
inhibition of cytochrome bc_1, **126**, 270
Dimerization
neuropeptides, **103**, 49
Dimethyl adipimidate
synthesis of chemiluminogenic protein conjugates, **133**, 557
Dimethylallylpyrophosphate
purification, **110**, 142
synthesis, **110**, 136, 341
Dimethylallylpyrophosphate:5′-AMP dimethylallyltransferase
tobacco crown gall tumor
assay, **110**, 344

properties, **110**, 346
purification, **110**, 343, 346
Dimethylallylpyrophosphate:L-tryptophan dimethylallyltransaferase, *see* Tryptophan dimethylallyltransferase
Dimethylallyl*trans*transferase
chicken liver
assays, **110**, 179
purification, **110**, 180
eukaryotic
assay, **110**, 145
interconvertible forms, **110**, 149
photoaffinity labeling, **110**, 128
photolabile substrate analogs
preparation, **110**, 125
properties, **110**, 128
properties, **110**, 151
purification, **110**, 146, 150
human liver
assay, **110**, 156
properties, **110**, 161
purification, **110**, 159
Micrococcus luteus
assay, **110**, 189
properties, **110**, 191
purification, **110**, 189
pumpkin fruit
assays, **110**, 167
properties, **110**, 170
purification, **110**, 169
tomato fruit chromoplasts
assay, **110**, 213
extraction and partial purification, **110**, 217
properties, **110**, 219
yeast
assays, **110**, 179
purification, **110**, 183
7-Dimethylaminonaphthalene-1,2-dicarbonic acid hydrazide
in cellular chemiluminescence assays, **133**, 502
1,5-Dimethylaminonaphthoyl
as fluorescent probe for ADP/ATP carrier protein and uncoupling protein, **125**, 618
Dimethylglycine dehydrogenase
rat liver
assay, **122**, 256
properties, **122**, 259
purification, **122**, 257
ε-*N*-Dimethyl-L-lysine
chromatographic properties, **123**, 294
^{14}C-labeled, synthesis, **123**, 293
5,5-Dimethyl-1-pyrroline *N*-oxide
chemistry, **105**, 200
in HPLC–electrochemical detection of oxygen free radicals, **105**, 233
preparation, storage, and use, **105**, 232
synthesis, **105**, 201
4,4-Dimethylsterols
radiolabeled
incubation with marine invertebrates and plant tissues, **111**, 320
preparation, **111**, 316
Dimyristoyl phosphatidylcholine
volume-perturbation kinetic calorimetry, **130**, 536
Dinitrophenol
specific hybridomas, production, **121**, 261
2,4-Dinitrophenylhydrazine
in quantitation of protein oxidative modification, **107**, 374
Dinoflagellates
bioluminescence
particulate activity, **133**, 310
soluble components, **133**, 312, 318, 324
cell culture and harvesting, **133**, 309
1,1′-Dioctadecyl-3,3,3′,3′-tetramethylindocarbocyanine perchlorate
fluorescent labeling of lipoproteins, **129**, 562
7,8-Diol
as chemiluminescent probes of singlet oxygen
emission spectra, **133**, 577
in hepatic microsomal extracts, **133**, 580
properties, **133**, 572
1,2-Dioxetanes
as labels for thermochemiluminescent immunoassays
applications, **133**, 538
principles, **133**, 531
synthesis, **133**, 534
Dioxygen
binding to hemoproteins, resonance Raman spectroscopy, **130**, 383
detection, **105**, 8

reactions with paraquat radical, **105**, 525
reduction products, **105**, 4
Dipalmitoyl phosphatidylcholine
bilayer water content, analysis by neutron diffraction and X-ray techniques, **127**, 226
volume-perturbation kinetic calorimetry, **130**, 536
Dipeptidase
rat kidney
assays, **113**, 472
properties, **113**, 480
purification, **113**, 475
substrate specificity, **113**, 482
Dipeptide permease
bacterial, deficient mutants, isolation, **125**, 370
Dipeptides
inhibition of γ-glutamyl transpeptidase, **113**, 425
as substrate for γ-glutamyl transpeptidase, **113**, 426
valine-containing, in selection of bacterial mutants deficient in dipeptide permease, **125**, 370
Dipeptide system
for *in vitro* determination of chloroplast gene translation initiation sites, **118**, 309
Diphenylhydantoin, *see* Phenytoin
Diphenylisobenzofuran
in assay for oxygen radical generation by prostaglandin H synthase, **105**, 414
2,2-Diphenylpropylamine
–Sepharose
applications, **133**, 105
binding to luciferase, **133**, 102
synthesis, **133**, 101
Diphthamide
in elongation factor 2
ADP-ribosylation by diphtheria toxin, **106**, 379, 392
biosynthesis *in vitro*, **106**, 388
demonstration, **106**, 395
properties, **106**, 385
purification from yeast, **106**, 381
Diphtheria toxin
A chain
coupling to antibodies, **112**, 218
isolation, **112**, 212
radiolabeling, **112**, 213
ADP-ribosylation of elongation factor 2, **106**, 379, 392
–antibody conjugates
biological evaluation, **112**, 221
cytotoxicity, **112**, 222
physicochemical characterization, **112**, 220
preparation, **112**, 213
purification and storage, **112**, 219
coupling to antibodies
with chlorambucil *N*-hydroxysuccinimidyl ester, **112**, 214
with *N*-hydroxysuccinimidyl ester of iodoacetic acid, **112**, 217
with *N*-succinimidyl-3-(2-pyridyldithio)propionate, **112**, 215
handling, safety considerations, **112**, 224
purification, **112**, 209
radiolabeling, **112**, 213
Diploptera punctata, *see* Cockroach
3,3′-Dipropylthiodicarbocyanine
as indicator of membrane potential
applications, **124**, 125
general methodology, **124**, 124
Diquat
mutagenicity testing with *Salmonella typhimurium* TA102, **105**, 250
Displacement chromatography
proteins, **104**, 113
with ampholytes as displacers, **104**, 115
with carboxymethyldextran displacers, **104**, 118
Distillation
in preparation of high-purity water, **104**, 398
removal capabilities and limitations, **104**, 400
steam, *see* Steam distillation
Disuccinimidyl suberate
in affinity crosslinking of receptors for insulin and insulin-like growth factors, **109**, 180
Disulfides
interferon αA, irreversible derivatization, **119**, 253
protein
conformational forces in formation, **131**, 104

enzymatic formation and isomerization, **107**, 281
enzymatic reduction–oxidation by thioredoxin, **107**, 295
formation, breakage, and rearrangement, **131**, 84
formation pathway
chemistry, **107**, 307
reconstruction, **107**, 328
intermediate species, trapping, **107**, 315; **131**, 87
intermolecular reagents, **131**, 100
nonenzymatic formation and isomerization, **107**, 301
as probe of protein folding pathway, **131**, 89
trapped intermediate species
analytical techniques, **107**, 318
identification and characterization, **107**, 323
–thiol exchange
differential effects on enzyme activities, **107**, 336
kinetic competence, **107**, 345
in redox control of enzyme activities, **107**, 330
–thiol ratio, variation *in vivo*, **107**, 332
Diterpenes
in termite soldiers
distribution and importance, **110**, 417
identification, **110**, 421
isolation, **110**, 419
variation, quantification, **110**, 424
5,5′-Dithiobis(2-nitrobenzoic acid)
inhibition of cytochrome bc_1, **126**, 271
Dithionite
in bioluminescence assay for aldehydes, **133**, 190
reduction of cytochrome *b*-245, spectrophotometric assay, **132**, 385
Dithiothreitol
oxidized, in disulfide bond formation, **107**, 313
Dityrosine
characterization, **107**, 384
production *in vitro*, **107**, 379
and trityrosine, halogenated, separation, **107**, 427
DNA
A octamer, constrained–restrained least-squares structure refinement, **115**, 294
aqueous solutions, computer simulations
ion positions, **127**, 128
water structure, **127**, 128
bacteriophage λ, operator–protein interactions, DNase footprint titrations, **130**, 132
cauliflower mosaic virus, cloning vectors
construction, rationale for, **118**, 661
introduction and expression in plants, **118**, 660, 663
preparation, **118**, 659
chloroplast
from cybrid plants, restriction endonuclease analysis, **118**, 605
identification in mtDNA preparations, **118**, 449
identification of promoter sequences for RNA polymerase, **118**, 267
isolation, **118**, 167
structural analysis by restriction site mapping, **118**, 174
complementary
aequorin
construction, **133**, 291
expression in *Escherichia coli* extracts, **133**, 294
identification, **133**, 293
apoaequorin
construction, **133**, 301
identification, **133**, 306
screening, **133**, 303
apolipoprotein A-I, isolation and characterization, **128**, 713
apolipoprotein A-II
preparation and sequence analysis, **128**, 749
screening, **128**, 746
apolipoprotein C-II, isolation, **128**, 789
apolipoprotein C-III, isolation and characterization, **128**, 713
apolipoprotein E, cloning and characterization, **128**, 801
apolipoprotein-specific
hybridization assay with, **128**, 680
synthesis, **128**, 672

for detection of viral RNAs
hybridization techniques with, **118**, 732
preparation, **118**, 723
firefly luciferase
construction, **133**, 5
expression in *Escherichia coli*, **133**, 10
screening, **133**, 8
to Ia antigens
cloning, **108**, 577
identification, **108**, 575
synthesis, **108**, 573
infectious viral clones
production, **118**, 707
properties, **118**, 705
in vitro transcription, **118**, 712
interferon-specific
hybridization *in situ* with, **119**, 478
manipulation, **119**, 240
preparation, **119**, 475
to low abundance red light-regulated sequences, identification, **118**, 369
low-density lipoprotein receptor, cloning, **128**, 895
neuropeptide, inserts in vaccinia virus vector
construction, **124**, 298
detection, **124**, 300
expression analysis, **124**, 305
purification, **124**, 301
structural analysis, **124**, 303
transcription, **124**, 303
neuropeptide precursor, detection with synthetic oligonucleotides, **124**, 309
phytochrome, identification, **118**, 369
^{32}P-labeled, synthesis techniques, **109**, 580
radiolabeled probes, in detection of neuropeptide mRNA, **124**, 497, 510, 534
cosmid clones, apolipoprotein E gene-containing, characterization, **128**, 818
double helices, intermolecular forces, direct measurement by osmotic stress, **127**, 414
elongated, molecular weight determination by viscoelastometry, **117**, 94
hydrated, calorimetric analysis, **127**, 149
insertion in apolipoprotein A-I gene, familial segregation, **128**, 725
internally repeated sequences
detection, **128**, 774, 777
duplication history, inference, **128**, 781
plant
hybridization to membrane filters, **118**, 66
isolation from purified nuclei, **118**, 57
restriction analysis, **118**, 65
restriction site polymorphisms, as genetic markers, **118**, 82
variation in repeated sequences, quantitative analysis, **118**, 76
plant mitochondrial
cloning, **118**, 458, 476
from cybrids, restriction endonuclease analysis, **118**, 609
examination for plastid DNA contamination, **118**, 449
gene-containing restriction fragments, identification, **118**, 473
isolation, **118**, 439, 446, 455, 472
physical mapping, **118**, 456
protein-coding open reading frames, identification, **118**, 479
–protein complexes, preparation for electron microscopy, **118**, 485
recombinant cosmid clones
integrity, analytical techniques, **118**, 467
preparation, **118**, 463
restriction mapping, **118**, 466
restriction mapping, **118**, 456, 458
sequence reiterations, mapping, **118**, 468
sequencing strategies, **118**, 477
plasmid
in labeling of interferon with [^{35}S]methionine, **119**, 293
structure determination by transient electric birefringence, **117**, 206
transfection and amplification in CHO cells, **119**, 398
transformation of plant protoplasts, **118**, 581

restriction length polymorphisms
detection, **128**, 730
as genetic markers, **128**, 735
Rhizobium
deletion creation with Tn*10*, **118**, 653
isolation, **118**, 528
recombinant, mini-prep procedures, **118**, 526
site-specific insertion of selective, markers, **118**, 649, 656
Tn*5* marker exchange methods, **118**, 522
transposon mutagenesis, **118**, 644
Rous sarcoma virus, as expression vehicle for human interferon β, **119**, 383
synthesis in lymphocytes, hybridoma supernatant effects, *Mycoplasma* interferences, **121**, 481
synthetic duplexes, in site-specific mutagenesis of bacterio-opsin gene, **125**, 206
DNA-directed RNA polymerase
Escherichia coli, phage-induced ADP-ribosylation, **106**, 418
promoter sequences for, identification in chloroplast extract, **118**, 267
DNA nucleotidylexotransferase
in assay for
polypeptide β_1, **116**, 247
thymosin α_1, **116**, 247
thymosin β_4, **116**, 253
thymosin fraction 5 and 5A, **116**, 231
DNA polymerase
chloroplast
assay, **118**, 187
properties, **118**, 192
purification, **118**, 188
DNA polymerase α
spinach
assay, **118**, 98
properties, **118**, 102
purification, **118**, 99
DNase, *see* Deoxyribonuclease I
Dodecyl-β-D-maltoside
binding to cytochrome-*c* oxidase, measurement, **126**, 51
Dogfish
isolation of erythrocyte marginal bands, **134**, 242, 245
Dolichol
assays, **111**, 209
isolation from animal tissues, **111**, 207
radiolabeled, preparation, **111**, 203
Dolichol esterase
bovine brain
assay, **111**, 480
properties, **111**, 482
Dolichol kinase
bovine brain
assay, **111**, 471
membrane-bound, properties, **111**, 473
Dolichyl-phosphatase
bovine brain
assay, **111**, 477
properties, **111**, 478
Dolichyl phosphate
assays, **111**, 214
isolation, **111**, 213
radiolabeled, preparation, **111**, 204
Dolichyl phosphate mannose
enzymatic synthesis, **123**, 64
separation, **123**, 65
Dolichyl phosphate phosphatase, *see* Dolichyl-phosphatase
Dolichyl-pyrophosphate phosphatase
bovine brain
assay, **111**, 548
properties, **111**, 551
solubilization, **111**, 550
Dopa
colorimetric assays, **107**, 399
isolation techniques, **107**, 405
Dopamine
delivery systems, analytical techniques, **112**, 392
hypothalamic, secretion into hypophysial portal blood, **103**, 607
control, **103**, 612
differential release into various portal vessels, **103**, 610
effect of aging, **103**, 616
radioligands, in measurement of dopamine receptors in anterior pituitary, **103**, 577
Dopamine receptors
in anterior pituitary, direct radioligand measurement
on dispersed cells, **103**, 585
in membranes, **103**, 579

in solubilized preparations of membranes, **103**, 587
Dose–response curves
for interferon induction
analysis, **119**, 112
generation, **119**, 107
for plant IVR on tobacco mosaic virus replication, **119**, 733
Dosimetry
and pulse radiolysis, **105**, 175
Dot immunobinding assay
monoclonal antibodies, **121**, 484
Doxorubicin
incorporation into hydrophilic albumin microspheres
assessment, **112**, 32
effect of protein concentration, **112**, 40
–polylysine conjugates
preparation, **112**, 274
properties, **112**, 277
reductively activated, generated oxygen radicals, ethidium binding assay, **105**, 532
release from albumin microspheres *in vitro*, **112**, 22
uptake by hydrophilic albumin microspheres, **112**, 21
Draft tube fermenter
for cultivation of agarose-entrapped hybridoma cells, **121**, 358
Drosophila melanogaster, *see* Fly
Drug delivery systems
albumin microsphere, **112**, 3
conjugate, **112**, 207
enzyme-controlled, **112**, 484
microencapsulation, **112**, 3
microsealed, fabrication and performance, **112**, 461
Ocusert ocular, structure, function, and release rates, **112**, 509
polymer, **112**, 399
prodrug, **112**, 309
Drug design
computationally directed biorational, peptides
case studies, **124**, 7, 13
conformational energy methods, **124**, 4
design methods, **124**, 7
Dye marking
neuroendocrine cells *in vitro*, **103**, 138
Dyes, *see also specific dyes*
pH-sensitive, in measurement of proton translocation in model membranes, **127**, 476
Dynein
bovine sperm
assay, **134**, 321
characterization, **134**, 322
extraction, **134**, 321
Chlamydomonas flagella
analytical techniques, **134**, 297
purification, **134**, 293
sea urchin egg cytoplasm
assays, **134**, 327, 339
properties, **134**, 334, 350
purification, **134**, 329, 344
Tetrahymena thermophila cilia
characterization, **134**, 313
isolation, **134**, 309
–microtubule complex, formation, **134**, 316
Dynein ATPase
assays, **134**, 301, 321, 327, 339
inhibition by vanadate, **134**, 485
kinetic analysis
chemical quench-flow methods, **134**, 692
stopped-flow methods, **134**, 678
Dysopsonins
overview, **132**, 306

E

EBV, *see* Epstein–Barr virus
Ecdysone
^{3}H-labeled, purification, **111**, 454
Ecdysone 3-epimerase
Manduca sexta
assay, **111**, 439
isolation, **111**, 438
properties, **111**, 440
substrate specificity, **111**, 440
Ecdysone 20-monooxygenase
Manduca sexta
assay, **111**, 455
isolation, **111**, 457
properties, **111**, 458
Ecdysone oxidase
blowfly
assays, **111**, 420

characterization, **111**, 423
isolation, **111**, 422
products, **111**, 426
substrate specificity, **111**, 425
Ecdysteroid carrier proteins
arthropod
assay, **111**, 430
irradiation, **111**, 432
isolation, **111**, 435
properties, **111**, 436
Ecdysteroid conjugates
enzymatic hydrolysis, **111**, 417
extraction and purification, **111**, 412
identification, **111**, 413
Ecdysteroid receptors
K_c cell
characteristics, **111**, 465
preparation, **111**, 462
tissue sources, **111**, 459
Ecdysteroids
biosynthesis and metabolism
experimental approaches, **111**, 384
overall pathways, **111**, 377
free and conjugated
mass spectrometry, **111**, 401
NMR spectroscopy, **111**, 405
purification and analysis, **111**, 387, 391
Edetic acid
in elution of absorbed cells from monolayers, **108**, 131
EDTA, *see* Edetic acid
EEDQ, *see* *N*-Ethoxycarbonyl-2-ethoxy-1,2-dihydroquinoline
Eggplant
purification of mitochondrial DNA, **118**, 439
Eggs, *see also* Oocytes
avian, isolation of riboflavin-binding protein, **122**, 227
sea urchin
fertilized, isolation of spindles, **134**, 194
unfertilized
induction of monasters and cytasters, **134**, 196
preparation for isolation of tubulin, **134**, 131
purification of
cytoplasmic dynein, **134**, 329, 344
microtubules and microtubule-associated proteins, **134**, 119
trout, fertilized, occurrence of Glu–Lys crosslink in chorion, **107**, 256
Elastase
crystal intensity, multiwire area detector diffractometry, **114**, 470
proteolysis of tubulin, **134**, 182
Electrical capacitance
in measurement of water penetration depth into lipid bilayers, **127**, 514, 515
Electrical conductivity
water in ice phases, **127**, 317
Electroblotting
lipoproteins on gradient gels, **128**, 430
monoclonal and site-directed polyclonal antibodies to *lac* permease of *Escherichia coli*, **125**, 459
proteins, for detection of antibodies, **121**, 501
Electrochemical detection
in catecholamine assay, **103**, 470
with liquid chromatography, pterins in urine and tissue, **122**, 301
Electrodes
in electrophysiological assay of neuronal cells, **103**, 96
for intracellular recording in dissociated tissue culture, **103**, 122, 129; **124**, 219
oxygen, *see* Oxygen electrodes
for patch recording, **124**, 231
pH-sensitive, in measurement of proton flux in model membranes, **127**, 474
in proton motive force estimations across membranes, **127**, 547
in pulse-driven ATP synthesis, **126**, 640
for single-channel electrophysiological analysis, **103**, 151
Electromyographic assay
thymopoietin, **116**, 280
Electron density maps
fast Fourier transform calculations
algebra for, **115**, 328
from diffraction patterns, **115**, 324
organization of, **115**, 333
high-resolution, interpretation, **115**, 199
low-resolution, interpretation, **115**, 189

medium-resolution, interpretation, **115**, 194
noisy, molecular boundary determination, **115**, 99
noncentrosymmetrical, phasing, **115**, 87
phase-refined, calculation, **115**, 114
and physical models, optical matching
construction and use, **115**, 149
corrections to original citations, **115**, 151
history, **115**, 145
proteins, automated analyses
applications, **115**, 216
methods, **115**, 207
review, **115**, 221
Electron density modification
for phase refinement/phase extension
applications, **115**, 77
cyclical nature, **115**, 82
function selection, **115**, 78
in noncentrosymmetrical difference maps, **115**, 87
phase changes, **115**, 86
with proteins, **115**, 86
refinement progress, **115**, 83
requirements, **115**, 86
Electron microscopy
antibody inhibition of myosin bipolar filament assembly, **134**, 444
antibody–myosin complexes, **134**, 440
with autoradiography, hepatocyte subcellular peptide hormone receptors, **109**, 229
Chlamydomonas basal bodies, **134**, 260
cytochrome-*c* oxidase reconstituted in lipid vesicles, **126**, 85
F_1-ATPase single molecules and crystals, **126**, 434
follicular cells of anterior pituitary and pars tuberalis, **124**, 248
freeze-fractured and etched lipoproteins, **128**, 461
frozen hydrated specimens, **127**, 719
horseradish peroxidase in neuroendocrine cells, **103**, 193, 199, 207
horseradish peroxidase uptake by leukocytes, **108**, 340
immunostained kinetochore–centromere fraction from metaphase chromosomes, **134**, 271
low-density lipoprotein receptors, **129**, 201
microtubule–neurofilament networks with affinity grids, **134**, 553
NADH dehydrogenase two-dimensional crystals, **126**, 345
negatively-stained lipoproteins, **128**, 442
artifacts, **128**, 450
comparison with physical–chemical techniques, **128**, 455
neuropeptide-secreting cells, **124**, 463
occupied insulin receptors with biologically active monomeric ferritin–insulin as marker, **109**, 200
peptide–lipid complexes, **128**, 636
in phagosome–lysosome fusion assay, **132**, 260
plant mtDNA–protein complexes, preparation methods, **118**, 485
plasmacytoma and lymphoblastoid cells, **121**, 125
receptor-mediated binding and internalization of human interferon, **119**, 337
receptosomes from human KB cells, **109**, 268
triple-labeled neuroendocrine cells, **103**, 143
ubiquinol–cytochrome-*c* reductase membrane crystals, **126**, 194
Electron paramagnetic resonance
calmodulin, **102**, 103
in measurement of H^+/OH^- fluxes across phospholipid vesicles, **127**, 502
Electron spin resonance
ADP/ATP carrier protein, **125**, 638
applications of nitroxides, **105**, 217
in assay of *in situ* radicals, **105**, 416
free radicals
in lyophilized tissues, **105**, 454
in wet and frozen tissues, **105**, 451
in measurement of
cell water viscosity, **127**, 366
hydroxyl radical production by neutrophils, **105**, 395
melanin free radicals, **105**, 479
for radical detection, spin destruction methods, **105**, 215

Electron transfer reactions
pulse radiolysis study, **105**, 179
Electron transport
generated proton motive force, critical assessment of estimations, **127**, 538
Electrophoresis, *see also specific electrophoretic techniques*
antigen-specific, in lymphocyte separation, **108**, 189
carotenoproteins, **111**, 223
O-phosphates, **107**, 13
in separation of
anterior pituitary cells, **103**, 277
lymphocytes, **108**, 181
Electrophoretic light scattering
in analysis of membrane surface charge density, **127**, 691
Electrophysiology
intracellular, recording in dissociated tissue culture
cell impalement techniques, **103**, 124
culture conditions, selection, **103**, 116
electrodes for, **103**, 122
electronics for, **103**, 119
grounding and interference in, **103**, 121
iontophoresis techniques, **103**, 124
micromanipulators for, **103**, 118
microperfusion techniques, **103**, 125
patch electrodes for, **103**, 129
recording solutions for, **103**, 116
staining techniques, **103**, 127
support equipment for, **103**, 116
vibration control, **103**, 117
voltage clamp methods, **103**, 128
neuroendocrine cells *in vitro*, recording and data analysis, **103**, 137
neurohormonal communication in mammals, assay, **103**, 93
single-channel
analytical patch clamp technique, **103**, 147
data analysis, **103**, 163
Electrostatic interactions
in proteins
calculation, **130**, 413
effect on structural stability, **130**, 437
ELISA, *see* Enzyme-linked immunosorbent assay
Ellipses
in position-sensitive area detectors
location determination, **114**, 536
parameters, **114**, 538
Elongation factors
EF-1, from plants
assays, **118**, 144
properties, **118**, 152
purification, **118**, 143, 145
EF-2, diphthamide in
ADP-ribosylation by diphtheria toxin, **106**, 379, 392
biosynthesis *in vitro*, **106**, 388
demonstration, **106**, 395
properties, **106**, 385
purification from yeast, **106**, 381
Euglena gracilis chloroplast
assay, **118**, 297
properties, **118**, 308
purification, **118**, 296, 305
spinach chloroplast
assays, **118**, 297
properties, **118**, 300, 303
purification, **118**, 296, 298
wheat germ
assays, **118**, 114, 135
functional properties, **118**, 127
physical properties, **118**, 124
purification, **118**, 110, 116, 130
Embedding
cytoskeletal antigens in cell monolayers, **134**, 596, 597
Embryo
chicken
bursectomy methods
hormonal, **108**, 5
surgical, **108**, 6
primary cultures
interferon induction, **119**, 120, 124
preparation, **119**, 115
sea urchin
culturing conditions, **134**, 206
mitotic apparatus
metaphase, isolation, **134**, 209
native, characterization, **134**, 212
Enamines
as amine prodrugs, **112**, 359
Encephalomyocarditis virus
in antiviral assay of interferons, **119**, 534

Endocytosis
asialoglycoproteins, postinternalization events, **109**, 242
interferon, receptor-mediated, measurement techniques, **119**, 332
low-density lipoproteins
defective cell mutants, selection methods, **129**, 227
receptor-mediated, visualization, **129**, 212
Endopolygalacturonase, *see* Polygalacturonase
Endorhamnosidase
P22 tailspike, temperature-sensitive folding mutants
characterization, **131**, 260
folding pathway, **131**, 256
isolation, **131**, 258
mutational analysis, **131**, 252
properties, **131**, 255
purification, **131**, 259
β-Endorphin
antibody production, **103**, 548
chromatographic separation, **103**, 556
extraction techniques, **103**, 552
radioimmunoassays, **103**, 559
radiolabeled, preparation, **103**, 550
Endorphins
immunocytochemistry, **103**, 670
Endosomes
human KB cell
characterization, **109**, 265
isolation, **109**, 258
low-density, from rat liver
biochemical properties, **109**, 252
depleted Golgi fractions, preparation, **109**, 256
preparation, **109**, 247
subcellular markers for, current status, **109**, 255
subfractionation, **109**, 252
Endothelial cell growth supplement
promotion of hybridoma cell growth, **121**, 293
Endothelial cells
human vascular
cytotoxic effects of neutrophil-generated free radicals, assay, **105**, 381
derived B-cell hybridoma growth promoters
comparison with other growth promoters, **121**, 12
overview, **121**, 10
preparation, **121**, 11
enzymatic generation of free radicals, assay, **105**, 383
preparation, **105**, 379
Endotoxins
chemiluminescent detection in macrophages, **133**, 515
Energy profiles
gramicidin A ion channels, computational methods, **127**, 250
Energy transfer
isochromic, thermochemiluminescent labeling, **133**, 544
Enkephalin receptors
image-intensified fluorescence microscopy, **103**, 219
Enkephalins
antibody production, **103**, 548
chromatographic separation, **103**, 556
computationally directed biorational drug design, **124**, 13
extraction techniques, **103**, 552
immunocytochemistry, **103**, 670
radioimmunoassays, **103**, 559
radiolabeled, preparation, **103**, 550
–Texas Red conjugate
preparation, **103**, 220
properties, **103**, 221
Enterotoxin A
staphylococcal
induction of
human interferon γ, **119**, 89
murine interferon γ in cultured spleen cells, **119**, 146
preparation, **119**, 145
Enterotoxin B
staphylococcal
induction of interferon γ production by leukocytes, **119**, 85
and OK-432, induction of human interferon γ, **119**, 93
Enterotoxins
Escherichia coli, ADP-ribosyltransferase assays, **106**, 414

demonstration, **106**, 413
effect on adenylate cyclase system, **106**, 411
Enthalpies
for protein–protein and protein–ligand interactions, determination, **130**, 64
Enucleation
neutrophils, **132**, 251
Enzymatic hydrolysis
glycoproteins, **106**, 69
NADase-catalyzed, spectrophotometric assay, **122**, 169
native and soluble proteins, **106**, 69
pantetheine
assays, **122**, 36
visualization on polyacrylamide gels, **122**, 38
polyprenol pyrophosphates, **110**, 153
wool and morphological components, **106**, 61
extent and distribution of diagnostic components, **106**, 62
inhibitors originating from substrate preparations, **106**, 67
Enzyme immunoassay
active center-based, with bacterial luciferase, **133**, 248
antigen-capture, for monoclonal antibodies
comparison with competitive inhibition assay, **121**, 545
methodology, **121**, 543
competitive inhibition, for monoclonal antibodies
comparison with antigen-capture assay, **121**, 545
methodology, **121**, 542
optimization, **121**, 543
S-100 protein, **102**, 259
solid-phase, with avidin–biotin, for epitope analysis, **121**, 463, 466
ultrasensitive
application to biological samples, **103**, 428
with sample purification, **103**, 431
without sample purification, **103**, 430
general considerations in, **103**, 413
history and development, **103**, 409
optimal conditions, determination, **103**, 420
Enzyme immunofiltration assay
for antibody specificity determination, **121**, 529
for Ig class and concentration determination, **121**, 533
overview, **121**, 527
with staining, overview, **121**, 529
Enzyme-linked immunosorbent assay
acetylcholine receptor antibodies in hybridoma supernatants, **121**, 103
antibodies to cellular and soluble antigens, **121**, 786
apolipoprotein (a)-specific antigen in plasma, **129**, 173
apolipoproteins, **129**, 428
calcium-binding protein, **102**, 294
H_2O_2 production by macrophages, **132**, 414
hybridoma antibody isotypes, **116**, 154
IgA, monoclonal hybridoma-produced, **121**, 48
IgM in human × human hybridoma supernatants, **121**, 115
immunoglobulin allotypes, **116**, 166
immunoglobulin isotypes, **116**, 152
interleukin 2, **116**, 501
interleukin-2 antibodies, **116**, 507
α_2-macroglobulin antibodies in hybridoma supernatants, **121**, 64
monoclonal antibodies produced by hybridomas, **121**, 32
in monoclonal antiidiotypic antibody screening, **116**, 185
nitroblue tetrazolium reduction by macrophages, **132**, 417
plant antiviral protein, **119**, 737
plant viruses, **118**, 742, 771
serum retinol-binding protein, **123**, 85
superoxide production by macrophages, **132**, 410
tobacco mosaic virus multiplication in protoplasts, **119**, 748
vitamin D-induced Ca^{2+}-binding protein, **123**, 154
Enzymes, *see also specific enzymes*
active
computer-controlled scanning gel chromatography, **117**, 116

hydrodynamic properties, determination by sedimentation, **117**, 97
affinity chromatography, **104**, 21, 22
buffers for
broad-range, preparation, **104**, 410
deuterated, preparation and application, **104**, 412
effect of organic solvents, **104**, 411
at low temperature, **104**, 412
metal ion content, control, **140**, 409
in NMR studies, **104**, 414
preparation factors, **104**, 408
selection, **104**, 404
in gels
activity retention, **104**, 417
localization, **104**, 419
in modification of human and murine interferons, **119**, 234
multienzyme systems incorporated in artificial cells
applications, **112**, 196
cofactor recycling in, **112**, 200
preparation, **112**, 196
regulation by thiol/disulfide exchange, **107**, 330
Eosinophils
human, purification, **132**, 232, 241
Eosin Y
in neutrophil viability assessment, **132**, 96, 136
Epidermal growth factor
effect on measurement of interferon antiproliferative effects, **119**, 645
radioiodinated, reaction with receptor
in detergent-solubilized preparations, **109**, 108
on intact cells, **109**, 102
in membrane fractions, **109**, 107
Epidermal growth factor receptors
associated kinases
ATP-binding sites in, analysis with 5′-*p*-fluorosulfonylbenzoyl adenosine, **109**, 816
identification as EGF receptor, **109**, 824
binding assays, **109**, 101
Epidermis, *see* Skin
Epinephrine
oxidation to adrenochrome, in detection of superoxide anion, **105**, 395, 430
Epithelial cells
binding of [^{125}I]EGF, assay, **109**, 102
bovine corneal, production of extracellular matrix, **103**, 288
Epitope ratio analysis
cell-surface antigens, **121**, 797
Epitopes
apolipoprotein
assignment, **128**, 535, 542
linkage to protein function, **128**, 551
hidden, on Ig light chains, recognition, **116**, 117
hololipoprotein, expression, **128**, 545
on human chorionic gonadotropin subunits, determination with monoclonal antibodies, **109**, 648
on hybridoma-secreted antibodies, specificity, **121**, 203
solid-phase enzyme immunoassay with avidin–biotin, **121**, 463, 466
solid-phase radioimmunoassay, **121**, 461
Epoxidation
7,8-dihydroxy-7,8-dihydrobenzo[*a*]pyrene during lipid peroxidation, detection, **105**, 347
Epoxide hydrolase
vertebrate and lepidopteran, assays
with cholesterol epoxide, **111**, 309
with juvenile hormone, **111**, 305
with *trans*- and *cis*-stilbene oxide, **111**, 308
Epoxyfarnesoic acid methyltransferase
Hyalophora cecropia
preparation, **111**, 541
product isolation and analysis, **111**, 543
Epoxyfarnesyl diazoacetate
in photoaffinity labeling of juvenile hormone-binding proteins, **111**, 516
criteria for binding specificity, **111**, 511
properties, handling, and repurification, **111**, 514
EPR, *see* Electron paramagnetic resonance
Epstein–Barr virus
immmortalization of B cells
application to human hybridoma production, **121**, 154
overview, **121**, 141
procedure, **121**, 145

induced nuclear antigen, in detection of lymphoblastoid cells, **121**, 125

transformation of B cells, **121**, 170

Equilibrium dialysis

in measurement of

fatty acid binding to serum albumin, **128**, 328

nucleotide binding to uncoupling protein, **126**, 501

Equilibrium partitioning

in measurement of

fatty acid binding to serum albumin, **128**, 328

ligand binding to proteins, **117**, 342

Equine chorionic gonadotropin, see Pregnant mare serum gonadotropin

Erwinia herbicola

ice nucleation in plant leaves, characteristics, **127**, 731

Erythorbic acid

HPLC, **122**, 3

Erythrocytes

bovine

in indirect rosetting assay for lymphocytes, **121**, 727

purification of

Cu,Zn-superoxide dismutase, **105**, 89

skeletal protein brain analogs, **134**, 55

chicken, purification of

ferrochelatase, **123**, 403

uroporphyrinogen decarboxylase, **123**, 415

coating of culture surfaces for phagocytosis studies, **132**, 213

cytotoxic effects of chloramines, **132**, 590

human

Ca²⁺ transport

assay in inside-out vesicles, **102**, 54

calmodulin effects, **102**, 59

coupling to immunoglobulins, **121**, 550

cytolysis, continuous assay, **132**, 491

and desensitized turkey, fusion

desensitization in normal and fused cells, **109**, 374

kinetics of cAMP production after, **109**, 373

methodology, **109**, 367

transfer of β-adrenergic receptor during, **109**, 371

in monoclonal antibody assays, **121**, 552, 553, 554

preparation of inside-out vesicles, **102**, 49

purification of

D-aspartic acid β-methyl ester from carboxyl methylated proteins, **106**, 330

Cu,Zn-superoxide dismutase, **105**, 89

Nₛ and Nᵢ coupling proteins, **109**, 446

protein methylase II, **106**, 304

removal from phagocyte preparations, **132**, 229

intracellular viscosity, analysis by rotational motion of spin labels, **127**, 248

malonaldehyde detection by HPLC, **105**, 326

nonmammalian vertebrate

cytoskeletal system, **134**, 232

isolation of marginal band, **134**, 235, 250

ovine

coupling to

haptenated (Fab′)₂ fragments, **121**, 345

protein A, **121**, 350

ligand-coated, preparation, **132**, 208

magnetic separation with polyacrolein-activated magnetic microspheres, **112**, 162

protein A-sensitized, in agglutination assay for immunoglobulins, **121**, 556

purification of glutathione peroxidase, **107**, 593

porcine, purification of skeletal protein brain analogs, **134**, 55

protein A-coated

in detection and quantitation of surface IgG, **108**, 411

preparation, **108**, 409

rat, purification of

γ-glutamylcysteine synthetase, **113**, 391

glutathione peroxidase, **107**, 610

turkey, purification of arginine-dependent ADP-ribosyltransferases, **106**, 433, 436
Erythro-9-[3-(2-hydroxynonyl)]adenine
applications, **134**, 481
cytotoxicity, **134**, 481
synthesis, **134**, 479
Escherichia coli
active transport, binding protein-dependent
protein components, **125**, 281, 289
regulation, **125**, 288
system types, **125**, 279
alanine carrier protein
purification, **125**, 486
reconstitution in proteoliposomes, **125**, 490
transport assay, **125**, 490
arabinose transport system, H^+-linked
assay
in intact cells, **125**, 388
in right-side out subcellular vesicles, **125**, 399
associated membrane proteins, identification
by gel electrophoresis, **125**, 411, 419
after radiolabeling, **125**, 404, 417
genetics, **125**, 403
properties, **125**, 389
reconstitution into liposomes, **125**, 425
-based system, for *in vitro* determination of chloroplast gene translation initiation sites, **118**, 309
chemotaxis-related carboxymethyltransferase, properties and assays, **106**, 310
cloned genes from bioluminescent bacteria
expression optimization, **133**, 76
light emission, **133**, 75
mRNA analysis, **133**, 78
screening, **133**, 71
transfer methods, **133**, 70
cytotoxic effects of chloramines, **132**, 592
damage induced by oxygen free radicals, assessment methods, **105**, 410
ecf mutants
isolation, **125**, 181
properties, **125**, 182
revertants, isolation, **125**, 182
eup mutants
generation and screening, **125**, 188
mapping, **125**, 189
properties, **125**, 190
storage, **125**, 190
expression of cloned genes for
apoaequorin, **133**, 294
firefly luciferase, **133**, 10
interferon α, **119**, 364, 434, 449, 471
interferons with lambda P_L promoter, **119**, 366, 376
ribulose-bisphosphate carboxylase, **118**, 420
ribulose-bisphosphate carboxylase large subunit, **118**, 431
galactose transport system, H^+-linked
assay
in intact cells, **125**, 388
in right-side out subcellular vesicles, **125**, 399
associated membrane proteins, identification
by gel electrophoresis, **125**, 411, 419
after radiolabeling, **125**, 404, 417
genetics, **125**, 401
properties, **125**, 389
reconstitution into liposomes, **125**, 425
glutamine synthetase, regulation, **107**, 185
glycerol facilitator, optical assay, **125**, 468
H^+-ATPase mutants, genetic analysis, **126**, 558, 595
intracellular pH
homeostasis mechanism, **125**, 347
measurement, **125**, 343
steady-state conditions and pH_0 transients, **125**, 345
ion extrusion systems
antiporter activity, *in vitro* measurements, **125**, 334, 336
energy coupling *in vivo*, determination, **125**, 331
lac permease
characterization, **125**, 440
monoclonal antibodies, preparation, **125**, 453
purification, **125**, 430

reconstitution into proteoliposomes, **125**, 440
site-directed polyclonal antibodies, preparation, **125**, 453
lac permease mutants
assays, **125**, 227
gene cloning and orientation in recombinant plasmids, **125**, 227
production by oligonucleotide-directed site-specific techniques, **125**, 217, 228
purification, **125**, 218
screening, **125**, 224
sequencing, **125**, 225
malE mutants, Ca^{2+}-treated
maltose-binding protein entry and function, influencing factors, **125**, 296
maltose transport characteristics, **125**, 297
melibiose transport system
assay, **125**, 379
cation cotransport, measurement, **125**, 380
reconstitution, **125**, 385
membrane vesicles
binding protein-dependent glutamine transport, analysis, **125**, 306
preparation, **125**, 305, 334
mobilizable plasmids
application as broad host range vectors, **118**, 657
mobilization in gram-negative bacteria, **118**, 641, 654
transposon carrier
random mutagenesis of gram-negative bacteria, **118**, 644
site-specific insertion in gram-negative bacteria, **118**, 649, 656
outer membrane, permeability measurement, **125**, 266
peptide transport
deficient mutants, isolation, **125**, 366
measurement, **125**, 370
periplasmic binding protein-dependent transport systems, Ca^{2+}-dependent reconstitution
applicability for whole cells, **125**, 300
assays, **125**, 295
influencing factors, **125**, 296
in spheroplasts and treated cells, **125**, 291
plasmid vectors, viral DNA-containing
construction, rationale for, **118**, 661
introduction and expression in plants, **118**, 660, 663
preparation, **118**, 659
polyhydric alcohol transport
complementation assays, **125**, 479
defective mutants, selection, **125**, 476, 477, 478
transport assays, **125**, 483
protein export, role of protonmotive force, **125**, 133
protein localization with pIN III secretion cloning vectors, **125**, 138
purification of
adenylosuccinate synthetase, **104**, 109
amidophosphoribosyltransferase, **113**, 270
aminoacyl-tRNA synthetase, **118**, 218
asparaginase, **113**, 614
aspartate ammonia-lyase, **113**, 620
L-aspartate α-decarboxylase, **113**, 591
aspartate kinases, **113**, 597
aspartate-semialdehyde dehydrogenase, **113**, 600
carbamyl-phosphate synthetase, **113**, 309
CTP synthetase, 283
cytochrome *d* terminal oxidase complex, **126**, 89, 97
cytochrome-*o* oxidase, **126**, 115
cytochrome *o* terminal oxidase complex, **126**, 95
F_0 portion of ATP synthase, **126**, 569, 588
F_1 subunits of ATP synthase, **126**, 590
F_1F_0 portion of ATP synthase, **126**, 583
fumarate reductase, **126**, 381, 385
glutamate-aspartate transaminase, **113**, 84, 86
L-glutamate decarboxylase, **113**, 12
glutamate synthase, **113**, 328
glutamine-binding protein, **125**, 303
glutamine synthase, **113**, 221
glutamyl-tRNA synthetase, **113**, 43, 47, 57
glutaredoxin, **113**, 528, 535
GMP synthetase, **113**, 274

H^+-ATPase, **126**, 545
inosine 5′-monophosphate dehydrogenase, **104**, 109
D-lactate dehydrogenase, **126**, 374
L-leucyl-tRNA:protein transferase, **106**, 200
methionine sulfoxide reductase, **107**, 355
methionyl-tRNA formyltransferase, **106**, 144
NAD synthetase, **113**, 298
porin for crystallization, **125**, 318
recombinant interferon $\alpha 2$, **119**, 168
recombinant interferon αA, **119**, 154
recombinant interferon β, **119**, 178, 183
recombinant interferon γ, **119**, 204
regulatory protein P_{II}, **113**, 228
uridylyltransferase and uridylyl-removing activities, **113**, 231
S-30 extract, in labeling of interferon with [^{35}S]methionine, **119**, 292
sugar transport mutants, isolation and characterization with transposons, **125**, 157
synthesis of radiolabeled glutathione, **113**, 567
temperature-sensitive mutant, in phagocytic killing assay, **132**, 525
toxin ADP-ribosyltransferase
assays, **106**, 414
demonstration, **106**, 413
effect on adenylate cyclase system, **106**, 411
xylose transport system, H^+-linked
assay
in intact cells, **125**, 388
in right-side out subcellular vesicles, **125**, 399
associated membrane proteins, identification
by gel electrophoresis, **125**, 411, 419
after radiolabeling, **125**, 404, 417
genetics, **125**, 404
properties, **125**, 389
ESR, *see* Electron spin resonance
Esterase
juvenile hormone-associated
metabolite identification, **111**, 488
partition assay, **111**, 490
TLC assay, **111**, 493
in phagocyte staining, **132**, 134, 175
Esterification
cholesterol in plasma, in assay for lecithin–cholesterol acyltransferase, **129**, 775
Estradiol
plasma, chemiluminescence immunoassay, **133**, 398
transdermal application, rate-controlled system design, **112**, 453
Estriol
releasing intrauterine device, structure, function, and release kinetics, **112**, 508
Estrogen
in biological samples, bioluminescent assay, **133**, 209
function in isolated luteal cells, **109**, 304
Ethane
exhaled
collection, **105**, 306
concentration, **105**, 308
gas chromatography, **105**, 309
as index of lipid peroxidation, advantages and disadvantages, **105**, 310
from isolated hepatocytes and subcellular organelles, assay, **105**, 315
from isolated perfused liver
collection, **105**, 312
gas chromatography, **105**, 314
production rates, calculation, **105**, 314
Ethanol
acidic, in fractionating precipitation of leukocyte interferon preparations, **119**, 28
in biological samples, continuous-flow assay, **133**, 238
microsomal oxidation, role of oxidant radicals, **105**, 516
precipitation of calmodulin, **102**, 16
Ethanolamine
as supplement in preparation of SFH medium, **121**, 282
Ethene
formation from 4-methylthio-2-oxobutanoic acid, in hydroxyl radical detection, **105**, 17
1,N^6-Ethenoadenine dinucleotide
NADase-catalyzed reactions, spectrophotometric assay, **122**, 169
Ethidium, *see* Homidium

Ethidium bromide
–acridine orange, in neutrophil viability assessment, **132**, 96, 138
N-Ethoxycarbonyl-2-ethoxy-1,2-dihydroquinoline
in coupling monoclonal antibodies to microspheres, **112**, 79
effect on mitochondrial ADP–ATP carrier protein binding, **125**, 663
modification of F_1-ATPase carboxyl groups, **126**, 719, 755
7-Ethoxycoumarin *O*-deethylase
assay for analysis of interferon effects on hepatic drug metabolism, **119**, 724
Ethylcellulose
ferromagnetic mitomycin C-containing microcapsules
clinical applications, **112**, 149
preparation, **112**, 144
mitomycin C-containing microcapsules
clinical applications, **112**, 149
intraarterial targeting, **112**, 146
preparation, **112**, 140
peplomycin-containing microspheres
clinical applications, **112**, 149
intraarterial targeting, **112**, 146
preparation, **112**, 146
Ethyl chloroformate
in coupling of *p*-hydroxyphenylacetic acid to small oligopeptides, **103**, 438
Ethylene
production, in measurement of hydroxyl radical production by neutrophils, **105**, 395
–vinyl acetate copolymers, as macromolecule delivery systems, **112**, 400
Ethylene glycol bis(succinimidyl succinate)
in crosslinking of neuropeptides, **103**, 53
1-Ethyl-2-(3-(1-ethyl-naphtho[1,2-*d*]thiazolin-2-ylidene)-2-methylpropenyl)naphthol[1,2-*d*]thiazolium bromide
in staining of phosphoproteins in gels, **104**, 453
N-Ethylmaleimide
in chloroplast phosphate transport protein labeling, **125**, 718
effect on mitochondrial ADP–ATP carrier protein binding, **125**, 663
in H^+–sugar transport protein identification, **125**, 404, 409
in membrane phosphate transport protein identification, **125**, 718
Ethyl methane sulfate
in preparation of *Rhizobium* H_2 uptake negative mutants, **118**, 529
Ethylmorphine *N*-demethylase
assay for analysis of interferon effects on hepatic drug metabolism, **119**, 722
N-Ethyl-*N*-nitroso-*p*-toluenesulfonamide
in diazoethane synthesis, **126**, 622
synthesis, **126**, 620
N-Ethyl-5-phenylisoxazolium 3′-sulfonate, *see* Woodward's reagent K
Euglena gracilis
isolation of chloroplasts, **118**, 256
purification of chloroplast elongation factor Tu, **118**, 296, 305
EXAFS, *see* Extended X-ray absorption fine structure
Excitotoxins
induction of axon-sparing lesions in hypothalamus, **103**, 381, 393
as neuroendocrine probes, **103**, 379
direct intracranial injection techniques, **103**, 391
provocative approach, basis for, **103**, 383
role of blood–brain barrier and age, **103**, 383
selection, **103**, 383
systemic ablative approach, **103**, 388
systemic provocative approach, **103**, 389
Exocytosis
by neutrophils, assessment methods, **132**, 267
Extended X-ray absorption fine structure
in measurement of metal–ligand distances
applications, **117**, 424
data interpretation, **117**, 443
historical overview, **117**, 414
theoretical considerations, **117**, 417
Extraction
adenine nucleotides from cells, **133**, 32
ATP from microbial and somatic cells, **133**, 14

carotenoids, **105**, 156; **111**, 120
carotenoproteins, **111**, 219, 239, 242,
　　244
catecholamines and neuropeptides from
　　tissues and incubation media, **103**,
　　498
chloramines, **132**, 582
chloroplast and cytoplasmic tRNAs from
　　polyacrylamide gel, **118**, 225
chloroplast polypeptides from leaves,
　　118, 364
corticotropin-releasing factor from tis-
　　sues and biological fluids, **103**, 575
in determination of protein phosphorami-
　　dates, **107**, 35
DNA–protein complexes from plant
　　mitochondria, **118**, 486
β-endorphin from various tissues, **103**,
　　552
enkephalins from various tissues, **103**,
　　552
fluorescent damage products of lipid
　　peroxidation from tissue samples,
　　105, 338
isoprenoids, **111**, 152, 154
microtubule-associated proteins from
　　cultured cells, **134**, 139
neuropeptides from tissues, **103**, 72
osteocalcin from bone powder, **107**,
　　520
phospholamban from sarcoplasmic
　　reticulum, **102**, 270
plant cell wall components, **118**, 8
plant prolyl hydroxylase for *in vitro*
　　assay, **107**, 362
radioactive sterol metabolites, **111**, 325
sulfated proteins with phenol, **107**, 208
ubiquinones from biological samples,
　　105, 149
vitamin E from blood and tissues, **105**,
　　136, 139, 140
Eye
　bovine, purification of interstitial retinol-
　　binding protein, **123**, 102
　human, occurrence of Glu–Lys crosslink
　　in lens tissue, **107**, 255
　retina, *see* Retina
Ezrin
　intestinal microvillus, purification, **134**,
　　34

F

FACS, *see* Fluorescence-activated cell
　　sorter
Factor XIII
　human
　　assay, **113**, 364
　　properties, **113**, 369
　　purification, **113**, 366
FAD (flavin adenine dinucleotide)
　^{3}H-labeled, in FAD synthetase assay,
　　122, 239
　luminometric assay
　　associated inactivation of apo-D-
　　amino-acid oxidase, **122**, 190
　　methodology, **122**, 186
FAD synthetase, *see* FMN adenylyltrans-
　　ferase
Falling-ball viscometry
　in analysis of microtubule–neurofilament
　　networks, **134**, 549
　in assay for gelsolin, **134**, 4
　principle, **134**, 547
Farnesyl pyrophosphate
　purification, **110**, 143
　synthesis, **110**, 137
Farnesylpyrophosphate synthetase, *see*
　　Dimethylallyl*trans*transferase
Farnesyl*trans*transferase
　porcine liver
　　assay, **110**, 185
　　properties, **110**, 187
　　purification, **110**, 185
　pumpkin fruit
　　assays, **110**, 167
　　properties, **110**, 170
　　purification, **110**, 169
Fascin
　extraction from actin gels, **134**, 17
　isolation from sea urchin egg, **134**, 15
Fast atom bombardment mass spectrometry
　short-chain acylcarnitines, **123**, 268, 271
Fast Blue
　in protein immunostaining on nitrocellu-
　　lose, **121**, 850
　as retrograde tracer for neuropeptide-
　　specific pathways, **103**, 665
Fatty-acid reductase
　Photobacterium phosphoreum
　　assays, **133**, 173

properties, **133**, 177
purification, **133**, 175
Fatty acids
binding to serum albumin
data analysis, **128**, 333
measurement, **128**, 327
structure effects, **128**, 335
diene, autoxidation chemistry, **105**, 273
hydroxy, in detection of lipid peroxidation products, **105**, 292
–serum albumin solutions, preparation, **128**, 321
triene and tetraene, autoxidation chemistry, **105**, 276
Fatty alcohols
analysis by bacterial bioluminescence assay, **133**, 194
Fc receptors
phagocytic, for opsonic immunoglobulins detection and quantitation, **132**, 291
overview, **132**, 290
T-cell, detection and characterization, **116**, 406
Ferricytochrome *c*
reduction, in assay for
superoxide anion, **105**, 14, 365, 372, 375, 394, 432
superoxide dismutase, **105**, 101
Ferritin
–amatoxin conjugates
biological properties, **112**, 233
preparation by mixed anhydride coupling, **112**, 231
–IgG conjugates
applications, **121**, 776
preparation, **121**, 774
–insulin conjugates, biologically active monomeric
in electron microscopic analysis of insulin receptors, **109**, 200
preparation and characterization, **109**, 189
–low-density lipoprotein conjugates, preparation, **129**, 203
Ferrochelatase
bacterial
assay, **123**, 409
properties, **123**, 413
purification, **123**, 410
bovine liver
assay, **123**, 402

properties, **123**, 405
purification, **123**, 403
chicken erythrocyte
assay, **123**, 402
purification, **123**, 403
murine liver
assay, **123**, 402
purification, **123**, 403
Ferroelectric compounds
proton motion, **127**, 279
α-Fetoprotein
antibody dilution analysis, **121**, 476
immunochemiluminometric assay, **133**, 381
Fetuin
–amatoxin conjugates
biological properties, **112**, 233
preparation by mixed anhydride coupling, **112**, 231
galactose-terminated, –pepstatin conjugates, preparation and applications, **112**, 300
Fibroblasts
binding of [^{125}I]EGF, assay, **109**, 102
chicken embryo, expression of human interferon β with Rous sarcoma viral genome, **119**, 385
equine dermal
cultivation, **119**, 130
interferon induction in, **119**, 132
guinea pig embryo
culture techniques, **119**, 126
interferon induction in, **119**, 127
human
antagonistic effects of interferons and growth factors, **119**, 658
bradykinin-stimulated prostaglandin production
analytical chromatographic techniques, **109**, 493
effects of inhibitors of arachidonate release and metabolism, **109**, 486
radioimmunoassay, **109**, 485
cultured, purification of low-density lipoprotein receptors, **109**, 409, 413
in interferon antiviral assay, **119**, 534
low-density lipoprotein receptors
endocytosis, visualization, **129**, 212
intracellular distribution, **129**, 209

surface membrane distribution, **129**, 207
in proliferation assay for interleukin 1, **116**, 461
quantification of secretory transcobalamin II, **123**, 47
visualization of low-density lipoprotein pathway, **129**, 201
-like cell lines from adipocytes with redifferentiation ability
adipose conversion, **109**, 383
characterization, **109**, 384, 385
establishment and handling, **109**, 378
murine
DNA-transfected, selection by FACS, **108**, 238
growth inhibition by interferon, measurement
effects of growth factors, **119**, 645
general methods, **119**, 643
in interferon antiviral assay, **119**, 534
protein synthesis and cell growth, inhibition by cordycepin analog of 2,5-p_3A_n, **119**, 668
purification of
colony-stimulating factor 1, **116**, 571
dsRNA-dependent protein kinase, **119**, 499
sedimentation profile, **108**, 85
transformants with enhanced sensitivity to human interferon
preparation, **119**, 599
screening, **119**, 605
selection, **119**, 603
viability assay, **132**, 488
murine embryo, monolayers
lymphocyte separation on, **108**, 130
preparation, **108**, 126
rat cartilage, as feeder cells for hybridoma production, **121**, 54
Fibronectin
activation of phagocytosis-promoting capacity of C3 receptors, **132**, 218
opsonic activity, assays, **132**, 337
purification from plasma, **132**, 335
role in opsonization, **132**, 303
Ficoll solutions
for crystalline protein density measurements, **114**, 187
Filaments
amyloid, from human brain
assays, **134**, 391
purification, **134**, 399
structure, **134**, 389
glial, protein separation from neurofilament triplet, **134**, 382
paired helical, from human brain
assays, **134**, 391
purification, **134**, 393
structure, **134**, 389
Filtration
cascade-type, in production and recovery of human immune interferon, **119**, 77
depth, in preparation of high-purity water, **104**, 393
removal capabilities and limitations, **104**, 400
gel, *see* Gel filtration
macroporous membrane, in preparation of high-purity water, **104**, 401
removal capabilities and limitations, **104**, 400
microporous, in analysis of membrane transport in rickettsiae, **125**, 258
Fimbrin
intestinal microvillus, purification, **134**, 26, 32
Firefly luciferase, *see Photinus*-luciferin 4-monooxygenase (ATP-hydrolysing)
Flagella
alkalophilic *Bacillus*, isolation, **125**, 591
Chlamydomonas
dynein purification, **134**, 293
fractionation, **134**, 289
isolation, **134**, 287
removal, **134**, 259
Flagellar motors
proton-driven bacterial
characterization, **125**, 572
energetic analysis, **125**, 579
reconstitution, **125**, 571
sodium-driven, in alkalophilic *Bacillus*, motility measurements, **125**, 584, 586
Flavin
and analogs, HPLC separation, **122**, 200
covalent attachment to flavoproteins
occurrence, **106**, 369
qualitative identification and assay, **106**, 373

synthesis, **106**, 377
redox state in intact tissues, measurement, **123**, 311
Flavin adenine dinucleotide, *see* FAD (flavin adenine dinucleotide)
Flavin 4a-hydroxide
luciferase-bound, formation and properties, **133**, 146
Flavin mononucleotide, *see* FMN (riboflavin 5′-phosphate)
Flavin semiquinone
luciferase neutral
formation and properties, **133**, 141
reaction with superoxide, **133**, 144
Flavobacterium lutescens
purification of L-lysine 6-aminotransferase, **113**, 96
Flavokinase, *see* Riboflavin kinase
Flavoproteins
flavin covalent attachment
occurrence, **106**, 369
qualitative identification and assay, **106**, 373
synthesis, **106**, 377
iron–sulfur, NADH dehydrogenase (ubiquinone)
properties, **126**, 364
resolution, **126**, 363
NADPH oxidase-associated
characterization, **132**, 362
preparation, **132**, 358
Flow cell
construction and use, **114**, 141
Flow cytometry
in analysis of neuroendocrine function, **103**, 227
basic principles, **103**, 228
FACS, *see* Fluorescence-activated cell sorter
large granular lymphocyte subsets, **132**, 452
in measurement of phagocytosis, **132**, 183
T-cell subsets in blood, **121**, 751
Fluorescamine
in detection of *p*-aminobenzoylpoly(γ-glutamates), **122**, 332
in measurement of peptide uptake by bacteria, **125**, 375
Fluorescein
as staining reagent in FACS, **108**, 208
Fluorescein diacetate
in neutrophil viability assessment, **132**, 96, 137
Fluorescein isothiocyanate
in phagocytic assay by fluorescence quenching, **132**, 200
in staining of glycoproteins in gels, **104**, 449
Fluorescence
apolipoproteins
properties, **128**, 517
in solution, **128**, 383
bacterial luciferase 4a-hydroperoxyflavin intermediates, properties, **133**, 135
α-bungarotoxin–luciferin conjugates, properties, **133**, 45
catecholamine neurotransmitters, in histochemical analysis, **103**, 620
dichlorofluorescein, in assay for picomole levels of lipid hydroperoxides, **105**, 352
in phagosome–lysosome fusion assay, **132**, 260
quenching, in assay of phagocytosis, **132**, 199
T-cell subsets labeled with fluorochromes, relative intensity, **121**, 751
tryptophan, peptide–lipid complexes, **128**, 638
Fluorescence-activated cell sorter
in analysis of interferon effects on histocompatibility antigen expression, **119**, 690
analytical data and data processing, **108**, 219
fluorescence analysis, **108**, 212
in gene amplification studies, **108**, 237
in immunochemical and enzymological research, **108**, 239, 240
instrumentation for
historical review, **108**, 199
operation and measurement, theoretical description, **108**, 200
light scatter analysis, **108**, 205
in lymphoid cell analysis, **108**, 232
in measurement of dextran uptake by leukocytes, **108**, 345
in quantitation of epidermal Langerhans cells, **108**, 359
in separation of
anterior pituitary cells, **103**, 287

thymocytes with peanut agglutinin, **108**, 178
sorting procedures, **108**, 225
staining reagents and procedures, **108**, 208
in suppressor T-cell analyses, **116**, 319
Fluorescence correlation spectroscopy
applications, **130**, 457
data analysis and interpretation, **130**, 482
fluorescent probes for, **130**, 478
instrumentation, **130**, 459
theoretical description, **130**, 467
Fluorescence lifetimes
measurement with synchrotron radiation, **130**, 484
Fluorescence microscopy
cellular binding and uptake of lipoproteins, **129**, 563
dextran uptake by leukocytes, **108**, 343
digital imaging, *see* Digital imaging fluorescence microscopy
image-intensified, enkephalin receptors, **103**, 219
Leishmania amastigotes in macrophages, **132**, 618
low-density lipoprotein receptors, **129**, 201
lymphocyte surface immunoglobulins stained with fluorescent antibodies, **108**, 423
T-cell subsets in blood, **121**, 755
Fluorescence photobleaching recovery
applications, **130**, 457
data analysis and interpretation, **130**, 482
fluorescent probes for, **130**, 478
instrumentation, **130**, 459
theoretical description, **130**, 467
Fluorescence polarization
macromolecules, in high-pressure bomb
applications, **130**, 516
instrumentation, **130**, 504
polarizer alignment, **130**, 505
standards, **130**, 506
viscosity effects, **130**, 515
Fluorescence spectroscopy
calmodulin, **102**, 88
ligand–protein interactions, **117**, 385
data analysis, **117**, 403
sources of variation, **117**, 408
theoretical considerations, **117**, 401
macromolecule structural changes
frequency-domain measurements, **131**, 540
lifetime-domain measurements, **131**, 532
time-domain measurements, **131**, 520
Fluorescent analogs
characterization, **134**, 494
conjugation to protein, **134**, 492
preparation, **134**, 488
Fluorescent labeling
actin, **134**, 508
α-actinin, **134**, 510
calmodulin, **102**, 6, 150
lipoproteins, **129**, 848
peptide hormones for receptor photoaffinity labeling, **109**, 143
vinculin, **134**, 510
Fluoride
binding to hemoproteins, resonance Raman spectroscopy, **130**, 393
Fluorochromes
optimal combinations for T-cell epifluorescence microscopy, **121**, 750
Fluorodeoxyuridylate
in chain length determination of tissue folate polyglutamate, **122**, 313
1-Fluoro-2,4-dinitrobenzene
modification of F_1-ATPase tyrosine residues, **126**, 728
Fluorography
chloroplast 32-kDa membrane protein, **118**, 391
for detection of radioactivity in gels, **104**, 460
LHRH precursor, **124**, 331
in tissue folate polyglutamate chain length determination, **122**, 316
Fluorometric titration
urinary riboflavin with apoflavoprotein, **122**, 234
Fluorometry
actin filaments, **134**, 22
in analysis of
interferon effects on histocompatibility antigen expression, **119**, 690
membrane surface charge density, **127**, 684
in assay for
avidin, **122**, 67
biotin, **122**, 67

gelsolin, **134**, 4
5-methyl-L-tetrahydrofolate, **122**, 384
relative membrane potential, **103**, 241
carboxyfluorescein leakage from liposome–lipoprotein complex, **128**, 662
in detection of lipid peroxidation products, **105**, 288
malonaldehyde in biological samples, **105**, 303
in measurement of
ADP-ribose *in vivo* levels, **106**, 483
intracellular pH, **125**, 343
redox state of nicotinamide nucleotides and flavins in intact tissues, **123**, 312
Fluorophores
blue- and yellow-emitting, microspectrophotometric analysis *in situ*, **105**, 342
formed during lipid peroxidation
extraction, **105**, 338
spectrophotometric analyses, **105**, 339
for image-intensified fluorescence microscopy, choice, **103**, 219
5′-*p*-Fluorosulfonylbenzoyl adenosine
^{3}H-labeled
labeling of F_1-ATPases, **126**, 746
synthesis, **126**, 743
preparation, **109**, 818
as probe of ATP-binding sites in hormone receptor-associated kinases, **109**, 819
5-Fluorouracil
incorporation into hydrophilic albumin microspheres
assessment, **112**, 32
effect of protein concentration, **112**, 40
Fly
imaginal disks, demonstration of ecdysteroid receptors, **111**, 462
purification of *N*-acetyltransferase, **103**, 489
salivary glands, preparation of ecdysteroid receptors, **111**, 461
FMN (riboflavin 5′-phosphate)
^{14}C-labeled, in flavokinase assay, **122**, 238
reduced, in injection assay for bacterial luciferase, **133**, 99
FMN adenylyltransferase
radiosubstrate HPLC assay, **122**, 239
Fodrin
mammalian brain
denatured form, purification, **134**, 51
native form, purification, **134**, 44
Folate
transport protein, purification from
L1210 cells, **122**, 267
Lactobacillus casei, **122**, 261
transport-defective *Lactobacillus casei*, **122**, 266
Folate-binding protein:cytosol I
rat liver
assay, **122**, 251
properties, **122**, 254
purification, **122**, 252
Folate oligoglutamate:amino acid transpeptidase
rat liver
activity versus pteroylpolyglutamate concentration, **122**, 370
assay, **122**, 367
purification, **122**, 369
Folate polyglutamate
chain length determination by electrophoresis, **122**, 313
Folic acid
naturally occurring derivatives
extraction from liver, **122**, 271
HPLC, **122**, 271
microbiological assay, **122**, 270
Follicles
guinea pig hair, purification of transglutaminase, **113**, 373
Follicle-stimulating hormone
assembly
detection, **109**, 737
methods for, **109**, 747
deglycosylated
chromatographic purification, **109**, 729
preparation
with anhydrous HF, **109**, 726
with trifluoromethanesulfonic acid, **109**, 733
disassembly
detection, **109**, 737
methods for, **109**, 739

Follicular cells
bovine anterior pituitary and pars tuberalis
culture technique, **124**, 246
ultrastructural analysis, **124**, 248
Follitropin, *see* Follicle-stimulating hormone
Folylpolyglutamates
chain length identification, **122**, 323
cleavage methods, **122**, 324
extraction, **122**, 323
synthesis, **113**, 169
tetrahydro derivatives, preparation and purification, **113**, 180
Folylpolyglutamate synthase
bacterial
assay, **122**, 350
properties, **122**, 356, 359
purification, **122**, 351, 353
porcine liver, properties, **122**, 359
Foot-and-mouth disease virus
induction of bovine interferon β, **119**, 212
Formaldehyde
in albumin microsphere preparation, **112**, 12
aqueous, structural analysis with proximity criterion, **127**, 33
Formate dehydrogenase
Wolinella succinogenes
assays, **126**, 397
properties, **126**, 391
purification, **126**, 389
reconstitution, **126**, 392
Formylmethionyl-tRNA
binding to methionyl-tRNA formyltransferase, stoichiometry, **106**, 151
(6*S*)-5-Formyltetrahydrofolate
labeled at high specific activity
with ^{14}C, preparation, **122**, 311
with ^{3}H, preparation, **122**, 310
10-L-Formyltetrahydrofolate
binding to methionyl-tRNA formyltransferase, stoichiometry, **106**, 149
Formyltetrahydromethanopterin
preparation and properties, **122**, 420
reverse-phase HPLC, **122**, 422
Fourier refinement
$F_o - F_c$, in crystal structures, **115**, 235
minimum variance, for protein phases
computer programs for, **115**, 21
correlations, **115**, 20
figure of merit, **115**, 18
theoretical background, **115**, 16
Fourier transform
fast, calculations for electron density maps
algebra for, **115**, 328
from diffraction patterns, **115**, 324
organization of, **115**, 333
Fourier transform infrared spectroscopy
bacteriorhodopsin carboxyl groups, protonation studies, **127**, 659
proton transport in membrane proteins, detection methods, **127**, 344
resolution-enhanced
applications, **130**, 302, 305
aqueous protein solutions, **130**, 300
solvent denaturation studies, **130**, 308
theoretical description, **130**, 292
retinoids, **123**, 92
Fractionation
bacterial cell lysate with ammonium sulfate, **133**, 104
Chlamydomonas cells, in isolation of basal bodies, **134**, 259
digitonin, in localization of superoxide dismutase in rat liver, **105**, 112
immunoaffinity, poly(ADP-ribosyl)ated domains of chromatin, **106**, 518
lipoproteins after sequential flotation ultracentrifugation, **128**, 166
Percoll gradient, in isolation of nonlysosomal non-Golgi intermediate vesicles for analysis of peptide hormone receptors, **109**, 227
subcellular, in localization of superoxide dismutase in rat liver, **105**, 106
Free radicals
assay by ESR, **105**, 416
in lyophilized tissues, ESR study, **105**, 454
in melanin
magnetic parameters and spin concentrations, **105**, 479
reaction kinetics and mechanisms, **105**, 483
structure determination, **105**, 482

oxidative
as probe in study of
active sites in enzymes and viruses, **105**, 494
structural aspects of enzymes and proteins, **105**, 499
radiolytic generation, **105**, 491
secondary, reaction selectivity, determination, **105**, 492
transient absorption spectra, as indicators of sites of attack, **105**, 493
role in carcinogenesis, analysis with *in vitro* cell cultures, **105**, 464
in wet and frozen tissues, ESR study, **105**, 451
Freeze-etching
cryofixed lipoproteins, **128**, 459
Freeze-fracturing
cryofixed lipoproteins, **128**, 459
and deep etching, quick frozen lung slices, **134**, 605
in evaluation of intracellular ice damage, **127**, 704
microtubules, **134**, 627, 631
Freezing
cytotoxic T-cell clones, **132**, 473
hybridoma cells, **103**, 467; **121**, 291, 419
quick, *see* Quick freezing
water in plants, NMR analysis, **127**, 761
Freezing front technique
in cell surface tension determination
methodology, **132**, 82
overview, **132**, 44
French press
in subfractionation of thylakoid membranes, **118**, 326
Fructose-1,6-bisphosphatase
binding to AMP, assay by equilibrium gel penetration, **117**, 345
FTIR, *see* Fourier transform infrared spectroscopy
Fucosterol 24,28-epoxide
synthesis, **111**, 348
Fumarate reductase, *see* Succinate dehydrogenase
Fungi, *see also specific fungi*
killing by phagocytes, assay, **132**, 523
Funiculosin
inhibition of cytochrome bc_1, **126**, 268
Furans
in singlet oxygen trapping, **105**, 40, 396
Furosine
in detection of nonenzymatic protein glycosylation, **106**, 87
Fusion
bacterial spheroplasts with plant protoplasts, **118**, 585
bovine lymphocytes with murine hybridoma, **121**, 245, 261, 262
cellular, virus-induced, analysis of interferon effects, **119**, 702
EBV-transformed B cells with human–mouse hybridoma, **121**, 168
hybridomas with hybridomas, **121**, 214
lymphocytes with hybridomas, **121**, 212
lymphocytes with myeloma cells
frequency, **121**, 8
polyethylene glycol-mediated, **121**, 3
methodology, **121**, 12, 18, 29, 46, 56, 61, 98, 184, 196, 241, 294, 315
membranes, in analysis of hepatic glucagon receptor function, **109**, 346
mitochondrial inner membranes
assessment, **125**, 44
methodology, **125**, 42
normal human and desensitized turkey erythrocytes
desensitization in normal and fused cells, **109**, 374
kinetics of cAMP production after, **109**, 373
methodology, **109**, 367
transfer of β-adrenergic receptor during, **109**, 371
phagosome–liposomes, fluorescence assay, **132**, 257
plant protoplasts, **118**, 578
spleen cells with plasmacytoma cells, **121**, 177, 237

G

Galactose
H^+-linked transport system in *Escherichia coli*
assay
in intact cells, **125**, 388
in right-side out subcellular vesicles, **125**, 399

associated membrane proteins, identification
by gel electrophoresis, **125**, 411, 419
after radiolabeling, **125**, 404, 417
genetics, **125**, 401
properties, **125**, 389
reconstitution into liposomes, **125**, 425
-terminated fetuin, –pepstatin conjugates, preparation and applications, **112**, 300
β-Galactosidase
encapsulated in crystallized carbohydrate microspheres
biological activity, assay, **112**, 121
preparation, **112**, 120
in vitro release studies, **112**, 122
as endosome marker, determination, **109**, 267
in immunoassay of S-100 protein, **102**, 259
lacZ gene fusions, in analysis of transport proteins, **125**, 150
linked antibodies, in protein detection on nitrocellulose, **121**, 855
Galactosyltransferase, *see* N-Acetyllactosaminide α-1,3-galactosyltransferase
Gamma irradiation
in acrolein microsphere preparation, **112**, 151
for grafting polyacrolein onto preformed microspheres, **112**, 155
in poly(methyl methacrylate) nanoparticle preparation, **112**, 130
Ganglion
isolated cells from dispersed retinal cells, identification, **103**, 363
superior cervical, explant preparation, **103**, 336
Gas chromatography
acyclic isoprenoid hydrocarbons, **111**, 176
carbocyclase products, **110**, 409
ethane
from exhaled air samples, **105**, 309
from isolated hepatocytes and subcellular organelles, **105**, 315
from isolated perfused liver, **105**, 314
in identification of amino-terminal myristyl blocking groups in proteins, **106**, 208
isoprenoids, **111**, 149
pentane
from isolated hepatocytes and subcellular organelles, **105**, 315
from isolated perfused liver, **105**, 314
retinoic acid, **123**, 119, 122
short-chain acylcarnitines, **123**, 270
terpenoids, **111**, 176
Gas chromatography–mass spectrometry
bile acids, **111**, 63
β-carboxyaspartic acid, **107**, 573
cholesterol side-chain cleavage products, **111**, 355
free and conjugated ecdysteroids, **111**, 394
^{18}O content of phosphate derivatives, **126**, 618
short-chain acylcarnitines, **123**, 267
Gas–liquid chromatography
free and conjugated ecdysteroids, **111**, 394
phytosterol dealkylation products, **111**, 350
radioactive sterol metabolites, **111**, 330
sterols, **111**, 16
Gastric juice
human, purification of vitamin B_{12}-binding proteins, **123**, 32
Gastrin
radiolabeled
preparation, **109**, 58
in receptor binding assay, **109**, 59
Gastrin receptors
gastric mucosal, binding assay, **109**, 56
Gastrointestinal tract
prodrug absorption, improvement by enzyme targeting, **112**, 360
in situ methodology, **112**, 372
in vitro methodology, **112**, 370
Gelatin
preparation from rat skin collagen, **132**, 271
Gelatinase
human neutrophil, assay, **132**, 273
Gel chromatography
for determination of asymptotic boundary shapes, **130**, 14
IgG on Sephadex G-100, **121**, 634
in measurement of protein–ligand interactions, **117**, 346

nucleotide-binding uncoupling protein, **126**, 501
scanning, computer-controlled
applications, **117**, 137
design and operation, **117**, 117
historical development, **117**, 116
software for data analysis, **117**, 128
Gel electrophoresis
(ADP-ribosyl)$_n$ proteins, **106**, 466
analytical, in monitoring of somatomedin C purification, **109**, 804
apolipoprotein A-containing lipoproteins, **129**, 138
apolipoprotein E, **128**, 282, 825, 833
apolipoprotein E-containing lipoproteins, **129**, 160
apolipoproteins A-I, A-II, and A-IV, **128**, 244; **129**, 424
in assay for actin amino-terminal acetylation, **106**, 187
bispecific monoclonal antibodies, **121**, 224
calmodulin, **102**, 298
calmodulin-binding polypeptides, **102**, 205
carotenoproteins, **111**, 233
chlorophyll-binding proteins and apoproteins, **118**, 346
chloroplast and cytoplasmic tRNAs, **118**, 223
chloroplast envelope membranes, **118**, 323
chloroplast 32-kDa membrane protein, **118**, 388, 391
chloroplast *in vitro* transcripts, **118**, 261
class I MHC antigens, **108**, 512
covalently labeled hepatic glucagon receptors, **109**, 209
cytochrome *b*-245, **132**, 394
cytokeratin polypeptides, **134**, 361, 367
cytoplasmic dynein, **134**, 328
in detection of
glutamine synthetase adenylylation, **107**, 191
nonenzymatic protein glycosylation, **106**, 82
in determination of
chain length of tissue folate polyglutamate in ternary complex, **122**, 313
optimal pH conditions for ion-exchange chromatography on macroporous supports, **104**, 225, 229
flagellar dyneins, **134**, 302
fluorescently labeled cytoskeletal proteins, **134**, 511
GAT antigen-specific suppressor factors, **116**, 331, 334
L-glutamate decarboxylase, **113**, 6
H-2 antigens, **108**, 456
HLA antigens, **108**, 598
H^+–sugar transport proteins in *Escherichia coli*, **125**, 419
Ia antigens, **108**, 456, 540
IgD and fragments, **116**, 99
IgG-binding factor, **116**, 414
IgG after DEAE Affi-Gel Blue chromatography, **121**, 591
IgG idiotypes after hydroxylapatite chromatography, **121**, 619
IgG subclasses during peptic digestion, **121**, 655
Ig isotypes from bovine × murine hybridomas, **121**, 253
interferon αA labeled with [^{35}S]methionine, **119**, 287
interferon α labeled with ^{125}I-labeled Bolton–Hunter reagent, **119**, 278
interferon antagonist sarcolectin, **119**, 699
interferon β labeled with ^{35}S, **119**, 285
interferons from
human placenta, **119**, 548
transformed rat cells, **119**, 225
interleukin 1, **116**, 477, 487
interleukin 2, **116**, 518
isolated mitotic spindles, **134**, 221
LHRH precursor, **124**, 329
lipoprotein (a)-specific antigen in plasma, **129**, 169
lipoproteins with nondenaturing polyacrylamide gradient, **128**, 417
lymphotoxin, **116**, 443
in measurement of ligand–protein interactions, **117**, 388
membrane-bound inorganic pyrophosphatase, **126**, 543
multimolecular aggregates
apparatus for, **130**, 118
applications, **130**, 128

data analysis, **130**, 125
horizontal slab gels for, **130**, 117
voltage and temperature control, **130**, 123
myosin, **102**, 67
nitrogenase components I and II, **118**, 516, 517
osteocalcin, **107**, 523
5-oxo-L-prolinase components, **113**, 457
peptide–lipid complexes with nondenaturing polyacrylamide gradient, **128**, 637
phospholamban, **102**, 272
photolabeled insulin receptors, **109**, 178
plant mitochondrial RNA, **118**, 495
polyacrylamide
denaturing systems, **104**, 241
dopa proteins, **107**, 408
nondenaturing systems, **104**, 239
phosphoramidate-containing proteins, **107**, 32
proteins, **104**, 244
poly(ADP-ribosyl)ated chromatin, **106**, 517
polypeptides in KB cell receptosomes, **109**, 268
prolactin receptors, **109**, 160
proteins, for detection of antibodies on blots, **121**, 500
proteins synthesized by isolated chloroplasts, **118**, 293
protein subunits of cytochrome bc_1, **126**, 231
Qa and TL antigens, **108**, 554
recombinant human interferon αA, **119**, 158, 164
recombinant human interferon γ, **119**, 206
single-stranded mitochondrial RNAs, **118**, 504
T-cell suppressor factor for mixed leukocyte response, **116**, 424
TGAL antigen-specific helper factor, **116**, 352
thymopoietin, **116**, 285
ubiquitin, **116**, 285
urea-gradient, protein folding intermediates, **131**, 156
Gel filtration
apolipoprotein C, **128**, 290
batch, in determination of ligand–protein binding, **117**, 499
calcineurin on Sephadex G-200, **102**, 237
[^{14}C]pyridoxal 5′-phosphate, **122**, 102
crude membrane extracts, in separation of proteins and excess detergent, **104**, 323
cytochrome-*c* oxidase, **126**, 47, 157, 165
dopa, **107**, 405
enkephalins and β-endorphin, **103**, 556
GAT antigen-specific suppressor factors, **116**, 337
granulocyte colony-stimulating factor, **116**, 609
granulocyte–macrophage colony-stimulating factor, **116**, 594
growth hormone-releasing factor, **124**, 384
hepatic lipase, **129**, 723
IgA component chains, **116**, 64
IgA secretory component, **116**, 70
IgD, **116**, 96
IgG proteolytic fragments, **121**, 666, 669
IgM, **116**, 30
insulin-like growth factors, **109**, 783, 785, 794
interferon antagonist sarcolectin, **119**, 698
interferon γ, **119**, 197
interferon from transformed rat cells, **119**, 222
interleukin 1, **116**, 476, 485
interleukin 2, **116**, 532
interstitial retinol-binding protein, **123**, 110
LHRH secretion granules, **124**, 264
lipoproteins, **129**, 45
lumazine proteins, **133**, 158
mast cell growth factor, **116**, 558
membrane proteins, **104**, 336
mevalonate kinase, **110**, 74
migration inhibitory factor, **116**, 392
monoclonal IgE, **116**, 79
monoclonal Ig light chains, **116**, 106
myeloperoxidase, **132**, 374
neuropeptides, **103**, 73
plant IVR, **119**, 734
polyclonal IgE, **116**, 81
polypeptide β_1, **116**, 240

in protein transfer to bile salt micelles, **104**, 325
prothymosin α, **116**, 261
recombinant human interferon αA, **119**, 160
ribulose-bisphosphate carboxylase subunits, **118**, 413
RNase L, **119**, 496
selenoproteins, **107**, 606, 614
in separation of phospholipid vesicles and detergent micelles, **140**, 326
soluble immune response suppressor, **116**, 401
somatomedins, **109**, 777, 808
T-cell suppressor factor for mixed leukocyte response, **116**, 424
thymopoietin, **116**, 287
thymosin α_1, **116**, 236
thymosin β_4, **116**, 249
thymulin, **116**, 272, 273
ubiquinol–cytochrome-*c* reductase, **126**, 202
Gel overlays
for detection of
calmodulin-binding polypeptides, **102**, 205
protein–protein interactions, **134**, 557, 558, 559
Gels, *see also specific gels*
affinity, for prenyltransferase construction, **110**, 173
enzymes in
activity retention, **104**, 417
localization, **104**, 419
glycoproteins in, staining, **104**, 447
granulated
for high-resolution preparative isoelectric focusing
description, **104**, 263
load capacity, **104**, 265
protein detection techniques, **104**, 269
in preparative isotachophoresis, **104**, 297
phosphoproteins in, staining, **104**, 451
preparation for fluorography, **104**, 465
proteins in
blotting onto paper, **104**, 455
silver staining, **104**, 441
staining with Coomassie Blue, **104**, 439
rehydratable, for high-resolution preparative isoelectric focusing
description, **104**, 265
load capacity, **104**, 265
protein detection techniques, **104**, 269
Gelsolin
cytoplasmic and secreted
assays, **134**, 4
purification, **134**, 6, 7, 8
Gemini viruses
propagation in plants, **118**, 703
properties, **118**, 702
Gene fusions
lacZ, in analysis of transport proteins, **125**, 150
Gene mapping
apolipoproteins
by nucleic acid hybridization with metaphase chromosomes, **128**, 863
with somatic cell hybrids, **128**, 851
genes controlling immune responsiveness to human parathyroid hormone, **109**, 634
lipoprotein transport regulatory genes in murine model, **128**, 883
Lyt-1, Lyt-2, and Lyt-3 antigens, **108**, 667
nitrogen fixation genes from *Rhizobium* with Tn*5*, **118**, 519
Thy-1.1 and Thy-1.2 antigens, **108**, 642
Genes
alt, coliphage T4
encoded ADP-ribosyltransferase
assay conditions and kinetic properties, **106**, 426
mechanism, **106**, 424, 425
physiological function, **106**, 428
properties, **106**, 418
purification, **106**, 427
substrate specificity, **106**, 421
expression, **106**, 418
apolipoprotein
chromosomal location, **128**, 49
expression regulation, **128**, 53
organization and expression, **128**, 42
apolipoprotein A-I
DNA insertion, familial segregation, **128**, 725
isolation and characterization, **128**, 717

apolipoprotein A-I–C-III complex
linkage, **128**, 720
restriction length polymorphisms
analysis, **128**, 740
detection, **128**, 730
as genetic markers, **128**, 735
role in premature atherosclerosis, **128**, 721
apolipoprotein C-II
cDNA clones, isolation, **128**, 789
chromosomal localization, **128**, 796
genomic clones, isolation, **128**, 793
apolipoprotein E
in bacteriophage λ, screening, **128**, 813
in cosmid vector
characterization, **128**, 818
screening, **128**, 816
polymorphism
analytical techniques, **128**, 825
genetic basis, **128**, 844
transfected
expression analysis, **128**, 820
product characterization, **128**, 821
araE, for H^+–arabinose symport system, characterization, **125**, 403
for ATP synthase subunits from photosynthetic bacteria
cloning and sequencing, **125**, 231
transcriptional mapping, **125**, 245
bacterio-opsin
amber mutations, introduction by generalized mutagenesis, **125**, 193
deletion mutagenesis, **125**, 198
oligonucleotide-directed point mutagenesis, **125**, 202
site-specific mutagenesis, **125**, 206
bioluminescence-related, in marine bacteria
expression, **133**, 76
screening, **133**, 71
transfer to *Escherichia coli*, **133**, 70
chloroplast
expression in transcriptionally active extracts, **118**, 253
promoter identification in transcriptionally active extracts, **118**, 267
psbA-encoded product
identification, **118**, 384
structural characterization, **118**, 391
transcription *in vitro*, **118**, 232
translation initiation sites, determination with *in vitro* dipeptide system, **118**, 309
dihydrofolate reductase, amplication, FACS analysis, **108**, 237
direct transfer to plants by protoplast transformation, **118**, 668
energy coupling factor, *Escherichia coli* mutants
isolation, **125**, 181
properties, **125**, 182
revertants, isolation, **125**, 182
eup, *Escherichia coli* mutants
generation and screening, **125**, 188
mapping, **125**, 189
properties, **125**, 190
storage, **125**, 190
galP, for H^+–galactose symport system, characterization, **125**, 401
H^+-ATPase, *Escherichia coli* mutants
altered sequence, determination, **126**, 602
analysis with λ-*unc* transducing phages, **126**, 558
cloning, **126**, 599
mapping, **126**, 597
H_2 uptake, isolation from *Rhizobium*, **118**, 534
hydroxymethylglutaryl-CoA reductase, expression regulation, **128**, 59
Ia antigen, cloning, **108**, 567
interferon α
assay, **119**, 440
characterization, **119**, 363, 444
cloning, **119**, 442
expression in
Escherichia coli, **119**, 364, 437, 449
yeast, **119**, 416
expression vectors, construction, **119**, 437
identification, **119**, 467
isolation
from bovine hepatic DNA, **119**, 465
from lambda phage library, **119**, 434, 444
with short oligonucleotide probes, **119**, 359
mutagenesis *in vitro*, **119**, 403
interferon αA and α2, detection of base substitution between, **119**, 481

interferon αC
expression in *Escherichia coli*, **119**, 471
nucleotide sequence, **119**, 470
interferon β
expression in
CHO cells, **119**, 398
Escherichia coli with lambda P_L promoter, **119**, 366
mammalian cells with Rous sarcoma viral genome, **119**, 383
sequence encoding carbohydrate attachment site, *in vitro* mutagenesis, **119**, 399
interferon γ
expression in
Escherichia coli with lambda P_L promoter, **119**, 366, 376
eukaryotic cells, **119**, 456
yeast, **119**, 416
identification, **119**, 454
primary structure, **119**, 454
β-lactamase, cloning into pIN III-*ompA*, **125**, 143
lacY, *Escherichia coli*
cloning into M13 RF DNA, **125**, 220
oligonucleotide-directed site-specific mutagenesis, **125**, 214
lipoprotein
expression regulation, **128**, 53
organization and expression, **128**, 42
lipoprotein transport-regulating, murine model
identification, **128**, 879
phenotypic variation, **128**, 877
lux, in marine bacteria
cloning, **133**, 83
expression, **133**, 93
identification, **133**, 88
MHC antigen-encoding, in mice, overview, **108**, 437
β_2-microglobulin
allelic variants, **108**, 499
cloning, **108**, 501
expression regulation, **108**, 502
mod, coliphage T4
encoded ADP-ribosyltransferase
assay conditions and kinetic properties, **106**, 426
mechanism, **106**, 424, 425
physiological function, **106**, 428
properties, **106**, 419
purification, **106**, 427
substrate specificity, **106**, 423
expression, **106**, 419
neuropeptide
expression analysis
in discrete brain regions, **124**, 548
by *in situ* hybridization, **124**, 497, 510, 534
with vaccinia virus vector, **124**, 295
transfer methods
long-term assay systems, **124**, 282
transient assay systems, **124**, 287
organelle-encoded
expression analysis in cybrids, **118**, 603
transfer by protoplast fusion, **118**, 595
plant mitochondrial
cloning and sequencing strategies, **118**, 476
expression confirmation, **118**, 482
identification, **118**, 473, 479, 484
ribulose-bisphosphate carboxylase
cloning, **118**, 420
expression in *Escherichia coli*, analytical techniques, **118**, 425, 427
ribulose-bisphosphate carboxylase large subunit
cloning, **118**, 429
expression in *Escherichia coli*, **118**, 431
ribulose-bisphosphate carboxylase small subunit
cloning in phage λ, **118**, 397
expression, analytical techniques, **118**, 407
genomic clones
rapid gel analysis, **118**, 405
screening, **118**, 399
staphylococcal nuclease A, cloning, **125**, 146, 148
transfer in plants with Ti plasmid vectors, **118**, 627
xylE, for H^+–xylose symport system, characterization, **125**, 404
Genotoxicity
evaluation with bioluminescence test, **133**, 266

Gentamicin
induced test with luminous bacteria, **133**, 278
Geranylgeranyl pyrophosphate
^{14}C-labeled, preparation, **110**, 269
Geranylgeranylpyrophosphate synthetase, *see* Farnesyl*trans*transferase
Geranyl pyrophosphate
purification, **110**, 142
radiolabeled, chemical synthesis, **110**, 410
synthesis, **110**, 137
Geranylpyrophosphate synthetase, *see* Dimethylallyl*trans*transferase
Germination
aseptic, seeds, **118**, 540
Giemsa stain
in assay of *Toxoplasma* killing by macrophages, **132**, 634
Glass surfaces
macrophage separation on, **108**, 294
γ-Globulin
–amatoxin conjugates
biological properties, **112**, 230, 233
preparation, **112**, 227, 231
Glomerulosa cells
bovine adrenal
cytosolic Ca^{2+} changes, monitoring, **124**, 127
membrane potential changes, monitoring, **124**, 127
preparation, **124**, 126
rat adrenal, preparation, **109**, 121
Glucagon
effects on
Ca^{2+} fluxes
in isolated hepatocytes, **109**, 554, 555
in perfused liver, **109**, 553
phosphorylase in hepatocytes, **109**, 557
monoiodinated
activation of adenylyl cyclase, **109**, 11
binding to hepatic membranes, **109**, 8, 208
crosslinking to hepatic membrane receptors, **109**, 209
preparation, **109**, 5
purification, **109**, 6
storage, **109**, 8
Glucagon receptors
hepatic
adenylyl cyclase assay, **109**, 11
covalent labeling, **109**, 207
direct *in vitro* assay, **109**, 3
functional analysis by membrane fusion, **109**, 346
wheat germ lectin–Sepharose assay, **109**, 13
induction in transformed MDCK cells
applications, **109**, 360
conditions for, **109**, 357
inhibition of, **109**, 359
Glucagon responsiveness
MDCK cells, **109**, 360
SV40 *tsA* mutant-transformed fetal hepatocytes, **109**, 394
Glucocorticoids
effect on interferon-mediated indoleamine 2,3-dioxygenase induction, **105**, 69
Glucosamine
^{35}S-labeled, in radiolabeling of interferon β, **119**, 401
Glucosamine-6-phosphate synthase
rat liver
assay, **113**, 278
properties, **113**, 281
purification, **113**, 279
Glucose
oxidation, assay for assessment of phagocytic function, **132**, 115, 155
transport in chloroplast envelope, **125**, 714
Glucose oxidase
controlled polymer system, preparation and characterization, **112**, 489
–lactoperoxidase, in ^{125}I labeling of interferons, **119**, 263
in peptide radiolabeling, **124**, 24
Glucose-6-phosphate dehydrogenase
assay for assessment of phagocytic function, **132**, 121, 161
α-Glucosidase
localization in gels by autochromic method with sandwich-type incubation, **104**, 439
Glucosylation
apolipoproteins, immunochemical isolation and identification, **128**, 354

low-density lipoproteins for catabolism studies, **129**, 583
nonenzymatic, lysine residues in albumin, **106**, 88
detection and quantitation, **106**, 95
dissociation kinetics, measurement, **106**, 93
kinetics, determination, **106**, 90
measurement, **106**, 90
Glucosylguanidines
synthesis, **106**, 401
D-Glucuronic acid
attached to cutinase N-terminus by amide linkage
detection, **106**, 210
occurrence, **106**, 216
β-Glucuronidase
assay for assessment of phagocytic function, **132**, 119, 160
human neutrophil, assay, **132**, 274
Glucuronoarabinoxylans
extraction and isolation, **118**, 17
Glutaconyl-CoA decarboxylase
Acidaminococcus fermentans
assay, **125**, 550, 553
properties, **125**, 556
purification, **125**, 552
sodium transport activity, reconstitution into phospholipid vesicles, **125**, 555
Glutamate
enzymatically synthesized
assay with glutamate dehydrogenase, **113**, 266
radiolabeled, filter binding assay, **113**, 267
excitotoxic properties in neuroendocrine research, **103**, 387
fluorometric assay, **113**, 245
D-Glutamate
enzymatic cyclization reaction, **113**, 62
and L-glutamate mixtures, analysis with L-glutamate decarboxylase, **113**, 61
L-Glutamate
optically pure
and D-glutamate mixtures, analysis with L-glutamate decarboxylase, **113**, 61
purification, **113**, 60
radiolabeling, **113**, 61
Glutamate-alanine transaminase, *see* Alanine aminotransferase
D-Glutamate-D-amino acid transaminase
bacterial
assays, **113**, 109, 110
properties, **113**, 112
purification, **113**, 110
L-Glutamate-L-amino acid transaminase
assay and purification, overview, **113**, 64
Glutamate-γ-aminobutyrate transaminase
assays, **113**, 81
natural occurrence, **113**, 80
Glutamate–ammonia ligase
adenylylation, detection methods, **107**, 187
Escherichia coli
catalytic cycle and topographic studies, **113**, 239
purification, **113**, 221
regulation
closed bicyclic cascade, **113**, 213
feedback inhibition, **113**, 235
overview, **107**, 185
oxidative inactivation, **113**, 237
porcine muscle
assay, **113**, 199
inhibitors, **113**, 205
kinetic properties, **113**, 203
physiological role, **113**, 211
purification, **113**, 200
rat liver
assay, **113**, 187
mechanism of action, **113**, 192
properties, **113**, 190
purification, **113**, 188
rat muscle
assay, **113**, 199
inhibitors, **113**, 205
kinetic properties, **113**, 203
physiological role, **113**, 211
purification, **113**, 201
species separation, **107**, 192
Glutamate-aromatic amino acid transaminase, *see* Aromatic-amino-acid aminotransferase
Glutamate-aspartate transaminase, *see* Aspartate aminotransferase
Glutamate-branched-chain amino acid transaminase, *see* Branched-chain amino-acid aminotransferase

Glutamate carboxylase
assays, **107**, 552
problems in, **107**, 561
influencing factors, **107**, 554
inhibitors and stimulators, **107**, 558
preparation techniques, **107**, 553, 562
species, tissue, and subcellular localization, **107**, 560
structural stability, **107**, 559
D-Glutamate cyclase
murine kidney, in purification of L-glutamate, **113**, 60
Glutamate–cysteine ligase
rat erythrocyte, purification, **113**, 391
rat kidney
assays, **113**, 380
inhibition, **113**, 385
mechanism of action, **113**, 389
properties, **113**, 383
purification, **113**, 381
Glutamate decarboxylase
bacterial
assay, **113**, 11
properties, **113**, 12
purification, **113**, 12
pyridoxal phosphate-binding site, isolation, **106**, 130
rat brain
assay, **113**, 3
immunochemical characterization, **113**, 7
localization, **113**, 8
purification, **113**, 4
in separation of radiolabeled D- and L-glutamates, **113**, 61
Glutamate dehydrogenase
bovine liver
affinity precipitation, **104**, 367
in amidophosphoribosyltransferase assay, **113**, 266
assay, **113**, 25
properties, **113**, 18
purification, **113**, 26
Glutamate mutase, *see* Methylaspartate mutase
Glutamate-ornithine transaminase
assays, **113**, 77
natural occurrence, **113**, 76, 78
Glutamate-5-semialdehyde dehydrogenase
bacterial, assay, **113**, 117
γ-Glutamate semialdehyde dehydrogenase, *see* Glutamate-5-semialdehyde dehydrogenase
Glutamate synthase (NADH)
Saccharomyces cerevisiae
assay, **113**, 333
properties, **113**, 336
purification, **113**, 334
Glutamate synthase (NADPH)
Escherichia coli
assay, **113**, 328
properties, **113**, 331
purification, **113**, 328
Klebsiella aerogenes, **113**, 331
assay, **113**, 328
properties, **113**, 331
purification, **113**, 328
Glutamate-tRNA^{Gln} amidotransferase
Bacillus subtilis
assay, **113**, 303
properties, **113**, 305
purification, **113**, 304
Glutamate–tRNA ligase
Bacillus subtilis
assay, **113**, 51
properties, **113**, 53
purification, **113**, 51
Escherichia coli
assays, **113**, 44, 56
properties, **113**, 47, 58
purification, **113**, 43, 47, 57
and regulatory factor, copurification, **113**, 49
Glutamic acid
radiolabeled
in analysis of bacteriorhodopsin carboxyl groups by FTIR, **127**, 651
synthesis, **127**, 655
reactivity with heavy atoms, **114**, 151
L-Glutamic acid-activating enzyme
membrane-bound
assays, **113**, 149
product characterization, **113**, 156
properties, **113**, 165
purification, **113**, 150, 168
Glutamic acid γ-methyl esters
identification as product of bacterial chemotaxis-related methyltransferase, **106**, 316

N^{ε}-(γ-Glutamic)lysine
analytical techniques, **107**, 241
occurrence
in extracellular proteins, **107**, 251
in intracellular proteins, **107**, 252
Glutaminase
Acinetobacter glutaminsificans
assays, **113**, 257
properties, **113**, 261
purification, **113**, 259
porcine brain
assays, **113**, 242
properties, **113**, 253
purification, **113**, 249
porcine kidney
assays, **113**, 242
properties, **113**, 253
purification, **113**, 246
rat kidney
assays, **113**, 242
properties, **113**, 253
purification, **113**, 251
Glutamine
periplasmic binding protein, *Escherichia coli*
in analysis of glutamine transport in isolated membrane vesicles, **125**, 306, 308
assays, **125**, 304
purification, **125**, 303
reactivity with heavy atoms, **114**, 152
Glutamine amidotransferase
natural occurrence and properties, **113**, 263
Glutamine–phenylpyruvate aminotransferase
rat kidney
assay, **113**, 345
purification, **113**, 346
substrate specificity, **113**, 349
Glutamine–pyruvate aminotransferase
rat liver
assays, **113**, 338
isozymic forms, occurrence, **113**, 343
properties, **113**, 342
purification, **113**, 340, 342
Glutamine synthetase, *see* Glutamate–ammonia ligase
Glutamine transaminase K, *see* Glutamine–phenylpyruvate aminotransferase
Glutamine transaminase L, *see* Glutamine–pyruvate aminotransferase
γ-Glutamyl amino acids
separation, **113**, 564
synthesis, **113**, 555
L-γ-Glutamyl-L-α-aminobutyric acid
characterization, **113**, 563
formation, in assay for γ-glutamylcysteine synthetase, **113**, 380
radiolabeled, enzymatic synthesis, **113**, 560
synthesis, **113**, 556
γ-Glutamylcyclotransferase
murine kidney, inhibition *in vivo*, **113**, 580
rat kidney
assay, **113**, 439
properties, **113**, 443
purification, **113**, 439, 441
γ-Glutamylcysteine disulfide
characterization, **113**, 563
preparation, **113**, 558
radiolabeled, enzymatic synthesis, **113**, 560
reduction, **113**, 559
γ-Glutamylcysteine synthetase, *see* Glutamate–cysteine ligase
L-γ-Glutamyl-L-cystine
characterization, **113**, 563
synthesis, **113**, 562
L-γ-Glutamyl-L-glutamine
characterization, **113**, 563
enzymatic synthesis, **113**, 563
L-Glutamyl γ-methyl esters
occurrence in eukaryotic tissues, **106**, 343
γ-Glutamyltransferase
inhibition *in vivo*, **113**, 581
rat kidney
antibody preparation, **113**, 407
assays, **113**, 402, 404
biosynthesis and processing, **113**, 412
inhibition and regulation, **113**, 416, 435
mechanism, **113**, 421, 436
properties, **113**, 408, 414
purification, **113**, 405
specificity and partitioning, **113**, 427
γ-Glutamyl transpeptidase, *see* γ-Glutamyltransferase
Glutamyl-tRNA
as intermediate in yeast mitochondrial protein synthesis, **206**, 152

Glutamyl-tRNA synthetase, *see* Glutamate–tRNA ligase
L-γ-Glutamyl-L-tyrosine
characterization, **113**, 563
synthesis, **113**, 556
Glutaraldehyde
in albumin microsphere preparation, **112**, 12
in coupling monoclonal antibodies to microspheres, **112**, 81
in crystal density measurement, **114**, 182
in hydrophilic albumin microsphere preparation, **112**, 19, 31
neuropeptide conjugation methods, **103**, 449
Glutaredoxin
bovine thymus
assay, **113**, 526
properties, **113**, 539
purification, **113**, 537
Escherichia coli
assays, **113**, 526
properties, **113**, 532, 536
purification, **113**, 528, 535
Glutathione
characterization, **113**, 563
determination in biological samples
chromatographic methods, **113**, 552
enzymatic methods, **113**, 550
enzymatic synthesis, **113**, 567
metabolism, selective modification methods, **113**, 571
oxidized
in detection and measurement of drug-induced hydrogen peroxide formation, **105**, 508
in fixed-time assay for glutathione peroxidase, **105**, 117
in preparation of glutathione disulfide, **113**, 557
radiolabeling, **113**, 569
reduced
in fixed-time assay for glutathione peroxidase, **105**, 115
and glutathione disulfide, enzymatic assay, **105**, 448
transport studies, **113**, 582
Glutathione dehydrogenase (ascorbate)
spinach leaves
assay, **122**, 10
properties, **122**, 12
purification, **122**, 11
Glutathione disulfide
characterization, **113**, 563
determination in biological samples
chromatographic methods, **113**, 552
enzymatic methods, **113**, 550
preparation from glutathione, **113**, 557
release from cells and tissues, assays, **105**, 445
Glutathione peroxidase
assays, **107**, 594
continuous, **105**, 117
fixed-time, **105**, 115, 117
mammalian
catalyzed reactions, **113**, 490
functional role, **113**, 494
major types, **113**, 491
natural occurrence, **113**, 492
subcellular localization, **113**, 492
oxidized forms
differentiation, **107**, 597
preparation, **107**, 597
purification from
ovine blood, **107**, 593
rat erythrocytes, **107**, 611
rat liver, **107**, 610
selenocysteine identification, **107**, 589
selenopeptide preparation, **107**, 617
self-rotation function, **115**, 69
Glutathione reductase (NAD(P)H)
bovine liver
assay, **113**, 485
properties, **113**, 490
purification, **113**, 486
rat liver
assay, **113**, 485
properties, **113**, 490
purification, **113**, 487
yeast, in assay for glutathione and glutathione disulfide in biological samples, **113**, 550
Glutathione synthase
rat kidney
assay, **113**, 393
properties, **113**, 398
purification, **113**, 395
Glutathione synthetase, *see* Glutathione synthase

Glutathione transferase
human liver
assay, **113**, 500
properties, **113**, 503
purification, **113**, 501
mammalian
catalyzed reactions, **113**, 498
overview, **113**, 495
rat liver
assay, **113**, 504
properties, **113**, 507
purification, **113**, 505
rat, nomenclature, **113**, 497
rat testis
assay, **113**, 508
properties, **113**, 510
purification, **113**, 508
Glycerate
transport in chloroplast envelope, **125**, 712
Glyceride
–cysteine modification, in membrane penicillinase
identification criteria, **106**, 366
nature of, **106**, 365
sequence required for, **106**, 368
Glycerol
in biological samples, continuous-flow assay, **133**, 238
facilitated transport in *Escherichia coli*, optical assay, **125**, 468
^{3}H-labeled, in analysis of phospholipid turnover in cultured cells, **124**, 429
Glycerol-3-phosphate dehydrogenase (NAD^+)
localization in gels by simultaneous capture technique, **104**, 434
Glycine
carboxy-terminal, analysis in active ubiquitin, **106**, 258
zwitterion, hydration, Monte Carlo simulation with proximity criterion analysis, **127**, 37
Glycogen phosphorylase
^{31}P NMR studies, **107**, 75
Glycogen synthase kinase
calmodulin-dependent, properties and function, **107**, 90
properties and function, **107**, 94
type 3
assay, **107**, 117
purification, **107**, 127
Glycolate
transport in chloroplast envelope, **125**, 712
Glycolide
–lactide copolymers, preparation, **112**, 102
Glycolysis
assay for assessment of phagocytic function, **132**, 108, 145
Glycopeptides
antifreeze, in cold-water fish
detection, isolation, and purification, **127**, 294
effect on water, **127**, 302
interactions with ice, **127**, 297
α_2-Glycoprotein
human serum, role in opsonization, **132**, 304
Glycoprotein β-D-galactosyltransferase
in hepatic Golgi subfractions, assay, **109**, 222
Glycoproteins
as drug carriers, **112**, 299
in drug targeting, **112**, 306
enzymatic hydrolysis, **106**, 69
in gels, staining, **104**, 447
hydroxyproline-rich, structure, **107**, 369
IgA secretory component, isolation, **116**, 70
mass spectrometric analysis, **106**, 56
Glycosaminoglycans
interactions with lipoprotein lipase, **129**, 742
Glycosylation
apolipoproteins, **128**, 63
cell wall polysaccharides
composition analysis, **118**, 27
linkage analysis, **118**, 30
nonenzymatic protein, measurement, **106**, 77
T-cell suppressor factor for mixed leukocyte response, analysis, **116**, 425
trans-Glyoxal
hydration, Monte Carlo simulation with proximity criterion analysis, **127**, 37

GMP synthetase
Escherichia coli
assay, **113**, 273
properties, **113**, 277
purification, **113**, 274
GnRH, *see* Gonadotropin-releasing hormone
Gold
colloidal, *see* Colloidal gold
–low-density lipoprotein conjugates, preparation, **129**, 204
Golgi apparatus
hepatic
endosome-depleted fractions, preparation, **109**, 256
isolation, **129**, 285
isolation of
apolipoproteins, **129**, 290
nascent lipoproteins, **129**, 291
peptide hormone receptor-rich structures
characterization, **109**, 221, 226
preparation, **109**, 219, 225
subfractionation, **129**, 288
Gonadotropin
human chorionic, *see* Human chorionic gonadotropin
pregnant mare serum, *see* Pregnant mare serum gonadotropin
Gonadotropin-releasing hormone
antagonist analog, radiolabeling, **103**, 46
biotinylated
applications, **124**, 51
preparation, **124**, 49
luciferin-derivatized analogs
activity determination, **124**, 34
binding to anterior pituitary membranes, **124**, 32
bioactivity, **124**, 31
preparation, **124**, 30
in receptor ligand assays, advantages, **124**, 35
sensitivity, **124**, 31
photoreactive
characterization, **103**, 60
nonradioactive, applications, **103**, 63
photolysis reactions, **103**, 62
radiolabeled, applications, **103**, 66
synthesis, **103**, 59
radiolabeled
preparation and purification, **103**, 34
storage, **103**, 45
superactive agonist analogs, radiolabeled
preparation and purification, **103**, 39
quality assessment, **103**, 43
storage, **103**, 45
tritiated, preparation, **103**, 48
Gonadotropin-releasing hormone receptors
autoradiographic localization in brain, **124**, 604
localization and processing analysis, **103**, 67
photoaffinity inactivation, **103**, 63
photoaffinity labeling, **103**, 66
solubilization and assay, **124**, 168
soluble, binding characteristics, **124**, 171
Gonyaulax polyedra
bioluminescence
particulate activity, **133**, 310
soluble components, **133**, 312, 318, 324
cell culture and harvesting, **133**, 309
Grafting
murine skin
methodology, **108**, 23, 27
recipient preparation, **108**, 23
rejection appraisal, **108**, 26
sources and preparation methods, **108**, 21
radiation, in preparation of polyacrolein-activated hybrid beads, **112**, 155
thymic tissue in athymic animals, **108**, 16
Graft-versus-host reaction
in birds, assays, **108**, 35
inducer cells, sources, **108**, 29
in rodents
regional assays, **108**, 31
systemic assays, **108**, 30
Gramicidin A
ion channels, energy profiles, computational methods, **127**, 250
Granules
neutrophil, purification, **132**, 382
Granulocytes
human blood, separation, **108**, 91
progenitors, detection assay, **116**, 554

Graves' disease
IgG from patients, in assay of thyroid-stimulating antibody, **109**, 682
Grayanotoxin
interactions with ion channels in neuroendocrine tissue, analysis, **103**, 402
Grounding
for electrophysiological recording in dissociated tissue culture, **103**, 121
Growth factors, *see also specific growth factors*
fibroblast-derived, effect on measurement of interferon antiproliferative effects, **119**, 645
hybridoma
comparative effects, **121**, 181
preparation, overview, **121**, 177
Growth hormone, *see* Somatotropin
Growth hormone receptors
monoclonal antibodies to
in characterization of hepatic receptors, **109**, 701
identification of Ig class, **109**, 698
production, **109**, 694
screening assay, **109**, 697
specificity studies, **109**, 701
Growth hormone-releasing factor
antiserum preparation, **124**, 380, 391
assays, **124**, 371, 391, 396, 400
dynamics, analysis by push–pull perfusion technique, **103**, 176
extraction methods, **124**, 390
human pancreatic, microisolation, **103**, 87
Guaiacol
in assay for thyroid peroxidase, **107**, 449
Guanidine
specific ADP-ribosyltransferase
chromatographic assay with guanylhydrazone substrate, **106**, 405
spectrophotometric assay with guanylhydrazone substrate, **106**, 407
Guanidine hydrochloride
in extraction of recombinant human interferon γ from *Escherichia coli* cells, **119**, 206
protein denaturation curves
analysis, **131**, 272
determination, **131**, 267
stock solutions, preparation, **131**, 269
Guanidinium
uptake in hybrid cells, hormonal regulation, testing procedures, **109**, 338
Guanylate cyclase
mammalian leukocyte
characterization of formed product, **132**, 427
nonradiometric assay, **132**, 426
radiometric assay, **132**, 425

H

Halobacterium halobium
preparation of halorhodopsin-containing envelope vesicles, **125**, 511, 513
Halorhodopsin
-containing envelope vesicles
absorption spectrum, effects of chloride, **125**, 518
chloride transport assays, **125**, 515, 518
content estimation, **125**, 519
light-generated chloride gradients, determination, **125**, 516
membrane potential, measurement, **125**, 511, 514
pH changes, measurement, **125**, 513, 514
preparation, **125**, 511, 513
volume changes, **125**, 515
Hansenula schneggii, *see* Yeast
Haplopappus gracilis
root tip meristems, synchronization by amphidicolin, **118**, 93
appropriate conditions for, definition, **118**, 91
assay, **118**, 89
effectiveness of method, **118**, 96
Haptens
in biological samples, luminescence immunoassays, **133**, 354
Harvey murine sarcoma virus
transformed MDCK cells, induction of glucagon receptors
applications, **109**, 360
conditions for, **109**, 357
inhibition, **109**, 359
Heart
bovine, purification of
mitochondrial carnitine acyltransferases, **123**, 280

mitochondrial cytochrome bc_1, **126**, 183, 186
canine, sarcoplasmic reticulum calcium pump, covalent regulation, **102**, 262
chicken, isolation of mitochondria with collagenase, **125**, 17
guinea pig and rat, simultaneous determination of ubiquinones, **105**, 147
intact, redox state of nicotinamide nucleotides and flavins, measurement, **123**, 311
porcine, purification of cytochrome-*c* oxidase, **126**, 34
vitamin E levels, analytical techniques, **105**, 138
Heat capacity
apparent specific, for protein–protein and protein–ligand interactions, determination, **130**, 64
Helper factors
antigen-specific T-cell-derived, overview, **116**, 296
TGAL antigen-specific
assays, **116**, 347
biochemical characterization, **116**, 349
isolation, **116**, 349
sources, **116**, 341
Hemadsorption assay
solid-phase, for monoclonal antibodies, **121**, 553
Hemagglutination assay
for antibodies to rat MHC class I antigens, **121**, 395
for immunoglobulins in tissue culture supernatants, **121**, 556
passive, for immunoglobulin idiotypes, **116**, 179
passive, inhibition, in detection of
immunoglobulin idiotypes, **116**, 179
monoclonal antibodies, **121**, 554
reverse passive, for monoclonal antibodies, **121**, 552
in screening of antiallotypic antisera, **116**, 162
Hemagglutinin
antibodies, multivalent antibody radioimmunoassay, **121**, 428
Helix pomatia A, –Sepharose gels
in chromatographic separation of lymphocytes, **108**, 159
preparation, **108**, 155
Hematin
–agarose conjugates
in affinity chromatography of heme-binding proteins, **123**, 327
preparation, **123**, 324
as marker substrate in hydroperoxide detection, **105**, 298
Hematoporphyrin
–agarose conjugates
in affinity chromatography of heme-binding proteins, **123**, 327
preparation, **123**, 324
Heme-binding proteins
affinity chromatography
with hematin- and hematoporphyrin-agarose, **123**, 324
with hemin-agarose, **123**, 331
iron in, spectrophotometric determination, **123**, 320
Hemicelluloses
extraction and isolation, **118**, 17
Hemicentrotus pulcherrimus, *see* Sea urchin
Hemin
–agarose conjugates
in affinity chromatography of heme-binding proteins, **123**, 336
preparation, **123**, 333
properties, **123**, 334
Hemin dicyanide
binding-induced conformational changes, stopped-flow circular dichroism, **130**, 526
Hemocytometer
in lymphoreticular cell number estimation, **108**, 45
Hemodialysis
chronic, detection with cellular chemiluminescence, **133**, 499
Hemoglobin
dilution-induced conformational changes, stopped-flow circular dichroism, **130**, 529
hydrogen exchange, effect of ligand saturation, **131**, 506
low-resolution electron density map, interpretation, **115**, 189
nonenzymatic glycosylation, measurement, **106**, 77
time-resolved X-ray scattering, **134**, 675

Hemolymph
Locusta migratoria
photoaffinity labeling of juvenile hormone-binding proteins, **111**, 526
purification of 20-hydroxyecdysone-binding protein, **111**, 446
Manduca sexta, purification of juvenile hormone carrier protein, **111**, 484
Hemolytic assay
for monoclonal antibodies to soluble antigens, **121**, 788
Hemolytic plaque assay
allotype-bearing cells, overview, **116**, 172
for anti-hapten antibody-forming cell clones, **121**, 341
Hemopexin
purification with hematin-agarose resin, **123**, 324
Hemoproteins
ligand binding, resonance Raman spectroscopy, **130**, 363
Heparin
binding sites on lipoprotein lipase, identification, **129**, 742
Hepatocytes, *see also* Liver
mammalian, receptor recycling, assessment, **109**, 232
rat
apolipoprotein synthesis, **129**, 281
Ca^{2+} fluxes and phosphorylase activity, hormone effects, **109**, 554, 557
culture techniques, **129**, 274
isolated, alkane production assay, **105**, 311
isolation, **109**, 469
isolation of ^{32}P-labeled apolipoprotein B, **129**, 538
lipoprotein synthesis, **129**, 278, 515
peptide hormone receptor-rich structures
in Golgi subfractions, preparation and characterization, **109**, 219
in intact Golgi, preparation and characterization, **109**, 225
in nonlysosomal non-Golgi intermediate vesicles, preparation and characterization, **109**, 227
quantitative autoradiography, **109**, 229
phosphoinositides in
degradation in cell-free systems, hormone stimulation, **109**, 477
metabolic enzymes, assays, 474
radiolabeling and analysis, **109**, 469
plasma membrane phosphatidylinositol levels, direct chemical measurement, **109**, 504
SV40 virus *tsA*-mutant-transformed
establishment, **109**, 387
growth and maintenance, **109**, 395
metabolic and hormonal properties, **109**, 392
Heptadecylmercaptohydroxyquinoline quinone
inhibition of cytochrome bc_1, **126**, 265
Heptaprenylpyrophosphate synthetase
Bacillus subtilis
assay, **110**, 199
component separation, **110**, 202
properties, **110**, 203
purification, **110**, 200
1-Heptene
aqueous solutions, water structure, infrared spectroscopy, **127**, 109
Heptylhydroxyquinoline-*N*-oxide
inhibition of cytochrome bc_1, **126**, 270
Hevea brasiliensis
latex, 3-hydroxy-3-methylglutaryl-CoA reductase assay and properties, **110**, 40
Hexafluororiboflavin
phosphorylation, in preparation of riboflavin phosphates, **122**, 217
Hexane
in extraction of vitamin E, **105**, 139
Hexaprenylpyrophosphate synthetase
Micrococcus luteus B-P 26
assay, **110**, 193
properties, **110**, 197
purification, **110**, 194
resolution into essential components, **110**, 194
Hexokinase
cardiac mitochondrial, assay, **125**, 18
yeast
coupling to sulfonated support, **104**, 61
purification by Mg^{2+}-promoted binding to immobilized dyes, **104**, 110
β-Hexosaminidase, *see* β-*N*-Acetylhexosaminidase

Hexose 6-phosphate
–phosphate antiport, from *Streptococcus lactis*
anion exchange assay, **125**, 561
reconstitution in liposomes, **125**, 559
Hexoses
cell wall, colorimetric assay, **118**, 25
Hexylene glycol
calmodulin crystallization from, **102**, 144
–protein interactions, measurement, **114**, 71
High-density lipoproteins
binding activity, modulation, **129**, 645
binding assays, **129**, 650
charge heterogeneity, **128**, 440
conversions, **129**, 363
delipidation, **128**, 343, 348, 350
distribution, **128**, 423
interactions with liposomes
biological relevance, **128**, 655
effects of liposome composition, **128**, 648
intestinal epithelial cell
characterization, **129**, 531
isolation, **129**, 530
isolation, **128**, 649; **129**, 422
isolation of
apolipoprotein D, **128**, 298
serum amyloid A, **128**, 312
mesenteric lymph
biliary diversion effects, **129**, 527
characterization, **129**, 521
distribution, **129**, 520
metabolic origins, **129**, 360
metabolism, intestinal contribution, **129**, 534
phospholipid enrichment
incubation conditions, **128**, 650
molecular basis, **128**, 654
product separation and analysis, **128**, 651, 653
promotion, role of transfer proteins, **128**, 652
radiolabeling methods, **128**, 652
radioiodination, **129**, 425, 651
reconstituted into discoidal complexes, comparison with native form, **128**, 578
reconstitution, **128**, 553
structural analysis with crosslinking reagents, **128**, 620
High-performance liquid chromatography
in analysis of dopamine and testosterone delivery systems, **112**, 392
apolipoprotein C-II, **128**, 294
apolipoproteins, **128**, 339, 383
applications, **133**, 119
ascorbic acid, **122**, 3
azidobenzoyl insulins, **109**, 175
bacterial luciferase, **133**, 119
bacterial menaquinone mixtures, **123**, 251
bile acids, **111**, 51
bilirubin isomers, **123**, 389
biotin and analogs, **122**, 63
carotenoids, **105**, 157; **111**, 149, 189
in catecholamine assay, **103**, 470
C_{15}–C_{60} polyprenols synthesized in *Myxococcus fulvus* cell-free system, **110**, 301
columns
installation, **104**, 134
operation, **104**, 136
repair, **104**, 150
storage, **104**, 143
testing, **104**, 135
troubleshooting, **104**, 143
coproporphyrin isomers, **123**, 386
coumarin antagonists of vitamin K in blood, **123**, 223
cytokinins, **110**, 354
dehydroascorbic acid, **122**, 3
dehydroerythorbic acid, **122**, 3
3-dehydroretinal, **123**, 55
description, **133**, 110
7,8-dihydroxy-7,8-dihydrobenzo[*a*]pyrene metabolites, **105**, 350
diketogluconic acid, **122**, 3
diketogulonic acid, **122**, 3
dolichols and polyprenols, **111**, 202
ecdysone 3-epimerase, **111**, 440
ecdysteroids, free and conjugated, **111**, 392, 412
enkephalin peptides, **124**, 621
enkephalins and β-endorphin, **103**, 556
erythorbic acid, **122**, 3
exchangeable protons in protein structure, **131**, 471, 514
in FAD synthetase assay, **122**, 239
flavins and flavin analogs, **122**, 199
in flavokinase assay, **122**, 238

fluorescers, with chemiluminescence detection, **133**, 435
folic acid derivatives, **122**, 269
gel permeation, IgG, **121**, 612
glutathione in biological samples, **113**, 553
growth hormone-releasing factor, **124**, 386
in guanidine-specific ADP-ribosyltransferase assay, **106**, 407
25-hydroxyvitamin D_3 metabolites, **123**, 144
in identification of posttranslationally modified amino acids, **106**, 22
interleukin 1, **116**, 473
interleukin 3, **116**, 547
interstitial retinol-binding protein, **123**, 104
labeled protein fragments after amide proton exchange, **131**, 141
lipoproteins, **129**, 57
comparison with sequential ultracentrifugal flotation, **129**, 70
luteinizing hormone-releasing hormone peptidase products, **103**, 543
lysine and ornithine analogs and dansyl derivatives, **122**, 297
malonaldehyde in biological samples, **105**, 304, 319
microbore, biogenic amines, **124**, 402
mitochondrial F_1-ATPase, **126**, 481
monoclonal antibodies, **121**, 596
murine Ia antigens, **108**, 548
neuronal catecholamine content and biosynthesis, **103**, 500
neuropeptides, **103**, 23, 75
osteocalcin, **107**, 524
oxidatively modified proteins, **107**, 375
oxygen free radicals, with electrochemical detection, **105**, 23
O-phosphates, **107**, 13
phosphorylated basic amino acids, **107**, 24
phosphorylated β-subunit of insulin receptor, **109**, 620
phylloquinone in tissues, **123**, 242
plant antiviral protein, **119**, 743
plant oligoadenylates, **119**, 757
polypeptide β_1, **116**, 243
porphyrin methyl esters, **123**, 352
porphyrins and isomers with radial compression columns, **123**, 346
prostaglandins, **109**, 497
pump and column maintenance, **133**, 113
quinones
in determination of partition coefficients, **125**, 121
in plastoquinone biosynthesis, **110**, 316
radioactive acylcarnitines, **123**, 273
radioactive sterol metabolites, **111**, 327
recombinant human interferons, **119**, 242
reduced NAD, **122**, 153
retinal, **123**, 55
retinoic acid, **123**, 116, 123
retinoic acid isomers, **123**, 82
retinoids, **123**, 95
retinol, **123**, 68
retinol isomers, **123**, 78
retinol and α-tocopherol in plasma or serum, **123**, 215
retinyl esters, **123**, 68
riboflavin phosphates, **122**, 210, 212
selenocysteine in proteins, **107**, 582
short-chain acylcarnitines, **123**, 267, 271
solvent and sample preparation for, **133**, 112
somatomedins, **109**, 781, 810
sterols, **111**, 37
substance P, **124**, 339
system design, **121**, 597
thiamin and phosphate esters, **122**, 15
thymosin α_1, **116**, 237, 243
thymosin β_4, **116**, 250, 252, 266, 268, 269
thymosin β_9, **116**, 266
thymosin β_{10}, **116**, 268, 269
thymosin β_{11}, **116**, 268
thymosin fraction 5 and 5A, **116**, 225
uro- and coproporphyrin isomers, **123**, 387
uroporphyrin I and III, **123**, 385
vitamin D in human milk, **123**, 171
vitamin D metabolites, **123**, 129
vitamin E in tissues, **105**, 131
vitamin K in tissues, **123**, 235
warfarin in plasma, **123**, 225
High-pressure bomb
for fluorescence polarization studies
applications, **130**, 516
description, **130**, 495

handling, **130**, 504
instrumentation, **130**, 504
viscosity effects, **130**, 515
window birefringence correction, **130**, 507
High-pressure liquid chromatography, *see* High-performance liquid chromatography
Hinge protein
assays, **126**, 249
isolation, **126**, 238
properties, **126**, 241
relationship to cytochromes *c* and c_1, **126**, 245
Histidine
binding to hemoproteins, resonance Raman spectroscopy, **130**, 399
C-2 proton exchange rates, **131**, 478
in proteins, mixed-function oxidation, **107**, 370
reactivity with heavy atoms, **114**, 150
Histidine decarboxylase
pyridoxal phosphate-dependent, from *Morganella* AM-15
assay, **122**, 139
properties, **122**, 142
purification, **122**, 140
pyruvoyl-dependent, from *Lactobacillus* 30a
assays, **122**, 128
purification, **122**, 130
structure and properties, **122**, 132
Histidinoalanine
identification, **106**, 353
isolation from *Rangia cuneata*, **106**, 352
occurrence, **106**, 351
S^{β}-(2-Histidyl)cysteine
assay in peptides and proteins, **106**, 356
occurrence, **106**, 358
properties, **106**, 355
Histone 2A
–ubiquitin conjugate
branched structure, **106**, 248
chromosomal localization, **106**, 253
deconjugation, **106**, 260
functional role, **106**, 261
isolation from bovine chromatin, **106**, 241
primary structure, **106**, 241
properties, **106**, 239
terminal amino acid sequences, **106**, 245
Histone-lysine methyltransferase
currently known, **106**, 275
Neurospora crassa
assay, **106**, 275
properties, **106**, 279
purification, **106**, 278
Histones
ADP-ribosyl derivatives
isolation, **106**, 466
linkage via ester bond, evidence for, **106**, 456
preparation, **106**, 450
chromatography on hydroxyapatite columns, **117**, 378
33258 Hoechst
effect on lymphocyte depletion by 5-bromodeoxyuridine and light, **108**, 268
Homidium
binding assay for oxygen radicals generated from reductively activated adriamycin, **105**, 532
Homocysteic acid
excitotoxic properties in neuroendocrine research, **103**, 386
Homovanillic acid
in assay of phagocytic H_2O_2 production, **132**, 395
Hordeum vulgare, *see* Barley
Horseradish peroxidase
in assay for H_2O_2 production
by mitochondria, **105**, 434
by phagocytes, **132**, 395
catalyzed chemiluminescent reactions
enhancement of light emission
emission spectra, **133**, 346
reaction characteristics, **133**, 335
reactions, **133**, 332
enhancers
concentration effects, **133**, 344
identification, **133**, 336
pH effects, **133**, 343
specificity, **133**, 341
synergistic action, **133**, 340
mechanism, **133**, 332
conjugates, applications of enhanced assays, **133**, 351
luminescent assays, **133**, 331

in neuroendocrine cells
cytochemistry, **103**, 187
immunocytochemistry, **103**, 193
intracellular transport, **103**, 205
as pinocytic markers in leukocytes
applications, **108**, 339
merits, **108**, 337
HPLC, *see* High-performance liquid chromatography
Human chorionic gonadotropin
assembly
detection, **109**, 737
methods for, **109**, 747
crosslinking to luteal receptors, **109**, 203
deglycosylated
chromatographic purification, **109**, 729
preparation
with anhydrous HF, **109**, 726
with trifluoromethanesulfonic acid, **109**, 733
derivatization and radioiodination, **109**, 204
disassembly
detection, **109**, 737
methods for, **109**, 739
subunits, monoclonal antibodies to, **109**, 638
in determination of epitope relative orientation, **109**, 648
effect on hormone–receptor interaction, **109**, 641
production, **109**, 639
properties, subclass and affinity, **109**, 641
screening, **109**, 640
synergistic interactions, **109**, 653
Human chorionic gonadotropin receptors
–hormone interaction, effect of monoclonal antibodies, **109**, 641
luteal, photoaffinity labeling, **109**, 207
Human chorionic somatomammotropin
monoclonal antibodies to, affinity and specificity, **121**, 207
Hyalophora cecropia, *see* Silkmoth
Hybrid cells
hormonally responsive intact, construction by cell fusion
desensitization, **109**, 374
kinetics of cAMP production, **109**, 373
methodology, **109**, 367
transfer of β-adrenergic receptors, **109**, 371
human–rodent somatic
application in gene mapping, **128**, 858
clone panel construction, **128**, 857
establishment, **128**, 852
human chromosome content, **128**, 854
neuroblastoma × glioma
culture and handling, **109**, 321
generation and characteristics, **109**, 317
hormonal regulation of
adenylate cyclase in homogenates, testing procedures, **109**, 333
guanidinium ion uptake, testing procedures, **109**, 338
intracellular cAMP levels, testing procedures, **109**, 327
intracellular cGMP levels, testing procedures, **109**, 337
morphological differentiation, **109**, 325
(neuroblastoma × glioma) × neuroblastoma
culture and handling, **109**, 321
generation, **109**, 321
hormonal regulation of
adenylate cyclase in homogenates, testing procedures, **109**, 333
guanidinium ion uptake, testing procedures, **109**, 338
intracellular cAMP levels, testing procedures, **109**, 327
intracellular cGMP levels, testing procedures, **109**, 337
morphological differentiation, **109**, 325
Hybridization
aequorin cDNA with oligonucleotide probes, **133**, 303
apolipoprotein A-II cDNA with radiolabeled probes, **128**, 748
apolipoprotein C-II clones with radiolabeled probes, **128**, 789, 793, 796
apolipoprotein genes with radiolabeled probes, **128**, 870
apolipoprotein mRNA
with radiolabeled probes, **129**, 315
with single-stranded probes, **128**, 671
chloroplast DNA, in restriction site mapping, **118**, 175

dot-blot
in analysis of plant antiviral protein, **119**, 739
in assay for tobacco mosaic virus multiplication in protoplasts, **119**, 748
double- and single-stranded mitochondrial RNAs with radiolabeled probes, **118**, 504
in identification of cDNA clones to phytochrome and other low abundance red light-regulated sequences, **118**, 369
interferon mRNA with cDNA probes, **119**, 474
molecular, in measurement of mRNA concentration and half-life as function of hormone treatment, **109**, 584
neuropeptide mRNA with radiolabeled probes, **124**, 497, 510, 534, 554
neuropeptide precursor cDNAs with synthetic oligonucleotides, **124**, 309
plant mitochondrial DNA, in identification of gene-containing fragments, **118**, 473
plant mitochondrial RNA
with cloned probe, **118**, 496
by Northern method, **118**, 482
plant nuclear DNA to membrane filters, **118**, 66
artifacts and backgrounds in, **118**, 74
in quantitative analysis of variation in repeated sequences, **118**, 76
plasmid interferon αA and α2 DNAs to oligonucleotide probes, **119**, 483
RNA, for endocrine and neuroendocrine studies
cytoplasmic dot methods, **124**, 270
microanalytical dot blot methods, **124**, 272
mini-gel blot methods, **124**, 276
in screening genomic clones of ribulose-bisphosphate carboxylase small subunit genes, **118**, 399, 405
viral RNAs
with cDNA probes, **118**, 732
with radiolabeled probes, **118**, 721
Hybridomas
agarose-entrapped
cell concentration, **121**, 355
cell growth
control, **121**, 359
monitoring, **121**, 356
preparation, **121**, 354
antibody isotyping, **116**, 154
antibody production in SFH medium, **121**, 291
anti-erythrocyte
in cytotoxic drug screening, **121**, 768
isolation, **121**, 759
in modified plaque-reduction assay, **121**, 764
as target cells for cytotoxicity assays, **121**, 761
in tumor cell growth analysis, **121**, 766
antigen-specific
maintenance, **121**, 31
production, **121**, 18, 28, 193
screening, **121**, 23, 32
anti-human α_2-macroglobulin antibody-secreting
production, **121**, 61
screening, **121**, 63
anti-receptor antibody-secreting
cloning, **121**, 100
preparation, **121**, 98
screening
for anti-idiotypic clones, **121**, 101
for anti-receptor clones, **121**, 102
B-cell, growth promoters, **121**, 9
bovine × murine
chromosome number determination, **121**, 260
preservation and application, **121**, 262
production, **121**, 245, 261, 262
screening, **121**, 247
chromosome analysis, **121**, 5
clonal isolation by single-cell isolation, **121**, 332
cloning in SFH medium, **121**, 290
cultivation in
cytostat, **121**, 360
SFH medium, **121**, 286
cytotoxic T-cell
applications, **132**, 476
generation, **132**, 470
fluorescence-activated cell sorting, **108**, 232

freezing
  in 96-well microculture plates, **121**,
    419
  in SFH medium, **121**, 291
growth-promoting factors, comparison,
  121, 181
growth promotion with endothelial cell
  growth supplement, **121**, 293
H-1, secreting multiple IgG idiotypes,
  generation, **121**, 616
hamster × murine
  cloning, **121**, 242
  growth characteristics, **121**, 241
  production, **121**, 239
  screening, **121**, 242
human
  comparative feeder cell effects, **121**,
    298
  phenotypic analysis of available fusion
    partners, **121**, 120
  production with EBV-transformed B
    cells, **121**, 154
human × human
  cloning, **121**, 115
  overview, **121**, 111
  production, **121**, 114
  screening, **121**, 115
human × murine
  fusion with EBV-transformed B cells,
    121, 171
  growth
    in ascites fluid, **121**, 163
    in serum-free culture, **121**, 165
  overview, **121**, 162
  production, **121**, 169
  secretion of
    anti-A erythrocyte antibodies, **121**,
      171
    anti-Rh$_0$(D) antibodies, **121**, 172
human × (murine × human)
  production, **121**, 228
  screening, **121**, 231
hybrid
  production, **121**, 211
  screening, **121**, 216
IgE-secreting
  isolation of IgE, **116**, 78
  preparation, **116**, 77
interspecific, associated phenomena,
  121, 7

intraperitoneal injection for ascites
  formation, pristane effects, **121**, 375
intrasplenic injection for ascites formation, **121**, 381
  special applications, **121**, 384
macrophage, chemiluminescent bioassays for
  activation regulation, **133**, 513
  effect of pharmacological agents, **133**,
    524
  phagocytosis and opsonization, **133**,
    520
monoclonal antibody-producing
  cloning, **121**, 315
  production, **121**, 308
  screening, **121**, 317
monoclonal IgA-secreting
  production, **121**, 43
  screening, **121**, 48
monoclonality, assessment by Poisson
  statistical analysis, **121**, 412
murine
  cloning
    in 20-μl hanging drops, **121**, 327
    in ultra-low gelation temperature
      agarose, **121**, 323
  secreting reactive antibodies
    antibody screening assay, **103**, 462
    cloning, **103**, 466
    freezing and thawing, **103**, 467
    production, **103**, 462
murine × rat
  growth
    in animals, requirements and con-
      siderations, **121**, 387, 390
    in nude mice, **121**, 396
    in nude rats, **121**, 399
      irradiation and splenectomy
        effects, **121**, 402
  secreting antibodies to rat MHC class
    I antigens, production, **121**, 394
NS-1 and derivatives, cultivation in
  KSLM medium, **121**, 270
preparation of ascites fluid, **116**, 126
production
  fusion frequency, **121**, 8
  fusion protocol, **121**, 12
  with Percoll-separated spleen cells,
    121, 174
  role of polyethylene glycol, **121**, 3

in production of site-specific monoclonal antibodies to insulin
assay for antibody production, **109**, 709
growth as ascites, **109**, 711
preparation, **109**, 705
purification of
GAT antigen-specific suppressor factors, **116**, 329, 334
IgM from culture supernatants, **116**, 33
monoclonal Igs, **116**, 128
rat, Ig purification from supernatants, **121**, 643
rat × rat
IgA-secreting
cloning, **121**, 58
production, **121**, 53
screening, **121**, 57
from IR983F cell line
cloning, **121**, 238
production, **121**, 235
screening, **121**, 238
screening
by culture-well solid-phase radioimmunoassay, **121**, 448
by multivalent antibody radioimmunoassay, **121**, 425
with radiolabeled soluble antigens, **121**, 433
by slide immunoenzymatic assay, **121**, 514
secreting antibodies to weak immunogens, production, **121**, 183
stability of cell lines, **116**, 128
supernatants, screening
with diazobenzyloxymethyl paper, **121**, 491
Mycoplasma contamination effects, **121**, 481
suppressor T-cell
derived factors, characterization, **116**, 307
establishment, **116**, 314
production, **116**, 306
T-cell
interferon γ production, **119**, 53
preparation for IgG-binding factor production, **116**, 404
as source of TGAL antigen-specific helper factor, **116**, 344
Hydantoins
N-acyloxymethylation, in prodrug formation, **112**, 355
Hydrates
proton motion, **127**, 274, 277
Hydration
biomolecules, computer simulation with proximity criterion, **127**, 21
preferential, lipoproteins, determination, **129**, 44
protein crystals, determination from crystal density, **114**, 176
proteins
measurement by
NMR relaxation, **117**, 220, 254
small-angle X-ray scattering and velocity sedimentation, **117**, 247, 254
solid-phase studies, **127**, 284
Hydrocarbons
acyclic isoprenoid, *see* Acyclic isoprenoid hydrocarbons
polycyclic aromatic, *see* Polycyclic aromatic hydrocarbons
Hydrocortisone
release from urea-sensitive polymer system, **112**, 493
as supplement in preparation of SFH medium, **121**, 284
Hydrogen
bonding in water and aqueous solutions, low-frequency Raman scattering, **127**, 91
–sugar uptake in *Escherichia coli*, measurement, **125**, 380
transfer reactions, pulse radiolysis study, **105**, 179
translocation by reconstituted cytochrome oxidase, measurement, **126**, 15, 20
uptake negative mutants, *Rhizobium*
characterization, **118**, 532
isolation, **118**, 529
in isolation of *Hup* genes, **118**, 534
preparation, **118**, 529
screening, **118**, 531
Hydrogen bonds
proton polarizability
in biological systems, **127**, 445

measurement by infrared spectroscopy, **127**, 455
role in enzyme function, **127**, 450
Hydrogen fluoride
in deglycosylation of glycoprotein hormones, **109**, 726
Hydrogen isotope exchange
in analysis of membrane dynamics, principles, **127**, 630
in proteins
analytical techniques, **131**, 453, 514
applications, **131**, 502
mechanisms, **131**, 471, 494
methodology, **131**, 452, 510
multiple-site, data analyses, **131**, 480
ratio determination, **131**, 444
Hydrogen peroxide
assays, **105**, 15, 542
drug-induced formation, detection and measurement, **105**, 508
extracellular pool generated by neutrophils, assay, **132**, 402
in fixed-time assay for glutathione peroxidase, **105**, 117
mutagenicity, testing with *Salmonella typhimurium* TA102, **105**, 250
photogeneration in thylakoids, **105**, 425
production by
mitochondria, assay, **105**, 429
phagocytes, assays, **105**, 397; **132**, 114, 154, 155, 395, 414
scavenging in chloroplasts, **105**, 426
Hydrolase
affinity chromatography, **104**, 21, 22
localization in gels, **104**, 424
Hydrolysis
acid, *see* Acid hydrolysis
alkaline, *see* Alkaline hydrolysis
ATP by F_1-ATPase, equilibrium and rate constants, **126**, 608
cysteinylglycine, catalyzing microvillus membrane peptidases, **113**, 471
dopa proteins, **107**, 403
ecdysteroid conjugates, **111**, 418
enzymatic, *see* Enzymatic hydrolysis
neuropeptides, **103**, 24
poly(ortho ester) polymers, **112**, 425
Hydron
polymers, as macromolecule delivery systems
kinetic analysis, **112**, 414
preparation, **112**, 400
Hydroperoxides
effect on
Ca^{2+} retention by mitochondria, **105**, 438
redox state of mitochondrial pyridine nucleotides, **105**, 435
Hydrophobic chromatography
bacterial luciferase, **133**, 127
cytoplasmic dynein, **134**, 350
F_0 portion from ATP synthase, **126**, 572
granulocyte colony-stimulating factor, **116**, 608
interferon antagonist sarcolectin, **119**, 698
interferon β, **119**, 217
interleukin 1, **116**, 474
membrane proteins, **104**, 338
proteins with alkylagarose supports, **104**, 69, 90
Hydrophobic proteins
ion-exchange HPLC, **104**, 187
purification strategy incorporating detergent separation, **104**, 319
2-Hydroxyacetaldehyde
modification of low-density lipoproteins for catabolism studies, **129**, 585
Hydroxyaldehyde
measurement, in detection of lipid peroxidation products, **105**, 291
Hydroxyapatite
chromatographic columns
guidelines for use, **117**, 375
protein adsorption, **117**, 371
protein elution, **117**, 374
β-Hydroxyaspartic acid
in proteins, mass spectrometric identification, **106**, 49
4-Hydroxybenzoate polyprenyltransferase
rat liver
assay, **110**, 327
product analysis, **110**, 330
properties, **110**, 330
p-Hydroxybenzoic acid hydrazide
in colorimetric assay for reducing sugars, **118**, 26

6-Hydroxy-benzo[*a*]pyrene
in assay for hydrogen peroxide, **105**, 542
autoxidation, generated products, **105**, 543
oxygen uptake, measurement, **105**, 542
preparation, **105**, 540
redox cycles, **105**, 545
6-Hydroxybenzothiazole
enhancement of chemiluminescent assays for horseradish peroxidase, **133**, 335
m-Hydroxybiphenyl
in colorimetric assay for uronic acids, **118**, 26
3-Hydroxybutyrate dehydrogenase
Rhodopseudomonas spheroides, purification by affinity chromatography on immobilized dyes, **104**, 111
6-Hydroxydopamine
cytotoxic effects, assessment, **105**, 511
20-Hydroxyecdysone
binding protein from locust hemolymph
assay, **111**, 445
properties, **111**, 448
purification, **111**, 446
radiolabeled, biosynthesis, **111**, 443
Hydroxyethyldisulfide
in assay for glutaredoxins, **113**, 527
Hydroxyindole *O*-methyltransferase
purification from bovine pineal, **103**, 490
in radioenzymatic assay for serotonin, **103**, 491
Hydroxylation
phenol, in hydroxyl radical detection, **105**, 18
Hydroxyl radicals
in aqueous media
chemical properties, **105**, 211
generation methods, **105**, 210
physical properties, **105**, 210
assays, **105**, 16, 18; **132**, 117, 157
in biological systems, **105**, 214
detection, **105**, 51, 214
formation, role of iron salts, **105**, 52
generation by microsomes
assays, **105**, 519
reaction conditions, **105**, 518
history, **105**, 209
production by neutrophils, measurement, **105**, 395
rate constants, **105**, 213
rate measurements, **105**, 213
reactions, classification, **105**, 212
role in ethanol oxidation, **105**, 516
scavengers and reaction products, **105**, 517
spin trapping, **105**, 198
N-Hydroxymethylation
amides, in prodrug formation, **112**, 355
carbamates, in prodrug formation, **112**, 355
hydantoins, in prodrug formation, **112**, 355
imides, in prodrug formation, **112**, 355
Hydroxymethylglutaryl-CoA
radiolabeled, purification, **110**, 56
Hydroxymethylglutaryl-CoA reductase
gene expression, regulation, **128**, 59
in *Hevea brasiliensis* latex
assay, **110**, 44
characteristics, **110**, 46
collection, **110**, 41
isolation, **110**, 43
measurement, **110**, 45
human fibroblast, assay, **129**, 556
Manduca sexta
assay, **110**, 52
product isolation, **110**, 57
properties, **110**, 54
purification, **110**, 53
murine macrophage, assay, **129**, 556
pea seedlings
compartmentation, **110**, 27
isolation variables, **110**, 28
microsomal and plastid
assays, **110**, 30, 36
interrelations, **110**, 39
isolation, **110**, 34, 38
properties, **110**, 35, 39
regulation *in vitro*, analysis with oxysterols, **110**, 11
Hydroxymethylglutaryl-CoA reductase kinase
properties and function, **107**, 95
Hydroxymethylglutaryl-CoA synthase
chicken liver
assays, **110**, 20
properties, **110**, 24
purification, **110**, 22

2-Hydroxy-5-nitrobenzyl bromide
effect on mitochondrial ADP–ATP carrier protein binding, **125**, 663
p-Hydroxyphenylacetic acid
coupling to small oligopeptides with ethyl chloroformate, **103**, 438
Hydroxyproline
in colorimetric assay for cell wall polysaccharides, **118**, 26
-rich proteins, structure, **107**, 369
Hydroxyproline glycosides
biosynthesis, **106**, 525
functional role, **106**, 527
preparation and assay, **106**, 523
structure, **106**, 526
3β-Hydroxysteroid dehydrogenase
in identification of isolated Leydig cells, **109**, 285
20α-Hydroxysteroid dehydrogenase
in interleukin 3 assay, **116**, 541
N-Hydroxysuccinimidobiotin
in biotinylation of
monoclonal antibodies, **121**, 720
peptide hormones and other biological materials, **109**, 429
preparation, **109**, 422
N-Hydroxysuccinimidyl-4-azidobenzoate
^{14}C-labeled
binding to ATPase inhibitor peptide, **126**, 663
synthesis, **126**, 661
in crosslinking of bound monoiodinated glucagon to hepatic membranes, **109**, 209
in structural analysis of lipoproteins, **128**, 613
N-Hydroxysuccinimidyl 3-(2-pyridyldithio)propionate
in coupling monoclonal antibodies to microspheres, **112**, 81
β-Hydroxy-ε-*N*-trimethyl-L-lysine
chromatographic properties, **123**, 294
^{3}H-labeled, synthesis, **123**, 295
25-Hydroxyvitamin D_2
quantitation in human milk, **123**, 167
25-Hydroxyvitamin D_3
metabolic pathway *in vitro*, demonstration, **123**, 141
quantitation in human milk, **123**, 167
25-Hydroxyvitamin D 1α-hydroxylase
mammalian kidney
assay, **123**, 161
properties, **123**, 160
Hyperlipidemia
turbidimetric screening for, **129**, 99
Hyperlipoproteinemia
phenotype definitions and association with various disorders, **128**, 6
Hypertriglyceridemia
role of apolipoprotein A-I–C-III gene complex, **128**, 742
Hypochlorous acid
production by neutrophils, measurement, **105**, 396
Hypogammaglobulinemia
effect on opsonization, **132**, 314
Hypophysial portal system
releasing factor dynamics, problems in evaluation, **103**, 177
Hypophysis
median eminence, push–pull perfusion technique, **103**, 178
Hyporesponsiveness
to interferon, measurement with prostaglandins, **119**, 707
Hypothalamic explants
from rat brain
long-term cultivation, for analysis of peptide synthesis and secretion, **124**, 362
preparation, **124**, 361
Hypothalamoneurohypophysial complex
rat, preparation, **103**, 133
Hypothalamus
arcuate nucleus, neuroendocrine research with excitotoxins, **103**, 379
axon-sparing lesions
histological evaluation, **103**, 396
induction by excitotoxins, **103**, 393
catecholamines and neuropeptides in
extraction from tissue and incubation media, **103**, 498
HPLC with electrochemical detection, **103**, 500
steady-state and non-steady-state turnover, comparison, **103**, 506
clonal cell lines, **103**, 331

cultured cells, synthesis of luteinizing hormone-releasing hormone, **103**, 521
dopamine, secretion into hypophysial portal blood, **103**, 607
electrophysiological recordings of cells, **103**, 95, 101
enzymatic degradation of luteinizing hormone-releasing hormone
assays, **103**, 542, 546
physiological modeling, **103**, 540
products, HPLC, **103**, 543
long-term explant cultures in serum-supplemented medium, **103**, 315, 318
neuropeptide precursors
molecular-weight size-exclusion chromatography, **103**, 530
newly synthesized, reverse-phase HPLC, **103**, 535
synthesis rates, prediction methods, **103**, 524
tissue content after *in vitro* incubation, **103**, 526
tracer amino acids, central delivery, **103**, 527
nuclei for studies of, microdissection and extraction, **103**, 529
organ cultures, **103**, 314, 316
primary cell cultures in serum-supplemented medium, **103**, 322, 328
rat, isolation of
LHRH precursor mRNA, **124**, 320
somatostatin-28, **103**, 79
serum-free culture of dissociated cells, **103**, 329
tissue dissection and incubation, **103**, 495, 541
Hypothyroidism
secondary, detection by cellular chemiluminescence, **133**, 500
Hypusine
detection in translation initiation factor eIF-4D, **106**, 348
properties, **106**, 345
radiolabeled, production in growing lymphocytes, **106**, 346

I

Ibotenic acid
excitotoxic properties in neuroendocrine research, **103**, 384
Ice, *see* Water, frozen
ICR-170
in preparation of *Phycomyces* color mutants, **110**, 227
Ileum
canine, isolation of intrinsic factor–cobalamin receptor, **123**, 23
Imaginal disks
Drosophila melanogaster, demonstration of ecdysteroid receptors, **111**, 462
Imaging
human tumors in athymic mice, **121**, 811, 814
L-2-Imidazolidone-4-carboxylate
inhibition of 5-oxo-L-proline metabolism *in vivo*, **113**, 468
8α-*N*-Imidazolylriboflavin
ribityl side chain, ^{1}H NMR, **122**, 241
Imides
N-acylation, in amide prodrug formation, **112**, 357
N-acyloxymethylation, in prodrug formation, **112**, 355
N-Mannich bases
applications as prodrugs, **112**, 351
hydrolysis, **112**, 348
preparation, **112**, 348
Iminobiotin
in affinity purification of avidin, synthesis, **122**, 83
–avidin complexes, in purification of plasma membrane proteins, **122**, 90
in biotinylation of peptide hormones and other biological materials, **109**, 429
and derivatives, synthesis, **122**, 88
preparation, **109**, 423
Iminodiacetic acid
-activated Sepharose, in nickel chelate chromatography of human immune interferon, **119**, 201
Immobilization
enzymes, for bioluminescence assays, **133**, 218, 233, 239

Immune complexes
circulating, chemiluminescent assay, **133**, 522
coating of culture surfaces for phagocytosis studies, **132**, 212
Immune response
to antigens entrapped in crystallized carbohydrate spheres, assay, **112**, 124
cytokine–cell interactions, **116**, 367
effect of thymic ablation, **108**, 13
graft-versus-host, *see* Graft-versus-host reaction
to human parathyroid hormone, genetic control, **109**, 625
Immunity
adoptive transfer, *see* Adoptive transfer
Immunization
with parathyroid hormone-(1–34) *in vivo*, **109**, 632
with parathyroid hormone–thyroglobulin conjugate *in vivo*, **109**, 632
in production of monoclonal antibodies, **103**, 462
Immunoadsorption chromatography
interferon α, **119**, 28
recombinant interferon γ, **119**, 205
Immunoaffinity chromatography
apolipoprotein A-containing lipoproteins, **129**, 132
apolipoprotein C-III, in determination of specific activity, **129**, 457
apolipoprotein E-containing lipoproteins, **129**, 186
bovine interferon binding to anti-human interferon, **119**, 143
cytokinins, **110**, 352
with desalting, insulin-like growth factors, 787
GAT antigen-specific suppressor factors, **116**, 337
glycosylated apolipoproteins, **128**, 362
growth hormone-releasing factor, **124**, 385
HLA-DR antigens, **108**, 600, 609
Ia antigens, **108**, 537, 560
IgM, **121**, 623
interferon α, **119**, 43, 451, 546
interferon δ, **119**, 99
interleukin 1, **116**, 489
interleukin 2, **116**, 511
intrinsic factor–cobalamin receptor, **123**, 26
D-lactate dehydrogenase, **126**, 376
large granular lymphocyte subsets, **132**, 452
LHRH, **124**, 263
lipoprotein subpopulations, **128**, 431
lymphocytes, **108**, 139, 148
monoclonal antibodies, **121**, 643, 646, 649
murine class I MHC antigens, **108**, 511
somatomedin C, **109**, 815
T-cell subsets, **121**, 737
TGAL antigen-specific helper factor, **116**, 349
tubulin tyrosine ligase, **134**, 176
Immunoassay
apolipoprotein (a)-specific antigen in plasma, **129**, 174
application of bispecific monoclonal antibodies, **121**, 227
cell surface binding, with monoclonal antibodies, **121**, 714
chemiluminescent, *see* Chemiluminescent immunoassay
cooperative
methodology, **121**, 700
monoclonal antibodies for, screening procedures, **121**, 696
specificity analysis, **121**, 701
enzyme, *see* Enzyme immunoassay
hepatic lipase, **129**, 734
human chorionic gonadotropin with avidin–biotin complex, **133**, 286
luminescence, haptens and proteins, **133**, 354
microtubule-associated 210K protein, **134**, 150
platelet-derived growth factor, **109**, 758
precipitating, with monoclonal antibodies, **121**, 703
ribulose-bisphosphate carboxylase expressed in *Escherichia coli*, **118**, 425
superoxide dismutase, **105**, 95
thermochemiluminescent
applications, **133**, 548
thermostable solid phase materials for, **133**, 548

Immunoblotting
antibody-stained myosin polypeptides, **134**, 433
apolipoproteins, **128**, 246
cytokeratin polypeptides, **134**, 365
1,25-dihydroxyvitamin D_3 receptors, **123**, 205
lipoprotein subpopulations, **128**, 419
for production of monospecific antibodies, **134**, 467
Immunochemical techniques
application of bispecific monoclonal antibodies, **121**, 226
in assay for bispecific monoclonal antibodies, **121**, 217
with autoradiography
steroid hormone-receiving cells in CNS, **103**, 639
steroid hormones and neuropeptides in brain, **103**, 631
with horseradish peroxidase, in analysis of neuroendocrine cells, **103**, 193
in localization of
apolipoproteins, **129**, 298
catecholamine neurotransmitters, **103**, 623
double-labeled neuroendocrine cells *in vitro*, **103**, 139
endorphins and enkephalins, **103**, 670
neuropeptides, antibody preparation and use, **103**, 448
neuropeptide-secreting cells, **124**, 461
with monoclonal antibodies
overview, **121**, 563, 829
staining techniques, **121**, 574, 835
tissue fixation, **121**, 572, 834
tissue handling, **121**, 569, 831
related articles previously published in Methods in Enzymology
production of hybridomas, **121**, 871
screening and assessment of monoclonal antibodies, **121**, 872
selected applications of monoclonal antibodies, **121**, 873
Immunodecoration
chloroplast membrane proteins, **118**, 360
polypeptides during chloroplast development, **118**, 367
Immunodiffusion
in detection of lipoprotein (a)-specific antigen in plasma, **129**, 170
Ig isotypes from bovine × murine hybridomas, **121**, 252
in specificity analysis of anti-light chain antisera, **116**, 108
Immunoelectron microscopy
apolipoproteins, **129**, 304
F_1-ATPase subunits, **126**, 770
nascent lipoproteins, **129**, 294
Immunoelectrophoresis
apolipoprotein (a)-specific antigen in plasma, **129**, 172
Immunofluorescence microscopy
with agar overlay, ameboid cell contractile apparatus, **134**, 573
for calmodulin-binding site localization in fixed and living cells, **102**, 117
for calmodulin localization
in fixed tissue sections and cultured cells, **102**, 111
in unfixed frozen tissue sections, **102**, 122
cytoskeletal components, polyethylene glycol embedding protocol for, **134**, 580
kinetochore–centromere fraction from metaphase chromosomes, **134**, 271
β_2-microglobulin, **108**, 498
Immunofluorescent staining
in analysis of interferon effects on histocompatibility antigen expression, **119**, 689
in detection of macrophages, applications of monoclonal antibodies, **108**, 320
in flow cytometric analysis of neuroendocrine cell antigen expression, **103**, 238
immunoglobulin allotypes in cells, **116**, 170
in immunohistology with monoclonal antibodies, **121**, 838
indirect, striated flagellar root protein antibodies, **134**, 412
membrane antigens, **121**, 580
monoclonal antibodies, **121**, 574
monoclonal antiidiotypic antibodies, overview, **116**, 187

murine IgM, **121**, 628
in quantitation of epidermal Langerhans cells, **108**, 359
tissue calmodulin, **102**, 128
in tracing neuropeptide-specific pathways, **103**, 667
Immunoglobulin A
component chains, isolation, **116**, 62
dysopsonic function, overview, **132**, 306
enzymatic digestion, **116**, 52
Fab and Fc fragments
naturally occuring, **116**, 61
production, **116**, 60
metabolism in humans, **116**, 208
monoclonal, assays in hybridoma supernatants, **121**, 48, 57
properties, **116**, 37
purification
from ascites fluid, **116**, 138
from external secretions, **116**, 49
from serum, **116**, 44, 138
secretory component, isolation, **116**, 70
subclasses and allotypes
characterization, **116**, 39, 41
determination, **116**, 43
Immunoglobulin A_1 protease
bacterial
cleavage of IgA, **116**, 60
properties and assays, **116**, 55
purification, **116**, 58
Immunoglobulin-binding factors
for IgG
biological activities, **116**, 410
physicochemical characterization, **116**, 413
producing T cells, preparation, **116**, 404
production and preparation, **116**, 408
overview, **116**, 368
Immunoglobulin D
chain separation, **116**, 98
characterization, **116**, 99
Fab and Fc fragments, preparation, **116**, 98
function, **116**, 99
metabolism in humans, **116**, 208
murine, purification, **116**, 140
myeloma-associated
description, **116**, 96
isolation, **116**, 96
Immunoglobulin E
assays, **116**, 88
immunological properties, **116**, 86
metabolism in humans, **116**, 208
monoclonal, purification, **116**, 78
murine, purification, **116**, 139
physicochemical properties and structure, **116**, 82
polyclonal, purification, **116**, 81
receptor reconstitution into lipid vesicles, **104**, 343
Immunoglobulin G
anion-exchange HPLC, **121**, 600, 602, 606
antibodies to, multivalent antibody radioimmunoassay, **121**, 431
cation-exchange HPLC, **121**, 608
and C3b, cooperative effect on phagocytosis, **132**, 302
chain separation, **116**, 20
chemical typing, **116**, 22
coupling to fluorochrome for cellular allotype analysis, **116**, 170
enzymatic digestion, **116**, 13, 143
Fab and Fc fragments
crystallization, **116**, 195
preparation, **116**, 191
ferritin derivatives
applications, **121**, 776
preparation, **121**, 774
fragmentation, **121**, 803
gel permeation HPLC, **121**, 612
haptenated $(Fab')_2$ fragments
coupling to sheep erythrocytes, **121**, 345
preparation, **121**, 344
idiotypes, hydroxylapatite chromatography, **121**, 615
metabolism in humans, **116**, 208
monoclonal, purification, **121**, 653
murine, subclasses
enzymatic digestion, **121**, 654, 657, 659, 661
monitoring assays, **121**, 655
$F(ab')_2$ fragments, preparation, **121**, 657
properties, **116**, 3
purification, **116**, 5, 130, 148; **121**, 588, 631, 803

in radioimmunoassay of interferon effects on antigen expression, **119**, 686
radioiodination, **119**, 685; **121**, 247, 804
rat, subclasses
purification, **121**, 650
proteolytic fragments
preparation, **121**, 664, 668
purification, **121**, 666, 669
role in opsonization, **132**, 292
selenomethionine derivatives
applications, **121**, 776
preparation, **121**, 774
surface, on lymphocytes, detection and quantitation with protein A, **108**, 405
thyroid-stimulating, assays, **109**, 677
in tumor distribution studies, **121**, 808, 812
in tumor imaging, **121**, 811, 814
Immunoglobulin M
assays, **116**, 34; **121**, 628
chain separation, **116**, 35
coating of liposomes, **121**, 818, 823
metabolism in humans, **116**, 208
precipitation in agar, **121**, 627
properties, **116**, 26
purification, **116**, 29, 136; **121**, 623, 650, 819
role in opsonization, **132**, 294
subunits, preparation, **121**, 821
Immunoglobulins
allotypes
definition, **116**, 157
quantitative assays, **116**, 163
from antibody-producing cells, culture-well radioimmunoassay, **121**, 438
from bovine × murine hybridomas
affinity purification, **121**, 251
assays, **121**, 248, 264
characterization, **121**, 252
crystallization, **116**, 195
human, light chains
antisera to
immunodiffusion analysis, **116**, 108
preparation, **116**, 106
specificity, **116**, 110
conformation, analysis with antisera, **116**, 120
C and V fragments
antisera to, specificity, **116**, 120
preparation and analyses, **116**, 114
hidden determinants, recognition, **116**, 117
isolation, **116**, 102
purification, **116**, 106, 191
interactions with protein A, **108**, 470
intracellular, localization in lymphoid cell suspensions, **108**, 393
levels in nude mouse, **108**, 351
metabolic studies
data analysis techniques, **116**, 206
data interpretation problems, **116**, 209
general principles, **116**, 202
in humans, **116**, 208
specific methodology, **116**, 205
monoclonal, quantitation, **121**, 478
murine
chain reconstitution, **116**, 142
enzymatic digestion, **116**, 143
production, **116**, 125
purification, **116**, 124, 128, 191
structure and genetics, **116**, 121
opsonic, overview, **132**, 284
protein A-binding, removal from fetal bovine serum, **121**, 301
purification from antisera, **118**, 748
radioiodinated, preparation, **116**, 203
surface
detection, direct and mixed rosetting techniques, **108**, 386
ligand-induced patching and capping, **108**, 371
localization in lymphoid cell suspensions, **108**, 393
on lymphocytes
biosynthetic labeling, **108**, 427
detection with fluorescent antibodies, **108**, 413
isolation, **108**, 429
labeling by surface iodination, **108**, 426
in tissue culture supernatants, hemagglutination assay, **121**, 556
Immunoperoxidase technique
in assay for monoclonal antibodies, **121**, 575
indirect, in immunohistological staining, **121**, 838

Immunoprecipitation
apolipoproteins
assays, **129**, 264
direct methods, **129**, 258
indirect methods, **129**, 261
calmodulin, **102**, 109
in detection of
cell-surface receptor antibodies, **109**, 660
glutamine synthetase adenylylation, **107**, 191
1,25-dihydroxyvitamin D_3 receptors with monoclonal antibodies as probes, **123**, 202
H-2 antigens, **108**, 454
human T-cell antigens, **108**, 639
Ia antigens, **108**, 454, 536, 569
2-iminobiotinylated plasma membrane proteins, **122**, 90
insulin receptors, **109**, 606
LHRH precursor, **124**, 324
β_2-microglobulin, **108**, 497
monoclonal antibodies to platelet proteins in agarose, **121**, 712
murine class I MHC antigens, **108**, 511
phosphorylated insulin receptors, **109**, 615
Qa and TL antigens, **108**, 554
somatostatin, **124**, 340
substance P, **124**, 340
succinate dehydrogenase from *Bacillus subtilis*, **126**, 403
surface immunoglobulins, **108**, 429
thyroid-stimulating antibody, **109**, 685
Immunosorbent chromatography
with monoclonal antibodies, advantages, **104**, 385
proteins, **104**, 381
recombinant human interferon αA, **119**, 155
Impalement
cells in dissociated tissue culture, in electrophysiological recording, **103**, 124
Indoleamine 2,3-dioxygenase
assays, **105**, 63
physiological significance, **105**, 67
properties, **105**, 62
purification from rabbit small intestine, **105**, 62
reaction mechanism, role of superoxide anion, **105**, 65
L-Indoleamines
radiolabeled, in assay for indoleamine 2,3-dioxygenase, **105**, 63
Infection
detection with cellular chemiluminescence, **133**, 498
Inflammation
detection with cellular chemiluminescence, **133**, 496
Influenza virus
pulmonary infection, induction of indoleamine 2,3-dioxygenase, **105**, 67
Infrared spectroscopy
carbohydrate and water interactions with membranes, **127**, 698
ecdysteroid conjugates, **111**, 413
Fourier transform, *see* Fourier transform infrared spectroscopy
H–D exchange in protein conformational analysis, **131**, 454, 457
lysozyme films, hydration studies, **127**, 288, 291
proton polarizability of hydrogen bonds, **127**, 439
water
in ice phases, **127**, 313
in nonpolar solvents, **127**, 106
Infusion pumps
implantable
applications, **112**, 524
available devices, **112**, 523
implant procedures
in dogs, **112**, 541
in humans, **112**, 530
potential uses, **112**, 529
refill procedures, **112**, 539
in dogs, **112**, 541
in humans, **112**, 539
Inhibitors of virus replication
in plants
characteristics, **119**, 734
dose–response relationships, **119**, 733
effects on cucumber mosaic virus and potato virus X, **119**, 733
inhibitory potency, assay, **119**, 732
molecular weight estimation, **119**, 734
preparation, **119**, 731
purification, **119**, 731, 734

Initiation factors
eIF-4D, hypusine detection, **106**, 348
wheat germ
assays, **118**, 114, 135
functional properties, **118**, 127
physical properties, **118**, 124
purification, **118**, 110, 116, 130
Inorganic phosphate
in assay for mevalonate 5-pyrophosphate decarboxylase, **110**, 88
–ATP exchange, assay in vesicles containing ATP synthase subunits, **126**, 580
in quantification of *O*-phosphates in proteins, **107**, 15
Inorganic pyrophosphatase
bovine cardiac mitochondria, membrane-bound
purification, **126**, 449
reconstitution, **126**, 452
Rhodospirillum rubrum, membrane-bound
assays, **126**, 541
inhibition, **126**, 543
purification, **126**, 539
reconstitution, **126**, 540
substrate specificity, **126**, 544
Inosine 5′-monophosphate dehydrogenase
Escherichia coli, purification by affinity chromatography on immobilized dyes, **104**, 109
Inositol phospholipids
extraction from cultured cells, **124**, 434
separation and analysis, **124**, 436
Insertion compounds
proton motion, **127**, 281
Insulin
azidobenzoyl derivatives
preparation, **109**, 171, 172
purification, **109**, 171, 172, 175
receptor photolabeling on isolated cells, **109**, 175
binding data, nonlinear least-squares analysis, **117**, 322
biotinylated
applications, **124**, 54
preparation, **109**, 429; **124**, 50
–ferritin, biologically active monomeric
in electron microscopic analysis of insulin receptors, **109**, 200
preparation and characterization, **109**, 189
hydrated, X-ray and neutron diffraction analysis, **127**, 169
photoreactive
characterization, **103**, 60
nonradioactive, applications, **103**, 63
photolysis reactions, **103**, 62
radiolabeled, applications, **103**, 66
synthesis, **103**, 60
rate-controlled delivery by osmotic system, **112**, 474
–Sepharose, in affinity chromatographic purification of insulin receptor, **109**, 400, 403
site-specific monoclonal antibodies
affinity purification, **109**, 712
assays, **109**, 709, 718
production, **109**, 705
specificity analysis, **109**, 715
structure
description, **109**, 705
least-squares-refined, **115**, 114
as supplement in preparation of SFH medium, **121**, 282
Insulin-like growth factor
types I and II
amino acid analysis, **109**, 781
assays, **109**, 775, 782
purification
from human serum and plasma, **109**, 783, 793
from rat serum, **109**, 776, 781
Insulin-like growth factor receptors
types I and II, affinity crosslinking, **109**, 179
Insulin receptors
affinity crosslinking, **109**, 179
cell-surface, antibodies to
binding-inhibition assay, **109**, 657
immunoprecipitation assay, **109**, 660
insulin-like bioactivity, **109**, 663
human placental
purification, **109**, 399, 401
solubilized, assay, **109**, 401
localization and processing analysis, **103**, 69
occupied, electron microscopy, **109**, 200
phosphorylation
in cultured cells, 613
insulin dose–response curve, **109**, 617

in solubilized system, **109**, 616
tryptic peptide mapping by HPLC,
  109, 620
photoaffinity labeling, **103**, 66; **109**, 170
reconstitution into lipid vesicles, **104**,
  347
subunits, radiolabeled
  autoradiography and quantification,
    109, 608
  immunoprecipitation, **109**, 606
  preparation, **109**, 594
  solubilization and purification, **109**,
    602
turnover rates, measurement, **109**, 600
Interfaces
air–water, adsorbed apolipoprotein
  monolayers
  spreading methods, **128**, 392
  surface concentration measurement,
    128, 395
  surface potential measurement, **128**,
    397
  surface pressure measurement, **128**,
    390, 392
Interference
in dissociated tissue culture electrophysiological recording, **103**, 121
Interferon α
bovine
  antigenic cross-reactivity with human
    interferon, **119**, 143
  assay, **119**, 138
  cross-species antiviral activity, **119**,
    140
  encoding gene
    identification, **119**, 467
    isolation, **119**, 465
  induction with Sendai virus, **119**, 137
  molecular weights, **119**, 141
  production in
    bacteria, **119**, 471
    leukocytes, **119**, 138
  purification, **119**, 138
  recombinant bacterial, properties, **119**,
    473
color-change growth-inhibition assay,
  119, 579
equine
  cross-species antiviral activities, **119**,
    134

induction with
  Newcastle disease virus, **119**, 133
  poly(I) · poly(C), **119**, 133
human
  amino acid sequence, **119**, 6
  antigenic cross-reactivity with bovine
    interferon, **119**, 143
  antitumor effects
    assay with human tumor xenografts,
      119, 651
    effect of combining with chemother-
      apy, **119**, 655
    mechanisms, **119**, 652
  assays, **119**, 40, 418, 427
  binding to cells in monolayer culture,
    analytical techniques, **119**, 312
  encoding genes
    characterization, **119**, 363
    expression in *Escherichia coli*, **119**,
      364
    isolation with short oligonucleotide
      probes, **119**, 359
    mutagenesis *in vitro*, **119**, 403
  enhanced sensitivity of transformed
    NIH 3T3 cells, selection and
    screening, **119**, 597
  equilibrium binding assay, **119**, 313
  interactions with growth factors, assay
    in fibroblasts, **119**, 658
  from lymphoblastoid cells, large-scale
    production, **119**, 35
  from peripheral blood, large-scale
    production and recovery, **119**, 39
  purification, **119**, 28, 38, 41, 428
  radiation inactivation and target size
    analysis, **119**, 258
  radioiodination
    by chloramine-T method, **119**, 267
    with ¹²⁵I-labeled Bolton–Hunter
      reagent, **119**, 276
  recombinant
    amino acid composition, **119**, 11
    effect on primary tumor cells, mea-
      surement, **119**, 635
    HPLC analysis, **119**, 242
  subtype composition, effect of purifi-
    cation procedure, **119**, 27
  synthesis in yeast, **119**, 420
  yeast expression vectors, construc-
    tion, **119**, 418, 424

murine
  assay, **119**, 440
  encoding gene
    expression in *Escherichia coli*, **119**,
      437
    isolation from lambda phage library,
      119, 434
    expression vectors, construction, **119**,
      437
  induction in C-243 cells by Newcastle
    disease virus, **119**, 477
  mRNA, detection by *in situ* hybridiza-
    tion, **119**, 474
rat
  chromosomal gene
    cloning, **119**, 442
    encoding plasmid, construction, **119**,
      445
    expression in *Escherichia coli*, **119**,
      449
    isolation and characterization, **119**,
      444
  recombinant, purification, **119**, 451
  from various sources, amino acid se-
    quences, **119**, 13
Interferon $\alpha 2$
  human
    conjugation with colloidal gold, **119**,
      333
    ^{125}I-labeled
      binding to Daudi cells, **119**, 348
      –receptor complexes, extraction and
        separation, **119**, 349
    induced changes in histocompatibility
      antigens, measurement, **119**, 688
    and interferon αA, single base substi-
      tution between, detection, **119**,
      481
    and naturally derived rat interferon,
      structural homology, **119**, 227
    receptor-bound
      binding of radiolabeled monoclonal
        antibody to, **119**, 329
      temperature-induced decrease,
        analysis with antibody, **119**, 330
    receptor-mediated internalization and
      processing, measurement, **119**,
      335
    recombinant
      production, **119**, 167

properties, **119**, 173
purification, **119**, 168, 327
radiolabeling, **119**, 263, 327
Interferon αA
  human
    conjugation with colloidal gold, **119**,
      333
    dimers and higher oligomers, sand-
      wich radioimmunoassay, **119**, 588
    ^{125}I-labeled
      binding to suspension cultures, **119**,
        307
      preparation, **119**, 306
    and interferon $\alpha 2$, single base substitu-
      tion between, detection, **119**, 481
    receptor-mediated internalization and
      processing, measurement, **119**, 335
    recombinant
      analytical techniques and assays,
        119, 250
      induced tumor cell antigens, ra-
        dioimmunoassay, **119**, 682
      irreversible derivatization of disul-
        fides, **119**, 253
      production, characterization, and
        storage, **119**, 684
      purification, **119**, 153, 288
      reduction and oxidation, **119**, 251
      [^{35}S]methionine labeling *in vivo* in
        maxicells, **119**, 288
      sulfitolysis and reoxidation, **119**, 252
    relative activity on various cell lines,
      119, 18
Interferon β
  bovine
    production, **119**, 212
    purification, **119**, 213
  equine
    cross-species antiviral activities, **119**,
      134
    induction with
      Newcastle disease virus, **119**, 133
      poly(I) · poly(C), **119**, 132
  guinea pig
    assay, **119**, 128
    harvest and storage, **119**, 128
    induction *in vitro*, **119**, 127
  human
    antigenic cross-reactivity with bovine
      interferon, **119**, 143

assay, **119**, 372
chromatographic analysis, **119**, 401
effect on cellular differentiation in
    melanoma cells, measurement,
    119, 611
enhanced sensitivity of transformed
    NIH 3T3 cells, selection and
    screening, **119**, 597
expression in
    CHO cells, **119**, 398
    Escherichia coli with phage lambda
        P_L promoter, **119**, 366
    mammalian cells with Rous sarcoma
        viral genome, **119**, 383
expression vectors, construction, **119**,
    368, 385
gene sequence encoding carbohydrate
    attachment site, *in vitro* muta-
    genesis, **119**, 399
metabolic labeling, **119**, 400
radiation inactivation and target size
    analysis, **119**, 258
recombinant
    amino acid composition, **119**, 11
    characterization, **119**, 180
    properties, **119**, 190
    purification, **119**, 178, 183
^{35}S-labeled, in cultured diploid fibro-
    blasts
    induction and radiolabeling *in vivo*,
        119, 284
    purification and characterization,
        119, 285
murine
    antitumor effects
        assay with human tumor xenografts,
            119, 651
        mechanisms, **119**, 652
    induction by Newcastle disease virus,
        119, 477
    induction of polyamine-dependent
        protein kinase, **107**, 160
    mRNA, detection by *in situ* hybridiza-
        tion, **119**, 474
    from various sources, amino acid se-
        quences, **119**, 8
Interferon δ
    human
        antiviral assay, **119**, 98
        production in leukocytes, **119**, 98
        purification, **119**, 99

Interferon γ
    equine
        cross-species antiviral activities, **119**,
            134
        induction with phytohemagglutinin,
            119, 134
    human
        assays, **119**, 56, 65, 81, 94, 200, 372,
            418
        concentration, **119**, 58, 61
        expression in *Escherichia coli* with
            phage lambda P_L promoter, **119**,
            366, 376
        expression vectors, construction, **119**,
            368, 418
        induced by A-23187 and mezerein
            bioassay, **119**, 193
            purification, **119**, 193
            purity assessment, **119**, 197
        induced changes in histocompatibility
            antigens, measurement, **119**, 688
        induction with
            A-23187 and mezerein, **119**, 74
            combinations of T-cell mitogens,
                119, 93
            concanavalin A, **119**, 89
            ionophores, **119**, 69
            lentil lectin, **119**, 54
            phorbol esters, **119**, 48
            phytohemagglutinin, **119**, 48, 66, 78,
                85
            staphylococcal enterotoxin B, **119**,
                85, 89
            streptococcal preparation OK-432,
                119, 89
        interaction with lipopolysaccharide in
            macrophage activation, assay,
            133, 517
        monoclonal antibodies to, **119**, 584
        nickel chelate chromatography, **119**,
            201
        phosphorylated by cAMP-dependent
            protein kinase, **119**, 296
        ^{32}P-labeled
            binding to U937 cells, **119**, 317
            preparation, **119**, 316
        production in
            leukocytes, **119**, 56, 63, 72, 83, 88
            lymphocytes, **119**, 77
            peripheral blood mononuclear cells,
                119, 93

plateletpheresis residues, **119**, 50
T-cell lymphoma or hybridoma cell
   lines, **119**, 52
yeast, **119**, 420
purification, **119**, 59, 85, 90
radiation inactivation and target size
   analysis, **119**, 258
radiolabeling, **119**, 263, 281
recombinant
   amino acid composition, **119**, 11
   HPLC analysis, **119**, 242
   purification and characterization,
      119, 206
recovery of crude preparation, **119**, 78
sandwich radioimmunoassay, **119**,
   582
ultrafiltration, **119**, 69
murine
   induction and harvesting, **119**, 146
   interaction with lipopolysaccharide in
      macrophage activation, assay,
      133, 517
   mediated cytolysis, microassay, **119**,
      574
   production in spleen cells, **119**, 145
   purification, **119**, 146
rat
   characterization, **119**, 460
   chromosomal gene
      identification, **119**, 454
      primary structure, **119**, 454
   concentration and partial purification,
      119, 459
   expression in eukaryotic cells, **119**,
      456
   production in CHO cells, **119**, 458
from various sources, amino acid sequences, **119**, 10
Interferon receptors
bound human interferon
   antibody binding to, methodology,
      119, 326
   internalization and processing, mea-
      surement, **119**, 332
complexed with interferon α2, extraction
   with digitonin, **119**, 347
on human cells
   binding of ^{125}I-labeled interferons, **119**,
      342
   identification by chemical crosslinking,
      119, 342

Interferons
antigenic purity, determination, **119**, 569
bovine, antiviral assay with ovine and
   caprine cells, **119**, 554
caprine, antiviral assay with ovine and
   homologous cells, **119**, 554
chicken
   bioassay, **119**, 124
   inducers, **119**, 118
   induction by viral infection, **119**, 120
   processing to remove residual inducer,
      119, 121
designation, **119**, 19
encapsulated in crystallized carbohydrate microspheres
   biological activity, assay, **112**, 121
   preparation, **112**, 120
   in vitro release studies, **112**, 122
equine, antiviral assay with ovine and
   homologous cells, **119**, 554
fibroblast, *see* Interferon β
human
   antiviral assay, **119**, 534, 540, 554
   chemical modification
      during biosynthesis, **119**, 233
      by cDNA manipulation, **119**, 240
      conjugates with other substances,
         119, 239
      enzymatic methods, **119**, 234
      reactions affecting cysteine and
         cystine residues, **119**, 237
      reactions with chemicals, **119**, 236
   detection in
      amniotic fluid, **119**, 542
      clinical blood samples, **119**, 540
      placenta and fetal membranes, **119**,
         544
   effect on
      cellular differentiation of skeletal
         muscle cells, **119**, 619
      proliferative capacity and cloning
         efficiency of normal and leuke-
         mic progenitor cells, **119**, 629
      tobacco protoplasts, assay, **119**, 744
      virus-induced cell fusion, **119**, 702
   ^{125}I-labeled, in receptor studies, **119**,
      342
   purification, **119**, 545
immune, *see* Interferon γ
induction, dose–response curves
   analysis, **119**, 112

generation, **119**, 107
leukocyte, *see* Interferon α
mixed-type, antibody neutralization, **119**, 571
murine
  antiproliferative effect
    effects of growth factors, **119**, 645
    measurement, **119**, 643
  antiviral assay, **119**, 534
  chemical modification
    during biosynthesis, **119**, 233
    by cDNA manipulation, **119**, 240
    conjugates with other substances, **119**, 239
    enzymatic methods, **119**, 234
    reactions affecting cysteine and cystine residues, **119**, 237
    reactions with chemicals, **119**, 236
  effect on
    drug metabolism, measurement, **119**, 718
    phenylhydantoin metabolism, measurement, **119**, 715
  hyporesponsiveness to, measurement with prostaglandins, **119**, 707
  ^{125}I-labeled
    binding analysis, **119**, 323, 325
    preparation, **119**, 322
  induction
    measurement with prostaglandins, **119**, 707
    by poly ICL-CM dextran, **119**, 104
  induction of pulmonary indoleamine 2,3-dioxygenase, **105**, 69
  interactions with growth factors, measurement, **119**, 662, 664
  neutralization by antibodies, **119**, 558
  antibody titration in, **119**, 559, 568
  monoclonal antibodies in, characteristics, **119**, 565
  titer definition and computation, **119**, 562
overview, **116**, 366; **119**, 3
ovine, antiviral assay with caprine and homologous cells, **119**, 554
porcine, antiviral assay with ovine and caprine and homologous cells, **119**, 554
radiolabeled with [35S]methionine
  gel electrophoresis, **119**, 295
  production in cell-free DNA-dependent system, **119**, 292
rat
  assay, **119**, 221
  characterization, **119**, 225
  concentration and purification, **119**, 222
  effect on drug metabolism, measurement, **119**, 718
  and human interferon α2, structural homology, **119**, 227
  production, **119**, 220
reference antisera, **119**, 15
reference standards, **119**, 15
simian, induction by poly ICL-CM dextran, **119**, 104
tertiary structure, computer simulation, **127**, 134
Interleukin 1
  human
    assays, **116**, 460, 468
    biochemical characterization, **116**, 466
    conditioned medium with, preparation, **116**, 470
    production, **116**, 457
    purification, **116**, 463, 471
  murine
    antibodies, preparation and use, **116**, 488
    assays, **116**, 481
    production, **116**, 483
    purification, **116**, 485
  overview, **116**, 362
Interleukin 2
  human
    antibodies, preparation, **116**, 506
    assays, **116**, 494
    cellular receptor, properties, **116**, 522
    production, **116**, 503
    purification, **116**, 511
    structural characterization, **116**, 518
  induction assays, interleukin 1-dependent, **116**, 481
  murine
    assays, **116**, 526
    proliferative response of thymocytes, inhibition by interferon, **119**, 662
    properties, **116**, 534
    purification, **116**, 530
  overview, **116**, 363

Interleukin 3
assays, **116**, 541, 551
purification, **116**, 547
radioiodination, **116**, 551
Intermediate filaments
X-ray diffraction analysis, **134**, 645
Intestine
chicken, purification of
calcium-binding proteins, **102**, 291
microvillus cytoskeletal proteins, **134**, 24
rabbit, purification of
cytosolic protein binding 1,25-dihydroxyvitamin D, **123**, 186
indoleamine 2,3-dioxygenase, **105**, 62
rat, purification of mucosal Δ^1-pyrroline-5-carboxylate synthase, **113**, 118
Intrauterine device
estriol-releasing, structure, function, and release kinetics, **112**, 508
fabrication, **112**, 469
progesterone-containing, structure, function, and release kinetics, **112**, 505
Intrinsic factor
–cobalamin receptor, canine ileal
assay, **123**, 23
purification, **123**, 24
Inulin
–protein conjugates
pharmacological properties, **112**, 292
preparation, **112**, 286
therapeutic applications, **112**, 296
Invertebrates
marine, sterol biosynthesis and interconversion, analytical techniques, **111**, 311
Iodide
in thyroid hormone iodination and coupling, **107**, 483
Iodination
in assay for assessment of phagocytic function, **132**, 116, 156
enzymatic
proteins with thyroid peroxidase, **107**, 445
thyroid hormones, **107**, 476
nonenzymatic, thyroid hormones, **107**, 483
small oligopeptide haptens for radioimmunoassay, **103**, 435
Iodine
in thyroid hormone iodination and coupling, **107**, 483
Iodine monochloride
in thyroid hormone iodination and coupling, **107**, 483
Iodipine
^{125}I-labeled, labeling of calcium channels, **109**, 539
Iodoacetamide
irreversible derivatization of disulfides in interferon αA, **119**, 253
Iodoacetate
irreversible derivatization of disulfides in interferon αA, **119**, 253
Iodoacetic acid
N-hydroxysuccinimidyl ester, in antibody coupling to toxins, **112**, 217
5-Iodoacetylaminoethylamino-1-naphthalenesulfonic acid
-labeled calmodulin
applications, **102**, 149, 153
biological activity, **102**, 152
preparation, **102**, 150
properties, **102**, 156
Iodogen
in peptide radiolabeling, **124**, 24
Iodometry
in assay for lipid hydroperoxides, **105**, 295
chloramines, **132**, 582
Iodotyrosine deiodinase
assay, **107**, 489
properties, **107**, 496
purification from bovine thyroid, **107**, 492
Ion channels
anterior pituitary cell, activation, **124**, 237
gramicidin A, energy profiles, computational methods, **127**, 250
in neuroendocrine tissue, activating and inhibiting drugs
action characterization, **103**, 401
in analysis of channel role in receptor actions, **103**, 404
in assays of numbers of channels, **103**, 404

single
applications for recording, **124**, 190
electrophysiology, analytical patch clamp technique, **103**, 147
patch recording, **124**, 230
Ion-exchange chromatography
amino acid derivatives, **106**, 17
apolipoprotein A-containing lipoproteins, **129**, 143
D-aspartic acid β-methyl ester, **106**, 333
N^{ε}-(β-aspartyl)lysine, **107**, 259
azidobenzoyl insulins, **109**, 171, 172
bispecific monoclonal antibodies, **121**, 223
calmodulin-dependent protein phosphatase, **102**, 247
4-carboxyglutamate, **107**, 536
in γ-carboxyglutamic acid assay, **107**, 503
columns
cleaning, **104**, 188
elution, **104**, 179
loading capacity, **104**, 178
resolution, optimization, **104**, 183
retention, **104**, 170
selection, **104**, 173
selectivity, **104**, 185
switching, **104**, 189
[^{14}C]pyridoxal, **122**, 99
cyclic nucleotides, **132**, 430
cytochrome-*c* oxidase, **126**, 157, 163, 165
cytokeratin polypeptides, **134**, 369
cytoplasmic dynein, **134**, 349
in detection of nonenzymatic protein glycosylation, **106**, 80
dopa, **107**, 405
dopa-containing peptides, **107**, 409
γ-glutamyl amino acids, **113**, 564
glutathione and glutathione disulfide in biological samples, **113**, 553
glycoprotein hormone subunits, **109**, 739
granulocyte–macrophage colony-stimulating factor, **116**, 596
hepatic lipase, **129**, 723
hydrophobic proteins, **104**, 187
Ia antigens, **108**, 547
IgA J chain, **116**, 66
IgA secretory component, **116**, 71
IgD, **116**, 96
IgG, **116**, 9; **121**, 599
IgM, **116**, 30
interferon antagonist sarcolectin, **119**, 698
interferon γ, **119**, 59, 147, 196
interleukin 1, **116**, 475, 486
interstitial retinol-binding protein, **123**, 107
lumazine proteins, **133**, 158, 159
on macroporous supports, optimal pH conditions, determination
by electrophoretic titration curves, **104**, 225, 229
by retention mapping, **104**, 227, 229
mannose, dolichyl phosphate mannose, retinyl phosphate mannose, and retinyl phosphate, **123**, 65
mannose, retinyl phosphate mannose, mannose phosphate, and GDP-mannose, **123**, 66
mast cell growth factor, **116**, 557
membrane proteins, **104**, 336
methanopterin and tetrahydromethanopterin derivatives, **122**, 415, 419
monoclonal IgE, **116**, 79
nucleotide-binding uncoupling protein, **126**, 501
operating conditions, **104**, 181
O-phosphates, **107**, 11
phosphoramidate-containing proteins, **107**, 33
polyclonal IgE, **116**, 81
polypeptide β_1, **116**, 240
protein disulfide-isomerase, **107**, 289
proteins, sample preparation for, **104**, 189
recombinant human interferon αA, **119**, 160
reduced NAD, **122**, 153
riboflavin-binding protein, **122**, 227
riboflavin phosphates, **122**, 213
selenoproteins, **107**, 608
somatomedin C, **109**, 807
thymopoietin, **116**, 287
thymosin α_1, **116**, 235
thymosin β_4, **116**, 249
thymulin, **116**, 273
tubulin from sea urchin egg, **134**, 132
vitamin B_6 metabolic forms, **122**, 102
Ionomycin
induction of human immune interferon, **119**, 69

Ionophore A-23187
induction of human interferon γ, **119**, 69
and mezerein, induction of human interferon γ, **119**, 74, 193
Ion-pair chromatography
short-chain prenyl phosphates on HPLC column, **111**, 250
Ions
negative air
concentration measurements, **105**, 244
laboratory generation, **105**, 240
natural generation, **105**, 238
Iontophoresis
in dissociated tissue culture electrophysiological recording, **103**, 124
Ion transport
bacterial, in maintenance of intracellular pH, **125**, 349
extrusion systems in *Escherichia coli*
antiporter activity, *in vitro* measurements, **125**, 334
energy coupling *in vivo*, determination, **125**, 331
in lipid bilayers
effect of phase transitions, **127**, 484
EPR measurement, **127**, 502
related membrane proteins, analysis with planar bilayers, **127**, 486
Iron
^{59}Fe, macrophage-induced release from tumor cells, assay, **132**, 516
in heme proteins, spectrophotometric determination
with 1,10-phenanthroline, **123**, 322
with 4-(2-pyridylazo)resorcinol, **123**, 323
Iron salts
detection *in vivo*, **105**, 56
role in hydroxyl radical formation, **105**, 52
Iron–sulfur protein
from bovine cardiac mitochondria
purification, **126**, 212
reconstitution, **126**, 220
resolution, **126**, 218
magnetic circular dichroism, **130**, 286
NADH dehydrogenase (ubiquinone)
properties, **126**, 366
resolution, **126**, 366
Neurospora cytochrome reductase
isolation, **126**, 204
structural properties, **126**, 206
N-Isobutoxycarbonyl-2-isobutoxy-1,2-dihydroquinoline
modification of F_1-ATPase carboxyl groups, **126**, 719
Isodityrosine
in plant cell wall
chemical synthesis, **107**, 395
isolation and qualitative identification, **107**, 390
quantitative estimation, **107**, 394
Isoelectric focusing
apolipoprotein C, **129**, 451
apolipoprotein C-II, **128**, 292
apolipoprotein E, **128**, 276, 280, 825, 837
apolipoproteins A-I, A-II, and A-IV, **129**, 423, 424
in detection of nonenzymatic protein glycosylation, **106**, 81
GAT antigen-specific suppressor factors, **116**, 330, 334
high-resolution preparative
apparatus description, **104**, 258
gel matrices for, **104**, 262
load capacity, determination, **104**, 265
pH gradients, generation, **104**, 259
proteins, **104**, 269, 272
recovery, influencing factors, **104**, 271
resolution, optimization, **104**, 266
Ig isotypes from bovine × murine hybridomas, **121**, 258
interferon from transformed rat cells, **119**, 225
lipoproteins
on agarose gel, **128**, 435, 438
in liquid phase, **128**, 433
on polyacrylamide gel, **128**, 434
mast cell growth factor, **116**, 559
in measurement of ligand–protein interactions, **117**, 393
migration inhibitory factor, **116**, 393
myosin P-light chain, **102**, 69
narrow-range, somatomedins, **109**, 777
polypeptide β_1, **116**, 242
somatomedin C, **109**, 809, 815
thymosin α, **116**, 256
thymosin α_1, **116**, 242

thymosin β_4, **116**, 252
thymosin fraction 5 and 5A, **116**, 224
Isoelectric point
IgG-binding factor, estimation, **116**, 414
T-cell suppressor factor for mixed leukocyte response, **116**, 424
Isomerase
affinity chromatography, **104**, 21, 22
localization in gels, **104**, 434
Isomerization
enzymatic, protein disulfides, **107**, 281
nonenzymatic, protein disulfides, **107**, 301
prolines, detection with proteolytic enzymes, **131**, 107
all-*trans*-[11-^{3}H]retinoic acid to *cis* isomers *in vivo*, HPLC analysis, **123**, 84
Isomorphous replacement
for macromolecular phasing
applications, **115**, 100
methods, **115**, 4, 92, 96
molecular boundary determination, **115**, 99
phase derivations, **115**, 9
previous approaches and background, **115**, 90
sources of error, **115**, 11
statistical data in, **115**, 13
for protein phasing, by minimum variance Fourier coefficient refinement, **115**, 15
Isonitriles
–cellulose disks, in biotin/avidin assays, preparation, **122**, 74
Isopentenyl-diphosphate Δ-isomerase
chicken liver
assay, **110**, 93
properties, **110**, 98
purification, **110**, 94
Claviceps purpurea
assay, **110**, 93
properties, **110**, 98
purification, **110**, 96
Isopentenyl pyrophosphate
purification, **110**, 141
synthesis, **110**, 134
thin-layer chromatography, **111**, 248
Isopentenyl pyrophosphate isomerase
tomato fruit chromoplasts
assay, **110**, 212
extraction and partial purification, **110**, 217
properties, **110**, 219
Isoprenoid benzoates
reverse-phase HPLC, **111**, 254
synthesis, **111**, 253
Isoprenoid naphthoates
reverse-phase HPLC, **111**, 254
synthesis, **111**, 253
Isoprenoids
gas chromatography
applications, **111**, 176
direct injection technique, **111**, 160
identification techniques, **111**, 170
stationary phase considerations, **111**, 163
system considerations, **111**, 161
isolation methods, **111**, 150
Isoproterenol
responsiveness of MDCK cells
evaluation, **109**, 363
measurement, **109**, 361
Isopycnic centrifugation
phagocytes, **132**, 225, 237, 239, 241
Isotachophoresis
5-aminolevulinic acid derivatives, **123**, 370
preparative
applications, **104**, 299
basic principles, **104**, 282
in column apparatus with polyacrylamide gel, **104**, 285
cooling systems, **104**, 297
free-flow technique, **104**, 297
in glass tubes with polyacrylamide gel, **104**, 296
hollow cylinder technique, **104**, 296
in horizontal slab of granulated gel, **104**, 289
Isotope dilution assay
riboflavin, **122**, 221
Isotopes
radioactive, *see* Radioisotopes
Isotyping
interleukin-2 antibodies, **116**, 509
IVR, *see* Inhibitors of virus replication

J

Juvenile hormone
in assay for epoxide hydrolase, **111**, 305
binding proteins
binding assays, **111**, 499, 506
demonstration, **111**, 500
photoaffinity labeling with epoxyfarnesyl diazoacetate, **111**, 509
carrier protein from *Manduca sexta* hemolymph
assay, **111**, 482
properties, **111**, 486
purification, **111**, 484
radiolabeled, esterase activity
metabolite identification, **111**, 488
partition assay, **111**, 490
TLC assay, **111**, 493
type III, biosynthesis *in vitro*, radiochemical assay, **111**, 530
applications, **111**, 538
validation, **111**, 537

K

Kainic acid
excitotoxic properties in neuroendocrine research, **103**, 384
Kanamycin
resistance in transformed plants, assay, **118**, 638
α-Keto acid ω-amidase, *see* ω-Amidase
α-Ketoglutaramate
assay in biological samples, **113**, 356
α-Ketoglutarate
decarboxylation, in assay for prolyl hydroxylase, **107**, 365
radiolabeled, in preparation of isotopically labeled L-glutamate, **113**, 61
Ketomethylthiobutyric acid
ethylene production, in measurement of hydroxyl radical production, **105**, 395
Kidney
bovine, cell cultures
production of interferon β, **119**, 211
purification of interferon β, **119**, 213
canine, plasma membranes, parathyroid hormone receptor assay, **109**, 52
catalase measurement, **105**, 123
guinea pig and rat, simultaneous determination of ubiquinones, **105**, 147
mammalian, measurement of 25-hydroxyvitamin D 1α-hydroxylase, **123**, 159
porcine, purification of
cytochrome-*c* oxidase, **126**, 34
glutaminase, **113**, 246
pantetheine-hydrolyzing enzyme, **122**, 39
rat
calcitonin receptors in
binding assays, **109**, 44
regulation, **109**, 47
isolated perfused
25-hydroxyvitamin D_3 metabolic pathway, **123**, 141
preparation, **123**, 141
purification of
ε-alkyllysinase, **106**, 285
L-α-aminoadipate aminotransferase, **113**, 669
cysteine *S*-conjugate *N*-acetyltransferase, **113**, 517
glutaminase, **113**, 251
glutamine transaminase K, **113**, 346, 347
γ-glutamylcyclotransferase, **113**, 439, 441
γ-glutamylcysteine synthetase, **113**, 381
γ-glutamyl transpeptidase, **113**, 405
glutathione synthetase, **113**, 395
kynurenine aminotransferase, **113**, 669
membrane peptidases, **113**, 475, 479
5-oxo-L-prolinase, **113**, 447
separation of selenoproteins, **107**, 614
transplanted, leukocyte detection with cellular chemiluminescence, **133**, 499
vitamin E levels, analytical techniques, **105**, 138
Kidney bean
purification of
cytoplasmic aminoacyl-tRNA synthetase, **118**, 218
cytoplasmic tRNAs, **118**, 215

Kinetochores
–centromere fraction from HeLa cell metaphase chromosomes
antigen identification, **134**, 271, 277
immunofluorescent staining, **134**, 271
preparation, **134**, 270
Klebsiella
K. aerogenes, purification of
glutamate synthase, **113**, 328
oxaloacetate decarboxylase, **125**, 530
K. pneumoniae, purification of nitrogenase, **118**, 511
Kupffer cells
rat
collection, **108**, 286
fibronectin-mediated phagocytosis, *in vitro* assay, **132**, 345
separation, **108**, 291
Kynurenine aminotransferase, *see* Kynurenine–oxoglutarate aminotransferase
Kynurenine–oxoglutarate aminotransferase
rat kidney
assay, **113**, 667
properties, **113**, 672
purification, **113**, 669
yeast
assay, **113**, 91
properties, **113**, 93
purification, **113**, 92
L-Kynurenine transaminase, *see* Kynurenine–oxoglutarate aminotransferase

L

β-Lactam
in milk, lysis test, **133**, 276
role of α-aminoadipate, **113**, 657
β-Lactamase
gene cloning into pIN III-*ompA*, **125**, 143
in measurement of bacterial outer membrane permeability, **125**, 271
membrane, N-terminal glyceride-cysteine modification
identification criteria, **106**, 366
nature of, **106**, 365
sequence required for, **106**, 368
Lactate dehydrogenase
affinty HPLC, **104**, 218
in assay for assessment of phagocytic function, **132**, 121, 161
bovine heart
affinity precipitation, **104**, 366, 368
ion-exchange chromatography on macroporous supports, **104**, 229
bovine muscle, ion-exchange chromatography on macroporous supports, **104**, 229
Escherichia coli
assays, **126**, 371
immunoaffinity purification, **126**, 374
human neutrophil, assay, **132**, 275
rabbit muscle, in assay for D-glutamate-D-amino acid transaminase, **113**, 110
Lactide
–glycolide copolymers, preparation, **112**, 102
Lactobacillus
L. casei
in assay for folic acid derivatives, **122**, 270
growth in culture, **122**, 351
purification of
folate transport proteins, **122**, 261
folylpoly(γ-glutamate) synthase, **122**, 353
transport-defective, purification of folate transport proteins, **122**, 266
L. leichmanni, extraction of corrinoids, **123**, 16
L. plantarum, purification of undecaprenylpyrophosphate synthetase, **110**, 289
strain 30a
growth in culture, **122**, 129, 136
purification of
prohistidine decarboxylase, **122**, 135
pyruvoyl-dependent histidine decarboxylase, **122**, 130
Lactoferrin
in assay for assessment of phagocytic function, **132**, 123, 163
β-Lactoglobulin A
asymptotic mass migration boundary shapes
calculations, **130**, 8
experimental determinations, **130**, 12

Lactoperoxidase
–glucose oxidase, in ^{125}I labeling of interferons, **119**, 263
induced erythrocyte cytolysis, continuous assay, **132**, 491
iodination and coupling reactions, **107**, 476
dissociation, **107**, 483
product identification, **107**, 485
in radiolabeling of
murine interferon, **119**, 322
neuropeptides, **124**, 24
Lactuca sativa, *see* Lettuce
Langerhans cells
epidermal
isolation, panning technique, **108**, 360
quantitation, **108**, 359
surface markers, detection by immunochemical techniques, **108**, 683
Lasers
He–Ne, in transient electric birefringence system, **117**, 200
induced proton pulse, in measurement of proton translocation in membranes and proteins, **127**, 522
Nd:YAG, in UV resonance Raman spectroscopy, **130**, 334
Latex
Hevea brasiliensis
collection, **110**, 41
fractionation, **110**, 42
3-hydroxy-3-methylglutaryl-CoA reductase assay, **110**, 44
Lead
in synthesis of (2′-5′)-oligoadenylic acid tubericidin analog, **119**, 526
Least-squares analysis
minimization techniques in correlation function profile analysis
linear methods, **117**, 267
nonlinear methods
applications, **117**, 321
assumptions, **117**, 302, 334
description, **117**, 264, 301
for multiple data sets, **117**, 311
numerical procedures, **117**, 305
parameter estimation processes, **117**, 310, 312
Raman amide I and III spectra, **130**, 312
Least-squares refinement
constrained–restrained, proteins and nucleic acids
applications, **115**, 281, 308
difficulties in, **115**, 271
implementation, **115**, 279
mathematical description, **115**, 277
restraints and constraints in, **115**, 272
in crystallography
acceleration techniques, **115**, 36
application in powder diffraction, **115**, 40
atomic parameter methods, **115**, 27
chemical constraints, **115**, 28
chemical restraints, **115**, 29
full-matrix technique, **115**, 31
hardware requirements, **115**, 38
Newton–Raphson method, **115**, 25
nonlinear methods, **115**, 26, 30
primary goals, **115**, 29
simplex method, **115**, 24
standard deviation calculations, **115**, 29, 40
strategy, **115**, 39
Lecithin–cholesterol acyltransferase, *see* Phosphatidylcholine–sterol acyltransferase
Lectins
affinity chromatography, **104**, 54
coupling to polyacrylhydrazidoagarose, **104**, 18
as drug and enzyme carriers, **112**, 248
fluorescein isothiocyanate-labeled, in staining of glycoproteins in gels, **104**, 449
induction of human immune interferon production, **119**, 57
Leishmania sp.
acquisition, cultivation, and preparation, **132**, 606
parasitism of macrophages, analysis and applications, **132**, 614, 617, 620
Lemon
purification of carbocyclase, **110**, 413
Lesions
axon-sparing, *see* Axon-sparing lesions
Lettuce
chloroplasts, purification of violaxanthin deepoxidase, **110**, 311

Leucine-enkephalin
antiserum preparation, **124**, 619
radioimmunoassay, **124**, 620
radioiodination, **124**, 620
regulation of cAMP levels in hybrid cells, testing procedures, **109**, 327
separation, **124**, 621
Leucophaea maderae, *see* Cockroach
Leucyltransferase
assay, **106**, 198
deficient mutants, characterization, **106**, 204
purification from *Escherichia coli*, **106**, 200
substrate identification, **106**, 202
L-Leucyl-tRNA:protein transferase, *see* Leucyltransferase
Leukocytes
fluid pinocytosis, analysis with fluorescent dextrans and horseradish peroxidase, **108**, 336
mammalian, cyclic nucleotide levels, assay, **132**, 432
peripheral blood
histocompatibility antigen expression, measurement of interferon-induced changes, **119**, 688
interferon induction in, **119**, 133
isolation, **119**, 131
production of
interferon α, **119**, 39, 137
interferon δ, **119**, 97
interferon γ, **119**, 50, 56, 63, 72, 83, 88
purification of methionine sulfoxide reductase, **107**, 358
separation, **108**, 91
polymorphonuclear, *see* Neutrophils
Levamisole
in detection of soluble immune response suppressor, **116**, 397
Levonorgestrel
loading in poly(lactic/glycolic acid) polymers, **112**, 438
release from poly(ortho ester) polymer disks, **112**, 425
Leydig cells
isolated, from rodents
characterization, **109**, 281
functional properties, **109**, 287
identification, **109**, 284
preparation, **109**, 276
LHRH, *see* Luteinizing hormone-releasing hormone
Ligand-binding isotherms
linked to protein self-associations
data analysis, **117**, 503
measurement, **117**, 497
simulations, **117**, 505
Ligase
affinity chromatography, **104**, 21, 22
localization in gels, **104**, 434
Light
and 5-bromodeoxyuridine, in depletion of immunoreactive T cells, **108**, 263, 268
Light microscopy
in assessment of phagocytic function, **132**, 101, 138
horseradish peroxidase in neuroendocrine cells, **103**, 193, 198, 205
horseradish peroxidase uptake by leukocytes, **108**, 340
low-density lipoprotein receptors, **129**, 201
phase-contrast, *Leishmania* amastigotes in macrophages, **132**, 615, 620
triple-labeled neuroendocrine cells, **103**, 143
Light scattering
analysis in FACS, **108**, 205
apolipoproteins in solution, **128**, 384
dynamic, in measurement of bacterial motility, **125**, 586
electrophoretic, *see* Electrophoretic light scattering
forward and right-angle, in flow cytometric analysis of neuroendocrine cells, **103**, 230
small-angle, *see* Small-angle light scattering
Limiting dilution analysis
large granular lymphocytes, **132**, 454
in repetitive subcloning, statistical evaluation, **121**, 412
Linalyl pyrophosphate
radiolabeled, chemical synthesis, **110**, 410

Lincomycin
resistance in chloroplast mutants, selection for, **118**, 615, 619
Linewidth distribution
photon correlation spectroscopy data for macromolecular solutions and colloidal suspensions
cumulant analysis, **117**, 279
functional approximations, **117**, 290
histogram analysis, **117**, 287
multiexponential analysis, **117**, 281
regularized solution, **117**, 291
Linoleic acid
as supplement in preparation of SFH medium, **121**, 284
Lipase
localization in gels, **104**, 438
salt-resistant, in human liver and plasma
assays, **129**, 719, 732
characterization, **129**, 726
purification, **129**, 718
Lipemia
effects on postprandial lipoprotein metabolism, **129**, 477
Lipid bilayers
conformational water forces between layers, direct measurement, **127**, 357
intermolecular forces, direct measurement by osmotic stress, **127**, 412
phase transitions, effect on
H^+/OH^- permeability, **127**, 484
water permeability, **127**, 482
planar
proton translocation, measurement in model systems, **127**, 477
reconstitution of bacterial cytochrome oxidases, **126**, 123
water penetration depth
capacitance measurements, **127**, 514
combined methods, **127**, 515
X-ray diffraction measurements, **127**, 512
Lipid enrichment
mitochondrial inner membranes
assessment, **125**, 39
methodology, **125**, 36
Lipid hydroperoxides
detection, chemical methods, **105**, 293
measurement, in detection of lipid peroxidation, **105**, 288
picomole levels, detection by dichlorofluorescein fluorescent assay, **105**, 352
Lipid peroxidation
associated metabolism of hydrocarbons to carcinogenic derivatives, detection, **105**, 347
chemistry, **105**, 273
detection methods, overview, **105**, 283
fluorescent damage products
extraction, **105**, 338
spectrophotometric analyses, **105**, 339
monitoring
with ethane and pentane as indicators, **105**, 311
with exhaled ethane as indicator, **105**, 305
as source of singlet oxygen, **105**, 229
Lipid peroxides
plasma and serum, assay, **105**, 328
Lipids
–apolipoprotein complexes
chemical modification, **129**, 548
lipoprotein mass determination, **129**, 557
preparation, **129**, 546
conjugated dienes, spectrophotometric detection, **105**, 331
core, low-density lipoprotein, reconstitution, **128**, 609
interactions with synthetic peptide analogs of apolipoproteins, **128**, 636
in KB cell receptosomes, composition, **109**, 265
lipoprotein
apolipoprotein A-containing, composition, **129**, 142
circular dichroism, **128**, 519
enzymatic quantification, **129**, 101
order, β-carotene as probe, **128**, 524
staining, **128**, 418, 422
monolayers at air–water interface, apolipoprotein adsorption, **128**, 399
in phagocytes, staining, **132**, 134, 178
phase transitions, volume-perturbation kinetic calorimetry, **130**, 536
plasma transport, genetic regulation, analysis in murine model, **128**, 877
reconstitution of chloroplast CF_0-F_1, requirements, **126**, 519

serum
detection limits, **129**, 74
HPLC elution pattern, **129**, 67
synthetic, reactions with lecithin–cholesterol acyltransferase, **129**, 785
vesicles, preparation with alkyl glycoside detergents, **125**, 34
Lipofuscin
fluorescent, measurement, **105**, 486
identification, **105**, 484
microspectrophotometric analysis *in situ*, **105**, 342
quantification, **105**, 485
Lipogenesis
adipocyte, effect of photoreactive insulin covalent binding, **103**, 64
Lipopolysaccharides
as B-cell hybridoma growth promoters
comparison with other growth promoters, **121**, 12
overview, **121**, 9
preparation, **121**, 10
induction of pulmonary indoleamine 2,3-dioxygenase, **105**, 68
interaction with interferon γ, chemiluminescent assay, **133**, 517
stimulation of lymph node cells *in vitro*, **121**, 264
Lipoprotein (a)
apo(a) antigen
discovery, **129**, 167
identification, **129**, 169
isolation, **129**, 180
properties, **129**, 184
quantitation, **129**, 174
blood levels, influencing factors, **129**, 167
Lipoprotein lipase
bovine milk
assay, **129**, 692
in assay for very-low-density lipoprotein metabolism in whole plasma, **129**, 704
properties, **129**, 703
purification, **129**, 698
catalyzed reactions
kinetics, **129**, 757
substrates for, **129**, 746
discovery, **128**, 4
human plasma
assay, **129**, 692
properties, **129**, 703
purification, **129**, 698
mechanism, analytical techniques, **129**, 738
metabolic function, **128**, 17
structural domains, **129**, 741
Lipoprotein receptors
–lipoprotein interactions, analytical techniques, **129**, 542
low-density, *see* Low-density lipoprotein receptors
Lipoproteins
analytical techniques, **128**, 167
impact of technology, **128**, 144
apolipoprotein A-containing
characterization, **129**, 138, 143
isolation, **129**, 132
types of, **129**, 132
apolipoprotein B-containing
characterization, **129**, 126
metabolism, analytical techniques, **129**, 395
modification mechanisms, **129**, 124
apolipoprotein D-containing, isolation, **128**, 299
apolipoprotein E-containing
characterization, **129**, 162
delipidation, **128**, 274
isolation, **128**, 274; **129**, 146, 186
properties, **129**, 156
separation, **129**, 159
apolipoprotein F-containing, isolation and characterization, **128**, 305
assembly and secretion, **128**, 19, 66
binding determinants, mapping, **129**, 559
catabolism in liver
analytical techniques, **129**, 596, 599
pathways, evaluation, **129**, 609
role of
apolipoproteins, **129**, 607
parenchymal and nonparenchymal cells, **129**, 605
charge heterogeneity, **128**, 432
chemical composition, **128**, 11
chemical properties, **128**, 89
cholesterol-induced, **128**, 142
cholesteryl alkyl ether-labeled, analysis, **129**, 826

classification theory, **129**, 368
collection methods, **129**, 81
cryofixed
electron microscopy, **128**, 461
fracture and etching, **128**, 459
preparation, **128**, 458
replication, **128**, 461
degradation sites in tissues, assessment techniques
applications, **129**, 626
overview, **129**, 612
sucrose–cyanuric chloride method, **129**, 619
tyramine–cellobiose method, **129**, 622
delipidation methods, overview, **128**, 213
density classes, properties, **128**, 157
density gradient ultracentrifugation, **128**, 170
digital imaging fluorescence microscopy, **129**, 867
discovery, **128**, 3
distribution anomalies and consequences, **128**, 77
effect on hepatic lipase activity, **129**, 730
encoding genes
expression regulation, **128**, 53
organization and expression, **128**, 42
epitope expression, **128**, 545
fluorescence labeling, **129**, 562, 848
fractionation, **128**, 166
glucosylated
antibody preparation, **128**, 357
immunochemical identification, **128**, 359
preparation, **128**, 356, 358
hepatic biosynthesis, analytical techniques, **129**, 498
high-density, *see* High-density lipoproteins
HPLC elution patterns, **129**, 64
human phenotypes, **128**, 6
immunological properties, **128**, 122
interactions with liposomes
carboxyfluorescein leakage assay, **128**, 657
mechanisms, **128**, 668
interconversions, role of liver, **129**, 611
interstitial fluid
characterization, **129**, 667
composition, **129**, 660
synthesis, analytical techniques, **129**, 674
intraplasma conversions, **129**, 355
isoelectric focusing
on agarose gel, **128**, 435, 438
in liquid phase, **128**, 433
on polyacrylamide gel, **128**, 434
two-dimensional analysis, **128**, 440
isolation, **128**, 10, 151
lipid–apolipoprotein complexes
chemical modification, **129**, 548
preparation, **129**, 546
lipids in
circular dichroism, **128**, 519
enzymatic quantification, **129**, 101
ether analogs
biological applications, **129**, 824, 834, 842
as enzyme inhibitors, **129**, 842
syntheses, **129**, 817, 832, 835
mass determination, **129**, 557
order, β-carotene as probe, **128**, 524
lipid transport, comparison with albumin, **128**, 338
localization and identification, **128**, 418
low-density, *see* Low-density lipoproteins
metabolic theory, **129**, 369
metabolism
analytical techniques, **129**, 353
relevance of apolipoprotein solution properties, **128**, 386
in steroidogenic tissues, analysis, **129**, 683
molecular weight determination, **128**, 430
monoclonal antibodies
applications, **128**, 535
assays, **128**, 530
production, **128**, 529
purification, **128**, 534
nascent
characterization, **129**, 292, 293
intracellular localization, **129**, 294
isolation, **129**, 291
radiolabeling, **129**, 285
negatively stained
electron microscopy, **128**, 448
preparation, **128**, 443
NMR studies, **128**, 472
nomenclature and classification, **128**, 71

physical properties, **128**, 10, 108
physicochemical properties, **128**, 121
polyclonal antisera
applications, **128**, 542
production, **128**, 528
postprandial
isolation, **129**, 472
metabolism
analytical techniques, **129**, 470
biological significance, **129**, 477
effects of alcohol, **129**, 480
pathways, **129**, 474
preferential hydration, determination, **129**, 44
processing and catabolism, **128**, 27
quantitation, precipitation methods, **129**, 83
radiolabeling, **129**, 348, 545
–receptor interactions, assays, **129**, 552
recombinant, structural analysis with crosslinking reagents, **128**, 624
separation
analytical ultracentrifugation methods, **129**, 26
charge-dependent methods, **128**, 432
gel filtration methods, **129**, 45
by gradient gel electrophoresis, **128**, 417
HPLC methods, **129**, 57
comparison with sequential ultracentrifugal flotation, **129**, 70
sequential flotation ultracentrifugation, **128**, 155
species groups, identification, **128**, 78
species separation, **128**, 196
spectroscopic studies, **128**, 515
structural analysis
with crosslinking reagents, **128**, 613
immunochemical methods, **128**, 527
by zonal ultracentrifugation, **129**, 14
structural features, **128**, 12
relevance of apolipoprotein solution properties, **128**, 386
subspecies separation, **128**, 201
synthesis and intracellular targeting, **128**, 60
transport
genetic regulation, analysis in murine model, **128**, 877
mechanisms, **128**, 17
triglyceride-rich
delipidation, **128**, 289
isolation, **128**, 288
uptake by liver
analytical techniques, **129**, 591, 596, 599
receptor-dependent, evaluation, **129**, 601
vertical spin density gradient ultracentrifugation, **128**, 181
very-low-density, *see* Very-low-density lipoproteins
Liposomes
antibody-coated
preparation, **121**, 818, 823
targeting, **121**, 824
antibody-coated and drug-containing
application *in vivo*, **121**, 827
cytotoxicity *in vitro*, **121**, 825
integration into target cells, **121**, 826
cholesteryl alkyl ether-labeled
analysis, **129**, 832
preparation, **129**, 831
in delivery of plasmid DNA to plant protoplasts, **118**, 587
F_0F_1, electric pulse-driven ATP synthesis, **126**, 640
interactions with lipoproteins
carboxyfluorescein leakage assay, **128**, 657
mechanisms, **128**, 668
muramyl peptide-containing
activation of macrophages, **132**, 601
preparation, **132**, 597
phospholamban reinsertion into, **102**, 275
preparation, **121**, 822; **128**, 661
proton translocation, measurement
with pH-sensitive dyes, **127**, 476
with pH-sensitive electrodes, **127**, 474
reconstitution of
chloroplast phosphate transport protein, **125**, 724
cytochrome-*c* oxidase, **126**, 170
cytochrome *d* and *o* terminal oxidase complexes, **126**, 109
F_1F_0-ATPase, **126**, 555
HLA-DR antigens, **108**, 610
H^+-pumping transhydrogenase, **126**, 357

membrane-bound inorganic pyrophosphatase, **126**, 540
mitochondrial tricarboxylate carrier,
125, 694
phosphate–hexose 6-phosphate antiport, **125**, 559
photosystem I reaction center and
chloroplast coupling factor complex, **126**, 288
sugar transport activity of *Escherichia
coli*, **125**, 425
Lipoxygenase
assay, chemiluminescence method, **105**,
129
Liquid chromatography
with electrochemical detection, pterins
in urine and tissue, **122**, 301
rat interferon from transformed cells,
119, 223
Liquid scintillation counter
for cellular chemiluminescence measurement, **133**, 504
in detection of bacteria containing bioluminescent genes, **133**, 72
for leukocyte luminescence velocity
measurement, **133**, 460
Lithium
in analysis of phospholipid turnover in
cultured cells, **124**, 439
in solution, coordination number, **127**,
326
–sugar uptake in *Escherichia coli*, measurement, **125**, 382
Liver, *see also* Hepatocytes
bovine, purification of
ferrochelatase, **123**, 403
L-glutamate dehydrogenase, **113**, 26
glutathione reductase, **113**, 486
porphobilinogen synthase, **123**, 429
protein disulfide-isomerase, **107**,
288
catalase measurement techniques, **105**,
123
chicken embryo, purification of 5-amino-
levulinate synthase, **123**, 397
chicken, purification of
farnesylpyrophosphate synthetase,
110, 180
hydroxymethylglutaryl-CoA synthase,
110, 22

isopentenyl-diphosphate Δ-isomerase,
110, 94
mevalonate 5-pyrophosphate decarboxylase, **110**, 89
prenyltransferase, **110**, 148
guinea pig
determination of ubiquinones, **105**,
147
purification of transglutaminase, **113**,
361
hamster, preparation of postnuclear
membranes, **123**, 62
human
purification of
Cu,Zn- and Mn-superoxide dismutase, **105**, 89
glutathione transferase, **113**, 501
lipase, **129**, 724
prenyltransferase, **110**, 159
vitamin K assay by HPLC with UV
detection, **123**, 236, 242
lipoprotein catabolism
analytical techniques, **129**, 596, 599
pathways, evaluation, **129**, 609
role of
apolipoproteins, **129**, 607
parenchymal and nonparenchymal
cells, **129**, 605
lipoprotein uptake
analytical techniques, **129**, 591, 596,
599
receptor-dependent, evaluation, **129**,
601
role in lipoprotein interconversions,
129, 611
murine
detection of
oxidized and reduced phosphofructokinase, **107**, 347
pterins, **122**, 301
microsomes in
diphenylhydantoin metabolism,
interferon effects, measurement, **119**, 717
drug metabolism, interferon effects,
measurement, **119**, 721
isolation, **119**, 716, 720
purification of
ferrochelatase, **123**, 403
β_2-microglobulin, **108**, 495

peroxisomal carnitine acyltransferases, **123**, 277
porcine, purification of
cytochrome-*c* oxidase, **126**, 34
geranylgeranylpyrophosphate synthetase, **110**, 185
methylenetetrahydrofolate reductase, **122**, 376
mevalonate kinase, **110**, 60, 74
phosphomevalonate kinase, **110**, 82
prenyltransferase, **110**, 150
rabbit
growth hormone receptors
characterization, **109**, 701
purification, **109**, 694
purification of
L-arginyl-tRNA:protein transferase, **106**, 201
cytochromes *P*-450 active in bile acid synthesis, **111**, 372, 374
pyridoxamine (pyridoxine) 5′-phosphate oxidase, **122**, 116
rat
collection of macrophages, **108**, 286
digitonin fractionation, **105**, 112
extraction of
thiamin and phosphates, **122**, 17
vitamin E, **105**, 136
fibronectin-mediated phagocytosis, *in vitro* assay, **132**, 339
isolated perfused, alkane production assay, **105**, 311
lipoprotein synthesis analysis
in hepatocytes, **129**, 515
perfusion techniques, **129**, 509
slice techniques, **129**, 511
whole animal studies, **129**, 499
membrane glucagon receptors
assays, **109**, 3, 15
covalent labeling, **109**, 207
functional analysis by membrane fusion, **109**, 346
microsomes in
drug metabolism, interferon effects, measurement, **119**, 721
isolation, **119**, 720
perfused, Ca^{2+} fluxes, hormone effects, **109**, 553
peroxidizing microsomes, malonaldehyde measurement, **105**, 321
preparation of
endosome-depleted Golgi fractions, **109**, 256
growth hormone receptors, **109**, 694
low-density endosomes, **109**, 247
postnuclear membranes, **123**, 62
purification of
acetoacetyl-CoA synthetase, **110**, 6
N-acetylglutamate synthase, **113**, 29
N-acetyltransferase, **103**, 489
5-aminolevulinate synthase, **123**, 401
asparagine transaminase, **113**, 605
catechol *O*-methyltransferase, **103**, 485
cysteine-conjugate β-lyase, **113**, 512
cytochrome-*c* oxidase, **126**, 34
cytochromes *P*-450 active in bile acid synthesis, **111**, 368, 370
dimethylglycine dehydrogenase, **122**, 257
folate-binding protein:cytosol I, **122**, 252
folate oligoglutamate:amino acid transpeptidase, **122**, 369
folic acid derivatives, **122**, 271
glucosamine-6-phosphate synthase, **113**, 279
glutamine synthetase, **113**, 188
glutamine transaminase L, **113**, 340, 342
glutathione peroxidase, **107**, 610
glutathione reductase, **113**, 487
glutathione transferase isozymes, **113**, 504
α-keto acid ω-amidase, **113**, 353, 357
microsomal glutamate carboxylase, **107**, 553
poly(ADP-ribose) synthetase, **106**, 502
sarcosine dehydrogenase, **122**, 257
squalene epoxidase, **110**, 377
squalene and sterol carrier protein, **111**, 296
UDP-*N*-acetylglucosamine–lysosomal enzyme *N*-acetylglucosaminephosphotransferase, **107**, 168
quantitation of superoxide dismutase, **105**, 112

subcellular distribution of 4-hydroxy-benzoate polyprenyltransferase, **110**, 330
subcellular fractionation, **105**, 106
superoxide anion production, measurement, **105**, 373
thiol/disulfide ratio, variation *in vivo*, **107**, 332
trout, purification of thymosin β_{11}, **116**, 268
vitamin E levels, analytical techniques, **105**, 138
Lobster
carapace, extraction of carotenoproteins, **111**, 219, 244
Locust
hemolymph
photoaffinity labeling of juvenile hormone binding proteins, **111**, 526
purification of 20-hydroxyecdysone binding protein, **111**, 446
Locusta migratoria, see Locust
Loligo pealii, see Squid
Lolium multiflorum, see Ryegrass
Long-chain fatty-acyl-CoA synthetase
cardiac mitochondrial, assay, **125**, 18
Low-density lipoprotein receptors
acetylated
activity assays, **129**, 222
isolation, **129**, 219
ligand blotting, **129**, 224
sources, **129**, 217
antibody preparation, **129**, 205
assays, **109**, 414
bovine, cDNA cloning, **128**, 895
interactions with very-low-density lipoproteins, **129**, 331
intracellular, visualization, **129**, 209
mediated endocytosis, visualization, **129**, 212
probes for, preparation, **129**, 202
properties, **109**, 416
purification from
bovine adrenal cortex, **109**, 410
cultured cells, **109**, 413
surface membrane, visualization, **129**, 207
Low-density lipoproteins
catabolism
analytical techniques, **129**, 567
receptor-independent
characteristics, **129**, 586
measurement, **129**, 570, 576
receptor-mediated, measurement, **129**, 570
delipidation, **128**, 252
endocytosis, defective cell mutants, selection methods, **129**, 227
–ferritin conjugates, preparation, **129**, 203
fluorescent, preparation, **129**, 202
–gold conjugates, preparation, **129**, 204
hydrophobic core extraction and reconstitution, **128**, 609
isolation, **128**, 250; **129**, 329
isolation of apolipoprotein B, **128**, 585
metabolic origins, **129**, 355
metabolism, defective cell mutants
immobilization, **129**, 241
induction, **129**, 240
screening, **129**, 245
reconstituted complexes
biological characteristics, **128**, 606
physical characteristics, **128**, 600
preparation, **128**, 590
storage, **128**, 252
structural analysis with crosslinking reagents, **128**, 623
Luciferase
bacterial, *see* Alkanal monooxygenase (FMN-linked)
dinoflagellate
assays, **133**, 312
properties, **133**, 316
purification, **133**, 313
firefly, *see Photinus*-luciferin 4-monooxygenase (ATP-hydrolysing)
Luciferin
binding protein, dinoflagellate
assays, **133**, 324
properties, **133**, 326
purification, **133**, 324
–α-bungarotoxin complexes
activity assay, **133**, 45
fluorescent properties, **133**, 45
membrane binding properties, **133**, 47
preparation, **133**, 43
dinoflagellate
assays, **133**, 318
properties, **133**, 321

purification, **133**, 319
firefly, commercial sources, **133**, 65
–GnRH analog conjugates
activity determination, **124**, 34
binding to anterior pituitary membranes, **124**, 32
bioactivity, **124**, 31
preparation, **124**, 30
in receptor ligand assays, advantages, **124**, 35
sensitivity, **124**, 31
Lucigenin
in cellular chemiluminescence assays, **133**, 501
as chemiluminigenic probe for phagocyte oxygenation activity
applications, **133**, 476, 480
azide effects, **133**, 469
comparison with luminol, **133**, 466
concentration effects, **133**, 465
temperature effects, **133**, 485
Lumazine proteins
Photobacterium strains
assay, **133**, 151
properties, **133**, 167
purification, **133**, 158, 164
Luminescence
in singlet oxygen detection, **105**, 42
Luminol
in cellular chemiluminescence assays, **132**, 545; **133**, 501
as chemiluminigenic probe for phagocyte oxygenation activity
applications, **133**, 476, 480
azide effects, **133**, 469
comparison with lucigenin, **133**, 466
concentration effects, **133**, 461
temperature effects, **133**, 485
Luminometers
for cellular chemiluminescence measurement, **133**, 505
Lung
murine
collection of macrophages, **108**, 278
purification of granulocyte–macrophage colony-stimulating factor, **116**, 592
rabbit, collection of macrophages, **108**, 278; **119**, 678
rat, collection of macrophages, **119**, 678
vitamin E levels, analytical techniques, **105**, 138
Luteal cells
isolated
applications, **109**, 310
biochemical/endocrine function, **109**, 302
hormone receptor function, **109**, 306
morphological aspects, **109**, 301
preparation, **109**, 299, 307
Luteinizing hormone
assembly
detection, **109**, 737
methods for, **109**, 747
deglycosylated
chromatographic purification, **109**, 729
preparation, **109**, 726, 733
disassembly
detection, **109**, 737
methods for, **109**, 739
function in isolated luteal cells, **109**, 302
release in perifused pituitary cells, **124**, 79
Luteinizing hormone receptors
in isolated luteal cells, functional analyses, **109**, 306
Luteinizing hormone-releasing hormone
antagonists, synthesis, **103**, 27
biosynthesis
in vitro studies, **103**, 521
in vivo studies, **103**, 519
enzymatic degradation by hypothalamic tissue
assays, **103**, 542, 546
physiological modeling, **103**, 540
products, HPLC, **103**, 543
function in isolated luteal cells, **109**, 305
hypothalamic precursor form
detection, **124**, 329
encoding mRNA
cell-free translation, **124**, 322
isolation, **124**, 320
immunoprecipitation, **124**, 324
secretion
dynamics in model system, **124**, 467
measurement techniques, **124**, 469
secretion granules
isolation from brain, **124**, 257
lectin-binding sites, demonstration, **124**, 259

molecular aspects of peptide release, analytical techniques, **124**, 262
synthesis, **103**, 27
Lutropin, *see* Luteinizing hormone
Lyase
affinity chromatography, **104**, 21, 22
localization in gels, **104**, 434
Lycopersicon, *see* Tomato
Lymph nodes
bovine
antigen stimulation *in vitro*, **121**, 264
preparation for cell fusion, **121**, 264
murine
popliteal, in assay for graft-versus-host reactions, **108**, 33
preparation of single-cell suspensions, **108**, 45
removal methods, **108**, 43
rat
mesenteric, as lymphocyte source for hybridoma production, **121**, 55
popliteal, in assay for graft-versus-host reactions, **108**, 33
preparation of single-cell suspensions, **108**, 45
removal methods, **108**, 43
Lymphoblastoid cells
as human hybridoma fusion partners
comparison with plasmacytoma cells, **121**, 131
EBV-induced nuclear antigen assay, **121**, 125
morphology, **121**, 123
non-Ig-secreting, selection, **121**, 136
overview, **121**, 127
ultrastructural analysis, **121**, 125
in large-scale production of human interferon
enhancers, **119**, 37
induction with Sendai virus, **119**, 36
maintenance, **119**, 35
stimulators, **119**, 37
Lymphoblasts
human, purification of
HLA-A and HLA-B antigens, **108**, 589
HLA-DR antigens, **108**, 602, 608
lymphotoxin-containing supernatants, production and harvesting, **116**, 442
T-cell, detection of thymulin receptors, **116**, 275
Lymphocytes
B-cell, *see* B cells
from chronic lymphocytic leukemia patient, fusion with myeloma cells, **121**, 114
cord blood, separation by affinity chromatography on immobilized peanut agglutinin, **108**, 177
fusion frequency in hybridoma production, **121**, 8
as fusion partners in EBV–hybridoma technique, selection, **121**, 151
hybridomas, antigen-specific, production, **121**, 18
hybridoma supernatant effects, *Mycoplasma* interferences, **121**, 481
IM-9, insulin receptor subunits, radioactive labeling and turnover studies, **109**, 593
large granular
cytotoxicity assay, **132**, 454
limiting dilution analysis, **132**, 455
morphologic analysis, **132**, 448
natural killer activity, **132**, 453, 454
preparation, **132**, 447
progenitors, identification, **132**, 456
separation from T cells, **132**, 449
subsets, separation, **132**, 452
treatment with monoclonal antibody, **132**, 450
lymphoid
fusion with myeloma cells, **121**, 57
localization of surface and intracellular immunoglobulins, **108**, 393
production of migration inhibitory factor, **116**, 380
purification of
class I MHC antigens, **108**, 506
Qa and TL antigens, **108**, 549
separation
by free-flow electrophoresis, **108**, 181
on Sephadex G-10 column, **108**, 303
single-cell suspensions
counting and size analysis, **108**, 45, 46, 76
preparation, **108**, 45, 75; **121**, 55
viability, determination, **107**, 48
mixed cultures, preparation for generation of cytotoxic activity, **132**, 438

monoclonal antibody-labeled, indirect rosetting assay
applications, **121**, 731
methodology, **121**, 726
peripheral blood
cryopreservation technique, **108**, 363
detection, rosetting techniques
with antibody-coated latex spheres, **108**, 56
antigen-specific, **108**, 58
direct and mixed antiglobulin methods, **108**, 386
generalized, **108**, 51
with tumor cells, **108**, 60
FACS analysis, **108**, 237
isolation, **108**, 51, 95, 148, 242
production of
interferon γ, **119**, 77
migration inhibitory factor, **116**, 381
radiolabeled hypusine, **106**, 346
recovery of crude interferon preparations from, **119**, 78
separation, **108**, 91
by antigen-specific electrophoresis, **108**, 189
with carbonyl iron, **108**, 307
by immunoadsorbent affinity chromatography, **108**, 139
on lectin–Sepharose gels, **108**, 153
by mixed rosetting with soybean agglutinin, **108**, 176
rosetting techniques, **108**, 61
on Sephadex G-10 column, **108**, 303
serum, equilibrium density separation on albumin and metrizamide, **108**, 102
splenic
biosynthetic labeling of H-2 and Ia antigens, **108**, 453
centrifugation for hybridoma production, **121**, 175
culture procedure for hybridoma production, **121**, 28
detection
with antibody-coated latex spheres, **108**, 56
hot antigen suicide technique, **108**, 255
hot pulse technique, **108**, 259
rosetting techniques, **108**, 51, 58
with tumor cells, **108**, 60
FACS analysis, **108**, 232
fusion with
hybridoma cells, **121**, 213
IR983F cells, **121**, 237
myeloma cells, **121**, 12, 31, 46, 57, 62, 241, 315
fusion protocol for preparation of antibodies to weak immunogens, **121**, 184
antigen-specific hybridomas, **121**, 197
induction assays with thymopoietin and ubiquitin, **116**, 283
induction of 20α-hydroxysteroid dehydrogenase by interleukin 3, **116**, 544
isolation of Ia antigens, **108**, 527, 536
preparation, **108**, 51; **121**, 62, 312
separation
by affinity chromatography on protein A gels, **108**, 132
on antibody-coated plates, **108**, 118, 124
with carbonyl iron, **108**, 307
by free-flow electrophoresis, **108**, 181
by immunoadsorbent affinity chromatography, **108**, 139
rosetting techniques, **108**, 61
by selective agglutination with soybean agglutinin, **108**, 175
on Sephadex G-10 column, **108**, 303
subpopulations
electrophoretic separation, **108**, 181
purification with antibody and complement, **108**, 249
surface immunoglobulins
biosynthetic labeling, **108**, 427
detection
with fluorescent antibodies, **108**, 413
with protein A, **108**, 405
isolation, **108**, 429
labeling by surface iodination, **108**, 426
T-cell, *see* T cells
thoracic duct
collection methods, **108**, 270
fusion with myeloma cells, **121**, 57
preparation for production of IgA-secreting hybridomas, **121**, 56

thymic, separation
in fluorescent activated cell sorter
with peanut agglutinin, **108**, 178
by free-flow electrophoresis, **108**, 181
on nylon wool columns, **108**, 298
by selective agglutination with peanut
agglutinin, **108**, 172, 174
Lymphokines
antigen-nonspecific, overview, **116**, 357
antigen-specific, overview, **116**, 295
Lymphotoxin
human
amino acid sequence homology with
tumor necrosis factor, **116**, 456
bioassay, **116**, 442
characterization, **116**, 445
purification, **116**, 443
supernatants with, production and
harvesting, **116**, 442
overview, **116**, 371
Lymph vessels
canine
peripheral
collection, **129**, 663
lipoproteins in, characterization,
129, 667
skeletal muscle, collection, **129**, 665
Lyophilization
lipoproteins, **128**, 216, 219
Lysine
in albumin, nonenzymatic glucosylation,
106, 88
detection and quantitation, **106**, 95
dissociation kinetics, measurement,
106, 93
kinetics, determination, **106**, 90
measurement, **106**, 90
F_1-ATPase
chemical modification, **126**, 728
radiolabeling, **126**, 751
methylated, in myosin
biosynthesis, analytical techniques,
106, 290, 293
determination, **106**, 289
functional and research role, **106**, 292
peptides with, isolation strategy, **106**,
295
tissue-specific occurrence, **106**, 287
protein-bound, enzymatic methylation
and demethylation, **106**, 274

reactivity with heavy atoms, **114**, 150
synthesis and breakdown, role of α-
aminoadipate, **113**, 645
L-Lysine 6-aminotransferase
Flavobacterium lutescens
proton abstraction, stereochemistry,
113, 100
purification, **113**, 96
subunit structure, **113**, 98
L-Lysine transaminase, *see* L-Lysine 6-
aminotransferase
Lysis test
β-lactam in milk, **133**, 276
Lysolecithin acyltransferase, *see* 1-Acyl-
glycerophosphocholine acyltransferase
Lysosomes
fusion with phagosomes, fluorescence
assay, **132**, 257
Lysozyme
active sites, determination by selective
oxidative free radical probes, **105**,
496
–amatoxin conjugates
biological properties, **112**, 233
preparation by mixed anhydride coupling, **112**, 231
assay for assessment of phagocytic
function, **132**, 120, 160
bacteriophage T4, folding and stability,
genetic analysis, **131**, 265
delivery systems, polymeric and copolymeric
kinetic analysis, **112**, 414
preparation, **112**, 400
hydration
analysis, solid-phase methods, **127**,
288
conformational effects, **127**, 382
hydrogen exchange, effect of inhibitor
binding, **131**, 503
papaya, chromatography on hydroxyapatite columns, **117**, 378

M

α_2-Macroglobulin
human, monoclonal antibody production
with conventional antiserum, **121**,
59

–ricin toxin conjugates, preparation and applications, **112**, 303
Macromolecules
backbone models, **115**, 391
conformational stability and transitions, **130**, 413
conformation, spectroscopic analysis, **130**, 185
crystallization
general principles, **114**, 112
heterogeneity, contributing factors, **114**, 120
influencing factors, **114**, 119
organic solvents in, **114**, 118
with polyethylene glycol, **114**, 120
precipitants in, **114**, 118
by variation of pH and temperature, **114**, 125
crystal structure, stereochemically restrained refinement
applications, **115**, 268
computer program design, **115**, 260
observational functions, **115**, 253
strategy, **115**, 264
weighting considerations, **115**, 263
elongated, molecular weight determination by viscoelastometry, **117**, 94
hydrated, NMR analysis, **127**, 154
hydration, theoretical approaches
applications, **127**, 382
computer simulations, **127**, 371
hydrodynamic properties, measurement by
active enzyme centrifugation, **117**, 97
difference sedimentation velocity, **117**, 28
molecular weight, determination by radiation inactivation, **117**, 65
multimolecular aggregates, characterization by gel electrophoresis, **130**, 116
phasing by isomorphous replacement
methodology, **115**, 4
phase derivations, **115**, 9
sources of error, **115**, 11
statistical data in, **115**, 13
polydisperse solutions and colloidal suspensions, correlation function profile analysis, **117**, 256
sedimentation equilibrium in air-turbine ultracentrifuge, **117**, 3
self-associations, analytical techniques, **130**, 3
solvents around, density and location measurements by neutron scattering, **127**, 619
structural analysis
by fluorescence spectroscopy, **131**, 518
by transient electric birefringence, **117**, 198
structural assemblies, analytical techniques, **130**, 3
Macrophage-arming factor
induced specific macrophage cytotoxicity, assays, **132**, 531
Macrophage cytotoxic factor
assay, **132**, 552
production, **132**, 549
Macrophages
activated
cytotoxic effect on tumor cells, assays, **132**, 516
ingestion of *Plasmodium*, assay, **132**, 642
killing of *Plasmodium*, assay, **132**, 642
preparation, **132**, 640
production of cytotoxic factor, **132**, 549
alveolar
collection methods, **108**, 278
depletion with silica and asbestos, **108**, 325
preparation, **132**, 630, 632
bone marrow-derived
preparation, **116**, 585
production of H_2O_2, assay, **132**, 397
colony-stimulating factor binding, analytical techniques, **116**, 582
cultured, number determination, **116**, 587
derived B-cell hybridoma growth promoters
comparison with other growth promoters, **121**, 12
overview, **121**, 10
preparation, **121**, 10
hepatic
collection, **108**, 286
separation, **108**, 291

hybridoma cells, chemiluminescent bioassays for
activation regulation, **133**, 513
effect of pharmacological agents, **133**, 524
phagocytosis and opsonization, **133**, 520
lymph node, separation on Sephadex G-10 column, **108**, 303
monoclonal antibodies
applications, **108**, 320
defined antigens, characterization, **108**, 314
preparation, **108**, 314
monocyte-derived
infection with *Leishmania*, analysis and applications, **132**, 617
preparation, **132**, 613, 639
murine P388D$_1$
production of interleukin 1, **116**, 483
propagation, **116**, 483
purification of interleukin 1, **116**, 485
normal and activated, killing of *Toxoplasma*, assay, **132**, 630
peripheral blood
depletion with silica and asbestos, **108**, 325
separation
with carbonyl iron, **108**, 307
on Sephadex G-10 column, **108**, 303
superoxide anion release, measurement, **105**, 365
peritoneal
activated, assays, **132**, 539, 544
activation, **132**, 509, 512, 601, 653
binding of
neurotensin and substance P, **132**, 333
(2′-5′)-oligoadenylate, **119**, 355
binding and lysis of tumor cells, assays, **132**, 559, 563, 565, 567
cocultivation with tumor cells, **132**, 511
collection methods, **108**, 275
cytotoxicity, assays, **132**, 458, 531, 601
feeder effect on tumor cells, **121**, 298
H_2O_2 production, microassay, **132**, 414
infection with *Leishmania*, analysis and applications, **132**, 614, 620
killing of schistosomula, assay, **132**, 656
monolayers
lymphocyte separation on, **108**, 130
preparation, **108**, 126
phagocytosis
effect of (2′-5′)-oligoadenylates, **119**, 678
fibronectin-mediated, assay, **132**, 340
zymosan particles, assay, **132**, 329
preparation, **116**, 388; **119**, 353, 678; **121**, 297; **132**, 259, 596, 610, 630, 639
reduction of nitroblue tetrazolium, microassay, **132**, 417
separation
on plastic and glass surfaces, **108**, 294
from tumor cells after cocultivation, **132**, 514
superoxide production, microassay, **132**, 410
phagocytic function, flow cytometric assays, **132**, 183, 186
progenitors, detection assay, **116**, 554
pulmonary
phagocytosis, effect of (2′-5′)-oligoadenylates, **119**, 678
preparation, **119**, 678
purification of gelsolin, **134**, 6
splenic
preparation, **132**, 639
separation
with carbonyl iron, **108**, 307
on plastic and glass surfaces, **108**, 294
on Sephadex G-10 column, **108**, 303
Madin–Darby bovine kidney cells
in monolayer culture, binding of interferon, **119**, 313
receptor-bound interferon, binding of monoclonal antibody, **119**, 326
Madin–Darby canine kidney cells
cAMP levels, radioimmunoassay, **109**, 362
cultivation, optimal conditions, **109**, 361
hormone responsiveness
evaluation, **109**, 363

measurement, **109**, 361
transformed, induction of glucagon receptors
applications, **109**, 360
conditions for, **109**, 357
inhibition of, **109**, 359
Magnesium
effect on phagocytosis, **132**, 97
Maize
isolation of
mitochondria, **118**, 471
mitochondrial DNA, **118**, 455, 472
mitochondrial DNA–protein complexes, **118**, 485
mitochondrial RNA, **118**, 472, 497
preparation of chloroplast transcription extracts, **118**, 234
Maize streak virus
propagation in plants, **118**, 703
Malate dehydrogenase
Rhodopseudomonas spheroides, purification, **104**, 111
Malonaldehyde
detection by HPLC, **105**, 304, 319
fluorometric determination, **105**, 303
spectrophotometric determination, **105**, 299
Malonyl-CoA
decarboxylation by methylmalonyl-CoA decarboxylase, assay, **125**, 541
Maltose
binding protein entry and function in *malE* mutants, influencing factors, **125**, 296
transport characteristics in reconstituted *malE* mutants, **125**, 297
Mammary gland
rabbit, preparation of
growth hormone receptors, **109**, 694
prolactin receptors, **109**, 157
Manduca sexta, *see* Tobacco hornworm
Manganese
binding to tRNA crystals, **114**, 165
Manganese dioxide
preparation, **122**, 98
in synthesis of [^{14}C]pyridoxal 5′-phosphate, **122**, 98
N-Mannich bases
as prodrugs
applications, **112**, 351
hydrolysis, **112**, 348
preparation, **112**, 348
Mannose
^{35}S-labeled, in radiolabeling of interferon β, **119**, 401
Mannose 6-phosphate
–ricin conjugates, preparation and applications, **112**, 304
Marginal band
erythrocyte, isolation
basic steps, **134**, 235
from dogfish, **134**, 241, 245
with proteases, **134**, 250
from salamander, **134**, 248
Mass spectrometry
bile acids, **111**, 63
in γ-carboxyglutamic acid assay, **107**, 505
ecdysteroid conjugates, **111**, 416
fast atom bombardment, *see* Fast atom bombardment mass spectrometry
free and conjugated ecdysteroids, **111**, 401
25-hydroxyvitamin D_3 metabolites, **123**, 150
in identification of
amino-terminal myristyl blocking groups in proteins, **106**, 207
posttranslationally modified amino acids, **106**, 29
retinoic acid, **123**, 120, 122
selenocysteine in proteins, **107**, 582
sterols, **111**, 27
vitamin D metabolites, **123**, 133
Mast cell growth factor
murine
assays, **116**, 554
properties, **116**, 563
purification, **116**, 557
Matrices
extracellular
culture of anterior pituitary cells, **103**, 289
production, **103**, 288
monolithic
applications, **112**, 515
drug delivery, **112**, 503
Maturation factors
B-cell, *see* B-cell maturation factors
Mechlorethamine
cytotoxicity, screening with anti-erythrocyte hybridoma, **121**, 768

Melanin
assay in human and murine melanoma cells, **119**, 614
free radicals in
magnetic parameters and spin concentrations, **105**, 479
reaction kinetics and mechanisms, **105**, 483
structure determination, **105**, 482
Melanotropin
–daunomycin conjugates, preparation, **112**, 259
–ouabain conjugates, preparation, **112**, 264
regulation of cAMP levels in primary glial cultures, analytical techniques, **109**, 344
γ_1-Melanotropin
related peptide, isolation from neurointermediate pituitary, **103**, 85
Melibiose
radiolabeled, preparation, **125**, 379
transport system in *Escherichia coli*
assay, **125**, 379
cation cotransport, measurement, **125**, 380
reconstitution, **125**, 385
Membrane potential
alkalophilic *Bacillus*, measurement, **125**, 591
in bioenergic systems, determination methods, **127**, 557
in *Escherichia coli ecf* mutants, measurement, **125**, 183
halorhodopsin-containing envelope vesicles
chloride-induced changes, **125**, 514
measurement, **125**, 511
neuroendocrine cells
fluorescent probes, **124**, 116
measurement, **124**, 223
reconstituted lipid bilayers containing cytochrome oxidases, generation, **126**, 131
reconstituted liposomes containing cytochrome *d* and *o* terminal oxidase complexes, formation, **126**, 109
relative, cytofluorometric assay, **103**, 241
small cells or vesicles, measurement, **125**, 47
in *Vibrio alginolyticus* cells, generation, **125**, 524
Membrane proteins
avidin–iminobiotin-based purification, **122**, 87
chloroplast, *psbA*-encoded
identification, **118**, 384
structural characterization, **118**, 391
crystallization with alkyl glycoside detergent as solubilizing agent, **125**, 34
erythrocyte, brain analogs, purification, **134**, 56
functional, solubilization
assay, **104**, 306, 311
buffer and temperature conditions, choice, **104**, 310
criteria for solubility, choice, **104**, 305
detergent choice, **104**, 306
functional characterization after, **104**, 315
optimization of protein-to-detergent ratio, **104**, 313
physical characterization after, **104**, 315
stabilization, **104**, 312
2-iminobiotinylated
detection, **122**, 93
preparation, **122**, 89
purification, **122**, 92
integral
fractionation, **104**, 335
isolation, **104**, 330
solubilization, **104**, 332
ion transport
analysis with planar bilayers, **127**, 486
measurement with optical probes, **127**, 746
occurrence of Glu–Lys crosslink, **107**, 253
reconstitution into lipid vesicles, **104**, 340
Membranes
anterior pituitary
binding of luciferin-derivatized GnRH analogs, **124**, 32
dopamine receptors, direct radioligand measurement, **103**, 579, 587

bacterial
binding protein-dependent active transport
protein components, **125**, 281, 289
regulation, **125**, 288
system types, **125**, 279
protein export, role of protonmotive force, **125**, 133
simultaneous determination of oxygen consumption and NAD(P)H oxidation, **122**, 168
solute transport mechanisms, **125**, 493
biological
binding of [^{125}I]EGF, assay, **109**, 107
changes during energization, **125**, 78, 80
detergent solubilization, **124**, 146
dynamic properties, analysis by hydrogen exchange kinetics, **127**, 630
energy transduction, chemiosmotic theory, **126**, 2
interactions with
small protein molecules, analysis, **125**, 81
water, NMR analysis, **127**, 156
intramembrane electrical field, relationship to surface potential, **125**, 76, 80
lipid peroxidation, detection methods, **105**, 283
N_s and N_i regulatory components, ADP-ribosylation by pertussis and cholera toxins, **109**, 566
processes affected by surface charges, characteristics, **125**, 63, 83
redox potentials, **125**, 75
structure and function, role of water, deuteration studies, **127**, 662
surface charge density, measurement by spectroscopic techniques, **127**, 678
surface ion concentration
relationship to electrical potential, **125**, 58, 73
and salt effect on surface reactions, **125**, 65, 68, 71
time-resolved X-ray scattering, **134**, 675
water content, quantitation by neutron diffraction and X-ray techniques, **127**, 217
chloroplast
NaBr-treated, preparation, **118**, 353
polypeptide synthesis during development, analytical techniques, **118**, 367
psbA gene-encoded protein
identification, **118**, 384
structural characterization, **118**, 391
purification of
photosystem II reaction center, **118**, 354
proteolipid, **126**, 520
quantitative estimation of protein subunits, **118**, 360
chloroplast envelope
characterization, **118**, 323
metabolite transport processes, **125**, 705
phosphate transport proteins
identification, **125**, 716
isolation, **125**, 723
preparation, **118**, 319
separation, **118**, 320
-coated reservoirs, applications, **112**, 505
electrical potential, *see* Membrane potential
erythrocyte, water channels
chemical modifications, **127**, 607
diffusional permeability measurements, **127**, 599, 601
osmotic permeability measurements, **127**, 603
volume changes, measurement, **127**, 605
Escherichia coli
fumarate reductase-enriched, isolation, **126**, 383
purification of
alanine carrier protein, **125**, 486
lac permease, **125**, 430
transport proteins
analysis with *lacZ* gene fusions, **125**, 150
mutant isolation and characterization with transposons, **125**, 157
fetal, detection of constitutively produced interferon, **119**, 544

fibroblast, visualization of low-density lipoprotein receptors, **129**, 207
hepatic
glucagon receptors
covalent labeling, **109**, 207
functional analysis by fusion to foreign adenylate cyclase, **109**, 346
postnuclear, preparation, **123**, 62
inner bacterial, active transport, coupling to outer membrane, **125**, 277
inner mitochondrial, fusion, **125**, 42
lipid bilayer
conformational water forces between layers, direct measurement, **127**, 357
intermolecular forces, direct measurement by osmotic stress, **127**, 400
ion permeability, measurement with optical probes, **127**, 746
phase transitions, effect on
H^+/OH^- permeability, **127**, 484
water permeability, **127**, 482
water penetration depth
capacitance measurements, **127**, 514
combined methods, **127**, 515
X-ray diffraction measurements, **127**, 512
lipid enrichment, **125**, 36
lymphocyte, detection of immunoglobulins, **108**, 393, 413
microsomal, interactions with water and carbohydrates
differential scanning calorimetry, **127**, 698
infrared spectroscopy, **127**, 698
mitochondrial
bound enzymes, analysis with *N,N'*-dicyclohexylcarbodiimide, **125**, 92
polypeptide uptake, role of protonmotive force, **125**, 130
preparation of ubiquinol–cytochrome-*c* reductase crystals, **126**, 192
superoxide anion production, measurement, **105**, 376
mycoplasma, transport processes, **125**, 259
neutrophil, purification, **132**, 382
outer bacterial
Ca^{2+}-dependent reconstitution
applicability for whole cells, **125**, 300
assays, **125**, 295
influencing factors, **125**, 296
in spheroplasts and treated cells, **125**, 291
permeability measurement
in estimation of lower limits, **125**, 277
by isotope equilibration, **125**, 266
with periplasmic hydrolytic enzymes, **125**, 271
with very hydrophobic compounds, **125**, 269
outer mitochondrial, diffusion channel crystalline arrays
characteristics, **125**, 601
mechanism of formation, **125**, 606
preparation, **125**, 597
structure, **125**, 607
Paracoccus denitrificans, isolation, **126**, 155
phospholipid oxidation, chemistry, **105**, 280
plasma, *see* Plasma membranes
polymeric
-coated reservoirs, drug release, **112**, 500
diffusion principles, **112**, 496
drug delivery rates and mechanisms, **112**, 500
enzyme-controlled, preparation and characterization, **112**, 484
proton translocation
description, barrier models, **127**, 465
energetics, theoretical calculation, **127**, 456
light-induced, correlations between structure and efficiency, **127**, 578
measurement
by displacement currents, **127**, 610
with laser-induced proton pulse, **127**, 522
with nitroxide spin probes, **127**, 738
proton transport, primary and secondary
energetics, **125**, 5, 9
energy conversion efficiency, **125**, 14

nonequilibrium thermodynamics, **125**, 12
pump stoichiometry, determination, **125**, 6, 11
proton tunneling
calculations, **127**, 81
coupling to molecular motions, **127**, 83
model, **127**, 79
solvent effects, **127**, 89
purple, *Halobacteria halobium*, proton transport, detection by FTIR, **127**, 344
retinal, α-bungarotoxin–luciferin binding, **133**, 47
rickettsiae, transport processes
analysis by microporous filtration, **125**, 258
bioenergetics, **125**, 259
skeletal muscle, α-bungarotoxin–luciferin binding, **133**, 47
surface potential, *see* Surface potential
synovial, *see* Synovial membranes
thylakoid, *see* Thylakoid membranes
vesicles, from *Escherichia coli*
Ca^{2+} uptake assay, **125**, 335
glutamine transport
assays, **125**, 306
properties, **125**, 306
preparation, **125**, 305, 334
Vibrio alginolyticus
preparation, **125**, 527
respiratory activities, **125**, 528
Membrane seal
in single-channel electrophysiological analysis, formation, **103**, 153
Menadione
induced superoxide formation, detection and measurement, **105**, 505
Menaquinone, *see* Vitamin K_2
Mercaptide
in assay of pantetheine enzymatic hydrolysis, **122**, 37
2-Mercaptoethanol
in detection of soluble immune response suppressor, **116**, 397
Mercury
binding to
proteins, **114**, 152
tRNA crystals, **114**, 166
Meristems
Haplopappus gracilis, synchronization by amphidicolin, **118**, 87
Meromyosin
heavy, *N*-ethylmaleimide-modified
applications, **134**, 476
preparation, **134**, 474
Metabolic chambers
for analysis of respiratory $^{14}CO_2$ production from radiolabeled substrates, **113**, 461
Metabolism, *see also* Catabolism
acetaminophen, alterations associated with prodrug use, **112**, 341
apolipoprotein B-containing lipoproteins, analytical techniques, **129**, 395
apolipoprotein C, analytical techniques, **129**, 449
apolipoprotein E, analytical techniques, **129**, 495
apolipoproteins
analysis with specific radioactivity data, **129**, 384
models, **129**, 375
apolipoproteins A-I, A-II, and A-IV
analytical techniques, **129**, 428
in dyslipidemic subjects, **129**, 438
in normal subjects, **129**, 435
aspirin, alterations associated with prodrug use, **112**, 343
cholesterol, models, **129**, 379
drugs, effects of interferon, measurement, **119**, 718
high-density lipoproteins, intestinal contribution, **129**, 534
lipoproteins
analytical techniques, **129**, 353
models, **129**, 369
in steroidogenic tissues, **129**, 683
phenacetin, alterations associated with prodrug use, **112**, 341
postprandial lipoproteins, **129**, 469
very-low-density lipoproteins, enzymatic assay in whole plasma, **129**, 704
Metal ions
content in buffers for enzymes, control, **104**, 409
trivalent, in staining of phosphoproteins in gels, **104**, 454

Metalloporphyrins
ligand binding, resonance Raman spectroscopy, **130**, 363
Metals
–ligand distances, measurement by EXAFS, **117**, 414
Metavinculin
chicken gizzard smooth muscle
assay, **134**, 76
purification, **134**, 74, 79
storage, **134**, 84
Methanol
aqueous, protein unfolding, **131**, 174
Methanopterin
identification, **122**, 415, 422
properties, **122**, 417
purification, **122**, 415
5,10-Methenyltetrahydrofolate cyclohydrolase
Clostridium formicoaceticum
assay, **122**, 386
properties, **122**, 390
purification, **122**, 388
5,10-Methenyltetrahydromethanopterin
identification, **122**, 415, 422
properties, **122**, 417
purification, **122**, 415
Methional
ethylene production, in measurement of hydroxyl radical production by neutrophils, **105**, 395
Methionine
reactivity with heavy atoms, **114**, 150
^{35}S-labeled, in radiolabeling of
interferon β, **119**, 284, 400
interferons in cell-free DNA-dependent system, **119**, 292
recombinant human interferon αA, **119**, 286
Methionine-enkephalin
antiserum preparation, **124**, 619
radioimmunoassay, **124**, 620
radioiodination, **124**, 620
separation, **124**, 621
synthesis, **103**, 29
Methionine sulfoxide
in neuropeptides, analytical techniques, **103**, 26
in proteins and peptides, enzymatic reduction, **107**, 352
Methionine sulfoxide reductase
distribution and physiological function, **107**, 359
Escherichia coli
properties and characteristics, **107**, 356
purification, **107**, 355
human leukocyte, purification, **107**, 358
Methionyl-tRNA
binding to methionyl-tRNA formyltransferase, stoichiometry, **106**, 151
Methionyl-tRNA formyltransferase
Escherichia coli
assay, **106**, 142
ligand binding stoichiometries, **106**, 148
properties, **106**, 146
purification, **106**, 144
specificity, **106**, 152
Methotrexate
lysine and ornithine analogs and dansyl derivatives
HPLC separation, **122**, 297
syntheses, **122**, 295
polyglutamate derivatives formed *in vivo*, chromatographic analysis, **122**, 339
–polylysine conjugates
preparation, **112**, 273
properties, **112**, 279
Methoxyl verapamil
interactions with ion channels in neuroendocrine tissue, **103**, 402
Methylamine–glutamate methyltransferase
bacterial
assays, **113**, 36, 37
properties, **113**, 39
purification, **113**, 38
N-Methyl aspartate
excitotoxic properties in neuroendocrine research, **103**, 385
β-Methylaspartate-glutamate mutase, *see* Methylaspartate mutase
Methylaspartate mutase
Clostridium tetanomorphum
assays, **113**, 121, 123
components S and E
interactions, **113**, 131
properties, **113**, 126, 130
purification, **113**, 124, 128

Methylation
calmodulin
electrophoretic assay, **102**, 159
radiometric assay, **102**, 160
enzymatic, protein-bound lysine residues, **106**, 274
lysine and histidine in myosin, tissue and developmental differences, **106**, 287
proteins
analytical techniques, **106**, 265
natural occurrence, **106**, 266
reductive, low-density lipoproteins for catabolism studies, **129**, 580
retinoic acid, **123**, 116
1-Methyl-3-(*N*-{β-[3,4-bis(pivalyloxy)-phenyl]carbamoyl})pyridinium iodide
synthesis, **112**, 392
Methylcellulose
hybridoma cloning in, **121**, 315
stock solution preparation, **121**, 310
Methylcrotonyl-CoA carboxylase synthetase
assay, **107**, 265
properties, **107**, 262, 274
purification, **107**, 271
reaction mechanism, **107**, 275
sequence homology with other synthetases, **107**, 264
Methylenetetrahydrofolate dehydrogenase ($NADP^+$)
chicken liver, in 5-methyl-L-tetrahydrofolate assay, **122**, 382
Methylenetetrahydrofolate reductase ($FADH_2$)
Clostridium formicoaceticum
assay, **122**, 393
properties, **122**, 397
purification, **122**, 395
Methylenetetrahydrofolate reductase (NADPH)
bovine liver, in 5-methyl-L-tetrahydrofolate assay, **122**, 382
porcine liver
assays, **122**, 373
properties, **122**, 377
purification, **122**, 376
5,10-Methylenetetrahydromethanopterin
preparation and properties, **122**, 421
reverse-phase HPLC, **122**, 422
Methylenetetrahydropteroylpolyglutamate
preparation and analysis, **122**, 337
Methyl ethyl ketone peroxide
mutagenicity, testing with *Salmonella typhimurium* TA102, **105**, 250
N-Methyl-L-glutamate synthase, *see* Methylamine–glutamate methyltransferase
Methyl green
in staining of phosphoproteins in gels, **104**, 451
3-Methylhistidine
in myosin
biosynthesis, analytical techniques, **106**, 290, 293
determination, **106**, 289
functional and research role, **106**, 292
peptides with, isolation strategy, **106**, 295
tissue-specific occurrence, **106**, 287
N^6-Methyl-lysine oxidase
rat kidney
assay, **106**, 283
properties, **106**, 286
purification, **106**, 285
Methylmalonyl-CoA decarboxylase
Veillonella alcalescens
assays, **125**, 541
properties, **125**, 543
purification, **125**, 542
sodium transport activity
properties, **125**, 546
reconstitution in proteoliposomes, **125**, 544
N-Methyl-*N'*-nitro-*N*-nitrosoguanidine
in preparation of *Phycomyces* color mutants, **110**, 226
3-Methyl-5-oxo-L-proline
inhibition of 5-oxo-L-proline metabolism *in vivo*, **113**, 468
2-Methyl-2,4-pentanediol, *see* Hexylene glycol
N-Methylphenazonium methosulfate
in reconstitution of H^+ translocation and photophosphorylation, **126**, 285
N-Methylphenylalanine
in proteins, mass spectrometric identification, **106**, 51

N-Methylpyridinium-2-carbaldoxime chloride
and dihydropyridine derivative, comparative effects in reactivation of cerebral enzymes, **112**, 381
4α-Methylsterols
radiolabeled
incubation with marine invertebrates and plant tissues, **111**, 320
preparation, **111**, 317
5-Methyl-L-tetrahydrofolate
enzymatic assay, **122**, 382
fluorometric procedure, **122**, 384
spectrophotometric procedure, **122**, 383
5-Methyltetrahydromethanopterin
preparation and properties, **122**, 422
reverse-phase HPLC, **122**, 422
Methyltetrahydropteroylpolyglutamate
preparation and analysis, **122**, 338
4-Methylthio-2-oxobutanoic acid
ethene formation, in hydroxyl radical detection, **105**, 17
4-Methylumbelliferyl phosphate
in protein immunostaining on nitrocellulose, **121**, 853
Metrizamide
density gradient media
in equilibrium density separation of lymphoid cells, **108**, 109, 116
osmolarity and pH control, **108**, 103, 106
preparation and calibration, **108**, 107
Mevalonate
identification as product of 3-hydroxy-3-methylglutaryl-CoA reductase, **110**, 57
metabolism, shunt pathway
alternate schemes, **110**, 102
compartmentation, **110**, 102
measurement, **110**, 103
reactions in, **110**, 100
in plasma and urine, radioenzymatic assay, **110**, 63
Mevalonate kinase
porcine liver
assays, **110**, 60, 72
inhibition, **110**, 77
properties, **110**, 76
purification, **110**, 60, 74
in radioenzymatic assay for plasma mevalonate, **110**, 63
reaction mechanism, **110**, 77
Mevalonate phosphates
thin-layer chromatography, **111**, 248
Mevalonate 5-phosphotransferase, *see* Mevalonate kinase
Mevalonate 5-pyrophosphate
synthesis, **110**, 86
Mevalonate-5-pyrophosphate decarboxylase
chicken liver
assays, **110**, 86
properties, **110**, 91
purification, **110**, 89
Mevalonic acid
radiolabeled
incubation with marine invertebrates and plant tissues, **111**, 320
preparation, **111**, 314
Mevastin
inhibitory effects on 3-hydroxy-3-methylglutaryl-CoA reductase, **110**, 55
Mezerein
and A-23187, induction of human interferon γ, **119**, 74, 193
Micelles
aqueous, formation, calorimetric analysis, **127**, 146
bile salt, protein transfer by gel filtration, **104**, 325
detergent
dimensions, **104**, 320
and phospholipid vesicles, separation by gel filtration, **104**, 326
physical properties, **124**, 139
mixed, in protein kinase C assay, **124**, 353
protein–detergent
detergent removal by ultrafiltration, **104**, 327
dimensions, **104**, 320, 323
Microaggregation
hormone receptors, analysis by ligand dimerization, **103**, 49
Microbial cells
ATP extraction, considerations, **133**, 20
Microcapsules, *see* Microspheres
Micrococcus luteus
cultivation, **110**, 193

purification of
geranylpyrophosphate synthetase, **110**, 189
hexaprenylpyrophosphate synthetase, **110**, 194
nonaprenylpyrophosphate synthetase, **110**, 207
Microfilaments
X-ray diffraction analysis, **134**, 648
β_2-Microglobulin
allelic variants, **108**, 499
detection methods, **108**, 497
expression regulation, **108**, 502
molecular cloning, **108**, 501
purification from murine liver, **108**, 495
Micromanipulators
for electrophysiological recording in dissociated tissue culture, **103**, 118
Micropunch technique
for brain nuclei sampling, **103**, 368
Microscope
inverted phase-contrast, for electrophysiological recording in tissue culture, **103**, 116
Microsomal aminopeptidase
rat kidney
assay, **113**, 474
properties, **113**, 480
purification, **113**, 479
substrate specificity, **113**, 482
Microsomes
hepatic
in assay of cholesterol acyltransferase, **111**, 286
γ-carboxyglutamic acid biosynthesis, **113**, 136
drug metabolism, interferon effects, measurement, **119**, 717, 721
hydroxyl radical scavengers and reaction products, **105**, 517
isolation of
apolipoproteins, **129**, 290
cytochromes *P*-450 active in bile acid synthesis, **111**, 368
nascent lipoproteins, **129**, 291
oxidant radical generation
assays, **105**, 519
reaction conditions, **105**, 518
peroxidizing, malonaldehyde measurement, **105**, 321
preparation, **111**, 367; **119**, 716, 720; **129**, 285
simultaneous determination of oxygen consumption and NAD(P)H oxidation, **122**, 166
subfractionation, **129**, 288
superoxide anion production, measurement, **105**, 375
Manduca sexta, isolation, **111**, 454
pea seedling, isolation of 3-hydroxy-3-methylglutaryl-CoA reductase, **110**, 34
Microspheres
agarose–polyacrolein
applications, **112**, 171
synthesis and properties, **112**, 169
albumin, *see* Albumin microspheres
crystallized carbohydrate, for slow release of proteins
biological activity of entrapped proteins, assay, **112**, 121
immunological response to entrapped antigens, assay, **112**, 124
preparation, **112**, 120
in vitro release studies, **112**, 122
ethylcellulose
mitomycin C-containing
clinical applications, **112**, 149
intraarterial targeting, **112**, 146
preparation, **112**, 140
peplomycin-containing
clinical applications, **112**, 149
intraarterial targeting, **112**, 146
preparation, **112**, 146
ferromagnetic ethylcellulose, mitomycin C-containing
clinical applications, **112**, 149
preparation, **112**, 144
fluorescent, in flow cytometric assay of phagocytosis, **132**, 185
latex, antibody-coated, in detection of cell surface markers, **108**, 56
magnetically responsive
applications, **112**, 17
polyacrolein-activated, for cell separation, **112**, 162
preparation, **112**, 17, 126
metallic poly(vinylpyridine), preparation and properties, **112**, 185
–monoclonal antibody conjugates
immunoreactivity, **112**, 84

preparation, **112**, 70
purification, **112**, 83
nylon, production and properties, **112**, 84
polyacrolein
antibody binding, **112**, 158
applications, **112**, 171
molecular structure and chemical analysis, **112**, 152
preparation, **112**, 151, 164
polyacrolein-activated hybrid
antibody binding, **112**, 158
preparation, **112**, 155
poly(lactide-*co*-glycolide), norethisterone-containing
characterization, **112**, 111
preparation, **112**, 102
poly(vinylpyridine) copolymer, preparation and properties, **112**, 186
poly(vinylpyridine) monomer
applications, **112**, 188
preparation, **112**, 177
properties, **112**, 176, 179
Microtiter plates
coating with antigens, **121**, 551
Microtubule-associated proteins
binding to neurofilaments, assay, **134**, 546
bovine brain
identification, **134**, 105
isolation, **134**, 111
and MAP 2, separation, **134**, 97
proteolytic digestion, **134**, 98
and tubulin, separation, **134**, 94
in cultured cells, direct identification by selective extraction, **134**, 139
mammalian brain
identification, **134**, 117
purification with preformed microtubules, **134**, 126
primate 210K
antibody preparation, **134**, 150
immunoassay, **134**, 152
purification, **134**, 150
sea urchin egg
identification, **134**, 117
purification with preformed microtubules, **134**, 126
Microtubules
assembly, time-resolved X-ray scattering, **134**, 666

axonemal, *Tetrahymena thermophila* cilia
complexing with dynein, **134**, 316
repolymerization, **134**, 314
bovine brain
purification, **134**, 90, 106
taxol-stabilized, reversible assembly, **134**, 115
component analysis, **134**, 143
in cultured cells
detergent extraction, **134**, 140
purification, **134**, 124
frozen-hydrated, cryo-electron microscopy, **134**, 612
HeLa cell, protein purification, **134**, 149
mammalian brain, cold-stable
isolation, **134**, 161
stability assay, **134**, 164
−neurofilament network
assays, **134**, 546
formation, **134**, 545
preformed, in purification of microtubule-associated protein, **134**, 126
sea urchin egg, purification, **134**, 119
single, video light microscopy, **134**, 561
taxol-stabilized, molecular analysis, **134**, 146
vitrification and freeze-fracturing, **134**, 627
X-ray diffraction analysis, **134**, 651
Migration inhibitory factor
assays, **116**, 382
characterization, **116**, 390
overview, **116**, 370
production, **116**, 380
Milk
bovine, vitamin K assay by HPLC with UV detection, **123**, 237, 243
human
purification of
IgA, **116**, 49
IgA secretory component, **116**, 71
vitamin D compounds, quantitation, **123**, 167
vitamin K assay by HPLC with UV detection, **123**, 237, 243
Mirrors
double-focusing, alignment method, **114**, 316

Mitochondria
brown fat, uncoupling protein
detection with fluorescent nucleotide analogs, **125**, 627
reconstitution in phospholipid vesicles, **127**, 772
cardiac
ADP–ATP carrier protein
active site labeling, **125**, 665
chemical modification, **125**, 659
detection with fluorescent nucleotide analogs, **125**, 620
fluorescent probes, **125**, 639
NMR and ESR studies, **125**, 631
transmembrane folding, analysis with pyridoxal phosphate, **125**, 650
enzyme assays, **125**, 18
hydrogen peroxide production, measurement, **105**, 432
isolation with collagenase, **125**, 19
membrane-bound enzymes, analysis with *N*,*N*′-dicyclohexylcarbodiimide, **125**, 92
oxygen consumption assays, **125**, 21
phosphate transport protein
isolation, **125**, 697, 702, 703
reconstitution, **125**, 699, 702, 703
purification of
adenine nucleotide carrier, **125**, 610
ATPase inhibitor, **126**, 460
carnitine acyltransferases, **123**, 280
complex I subunits, **126**, 362
cytochrome bc_1, **126**, 183, 225
F_0F_1-ATPase, **126**, 428
H^+-transporting ATP synthase, **126**, 417
membrane-bound pyrophosphatase, **126**, 447
oligomycin sensitivity-conferring protein, **126**, 455
succinate–cytochrome-*c* reductase, **126**, 214
transhydrogenase, **126**, 354
redox Bohr effects, measurement, **126**, 334
respiratory control, **125**, 22
superoxide anion production, measurement, **105**, 376, 429
hepatic
calcium retention, hydroperoxide effects, **105**, 438
electric pulse-driven ATP synthesis, **126**, 640
inner membranes
fusion, **125**, 42
lipid enrichment, **125**, 36
isolation, **105**, 436; **111**, 367
malonaldehyde measurement, **105**, 325
membrane-bound enzymes, analysis with *N*,*N*′-dicyclohexylcarbodiimide, **125**, 92
oxygen consumption and NAD(P)H oxidation, simultaneous determination, **122**, 167
purification of
5-aminolevulinate synthase, **123**, 395
cytochrome *P*-450 active in bile acid synthesis, **111**, 368
F_1-ATPase, **126**, 477
F_1-ATPase subunits, **126**, 485
F_0F_1-ATPase, **126**, 470
3-hydroxy-3-methylglutaryl-CoA synthase, **110**, 22
pyridine nucleotides, redox state, hydroperoxide effects, **105**, 435
tricarboxylate carrier protein
incorporation into liposomes, **125**, 694
purification, **125**, 692
reconstituted, properties, **125**, 695
isolated, membrane potential measurements
by ion distribution, **127**, 566
by lipophilic phosphonium cations, **127**, 567
Manduca sexta, isolation, **111**, 454
membrane proton transport, stoichiometry, **125**, 6
Neurospora
outer membrane channel crystalline arrays
characteristics, **125**, 601
mechanism of formation, **125**, 606
preparation, **125**, 597
structure, **125**, 607
ubiquinol–cytochrome-*c* reductase membrane crystals, preparation, **126**, 192

plant
controlled traits
identification in cybrids, **118**, 603
transfer by protoplast fusion, **118**, 595
DNA isolation, **118**, 439, 446, 455, 472
gradient purification, **118**, 491
isolation, **118**, 444, 472, 490
RNA isolation, **118**, 472, 492, 497
polypeptide uptake, role of protonmotive force, **125**, 130
Rhodopseudomonas sphaeroides, purification of cytochrome bc_1, **126**, 183
synaptosomal, membrane potential measurement by lipophilic phosphonium cations, **127**, 573
tumor, citrate transport
kinetics
determination, **125**, 674
temperature effects, **125**, 684
metabolite flux, determination, **125**, 685
yeast
Glu-tRNA identification as intermediate in protein synthesis, **106**, 152
purification of
ATPase inhibitor and stabilizing factors, **126**, 505
cytochrome bc_1, **126**, 173, 183
Mitomycin C
in ethylcellulose microcapsules
clinical applications, **112**, 149
intraarterial targeting, **112**, 146
preparation, **112**, 140
in ferromagnetic ethylcellulose microcapsules
clinical applications, **112**, 149
preparation, **112**, 144
mutagenicity, testing with *Salmonella typhimurium* TA102, **105**, 250
Mitotic index
plant cells, measurement, **118**, 91
Mitotic spindles
Chinese hamster ovary cell
isolation, **134**, 219
morphological analysis, **134**, 220
purification, **134**, 221
sea urchin egg, isolation, **134**, 194
sea urchin embryo
metaphase, isolation, **134**, 209
native, characterization, **134**, 212
yeast
isolation, **134**, 225
staining, **134**, 228
Models
animal, in analysis of
interferon antitumor effects, **119**, 649
plasma lipid transport regulatory genes, **128**, 877
aqueous solution structure, **127**, 3, 21
barrier, for proton translocation across membranes, **127**, 465
binding, nonlinear analysis, **126**, 733
closed bicyclic cascade, for regulation of glutamine synthase, **113**, 213
energy recycling, for generation of protonmotive force in anaerobic bacteria, **125**, 495
fatty acid binding to serum albumin, **128**, 333
ligand-binding isotherms linked to protein self-associations, **117**, 505
macromolecular
for backbone, **115**, 391
structure refinement, **115**, 227
metabolic
apolipoproteins, **129**, 375
cholesterol, **129**, 379
lipoproteins, **129**, 369
molecular
with flexible geometry, rationalization, **115**, 182
with ideal geometry, rationalization, **115**, 173
with interactive computer graphics program, **115**, 168
protein structure
and electron density maps, optical matching, **115**, 145
with interactive computer graphics program, **115**, 157
methods of coordinate measurement with Richards optical comparator, **115**, 154
velocity sedimentation of ligand-induced protein self-association, **117**, 485
Molecular sieve chromatography
apolipoprotein E, **128**, 275
glycosylated apolipoproteins, **128**, 362

methotrexate polyglutamate derivatives formed *in vivo*, **122**, 340
Molecular weight determination
apolipoprotein B-48, **128**, 269
apolipoprotein B-100, **128**, 260
bovine interferon α, **119**, 141
elongated molecules by viscoelastometry, **117**, 94
human interferons, **119**, 255
IgG-binding factor, **116**, 414
lipoproteins, **128**, 430
lymphotoxin, **116**, 445
macromolecules by radiation inactivation, **117**, 65
plant IVR, **119**, 734
polypeptide β_1, **116**, 243
protein crystals from crystal density, **114**, 176
proteins by sedimentation equilibrium, **117**, 10, 41
random coil polymers by size-exclusion chromatography, **117**, 142
T-cell suppressor factor for mixed leukocyte response, **116**, 424
Thy-1 antigens, **108**, 660
thymosin α_1, **116**, 243
thymosin β_4, **116**, 253
tumor necrosis factor, **116**, 453
Molybdenum cofactor
chemical assay, **122**, 408
enzymes with, occurrence, **122**, 400
reconstitution assay, **122**, 403
Molybdopterin
oxidized derivatives, in assay and detection of molybdenum cofactor, **122**, 408
Monasters
sea urchin egg, preparation, **134**, 196
$N^{\alpha B1}$-Monoazidobenzoyl insulin
in photolabeling of insulin receptors, **109**, 175
preparation, **109**, 171, 172
purification, **109**, 171, 172, 175
$N^{\varepsilon B29}$-Monoazidobenzoyl insulin
in photolabeling of insulin receptors, **109**, 175
preparation, **109**, 171, 172
purification, **109**, 171, 172, 175
Monochromators
in X-ray crystal diffractometry of large proteins, **114**, 390
Monoclonal antibodies
acetylcholine receptor
assay in hybridoma supernatants, **121**, 103
preparation, **121**, 97
specificity assay, **121**, 104
acridinium-labeled
preparation, **133**, 374
properties, **133**, 378
in affinity chromatography of T-cell subsets, **121**, 744
affinity constants, analysis by solution-phase radioimmunoassay, **121**, 468
antiallotypic, preparation, **116**, 160
antibacterial, slide immunoenzymatic assay, **121**, 514, 521
antigen-specific
passive hemagglutination inhibition assay, **121**, 554
reverse passive hemagglutination assay, **121**, 552
solid-phase hemadsorption assay, **121**, 553
antiidiotypic
idiotypic specificity, **116**, 188
production, **116**, 184
screening, **116**, 185
antireceptor
auto-antiidiotypic preparation strategy, **121**, 95
preparation for phagocytosis studies, **132**, 217
attachment to microspheres, **112**, 70
automated production in cytostat, **121**, 360
biotinylated
in membrane antigen immunoaffinity isolation, **121**, 724
preparation, **121**, 719
bispecific, from hybrid hybridomas
applications, **121**, 225
assays, **121**, 217
chain composition, **121**, 224
purification, **121**, 221
theoretical considerations, **121**, 211
calmodulin
for affinity chromatography, **102**, 107
for immunoprecipitation, **102**, 109
preparation, **102**, 105
in cell surface binding immunoassays, **121**, 714

to cellular and soluble antigens, assays, **121**, 783

for cooperative immunoassays, screening procedures, **121**, 696

coupling to activated dextran or agarose, **121**, 741

crosslinking to immobilized protein A or anti-Ig, **121**, 743

detection with tritiated antibodies, **121**, 674, 675

1,25-dihydroxyvitamin D_3 receptor
applications as probes, **123**, 201
preparation, **123**, 200

dilution analysis for affinity determination, **121**, 472

dot immunobinding assay, **121**, 484

enzyme-linked, in protein detection on nitrocellulose, **121**, 848

epitopes
solid-phase enzyme immunoassay with avidin–biotin, **121**, 463, 466
solid-phase radioimmunoassay, **121**, 461

to F_1-ATPase subunits
applications, **126**, 766, 774
preparation, **126**, 762, 771

generation by
immunization *in vitro*, **121**, 27
intrasplenic immunization, **121**, 33

L-glutamate decarboxylase
characterization, **113**, 8
production, **113**, 8

to gonadotropin subunits
in determination of epitope relative orientation, **109**, 648
effect on hormone–receptor interaction, **109**, 641
production, **109**, 639
properties, subclass and affinity, **109**, 641
screening methods, **109**, 640
synergistic interactions, **109**, 653

growth hormone receptor
in characterization of hepatic receptors, **109**, 701
identification of Ig class, **109**, 698
production, **109**, 694
screening assay, **109**, 697
specificity studies, **109**, 701

HPLC, **121**, 596

human
to A erythrocytes, preparation, **121**, 171
anti-Rh_0(D), preparation, **121**, 172
to chorionic somatomammotropin, affinity and specificity, **121**, 207
enzyme immunofiltration assay, **121**, 528
to HLA-DR, in immunoaffinity purification of antigen subsets, **108**, 600
production by EBV–hybridoma technique, **121**, 163
to T-cell antigens
applications, **108**, 637
production, **108**, 632

human–human
overview, **121**, 107
preparation, **121**, 114

from human × (murine × human) hybridomas, production, **121**, 230

from hybridoma-bearing nude mice and rats, analysis, **121**, 408

hybridoma-secreted
affinity constants, calculation, **121**, 205
antigen-capture enzyme immunoassay, **121**, 543
assays, **121**, 200, 204
competitive inhibition enzyme immunoassay, **121**, 542
epitope specificity, assay, **121**, 203
purification, **121**, 202
screening, **121**, 32, 491

immobilization on polymer matrix, **104**, 386

immunofluorescence assay, **121**, 580

immunohistochemical techniques with, **121**, 574, 828

immunoreactivity after radiolabeling, assay, **121**, 678

in immunosorbent chromatography, **104**, 385

in indirect rosetting assay for lymphocytes, **121**, 726

insulin site-specific
affinity purification, **109**, 712
assays, **109**, 709, 718
production, **109**, 705
specificity analysis, **109**, 715

interferon αA
preparation and isolation, **119**, 589

in sandwich radioimmunoassay, **119**, 590
interferon γ
in sandwich radioimmunoassay, **119**, 582
summary list, **119**, 584
to interferon-induced surface tumor antigens
preparation, **119**, 683
radioimmunoassay with, **119**, 686
interferon, neutralization characteristics, **119**, 565
labeling of T cells, **121**, 744
to *lac* permease of *Escherichia coli*
binding assays, **125**, 461
Fab fragment preparation, **125**, 458
preparation, **125**, 455
purification, **125**, 456
screening, **125**, 458
D-lactate dehydrogenase
in enzyme purification, **126**, 374
preparation, **126**, 372
purification, **126**, 373
lipoprotein
applications, **128**, 535
assays, **128**, 530
for mapping receptor binding determinants, **129**, 559
production, **128**, 529
purification, **128**, 534
to Lyt-1, Lyt-2, and Lyt-3 antigens, description, **108**, 667
α_2-macroglobulin
assays in hybridoma supernatants, **121**, 63
characterization, **121**, 67
production, **121**, 60
macrophage-specific
applications, **108**, 320
defined antigens, characterization, **108**, 314
preparation, **108**, 313
–microsphere conjugates
immunoreactivity, **112**, 84
purification, **112**, 83
murine
enzyme immunofiltration assay, **121**, 528
to IgM
applications, **121**, 623
production and purification from rat hybridomas, **121**, 623
to MHC antigens, applications in specificity analysis, **108**, 481
reacting with cytoplasm and surface of differentiated cells
availability, **103**, 468
production, **103**, 460, 467
screening assay, **103**, 465
to Thy-1.1 and Thy-1.2 antigens
applications, **108**, 647, 652
production, **108**, 645
myosin
binding analysis, **134**, 437
binding assays, **134**, 430, 431
immunoblotting, **134**, 433, 442
inhibitory properties, **134**, 444, 446, 449, 452
production and purification, **134**, 424
properties, **134**, 427
nutrient-coupled, to surface antigens, in positive cell selection, **121**, 771
to oligomycin sensitivity-conferring protein
interactions with F_0 and F_1, **126**, 458
in protein purification, **126**, 455
parathyroid hormone
production, **109**, 625
serum titers, determination, **109**, 631
to plant viruses
characterization, **118**, 773
production, **118**, 768
screening and serotyping assays, **118**, 771
in precipitating immunoassays, **121**, 703
to preselected protein regions, preparation and properties, **121**, 69, 88
production by
agarose-entrapped hybridoma cells, **121**, 352
ascites tumors, pristane effects, **121**, 375
hybridomas, **121**, 3, 291
to proteins, detection on blots with enzyme-conjugated anti-Ig antibodies, **121**, 502
purification, **116**, 191; **121**, 718
radioiodinated, in epitope ratio analysis of cell surface antigens, **121**, 799

rat
to Ia antigens, production, **108**, 559, 565
to MHC class I antigens
hemagglutination assay, **121**, 395
secreting hybridomas, production, **121**, 394
properties, **121**, 640
purification, **121**, 642, 643, 646, 649, 650
reaction with large granular lymphocytes, **132**, 450
to receptor-bound interferon
binding to bound interferon, **119**, 329
purification, **119**, 326
in quantitation of temperature-induced decrease in bound interferon, **119**, 330
radiolabeling, **119**, 327
to soluble immune response suppressor
applications, **116**, 397
preparation and properties, **116**, 396
to sperm whale myoglobin antigenic sites
binding specificity, assay, **121**, 76
class determination, **121**, 76
overview, **121**, 90
preparation, **121**, 74
to sperm whale myoglobin nonantigenic sites, preparation, **121**, 92
stepwise amplified immunoenzymatic staining, **121**, 855
T101, immunoreactivity test after radiolabeling, **121**, 682
in T-cell epifluorescence microscopy, **121**, 750
vitamin B_6
applications, **122**, 124
preparation, **122**, 122
to weak immunogens, preparation, **121**, 183
Monocytes
cryopreservation, **132**, 234, 241
culturing in suspension, **132**, 208, 243
cytotoxicity assays, **132**, 458
feeder effect on tumor cells, **121**, 298
isolation, **108**, 96, 242; **121**, 296; **132**, 207, 230, 237, 238, 632
production of
H_2O_2, assay, **132**, 397
interleukin 1, **116**, 457
purification of interleukin 1, **116**, 463
separation, **108**, 91
superoxide anion release, measurement, **105**, 365
5′-*R*,*S*-Monodeuteroriboflavin
ribityl side chain, ^{1}H NMR, **122**, 241
Monolayers
antibody-bound
lymphocyte separation on, **108**, 130
preparation, **108**, 128, 130
apolipoproteins at air–water interface
spreading methods, **128**, 392
surface pressure measurement, **128**, 390, 392
fibroblast
lymphocyte separation on, **108**, 130
preparation, **108**, 126
lectin-bound
lymphocyte separation on, **108**, 130
preparation, **108**, 128
lipids at air–water interface, apolipoprotein adsorption, **128**, 399
macrophage
lymphocyte separation on, **108**, 130
preparation, **108**, 126
poly(L-lysine)-bound
lymphocyte separation on, **108**, 130
preparation, **108**, 127
Mononuclear cells
production of
interferon γ, **119**, 93
interleukin 1, **116**, 457, 470
purification of interleukin 1, **116**, 463, 471
Monosodium urate
interaction with macrophages, chemiluminescent assay, **133**, 524
Monoterpene cyclase
plant and fungal
assays, **110**, 389
extraction and purification, **110**, 398
properties, **110**, 384
Morganella AM-15
growth in culture, **122**, 140
purification of pyridoxal phosphate-dependent histidine decarboxylase, **122**, 140
Mössbauer radiation
Rayleigh scattering, in analysis of protein structure dynamics, **131**, 598

Mössbauer spectroscopy
in analysis of
protein structure dynamics, **131**, 568
water mobility in proteins, **127**, 197
Mouse
nude, *see* Nude mouse
Müller cells
isolated from dispersed retinal cells, identification, **103**, 363
Mung bean
isolation of protoplasts, **118**, 575
Muramyl peptides
-containing liposomes
activation of macrophages, **132**, 601
preparation, **132**, 597
Muscle
fibers, time-resolved X-ray scattering, **134**, 675
skeletal, *see* Skeletal muscle
smooth, *see* Smooth muscle
–water interactions, NMR analysis, **127**, 158
Mustelus canis, *see* Dogfish
Mutagenesis
bacterio-opsin gene
generalized methods for introduction of amber mutations, **125**, 193
by introduction of small deletions, **125**, 198
oligonucleotide-directed methods, **125**, 202
site-specific methods, **125**, 206
CHO cells, conditions, **129**, 240
interferon α gene sequence *in vitro*, **119**, 403
interferon β gene sequence encoding carbohydrate attachment site *in vitro*, **119**, 399
lac permease of *Escherichia coli* by site-specific technique, **125**, 214
Phycomyces blakesleeanus, in production of color mutants, **110**, 226
transposon
random, in gram-negative bacteria, **118**, 644
Rhizobium DNA with Tn*5*, **118**, 522
site-specific, in gram-negative bacteria, **118**, 649, 656
Mutagenicity testing
oxygen free radicals in microbial systems
liquid incubation assay, **105**, 259
plate incorporation assay, **105**, 257
preincubation assay, **105**, 259
spot test, **105**, 259
with *Salmonella typhimurium* TA102, **105**, 251
Mutants
ACTH-resistant, from Y1 adrenal tumor cells
characteristics, **109**, 355
preparation, **109**, 353
selection, **109**, 354
amber, bacterio-opsin gene, production, **125**, 193
arginyltransferase-deficient, characterization, **106**, 204
ATPase, *Escherichia coli*
anomalous migration of ssDNA fragments, detection, **126**, 602
DNA cloning, **126**, 599
genetic analysis with λ-*unc* transducing phages, **126**, 558
identification of defective F_1 subunits, **126**, 593
mapping by genetic recombination, **126**, 597
nucleotide sequence, **126**, 602
carotene, *Phycomyces blakesleeanus*
applications, **110**, 235
conservation, **110**, 234
description, **110**, 223
detection, **110**, 220
genetic analyses, **110**, 229
genetic nomenclature and strain list, **110**, 237
preparation, **110**, 226
chlorina *f*2 of barley, isolation of chlorophyll-binding proteins, **118**, 348
chloroplast, isolation
via protoplasts, **118**, 619
via seed mutagenesis, **118**, 612
CHO cell defective in
low-density lipoprotein endocytosis, selection methods, **129**, 227
low-density lipoprotein metabolism
immobilization, **129**, 241

induction, **129**, 240
screening, **129**, 245
energy coupling factor, *Escherichia coli*
type I
isolation, **125**, 181
properties, **125**, 182
revertants, isolation, **125**, 182
type II, defects, **125**, 185
type III, defects, **125**, 186
energy uncoupled phenotype, *Escherichia coli*
generation and screening, **125**, 188
mapping, **125**, 189
properties, **125**, 190
screening, **125**, 188
storage, **125**, 190
H_2 uptake negative, *Rhizobium*
characterization, **118**, 532
isolation, **118**, 529
in isolation of *hup* genes, **118**, 534
preparation, **118**, 529
screening, **118**, 531
lac permease, *Escherichia coli*
assays, **125**, 227
cloning and orientation in recombinant plasmid, **125**, 227
production, **125**, 217, 228
purification, **125**, 225, 226
screening, **125**, 225
sequencing, **125**, 225
leucyltransferase-deficient, characterization, **106**, 204
malE, *Escherichia coli*, Ca^{2+}-treated
maltose-binding protein entry and function, influencing factors, **125**, 296
transport characteristics, **125**, 297
peptide transport-deficient, bacterial, isolation, **125**, 366
polyhydric alcohol transport-deficient, *Escherichia coli*, selection methods, **125**, 476, 477, 478
protein folding
mutational analysis, **131**, 252
temperature-sensitive
characterization, **131**, 260
folding pathway, **131**, 256
isolation, **131**, 258
properties, **131**, 255
purification, **131**, 259
sodium pump-defective, *Vibrio alginolyticus*
isolation, **125**, 521
sodium pump activities, **125**, 523, 525
sterol, yeast
in analysis of sterol metabolism, **111**, 341
characterization, **111**, 336
description, **111**, 333
isolation, **111**, 334, 339
succinate dehydrogenase-deficient, *Bacillus subtilis*
complementation analysis, **126**, 411
isolation, **126**, 409
sugar transport, *Escherichia coli*, isolation and characterization with transposons, **125**, 157
temperature-sensitive, *Escherichia coli*, in phagocytic killing assay, **132**, 525
temperature-sensitive, simian virus 40
transformed mammalian cells, general properties, **109**, 387
transformed rat fetal hepatocytes
establishment, **109**, 387
growth and maintenance, **109**, 395
metabolic and hormonal properties, **109**, 392
Mutations
single-base, in interferon αA and α2 genes, detection, **119**, 481
Mycoplasmas
cell water measurement, **125**, 264
contamination effects on screening of hybridoma supernatants, **121**, 481
growth in culture, **125**, 261
harvesting and washing, **125**, 263
transport processes, analytical techniques, **125**, 263
Myeloblasts
normal and leukemic, interferon effects, **119**, 629
Myeloperoxidase
antimicrobial activity, measurement, **105**, 399
in bromination of proteins, **107**, 441, 444
deficiency, detection with cellular chemiluminescence, **133**, 496

human leukocyte
assays, **132**, 274, 370
hemi-enzyme production, **132**, 374
isolation, **132**, 371
physicochemical properties, **132**, 377
purification, **132**, 372
in preparation of chloramines, **132**, 575
Myoglobin
bound water, mobility analysis, **127**, 203
conformation
dilution-induced changes, stopped-flow circular dichroism, **130**, 529
effects of hydration, **127**, 213
hierarchies, **127**, 211
free peptides
antibody responses to, analysis, **121**, 76
monoclonal antibodies to, preparation, **121**, 74
structure dynamics by Rayleigh scattering of Mössbauer radiation, **131**, 604
Myoinositol
^{3}H-labeled, in analysis of phospholipid turnover in cultured cells, **124**, 427
Myosin
antibody binding
competitive, **134**, 443
dissociation constant, determination, **134**, 437
localization, **134**, 440, 444
bipolar filament assembly, antibody inhibition, **134**, 444
coated beads, movement, antibody inhibition, **134**, 452
contraction of cytoplasmic extracts, antibody inhibition, **134**, 449
methylated lysines and 3-methylhistidine
biosynthesis, analytical techniques, **106**, 290, 293
determination, **106**, 289
functional and research role, **106**, 292
peptides with, isolation strategy, **106**, 295
tissue-specific occurrence, **106**, 287
movement in algal cells, assay
with coated beads, **134**, 531
video tracking device for, **134**, 538
P-light chain phosphorylation, quantitation, **102**, 65, 75
Myosin ATPase
inhibition by
antibodies, **134**, 446
vanadate, **134**, 485
kinetic analysis
chemical quench-flow methods, **134**, 692
stopped-flow methods, **134**, 678
Myosin light chain kinase
properties and function, **107**, 90
Myristyl groups
protein amino-terminal, identification, **106**, 205
Myxococcus fulvus
cell-free system, synthesis of C_{15}–C_{60} polyprenols, **110**, 300
Myxothiazol
inhibition of cytochrome bc_1, **126**, 259

N

NAD$^+$ (oxidized nicotinamide adenine dinucleotide)
bioluminescent assay, **122**, 150
continuous-flow assay with firefly luciferase, **133**, 237
extraction from tissue, **122**, 149
reversed affinity chromatography, **104**, 220
NADase, *see* NAD$^+$ nucleosidase
NAD glycohydrolase, *see* NAD$^+$ nucleosidase
NADH (reduced nicotinamide adenine dinucleotide)
bioluminescent assays, **122**, 150; **133**, 203, 244
continuous-flow assay with bacterial luciferase, **133**, 235
dry storage, **122**, 154
extraction from tissue, **122**, 149
formation or disappearance, simultaneous determination with oxygen consumption, **122**, 161
HPLC, **122**, 153
ion-exchange chromatography, **122**, 153
nucleotide desalting, **122**, 155
1,2-propanediol removal, **122**, 154

NADH dehydrogenase
mitochondrial, two-dimensional crystals
electron microscopy, **126**, 345
image analysis, **126**, 348
preparation, **126**, 345
NADH dehydrogenase (ubiquinone)
mitochondrial, resolution into polypeptides, **126**, 362
NADH oxidase
guinea pig neutrophil
assay, **132**, 366
properties, **132**, 368
purification, **132**, 367
human neutrophil
characterization, **132**, 359
preparation, **132**, 356
resolution into components, **132**, 358
in *Vibrio alginolyticus* membranes, assay, **125**, 528
NADH:ubiquinone oxidoreductase, *see* NADH dehydrogenase (ubiquinone)
NAD^+ nucleosidase
Bungarus fasciatus snake venom
assays, **122**, 169, 173
immobilization and transglycosidation, **122**, 178
properties, **122**, 176
purification, **122**, 175
NADPH (reduced nicotinamide adenine dinucleotide phosphate)
bioluminescent assay with coimmobilized enzymes, **133**, 203, 244
formation or disappearance, simultaneous determination with oxygen consumption, **122**, 161
NADPH–cytochrome reductase, *see* NADPH–ferrihemoprotein reductase
NADPH–ferrihemoprotein reductase
rat liver
assay, **110**, 375
purification, **110**, 376
NADPH oxidase
-containing particles in human neutrophils,
assay, **105**, 364
isolation, **105**, 363
$NAD(P)^+$ transhydrogenase
bovine cardiac mitochondria
assay, **126**, 354
purification, **126**, 354
reconstitution, **126**, 357
NAD synthetase
Escherichia coli
assay, **113**, 297
properties, **113**, 302
purification, **113**, 298
Saccharomyces cerevisiae
assay, **113**, 297
properties, **113**, 302
purification, **113**, 300
Nafcillin
in measurement of bacterial outer membrane permeability, **125**, 269
Naltrexone pamoate
loading in poly(lactic/glycolic acid) polymers, **112**, 438
Nandrolone
in hormonal bursectomy, **108**, 5
Nanospheres
polyacrolein, chromatographic characteristics, **112**, 156
Naphthalenesulfonamides
as calmodulin antagonists
biopharmacological assessment, **102**, 189
comparison with phenothiazines, **102**, 193
4-carbon, synthesis and properties, **102**, 202
6-carbon, synthesis and properties, **102**, 200
interaction with calmodulin, analytical techniques, **102**, 185
structure–activity relationships, **102**, 188
Naphthoyl-ADP
as fluorescent probe for mitochondrial ADP/ATP carrier protein, **125**, 640
Naphthoyl-ATP
as fluorescent probe for mitochondrial ADP/ATP carrier protein, **125**, 640
Naphthyl esterase
in identification of isolated Leydig cells, **109**, 286
β-Naphthyl phosphate
in protein immunostaining on nitrocellulose, **121**, 850

Narcissus pseudonarcissus, *see* Daffodil
National Bureau of Standards
reactor for neutron diffraction data collection and processing, **114**, 551
NBS, *see* National Bureau of Standards
Nematodes
growth in culture, **134**, 416
isolation of males, **134**, 417
major sperm protein purification, **134**, 418
sperm isolation, **134**, 418, 419
Neocarzinostatin, *see* Zinostatin
Nerve growth factor
iodinated
binding to cell-surface receptors, determination, **109**, 26
preparation, **109**, 22
Nerve growth factor receptors
cell-surface binding assay, **109**, 26
different types, detection, **109**, 32
Neryl pyrophosphate
radiolabeled, chemical synthesis, **110**, 410
Neuraminidase, *see* Sialidase
Neuroendocrine cells
excitotoxin application as probes, **103**, 379
function, flow cytometric analysis, **103**, 227
horseradish peroxidase in
cytochemistry, **103**, 187
intracellular transport, **103**, 205
in hypothalamoneurohypophysial complex
arginine vasopressin in, radioimmunoassay, **103**, 144
double-labeled, immunocytochemistry and identification, **103**, 139
electrophysiological recording and data analysis, **103**, 137
preparation, **103**, 133
single-labeled, localization, **103**, 138
triple-labeled, identification, **103**, 143
ion channel-interacting drugs, analytical application, **103**, 401
Neuroendocrine tissues
computer-controlled perifusion system
applications, **124**, 73
development, **124**, 67
intracellular Ca^{2+} fluorescent indicators, **124**, 116
membrane potential fluorescent indicators, **124**, 116
protein kinase C activation by diacylglycerols, **124**, 60
Neurofilaments
invertebrate, isolation, **134**, 373
mammalian
isolation, **134**, 376, 381
separation from glial filament protein, **134**, 382
subunit reassembly, **134**, 386
subunit separation, **134**, 384
–microtubule network
assays, **134**, 546
formation, **134**, 545
Neurohypophysis
portal blood, secretion of hypothalamic dopamine into, **103**, 607
Neurons
hypothalamic
clonal cell lines, **103**, 331
identification *in vivo*, **103**, 98
long-term explant cultures in serum-supplemented medium, **103**, 315, 318
organ cultures, **103**, 314, 316
primary cell cultures in serum-supplemented medium, **103**, 322, 328
serum-free culture of dissociated cells, **103**, 329
peptidergic, methodological considerations in culturing, **103**, 347
preparation for single-channel electrophysiological analysis, **103**, 150
steroid hormone-receiving
characterization by peptide content, **103**, 655
immunocytochemistry, **103**, 639
sympathetic
dissociation techniques, **103**, 337
tissue culture techniques, **103**, 335
tyrosine levels and specific activity, determination, **103**, 508
Neuropeptide receptors
in brain, autoradiographic localization, **124**, 590
membrane, solubilization and assay, **124**, 164

microaggregation, analysis by ligand dimerization, **103**, 54

Neuropeptides

analytical techniques, **103**, 23

biosynthesis, analytical techniques, **103**, 512, 521; **124**, 359

biotinylated

applications, **124**, 51

preparation, **124**, 48

colloidal gold-coupled, synthesis, **124**, 36

computationally directed biorational drug design, **124**, 3

conformation, computational principles, **124**, 3

divalent, synthesis, **103**, 49

encoding genes

expression analysis, **124**, 295, 497, 510, 534, 548

transfer methods, **124**, 282, 287

extraction, **103**, 72

handling and use, **103**, 29

in hypophysial portal blood, **103**, 607

hypothalamic, extraction from tissue and incubation media, **103**, 498

immunohistochemistry, antibodies for

application, **103**, 458

preparation, **103**, 448

luciferin derivatization for development of nonisotopic binding assays, **124**, 29

γ_1-melanotropin-like, isolation from neurointermediate pituitary, **103**, 85

ornithine decarboxylase applicability as marker, **103**, 590

precursor cDNA clones, detection with synthetic oligonucleotides, **124**, 309

precursors in hypothalamus

molecular-weight size-exclusion chromatography, **103**, 530

reverse-phase HPLC, **103**, 535

synthesis rates, prediction methods, **103**, 524

tissue content after *in vitro* incubation, **103**, 526

tracer amino acids, central delivery, **103**, 527

processing products, identification, **124**, 335

purification, **103**, 19, 73

radiolabeled, isolation, **124**, 27

radiolabeling

chloramine-T method, **124**, 22

conjugation methods, **124**, 26

direct radioiodination methods, **124**, 21

enzymatic method, **103**, 34; **124**, 24

iodogen method, **124**, 24

method selection, **103**, 34; **124**, 28

with radionuclides other than iodine, **124**, 28

related articles previously published in Methods in Enzymology

cell and tissue handling, **103**, 691

electrophysiology, **103**, 692

equipment and technology, **124**, 629

microscopy, **103**, 691

preparation of chemical probes, **124**, 629

quantitation of neuroendocrine substances, **124**, 629

radioimmunoassay and radioligand assay, **103**, 691

specific assays, **103**, 692

subcellular fractionation, **103**, 692

synthesis, **103**, 692

RNA dot and blot hybridization, **124**, 269

secretion

analysis with cultured hypothalamic explants, **124**, 359

measurement techniques, **124**, 466

reverse hemolytic plaque assay, **124**, 443

solid-phase synthesis, **103**, 4

specific pathways, tracing techniques, **103**, 663

and steroid hormones, simultaneous localization in brain, **103**, 631

Neurospora crassa

in assay and detection of molybdenum cofactor, **122**, 403

carotenoid synthesis in, **110**, 253

conidia, aging rate, assay, **105**, 263

growth in culture, **106**, 277; **123**, 303

isolation of mitochondrial outer membranes, **125**, 597

purification of

S-adenosylmethionine:ε-*N*-L-lysine methyltransferase, **123**, 306

histone-lysine methyltransferase, **106**, 278

ubiquinol–cytochrome-*c* reductase,
   126, 202
Neurotensin
   assay, **132**, 328
   binding studies, **132**, 333
   properties, **132**, 326
Neurotransmitters
   catecholamine
      fluorescence histochemistry, **103**, 620
      immunohistochemistry, **103**, 623
   ornithine decarboxylase applicability as
      marker, **103**, 590
Neutron diffraction
   aqueous solutions, structural study, **127**,
      13
   data collection and processing facility at
      NBS reactor, **114**, 551
   data processing, automated peak filtering
      procedure, **114**, 530
   proteins, **114**, 511
      conformational dynamics studies, **131**,
         433, 467, 495
      data collection and processing strat-
         egy, **114**, 527
      diffractometer for, **114**, 514
      position-sensitive detectors, **114**, 518
      reflection shape, **114**, 521
      solvent structure analysis, **127**, 329
   water
      around biomolecules, structural study,
         127, 162
      in membranes, quantitative study, **127**,
         217
      pure and in solution, structural study,
         127, 320, 325
Neutron scattering
   inelastic, lysozyme powders, hydration
      studies, **127**, 292
   small-angle, *see* Small-angle neutron
      scattering
   water in ice phases, **127**, 313
Neutrophils
   activation, **132**, 383
      drug inhibition studies, **105**, 392
      role of calcium and calmodulin, **105**,
         391, 392
   calmodulin in, assay and localization,
      105, 390
   cytotoxic effects of chloramines, **132**,
      587

dysfunctions, evaluation, **132**, 126
enucleation, **132**, 251
exocytosis, assessment methods, **132**,
   271
fractionation, **132**, 382
free radical generation
   assay, **105**, 383
   cytotoxic effects on endothelial cells,
      assay, **105**, 381
functional assessment, **132**, 100, 127, 138
guinea pig
   isolation from blood, **105**, 372
   purification of NADH oxidase, **132**,
      367
histochemical distinction from monocytes, **132**, 132, 171
human
   isolation, **105**, 359, 372, 380; **132**, 232,
      239, 270, 380, 381, 401; **133**, 459
   purification of
      cytochrome *b*-245, **132**, 384
      myeloperoxidase, **132**, 372
      NADPH oxidase, **105**, 363; **132**, 356
   oxidizing radical production, measure-
      ment, **105**, 361, 372, 393; **132**, 397,
      402
   oxygenation activity, chemiluminigenic
      probing, **133**, 456
   porcine, isolation, **132**, 381
   viability assessment, **132**, 96, 135
Newcastle disease virus
   induction of interferons, **119**, 127, 133,
      212
Nickel
   in solution, coordination number, **127**,
      325
Nickel chloride
   aqueous, structure determination by
      neutron diffraction, **127**, 13
Nicotiana tabacum, *see* Tobacco
Nicotinamide adenine dinucleotide
   oxidized, *see* NAD⁺ (oxidized nic-
      otinamide adenine dinucleotide)
   reduced, *see* NADH (reduced nic-
      otinamide adenine dinucleotide)
Nicotinamide adenine dinucleotide phosphate
   reduced, *see* NADPH (reduced nic-
      otinamide adenine dinucleotide
      phosphate)

Nicotinamide nucleotides
redox state in intact tissues, measurement, **123**, 311
N-Nicotinoyldopamine
synthesis, **112**, 390
Nifedipine
in calcium channel assays
basic binding study, **109**, 518
differential labeling in brain, skeletal muscle, and heart, **109**, 536
Nigrosin
in neutrophil viability assessment, **132**, 96, 136
Nimodipine
in calcium channel assays
basic binding study, **109**, 518
differential labeling in brain, skeletal muscle, and heart, **109**, 536
direct binding studies, **109**, 521
Nitella axillaris
dissection, **134**, 533
myosin movement on actin filaments, assay
with coated beads, **134**, 536
video tracking device for, **134**, 538
Nitrate reductase
apoprotein reconstitution, in assay of molybdenum cofactor, **122**, 403
Nitrendipine
in calcium channel assays
basic binding study, **109**, 518
differential labeling in brain, skeletal muscle, and heart, **109**, 536
Nitric oxide
binding to hemoproteins, resonance Raman spectroscopy, **130**, 377
p-Nitrobenzylidine aminoguanidine
in assay for guanidine-specific ADP-ribosyltransferase, **106**, 403
Nitroblue tetrazolium
in assay for
superoxide anion, **105**, 15, 394
superoxide dismutase, **105**, 95
in assessment of phagocytic function, **132**, 106, 113, 144, 152, 153
reduction by macrophages, microassay, **132**, 417
Nitrocellulose
ELISA techniques with, **118**, 761
immunostaining of proteins, **121**, 848
protein blotting onto, **104**, 458
Nitrogen
fixation genes from *Rhizobium*, Tn*5* mapping, **118**, 519
in solution, coordination number, **127**, 326
Nitrogenase
Klebsiella pneumoniae
assays, **118**, 514
properties, **118**, 518
purification, **118**, 516
Nitrogen mustard, *see* Mechlorethamine
Nitroglycerin
Synchron hydrogel delivery system, **112**, 517
transdermal delivery system, **112**, 460, 515
4-Nitroquinoline 1-oxide
in preparation of *Phycomyces* color mutants, **110**, 227
N-Nitroso-*N*-ethylurea
in chloroplast mutagenesis, **118**, 619
N-Nitroso-*N*-methylurea
in chloroplast mutagenesis, **118**, 612
Nitrous acid
in colorimetric assay for dopa and dopa proteins, **107**, 401
Nitroxides
in characterization of trapped radicals, **105**, 219
in free radical detection by ESR, **105**, 217
in oxygen addition reactions, **105**, 220
spin probes, for transmembrane proton movement
electrical potential measurements, **127**, 744
gradient measurements, **127**, 742
volume measurements, **127**, 740
NMR, *see* Nuclear magnetic resonance
Nocardiopsis dassonvillei
isolation of menaquinone mixtures, **123**, 252
all-*trans*-Nonaprenyl alcohol
preparation, **110**, 329
Nonaprenylpyrophosphate synthetase
Micrococcus luteus
assay, **110**, 206
properties, **110**, 208
purification, **110**, 207

Nonylhydroxyquinoline-*N*-oxide
inhibition of cytochrome bc_1, **126**, 270
Norethindrone
release from poly(ortho ester) polymer disks, **112**, 425
Norethisterone
in poly(lactic/glycolic acid) polymers, solubility effects, **112**, 438
in poly(lactide-*co*-glycolide) microcapsules
content determination, **112**, 111
incorporation methods, **112**, 107
release kinetics, **112**, 113, 115
19-Nortestosterone, *see* Nandrolone
Nuclear magnetic resonance
amide proton exchange, in detection of protein folding pathways, **131**, 141
calmodulin, **102**, 92
carbon-13
lipoproteins, **128**, 489
proteinase I, **107**, 176
ecdysteroids, free and conjugated, **111**, 405, 415
enzymes, buffers for, **104**, 414
erythrocyte diffusional permeability, in assay for water channels, **127**, 601
histidinoalanine, **106**, 353
phosphorus-31
ADP/ATP carrier protein, **125**, 634
lipoproteins, **128**, 475
in measurement of intracellular pH, **125**, 343
phosphorylated basic amino acids, **107**, 25
phosphorylated proteins, **107**, 36
proteinase I, **107**, 176
protein conformational dynamics, **131**, 309, 463, 495
proton
lipoproteins, **128**, 482
peptide–lipid complexes, **128**, 640
protein refolding at subzero temperatures, **131**, 179
protein unfolding in aqueous methanol, **131**, 175
ribityl side chain of riboflavin and ring-substituted analogs, **122**, 240
selenocysteine in proteins, **107**, 585
relaxation measurements, in determination of protein hydration, **117**, 220
solid-state, protein internal dynamics, **131**, 327
sterols, **111**, 29
vitamin D metabolites, **123**, 140
water in ice phases, **127**, 316
water interactions in biological systems, **127**, 151
water in plants during freezing process, **127**, 761
Nuclease
staphylococcal
electron density maps, interpretation, **115**, 194, 199
folding and stability, genetic analysis, **131**, 266
fragment complementing systems
folding kinetics and intermediates, **131**, 205
fragment exchange in, **131**, 207
preparation, **131**, 193
gene cloning, **125**, 146, 148
Nucleation
bacterial, ice in plant leaves
characteristics, **127**, 731
quantitation, **127**, 731, 732
tube nucleation test, **127**, 733
protein crystals, analytical techniques, **114**, 80
Nuclei
brain, micropunch sampling, **103**, 368
hypothalamic, microdissection and extraction, **103**, 529
plant, DNA isolation, **118**, 57
Nucleolytic enzymes
localization in gels, **104**, 436
Nucleosomes
chromatin poly(ADP-ribosyl)ation, **106**, 513
analytical immunofractionation methods, **106**, 518
electrophoretic analysis, **106**, 517
detection of ubiquitin–histone 2A conjugate, **106**, 253
Nucleotidase
assay for assessment of phagocytic function, **132**, 121, 162
5′-Nucleotidase
as endosome marker, determination, **109**, 267

Nucleotides
binding to
mitochondrial F_1-ATPase, nonlinear analysis, **126**, 733
uncoupling protein, **126**, 499, 501
coupling to polyacrylhydrazidoagarose, **104**, 18
desalting by reverse-phase HPLC, **122**, 155
Nucleotide sequence
apolipoprotein A-II cDNA, **128**, 751
murine interferon αA gene, **119**, 438
partial, rat interferon γ gene, **119**, 456
Nucleotide sequencing
ATPase mutant genes, **126**, 602
ATP synthase subunit genes, **125**, 231
plant mitochondrial DNA, **118**, 477
5′-Nucleotidyl-*O*-tyrosine bond
in glutamine synthetase, analytical techniques, **107**, 183
Nude mouse
applications in immunological research, **108**, 351
breeding systems, **108**, 350
housing systems, **108**, 349
human tumor xenografts, interferon effects
assay, **119**, 651
effect of combining with chemotherapy, **119**, 655
mechanism, **119**, 652
hybridoma-bearing, monoclonal antibodies, analysis, **121**, 408
Nude rat
hybridoma-bearing, monoclonal antibodies, analysis, **121**, 408
Nylon
in column separation of lymphoid cells, **108**, 298
encapsulated pharmaceuticals
applications, **112**, 100
production and properties, **112**, 84
enzyme immobilization for bacterial bioluminescence assay, **133**, 239

O

Oats
isolation of poly(A)$^+$ RNA, **118**, 372
Obituary
Sidney P. Colowick, **113**, xvii
Oil-droplet technique
for assay of phagocytosis, **132**, 192
Oil red O
–nitroblue tetrazolium, in assessment of phagocytic function, **132**, 106, 144
OK-432
induction of human interferon γ, **119**, 89, 93
Olefins
in measurement of singlet oxygen production by neutrophils, **105**, 396
Oleic acid
^{3}H-labeled, in analysis of phospholipid turnover in cultured cells, **124**, 429
(2′-5′)-Oligoadenylate
analogs
cordycepin, inhibition of protein synthesis and cell growth, **119**, 668
core, inhibition of murine Swarm chondrosarcoma growth *in vivo*, **119**, 671
effects on tobacco mosaic virus replication, **119**, 759
synthesis methods, **119**, 522
tubericidin, lead ion-catalyzed polymerization synthesis, **119**, 526
core oligonucleotide, enzymatic phosphorylation, **119**, 524
effects on
macrophages, assay, **119**, 676
tobacco mosaic virus replication, **119**, 759
enzymatic synthesis, **119**, 519
^{3}H-labeled
binding to macrophages, **119**, 355
preparation, **119**, 353
^{32}P-labeled
in binding assay for RNase L, **119**, 492
crosslinking to RNase L, **119**, 497
preformed, chemical modification, **119**, 522
Oligoadenylates
plant
enzymatic synthesis *in vitro*, **119**, 754
purification, **119**, 756
Oligoadenylate synthetase
from Ehrlich ascites tumor cells
assay, **119**, 518

partial purification, **119**, 518
synthesis of (2′-5′)(A)$_n$, **119**, 519
from tobacco plants
induction and preparation, **119**, 754
synthesis of oligoadenylates *in vitro*, **119**, 754
(2′-5′)-Oligoadenylate(trimer) receptors
macrophage, measurement, **119**, 351
Oligomycin
sensitivity-conferring protein
interactions with F_0- and F_1-ATPase, **126**, 458
purification, **126**, 455
radiolabeling, **126**, 729
Oligonucleotides
for apolipoprotein A-II cDNA screening
applications, **128**, 747
preparation, **128**, 746
crystalline, preparation of heavy-atom derivatives, **114**, 173
crystallization conditions, **114**, 168
mutagenic
phosphorylation, **125**, 219
preparation, **125**, 218
site-specific mutagenesis of *Escherichia coli lac* permease, **125**, 220
synthetic, as screening probes
applications, **124**, 315
in neuropeptide mRNA detection, **124**, 534
in neuropeptide precursor cDNA detection, **124**, 309
Oligopeptide permease
bacterial, deficient mutants, isolation, **125**, 367
Oligopeptides
radioimmunoassay, preparation of haptens and substrates for, **103**, 435
Oocytes, *see also* Eggs
Xenopus
in analysis of neuropeptide gene expression, **124**, 282
translation of T-suppressor factor mRNA, **116**, 324
Opines
in transformed plants, assay, **118**, 638
Opioid receptors
distribution, **124**, 175
ligand selectivity patterns, **124**, 174
purification, **124**, 189
selective isolation, **124**, 187
solubilization, **124**, 177
solubilized, identification, **124**, 185
Opsonins
overview, **132**, 282
Opsonization
defective, clinical consequences, **132**, 313
effective, measurement, **132**, 318
by fibronectin, assays, **132**, 337
macrophage, chemiluminescent assay, **133**, 520
Optical rotary dispersion
calmodulin, **102**, 86
Oral pellets
Theo-Dur Sprinkle, structure, function, and release rates, **112**, 512
Orcinol
in colorimetric assay for pentoses and uronic acids, **118**, 26
Orconectes limosus, *see* Crayfish
Organ cultures
apolipoprotein synthesis, immunochemical measurement, **129**, 254
hypothalamus, techniques for, **103**, 314, 316
Ornithine decarboxylase
assay methods, **103**, 592
mechanisms of stimulation, **103**, 601
regulatory properties of polyamine-dependent protein kinase, **107**, 154
α-Ornithine-α-ketoglutarate transaminase
natural occurrence, **113**, 78
δ-Ornithine-α-ketoglutarate transaminase
assays, **113**, 77
natural occurrence, **113**, 76
Orthophosphate
^{32}P-labeled, in analysis of phospholipid turnover in cultured cells, **124**, 425, 427
Oryza sativa, *see* Rice
Oscillation photography
Bragg reflection intensity, determination, **114**, 237
historical review, **114**, 237
with large unit cells, in X-ray diffraction data collection, **114**, 211
Oscilloscope
digitizing, in transient electric birefringence system, **117**, 205

Osmium
  binding to tRNA crystals, **114**, 163
Osmolarity
  effect on phagocytosis, **132**, 99
Osmosis
  reverse, in preparation of high-purity water, **104**, 398
Osmotic pumps
  ALZET
    applications, **112**, 474
    design, **112**, 470
  OSMET, applications, **112**, 483
Osmotic stress
  for measurement of intermolecular forces, **127**, 400
Osteocalcin
  4-carboxyglutamate in, quantitation, **107**, 533
  circulating levels, variation, **107**, 533
  isolation and purification, **107**, 518
  normalization parameters, **107**, 538
  quantitation by radioimmunoassay, **107**, 528
  structural properties, **107**, 527
  synthesis *in vitro*, **107**, 541
Ouabain
  –melanotropin conjugates, preparation, **112**, 264
Oudemansin
  types A and B, inhibition of cytochrome bc_1, **126**, 260
Ovalbumin
  –amatoxin conjugates
    biological properties, **112**, 233
    preparation by mixed anhydride coupling, **112**, 231
Ovarian cells
  rat, preparation for analysis of lipoprotein metabolism, **129**, 690
Ovary
  rat, preparation of free luteal cells, **109**, 307
Ovomucoid
  commercial, chromatography on hydroxyapatite columns, **117**, 377
Oxaloacetate
  transport in chloroplast envelope, **125**, 710
Oxaloacetate decarboxylase
  Klebsiella aerogenes
    assays, **125**, 531
    carboxyltransferase subunit
      assay, **125**, 535
      isolation, **125**, 536
    properties, **125**, 534, 539
    purification, **125**, 532
    reconstitution in proteoliposomes, **125**, 538
    sodium transport activity, assays, **125**, 536, 538
Oxazolidines
  as β-amino alcohol prodrugs, **112**, 359
Oxidation
  auto-, *see* Autoxidation
  glucose, assay for assessment of phagocytic function, **132**, 115, 155
  inhibition of carbamyl phosphate synthetase, **113**, 320
  interferon αA
    after reduction, **119**, 251
    after sulfitolysis, **119**, 252
  marker substrates, in hydroperoxide detection, **105**, 298
  mixed-function, histidine residues, **107**, 370
Oxidoreductase
  affinity chromatography, **104**, 21, 22
  coupled luminescent assay, **133**, 200
  localization in gels, **104**, 421
  –luciferase systems, stability, **133**, 203
  spectrophotometric assay, **133**, 201
2,3-Oxidosqualene
  radiolabeled
    incubation with marine invertebrates and plant tissues, **111**, 320
    preparation, **111**, 315
2,3-Oxidosqualene-cycloartenol cyclase
  plant, active cell-free preparations, isolation, **111**, 323
6-Oxo-benzo[*a*]pyrene radical
  characterization, **105**, 543
  preparation, **105**, 541
5-Oxo-L-prolinase
  inhibition *in vivo*, **113**, 580
  Pseudomonas putida
    assay, **113**, 452
    components β^5 and F^8
      properties, **113**, 456

purification, **113**, 454
separation, **113**, 457
stability, **113**, 455
rat kidney
assay, **113**, 447
mechanism, **113**, 451
properties, **113**, 449
purification, **113**, 447
substrate specificity, **113**, 450
5-Oxo-D-proline
enzymatic synthesis from D-glutamate, **113**, 62
5-Oxo-L-proline
metabolism *in vivo* and inhibition, **113**, 468
L-2-Oxothiazolidine-4-carboxylate
inhibition of 5-oxo-L-proline metabolism *in vivo*, **113**, 468
L-2-Oxothiazolidine-4-carboxylic acid
synthesis, **113**, 458
Oxygen
assays, **105**, 10, 129
in detection of paraquat-mediated oxy-radical formation, **105**, 525
effect on activity of cytochrome *d* and *o* terminal oxidase complexes, **126**, 109
enzymology, classification, **105**, 26
intermediates in cytochrome-*c* oxidase reaction
detection, **105**, 32
general features, **105**, 34
molecular, *see* Dioxygen
^{18}O, in phosphate derivatives, analysis by gas chromatography–mass spectrometry, **126**, 618
relationship to biology and Hund's maximum multiplicity rule, **133**, 450
singlet, *see* Singlet oxygen
Oxygenation
phagocyte activity, chemiluminigenic probing, **133**, 456
photosensitized, in singlet oxygen production, **105**, 37
Oxygen consumption
and NAD(P)H formation or disappearance, simultaneous determination, **122**, 161
Oxygen electrodes
calibration, **105**, 10
design, **105**, 8
operation, **105**, 9
Oxygen radicals
catalytic reactions, **105**, 25
enzymatic generation, assay, **105**, 383
enzymology, classification, **105**, 26
ESR detection, **105**, 215
formation, catalytic processes, **105**, 22
generated from activated adriamycin, assay, **105**, 532
generating systems
benzopyrene, **105**, 543, 545
deleterious effects, **105**, 48
generation methods, **105**, 255
extracellular, **105**, 405
intracellular, **105**, 407
HPLC–electrochemical detection, **105**, 231
induced microbial damage, assessment, **105**, 410
mutagenicity testing in microbial systems
liquid incubation assays, **105**, 259
plate incorporation assay, **105**, 257
preincubation assay, **105**, 259
spot test, **105**, 259
neutrophil-generated
assay, **105**, 383
cytotoxic effects on endothelial cells, assay, **105**, 381
noncatalytic reactions, **105**, 24
prostaglandin H synthase-generated, assays, **105**, 412
rate constants, **105**, 213
reactions, classification, **105**, 212
spin trapping, overview, **105**, 188
Oxygen uptake
assays for assessment of phagocytic function, **132**, 112, 147, 149
by 6-hydroxy-benzo[*a*]pyrene, measurement, **105**, 542
as lipid peroxidation detection method, **105**, 287
Oxyhemoglobin
human, asymptotic reaction boundaries, determination, **130**, 16
Oxysterols
in analysis of 3-hydroxy-3-methylglutaryl-CoA reductase regulation *in vitro*, **110**, 11

cytosolic binding protein
assays, **110**, 15
properties, **110**, 18
Oxytocin
biosynthesis, *in vivo* studies, **103**, 512
hypothalamic, measurement in tissue culture, **124**, 362

P

Pancreas
guinea pig
dispersed acini, cholecystokinin receptor assay, **109**, 64
preparation of
dispersed acinar cells, **109**, 289
dispersed acini, **109**, 291
human, isolation of growth hormone-releasing factors, **103**, 87
Pantetheine
hydrolyzing enzyme, porcine kidney
assays, **122**, 36
properties, **122**, 41
purification, **122**, 39
visualization on polyacrylamide gels, **122**, 38
Pantothenase
Pseudomonas fluorescens, in pantothenic acid assay, **122**, 33
Pantothenic acid
enzymatic assay, **122**, 33
Papain
digestion of
IgG and subclasses, **116**, 13, 143; **121**, 664
tubulin, **134**, 181
in isolation of intrinsic factor–cobalamin receptor, **123**, 25
Paper chromatography
isodityrosine, **107**, 392
phosphorylated basic amino acids, **107**, 27
radioactive acylcarnitines, **123**, 273
Paper electrophoresis
apolipoprotein E-containing lipoproteins, **129**, 148
inositol phospholipids, **124**, 436
isodityrosine, **107**, 391
phosphorylated basic amino acids, **107**, 27
Paracoccus denitrificans
cultivation, **126**, 155
purification of
cytochrome *b*, **126**, 326
cytochrome bc_1, **126**, 318
cytochrome-*c* oxidase, **126**, 156
ubiquinol oxidase, **126**, 307
Paraffin oil
in preparation of agarose matrix for hybridoma cell entrapment, **121**, 352
Parallax
in X-ray diffractometers, **114**, 435
Paraquat
generation of oxyradicals
detection and measurement, **105**, 525
methods, **105**, 523
mutagenicity, testing with *Salmonella typhimurium* TA102, **105**, 250
reactions with dioxygen, **105**, 525
Parathymosin α
biological activities, **116**, 265
properties, **116**, 264
purification, **116**, 264
tissue distribution, **116**, 264
Parathyroid hormone
bovine, radiolabeled
preparation, **109**, 49
purification, **109**, 50
in receptor binding assays, **109**, 52
human, immune response to, genetic control, **109**, 625
Parathyroid hormone receptors
in bone and kidney, binding assays, **109**, 52
Pars tuberalis
bovine, derived follicular cells
culture technique, **124**, 246
ultrastructural analysis, **124**, 248
Partial specific volume
proteins
in ammonium sulfate, calculation, **117**, 64
in concentrated salt and amino acid solutions, calculation, **117**, 60
in urea solutions, calculation, **117**, 53
Particles
colloidal gold, sizing, **124**, 39
phagocytosable, targeting, **112**, 305
viral, *see* Viral particles

Partition coefficients
ubiquinones, determination by HPLC, **125**, 125
Patch clamp
in analysis of single-channel electrophysiology, **103**, 147
Patch-clamp recording
anterior pituitary cells, **124**, 230
single channel, temperature control methods, **124**, 190
Patching
surface immunoglobulins, **108**, 371
Pea
isolation of
chlorophyll-binding proteins, **118**, 342
chloroplasts, **118**, 283
microsomal and plastid hydroxymethylglutaryl-CoA reductase, **110**, 27
nuclear DNA, **118**, 60
preparation of chloroplast transcription extracts, **118**, 234
Peanut agglutinin
in separation of lymphocytes
by affinity chromatography, **108**, 177
with fluorescent activated cell sorter, **108**, 178
by selective agglutination, **108**, 172, 174
Penicillin
in enrichment of *Rhizobium* H_2 uptake negative mutants, **118**, 529
Penicillinase, *see* β-Lactamase
Penicillus capitatus
purification of bromoperoxidase, **107**, 442
Pentadecylhydroxydioxobenzothiazole
inhibition of cytochrome bc_1, **126**, 264
1,4-Pentadiene fatty acids
assay, chemiluminescence method, **105**, 130
Pentane
from isolated hepatocytes and subcellular organelles, assay, **105**, 315
from isolated perfused liver
collection, **105**, 312
gas chromatography, **105**, 314
production rate, calculation, **105**, 314
Pentose
cell wall, colorimetric assay, **118**, 26
Peplomycin
encapsulated in ethylcellulose microcapsules
clinical applications, **112**, 149
intraarterial targeting, **112**, 146
preparation, **112**, 146
Pepsin
digestion of
IgA, **116**, 53
IgG and subclasses, **116**, 16, 143, 145; **121**, 654, 657, 659, 661, 665, 668, 801
Ig light chains, **116**, 114
Pepstatin
conjugates with galactose-terminated fetuin, preparation and applications, **112**, 300
Peptidase
LHRH-degrading, assay, **103**, 542
Peptide hormone receptors
in hepatic intracellular structures, analytical techniques, **109**, 219
photoaffinity labeling, general principles, **109**, 129
Peptide hormones
biotinyl derivatives
displacement from avidin, succinoyl-avidin, and streptavidin by biotin, **109**, 443
preparation, **109**, 429, 432
induced changes in hepatic phosphoinositide metabolism, analysis, **109**, 469
photolabels for
applications to peptide hormone–receptor systems, **109**, 149, 151
choice of, **109**, 131
cleavable, application, **109**, 141
design of, **109**, 134
equipment and procedures for photolysis, **109**, 145
radioactive and fluorescent, **109**, 143
receptor labeling, specificity, control, and analysis, **109**, 147
receptor binding, analysis in hepatic subcellular structures, **109**, 224, 227
Peptide mapping
actin, in assay for amino-terminal acetylation, **106**, 189
calmodulin, **102**, 305
interleukin 2, **116**, 520

murine class I MHC antigens, **108**, 516
murine Qa and TL antigens, **108**, 555
tryptic, phosphorylated β-subunit of insulin receptor, **109**, 620
Peptides
amino-containing, coupling to activated agarose, **104**, 18
anomalous mobility on HPLC, as signal of amino acid modification, **106**, 25
antifreeze, in cold-water fish
detection, isolation, and purification, **127**, 294
effect on water, **127**, 302
interactions with ice, **127**, 297
carboxy-terminal amides
biosynthesis, **106**, 220
detection, **106**, 219
occurrence, **106**, 218
dopa-containing, isolation, **107**, 409
interferon-related, HPLC analysis, **119**, 244
methionine sulfoxide in, enzymatic reduction, **107**, 352
model, proline isomerization, analysis by isomeric specific proteolysis, **131**, 117
modified amino acids in, mass spectrometric identification, **106**, 33
myosin, with methylated amino acids, isolation strategy, **106**, 295
neuroendocrine, *see* Neuropeptides
^{32}P-labeled, proteolytic removal from [^{32}P]phospholamban, **102**, 271
radiolabeled, uptake by bacteria, measurement, **125**, 370
selenocysteine-containing, *see* Selenopeptides
synthetic
apolipoprotein analogs
characterization, **128**, 637, 642, 644
design, **128**, 629
interactions with lipids, **128**, 636
synthesis, **128**, 632
myoglobin, antibody responses to, analysis, **121**, 76
phosphorylation, quantitative techniques, **107**, 139
as protein kinase substrates, preparation and assay, **107**, 136
UV resonance Raman spectroscopy, **130**, 344
Peptidylarginine deiminase, *see* Arginine deiminase
Perfusion
for continuous cytosolic Ca^{2+} measurement, **124**, 95
hepatocytes, methodology, **129**, 515
liver
in analysis of lipoprotein biosynthesis, **129**, 502
methodology, **108**, 286
micro-, in dissociated tissue culture electrophysiological recording, **103**, 125
push–pull
animal preparation for, **103**, 181
with intraatrial catheterization in freely behaving unanesthetized rat, **103**, 182
in median eminence, **103**, 178
protocol, **103**, 181
in tracing neuropeptide-specific pathways, **103**, 666
Perifusion
continuous, dispersed anterior pituitary cells
in analysis of hormone secretion, **124**, 79
comparison with static incubation techniques, **103**, 296
methodology, **103**, 298
neuroendocrine tissues, computer-controlled system
applications, **124**, 73
development, **124**, 67
Periodate
in detection of
nonenzymatic protein glycosylation, **106**, 86
vitamin D metabolites, **123**, 138
Periodic acid
and dansyl hydrazine, in staining of glycoproteins in gels, **104**, 448
–Schiff reagent, in phagocyte staining, **132**, 135, 179
Peritoneum
murine, collection of macrophages, **108**, 275, 294; **119**, 678

rabbit, collection of macrophages, **108**, 275
Periwinkle
purification of prolyl hydroxylase, **107**, 363
Permease
lac, from *Escherichia coli*
characterization, **125**, 440
and cytochrome *o*-containing liposomes, characteristics, **126**, 122
defective mutants
assays, **125**, 227
gene cloning and orientation in recombinant plasmids, **125**, 227
preparation by site-specific mutagenesis, **125**, 217, 228
purification, **125**, 225, 226
screening, **125**, 225
sequencing, **125**, 225
monoclonal antibodies, preparation, **125**, 453
purification, **125**, 430
reconstitution in proteoliposomes, **125**, 440
site-directed polyclonal antibodies, preparation, **125**, 453
Peroxidase
activation of macrophages, assays, **132**, 539, 544
–antibody conjugates
in localization of Igs in lymphoid cell suspensions, **108**, 397
preparation, **108**, 394
endogenous, blocking in immunohistological staining, **121**, 841
in estimation of lipid hydroperoxides, **105**, 296
in immunohistochemical assays with monoclonal antibodies, **121**, 576, 838
monoclonal antibodies, assay, **121**, 220
in phagocyte staining, **132**, 133, 171
stepwise amplified immunoenzymatic staining, **121**, 855
and substance P, bispecific monoclonal antibodies, assay, **121**, 217
Peroxidation
lipids, *see* Lipid peroxidation
Peroxisomes
hepatic, purification of carnitine acyltransferases, **123**, 277
Peroxyoxalate
chemiluminescence detection system for fluorescers, **133**, 439
Pertussis toxin
ADP-ribosylation of membrane N_s and N_i components, **109**, 566
ADP-ribosyltransferase
assays, **106**, 416
effect on adenylate cyclase system, **106**, 411
assay, **109**, 560
as probe for adenylate cyclase function, **109**, 563
purification, **109**, 561
Petunia
gene transfer with Ti plasmid vector, **118**, 627
isolation of protoplasts, **118**, 550, 573
plasmid-transformed
gene expression, analysis and verification, **118**, 637
selection and regeneration, **118**, 635
purification of
mitochondria for *in organello* protein synthesis, **118**, 444
mitochondrial DNA, **118**, 439, 447
regeneration from protoplasts, **118**, 575
pH
alkaline, sugar-promoted changes, measurement in *Escherichia coli*, **125**, 388
changes across lipid bilayers and membrane proteins, measurement with optical probes, **127**, 751
effect on
glutamate carboxylase, **107**, 556
horseradish peroxidase-catalyzed chemiluminescent reactions, **133**, 343
HPLC separation of bilirubin isomers, **123**, 392
phagocytosis, **132**, 100
protein electrostatic free energy, **130**, 444
protein electrostatic interactions, **130**, 421
protein stability, **130**, 441
proton conduction in proteins, **127**, 430
in estimation of H^+/OH^- permeability in phospholipid vesicles, **127**, 508

extracellular, in measurement of cytochrome oxidase H^+ pumping activity, **126**, 15
gradient generation
for high-resolution preparative isoelectric focusing, **104**, 259
in *Vibrio alginolyticus* cells, analysis, **125**, 524
halorhodopsin-containing membrane vesicles
chloride-induced changes, **125**, 514
light-induced changes, **125**, 513
intracellular, in bacteria
homeostasis
functional role, **125**, 337
mechanisms, **125**, 347
measurement, **125**, 343
regulation in
acidophilic strains, **125**, 354
alkalophilic strains, **125**, 361
steady-state conditions and pH_0 transients, **125**, 345
optimal, for ion exchangers on macroporous supports, determination, **104**, 223
variation in protein crystallization, **114**, 125
Phagocytes
chemiluminescence, assay, **132**, 505
functions, **132**, 7
ingested and adhering particles, distinction by fluorescence quenching, **132**, 198
killing of bacteria and fungi, assays, **132**, 521, 523, 525
metabolism, phagocytosis-induced changes, analytical methods, **132**, 220
monolayers, preparation for phagocytosis studies, **132**, 215
overview, **132**, 4
particle recognition and pathogenicity, **132**, 14
surface interactions, **132**, 10
surface tension determination, **132**, 22, 70
Phagocytosis
assay by
chemiluminescence, **132**, 498
flow cytometry, **132**, 183, 186
fluorescence quenching, **132**, 199
oil-droplet method, **132**, 192
associated respiratory burst, overview, **132**, 353
in cell surface tension determination, **132**, 47
dysfunction assessment, **132**, 126
effects of
cell surface charge and shape, **132**, 64
environmental factors, **132**, 97
fibronectin-mediated, assays, **132**, 337
macrophage, chemiluminescent assay, **133**, 520
measurement techniques, **132**, 100, 138
overview, **132**, 7
particle recognition and pathogenicity, **132**, 14
particles, targeting methods, **112**, 305
receptor-mediated
manipulation techniques, **132**, 217
measurement, **132**, 215
surface interactions, **132**, 10
yeast by macrophages, effect of (2′-5′)-oligoadenylates, **119**, 678
zymosan particles by macrophages, assay, **132**, 329
Phagosomes
fusion with lysosomes, fluorescence assay, **132**, 257
Phallotoxins
protein coupling
with carbodiimides, **112**, 236
with mixed anhydrides, **112**, 236
Pharmaceuticals
dextran and inulin conjugates
pharmacological properties, **112**, 292
preparation, **112**, 286
therapeutic applications, **112**, 296
nylon-encapsulated
applications, **112**, 100
production and properties, **112**, 84
Phase angles
ambiguity, resolution
applications, **115**, 100, 106
molecular boundary determination, **115**, 99
previous approaches and background, **115**, 90
procedures in, **115**, 92, 96
refinement/extension
by density modification, **115**, 77

by resolution of phase ambiguity, **115**, 100
Phaseolus aureus, see Mung bean
Phaseolus vulgaris, see Kidney bean
Phase partitioning
in purification of plasma membranes, **118**, 43
in subfractionation of thylakoid membranes, **118**, 326
Phasing
macromolecules, by isomorphous replacement, **115**, 4
noncentrosymmetrical difference maps, **115**, 87
proteins
by anomalous scattering, **115**, 41
by isomorphous replacement, **115**, 15
Phenacetin
prodrug structure, effect on drug metabolism, **112**, 341
1,10-Phenanthroline, *see* *o*-Phenanthroline
o-Phenanthroline
in spectrophotometric determination of iron in heme proteins, **123**, 322
Phenol
in extraction of
DNA–protein complexes from mitochondria, **118**, 486
sulfated proteins, **107**, 208
hydroxylation, in hydroxyl radical detection, **105**, 18
Phenol red
in color-change growth-inhibition assay for interferon, **119**, 579
Phenothiazines
as calmodulin antagonists, comparison with naphthalenesulfonamides, **102**, 193
interaction with calmodulin, analytical techniques, **102**, 171
–Sepharose conjugates
application in calmodulin purification, **102**, 29
calmodulin-binding capacity, estimation, **102**, 29
coupling estimation, **102**, 24
preparation, **102**, 18
synthesis and properties, **102**, 196
Phenylalanine
protein, aromatic ring flips and rotational motions, NMR spectroscopy, **131**, 317
L-Phenylalanine–α-keto-γ-methiolbutyrate transaminase
rat liver, in assay for glutamine transaminase K, **113**, 346
Phenylbutazone
in assay for oxygen radical generation by prostaglandin H synthase, **105**, 415
Phenylglyoxal
effect on mitochondrial ADP–ATP carrier protein binding, **125**, 661
in identification of chloroplast phosphate transport protein, **125**, 719
Phenylhydrazine
mutagenicity, testing with *Salmonella typhimurium* TA102, **105**, 250
Phenylmercuric acetate
stimulation of neutrophils, effect on chemiluminigenic response, **133**, 480
Phenylthiohydantoins
as HPLC probe for posttranslationally modified amino acids, **106**, 22
in mass spectrometric identification of posttranslationally modified amino acids, **106**, 32
Phenytoin
metabolism in hepatic microsomes, interferon effects, measurement, **119**, 715
Pheromones
insect
biosynthesis, enzyme assay, **133**, 197
extraction, **133**, 194
release rates, **133**, 195
Phorbol myristate acetate
effect on interferon γ production
in plateletpheresis residues, **119**, 50
in T-cell lymphoma or hybridoma cell lines, **119**, 52
Phosphate
counterion positions in hydrated DNA, computer simulation, **127**, 128
–hexose 6-phosphate antiport, from *Streptococcus lactis*
anion exchange assay, **125**, 561
reconstitution in liposomes, **125**, 559

inorganic, *see* Inorganic phosphate
side chains, hydrogen bonds, proton polarization, **127**, 445
transport protein
in chloroplast envelope
identification, **125**, 717
inhibitory effect of amino acid reagents, measurement, **125**, 718
isolation, **125**, 723
labeling with amino acid reagents, **125**, 718
properties, **125**, 708
reconstitution, **125**, 724
mitochondrial
isolation, **125**, 697, 702, 703
reconstitution, **125**, 699, 702, 703
O-Phosphate
in proteins
chemical stability to acid, **107**, 4
detection, **107**, 13
lability to base, **107**, 5
linkage characterization, enzymatic methods, **107**, 7
quantification, **107**, 15
release by partial acid hydrolysis, **107**, 10
semipurification, **107**, 11
Phosphate acetyltransferase
in determination of acetyl phosphate and acetyl-CoA, **122**, 44
Phosphatidate:CTP cytidyltransferase
rat liver, assay, **109**, 474
Phosphatidylcholine
–apolipoprotein complexes
biological characteristics, **128**, 606
characterization, **128**, 569
comparison with native forms, **128**, 578
physical characteristics, **128**, 600
preparation, **128**, 558, 590
transfer assay for plasma lipid transfer proteins I and II, **129**, 812
unilamellar vesicles
application in cholesterol efflux studies, **129**, 637
preparation, **128**, 555; **129**, 635
Phosphatidylcholine–sterol acyltransferase
activation by synthetic peptide analogs of apolipoproteins, **128**, 644
discovery, **128**, 4
human plasma
active site structure, **129**, 784
assays, **111**, 268; **129**, 773
characterization, **129**, 768
esterified cholesterol, lipoprotein origin, **111**, 270
isolation, **129**, 764
mechanisms, analytical techniques, **129**, 783
reactions with
native substrate, **129**, 787
synthetic lipid substrates, **129**, 785
utilization of cell membrane cholesterol, **111**, 271
metabolic function, **128**, 17
role in peripheral lymph, **129**, 676
Phosphatidylinositol
in hepatocyte plasma membranes, direct chemical measurement, **109**, 504
Phosphatidylinositol exchange enzyme
Mn^{2+}-activated, in rat liver, assay, **109**, 475
Phosphatidylinositol kinase
rat liver, assay, **109**, 476
Phosphoamino acid
^{32}P-labeled apolipoprotein B, analysis, **129**, 540
Phosphoarginine
preparation, **107**, 31
Phosphodiesterase
Ca^{2+}-dependent
application in calmodulin assay, **102**, 42
purification from bovine brain, **102**, 40
calmodulin-dependent, purification from bovine brain, **102**, 215
calmodulin-sensitive
assay, **102**, 172
preparation from rat brain, **102**, 171
snake venom
in analysis of poly(ADP-ribosyl)ated proteins, **106**, 508
assay, **106**, 507
Phosphoenolpyruvate
^{18}O-labeled
assay, **126**, 639
synthesis, **126**, 637

Phosphofructokinase
oxidized and reduced, detection in liver, **107**, 347
Phosphoglucomutase
^{31}P NMR studies, **107**, 72
6-Phosphogluconate
radiolabeled
preparation, **122**, 360
specific activity determination, **122**, 361
as substrate in dihydrofolate reductase assay, **122**, 363
6-Phosphogluconic acid
continuous-flow assay with bacterial luciferase, **133**, 235
3-Phosphoglycerate
transport protein in chloroplast envelope
identification, **125**, 717
inhibitory effect of amino acid reagents, measurement, **125**, 718
isolation, **125**, 723
labeling with amino acid reagents, **125**, 718
properties, **125**, 708
reconstitution, **125**, 724
Phosphohistidine
preparation, **107**, 30
Phosphoinositides
hepatic
degradation in cell-free systems, hormone stimulation, **109**, 477
metabolic enzymes, assays, **109**, 474
radiolabeling and analysis, **109**, 469
[^{3}H]myoinositol-labeled
chromatographic separation, **124**, 433
extraction from cultured cells, **124**, 433
Phospholamban
physicochemical properties, **102**, 273
[^{32}P]phosphopeptides, proteolytic removal, **102**, 271
purification, **102**, 272
reinsertion into liposomes, **102**, 275
solvent extraction, **102**, 270
Phospholamban kinase
–Ca^{2+}-ATPase–phospholamban complex, purification, **102**, 269
Phospholipase A_2
in ordering of mitochondrial outer membrane proteins, **125**, 599
Phospholipase C
phosphoinositide-specific, in rat liver, assay, **109**, 476
Phospholipids
–apolipoprotein complexes
application in cholesterol efflux studies, **129**, 637
preparation, **129**, 634
chromatographic separation, **124**, 431
enzymatic quantification, **129**, 120
extraction from cultured cells, **124**, 431
hydrated, calorimetric analysis, **127**, 147
interactions with apolipoproteins
enthalpy measurements, **128**, 367
helical amphipathic moment, **128**, 406
thermodynamics, measurement, **128**, 407
membrane
chemistry of autoxidation, **105**, 280
ion permeability, measurement with optical probes, **127**, 748
polyunsaturated, peroxidation detection methods, **105**, 283
turnover in cultured pituitary cells, measurement, **124**, 424
Phospholipid vesicles
composition, effects on interactions with lipoproteins, **128**, 648
and detergent micelles, separation by gel filtration, **104**, 326
enrichment of high-density lipoproteins
biological relevance, **128**, 655
incubation conditions, **128**, 650
molecular basis, **128**, 654
product separation and analysis, **128**, 651, 653
promotion, role of transfer proteins, **128**, 652
radiolabeling methods, **128**, 652
ion transport, measurement by EPR methods, **127**, 502
phosphatidylcholine, preparation, **128**, 555
with photosynthetic reaction center and cytochrome *bc*$_1$ hybrid system
construction, **126**, 303
quinone activity, **126**, 304
preparation, **128**, 648
reconstitution of
ADP/ATP carrier protein, **125**, 613, 633

chloroplast membrane proteolipid, **126**, 526
cytochrome-*c* oxidase, **126**, 14, 78
F_0F_1-ATPase, **126**, 605
glutaconyl-CoA decarboxylase Na^+ transport activity, **125**, 555
membrane-bound pyrophosphatase, **126**, 452
membrane proteins, **104**, 340
ubiquinol–cytochrome-*c* reductase, **126**, 206
uncoupling protein, **127**, 772

Phospholysine
preparation, **107**, 30

Phosphomevalonate
synthesis, **110**, 80

Phosphomevalonate kinase
porcine liver
assays, **110**, 78
properties, **110**, 84
purification, **110**, 82

Phosphoprotein phosphatase
calmodulin-dependent
antiserum production, **102**, 251
assay, **102**, 231
purification from bovine brain, **102**, 245
radioimmunoassay, **102**, 253
radioiodination, **102**, 251
classification, **107**, 103
identification, **107**, 118
phosphoprotein substrates, preparation, **107**, 108
types 1 and 2
assays, **107**, 110
isolation, **107**, 120
protein modulators, purification, **107**, 124
purification, **107**, 123
regulators, assay, **107**, 116
substrate specificity, **107**, 104
tissue distribution, **107**, 121

Phosphoproteins
in gels, staining, **104**, 451
phosphoramidate content, determination, **107**, 31

5′-Phosphopyridoxyl
proteins with
preparation, **122**, 121
in preparation of monoclonal antibodies to vitamin B_6, **122**, 122

Phosphoramidates
diagnostic tests, **107**, 29
in phosphoproteins, content determination, **107**, 31
in proteins, detection and characterization, **107**, 23

Phosphoribosylamine
enzymatically synthesized, assay, **113**, 268

5-Phosphoribosyl-1-pyrophosphate
assay, **113**, 269

Phosphoric acid
derivatization for analysis by gas chromatography–mass spectrometry, **126**, 619

Phosphorylase
hepatic, hormone effects, assessment, **109**, 550

Phosphorylase *a*
preparation, **107**, 108

Phosphorylase kinase
calmodulin-bound, quantitation, **102**, 220
and calmodulin subunits, interactions, **102**, 223
properties and function, **107**, 89
and troponin C, interactions, **102**, 225

Phosphorylase kinase *a*
preparation, **107**, 109

Phosphorylation
apolipoprotein B *in vivo*, **129**, 539
associated reactions, evaluation, **107**, 99
auto-, *see* Autophosphorylation
enzymatic, in synthesis of (2′-5′)-oligoadenylic acid analog, **119**, 524
insulin receptors
in cultured cells, **109**, 613
in solubilized system, **109**, 616
interferon γ by cAMP-dependent protein kinase, **119**, 296
myosin P-light chain, quantitation
in skeletal muscle, **102**, 75
in smooth muscle, **102**, 65, 71
oligonucleotides, for site-specific mutagenesis of *lac* permease, **125**, 219
oxidative, in mitochondria, inhibition with *N*,*N*′-dicyclohexylcarbodiimide, **125**, 92
photosynthetic, reconstitution
in estimation of proton motive force, **127**, 551

with photosystem I reaction center and CF_1CF_0, **126**, 285
proteins
detection in crude tissue extracts, **107**, 131
in intact cells and *in vivo*, **107**, 142
quantitative techniques, **107**, 139
riboflavin analogs, in preparation of riboflavin phosphates, **122**, 216
sarcoplasmic reticulum vesicles, **102**, 263
synthetic peptides, quantitative techniques, **107**, 139
Phosphoserine
codon usage in casein mRNA, **106**, 158
O^{β}-Phosphoseryl-tRNA
function, characterization, **106**, 157
structural features, **106**, 158
Phosphotransacetylase, *see* Phosphate acetyltransferase
Phosphotyrosine
in proteins, quantification, **107**, 18
in protein-tyrosine kinase reaction mixtures, quantification, **107**, 18
Phosvitin kinase
properties and function, **107**, 92
Photinus-luciferin 4-monooxygenase (ATP-hydrolysing)
cDNA
construction, **133**, 5
expression in *Escherichia coli*, **133**, 10
screening, **133**, 8
commercial reagents
applications, **133**, 62, 68
development, **133**, 51
evaluation, **133**, 53
suppliers, **133**, 53, 65
in continuous-flow bioluminescent assay for NAD, **133**, 237
effect on light production by α-bungarotoxin–luciferin conjugates, **133**, 45
Photoaffinity labeling
atractyloside-binding site on ADP/ATP carrier protein, **125**, 665
ecdysteroid carrier protein, **111**, 432
F_1-ATPase with
ADP analogs, **126**, 706
benzophenone-ATP, **126**, 678
divalent azido–ATP analog, **126**, 649
glucagon receptors in liver, **109**, 207
hormone receptors in neuroendocrine tissues, **103**, 66
insulin receptors, **109**, 170
juvenile hormone-binding proteins with epoxyfarnesyl diazoacetate, **111**, 509
luteal receptors with derivatized human chorionic gonadotropin, **109**, 207
peptide hormone receptors, general principles, **109**, 129
prenyltransferase with photolabile substrate analogs, **110**, 128
prolactin receptors, **109**, 161, 165
Photobacterium
P. leiognathi, purification of luciferase by HPLC, **133**, 119
P. phosphoreum
bioluminescence-related acyltransferase
assay, **133**, 183
properties, **133**, 187
purification, **133**, 185
bioluminescence-related genes, cloning in *Escherichia coli*, **133**, 70
fatty acid reductase
assays, **133**, 173
properties, **133**, 177
Photobleaching
fluorophores in digital imaging fluorescence microscopy, **129**, 860
Photography
for detection of luminescent reactions
applications, **133**, 413, 418
assays, **133**, 412
film types, **133**, 404
instrumentation, **133**, 405
in measurement of bacterial motility, **125**, 584
oscillation, *see* Oscillation photography
Photolabels
for peptide hormones
applications to hormone–receptor systems, **109**, 149, 151
cleavable, application, **109**, 141
choice of, **109**, 131
design, **109**, 134
equipment and methods for photolysis, **109**, 145
radioactive and fluorescent, **109**, 143
receptor labeling, specificity, control, and analysis, **109**, 147
Photolysis
hormone–receptor complexes, **103**, 62

peptide hormones for receptor photoaffinity labeling, **109**, 145
vacuum–UV, generation of superoxide in aqueous and ethanolic solutions, **105**, 81
Photometers
aequorin luminescence
assay and calibration methods, **124**, 106
design, **124**, 97
operation, **124**, 104
for detection of cellular chemiluminescence, **133**, 505
Photometry
in determination of oxygen consumption and NAD(P)H formation or disappearance, **122**, 161
Photomicrography
in immunocytochemical analysis of endorphins and enkephalins, **103**, 683
in immunohistological staining, **121**, 845
Photon correlation spectroscopy
polydisperse macromolecular solutions and colloidal suspensions
background, **117**, 256
experimental considerations, **117**, 259
linewidth distribution analysis, **117**, 278
mathematical techniques, **117**, 263
Photoreceptor cells
isolation from dispersed retinal cells, **103**, 363
Photosystem I
chlorophyll–protein complexes, isolation
from barley chlorina *f*2 mutant, **118**, 348
from pea, **118**, 342
reaction center complex
individual subunits
antibody preparation, **118**, 358
isolation, **118**, 358
quantitative estimation, **118**, 360
structure and function, **118**, 357
properties, **118**, 357
purification, **118**, 354; **126**, 286
reconstituted in liposomes
photophosphorylation in, **126**, 290
preparation, **126**, 288
proton translocation, assays, **126**, 288
synthesis during development, analytical techniques, **118**, 367
Photosystem II
chlorophyll–protein complexes, isolation
from barley chlorina *f*2 mutant, **118**, 348
from pea, **118**, 342
from spinach, **118**, 351
Phycobiliproteins
as staining reagent in FACS, **108**, 208
Phycomyces blakesleeanus
color, **110**, 220
color mutants
applications, **110**, 235
conservation, **110**, 234
description, **110**, 223
detection, **110**, 220
genetic analyses, **110**, 229
genetic nomenclature and strain list, **110**, 237
preparation, **110**, 226
standard growth conditions and variations, **110**, 222
Phylloquinone, *see* Vitamin K_1
Physical mapping
plant mitochondrial DNA, **118**, 456
Rhizobium nitrogen fixation genes with Tn*5*, **118**, 519
Phytic acid
in detection of nonenzymatic protein glycosylation, **106**, 83
Phytochrome
cDNA clones, identification, **118**, 369
Phytoene
enzymatic synthesis, assay, **110**, 209, 211
Phytoene synthetase
tomato fruit chromoplasts
extraction and partial purification, **110**, 216
properties, **110**, 218
Phytohemagglutinin
induction of interferon γ production, **119**, 66, 78, 85, 98, 134
Phytosterols
enzymatic dealkylation in insects, analytical techniques, **111**, 346
Pilocarpine
ocular delivery system, structure, function, and release rates, **112**, 509

Pineal gland
  bovine, purification of hydroxyindole O-
    methyltransferase, **103**, 490
Pinocytosis
  fluid, in leukocytes, analysis with fluo-
    rescent dextrans and horseradish
    peroxidase, **108**, 336
Pisum sativum, see Pea
Pituitary
  anterior, *see* Anterior pituitary
  bovine, isolation of γ_1-melanotropin,
    103, 85
  electrophysiological recordings of cul-
    tured cells, **103**, 102
  extraction of β-endorphin and enkepha-
    lins, **103**, 552
  human, distribution of CRF receptors,
    124, 567
  rat
    N^α-acetyltransferase, localization, **106**,
      175, 177
    CRF receptors, distribution, **124**,
      567
    neuropeptide gene expression, analy-
      sis by *in situ* hybridization, **124**,
      510
    neuropeptide receptors, localization,
      124, 600, 616
  short-term cell culture
    application for simultaneous hormone
      determination, **103**, 257
    cell adhesion and aggregation in, **103**,
      252
    cell density, effect on function, **103**,
      253
    cell dispersion and culturing condi-
      tions, **103**, 249
    effects of incubation length, **103**, 255
    media for, choice, **103**, 251
Placenta
  human
    detection of constitutively produced
      interferon, **119**, 544
    FACS analysis, **108**, 240
    purification of
      insulin receptors, **109**, 399, 401
      thioltransferase, **113**, 522
Plants, *see also specific plants*
  isolation of
    calmodulin, **102**, 11

cell wall, **118**, 3
chloroplast DNA, **118**, 167, 188
chloroplasts, general methodology,
  228, 188
cytokinins, **110**, 352
mitochondria, **118**, 490
mitochondrial RNA, **118**, 492
monoterpene and sesquiterpene cyclases, **110**, 400
plasma membranes, **118**, 41
plastoquinone biosynthesis, analytical
  techniques, **110**, 315
sterol biosynthesis and interconversion,
  analytical techniques, **111**, 311
vitamin K assay by HPLC with UV
  detection, **123**, 236, 242
Plant viruses, *see also specific viruses*
  detection
    with DNA probes, **118**, 723
    ELISA techniques, **118**, 742
    with RNA probes, **118**, 717
  monoclonal antibodies, production, **118**,
    768
  serotyping with monoclonal antibodies,
    118, 771
Plaque assay
  hemolytic, *see* Hemolytic plaque assay
  sequential, secretion from neuroendo-
    crine cells, **124**, 457
Plaque-forming cells
  screening with culture-well solid-phase
    radioimmunoassay, **121**, 448
Plaque reduction assay
  in screening of cells for sensitivity to
    interferon, **119**, 607
Plasma, *see also* Blood; Serum
  extraction of
    β-endorphin, **103**, 554
    enkephalins, **103**, 553
  guinea pig, human, and rat, simulta-
    neous determination of ubiquinones,
    105, 147
  human
    determination of apolipoprotein C
      levels, **129**, 448
    lecithin–cholesterol acyltransferase
      reaction and cholesterol transport,
      111, 267
    leukocyte-rich
      isolation of monocytes, **108**, 96

separation of mononuclear cells and granulocytes, **108**, 94
lipoprotein metabolism, assay with lipoprotein lipase, **129**, 704
measurement of growth hormone-releasing factor, **124**, 384
mevalonate assay, **110**, 63
purification of
catecholamines, **124**, 404
cholesteryl ester transfer activity, **111**, 275
coagulation factor XIIIa, **113**, 366
1,25-dihydroxyvitamin D, **123**, 179, 186
gelsolin, **134**, 8
hepatic lipase, **129**, 718
insulin-like growth factors, **109**, 793
lipid transfer proteins, **129**, 798
phosphatidylcholine–sterol acyltransferase, **129**, 764
platelet-derived growth factor, **109**, 759
somatomedin C, **109**, 805, 814
quantitation of 1,25-dihydroxyvitamin D, **123**, 190
reactions with lecithin–cholesterol acyltransferase, **129**, 787
vitamin K assay by HPLC with UV detection, **123**, 237
warfarin assay, **123**, 225
lipid peroxide assay, **105**, 328
rat, purification of
fibronectin, **132**, 335
selenoprotein P, **107**, 612
simultaneous determination of retinol and α-tocopherol by HPLC, **123**, 215
Plasma membranes
associated proteins, avidin–iminobiotin-based purification, **122**, 87
calvarial, calcitonin receptor assay, **109**, 44
hepatic
glucagon receptor assays, **109**, 3, 15
phosphatidylinositol levels, direct chemical measurement, **109**, 504
plant
markers for, **118**, 46
purification, **118**, 42
relative electrical potential, cytofluorometric assay, **103**, 241
renal
calcitonin receptor assay, **109**, 44
parathyroid hormone receptor assays, **109**, 52
synaptosomal, antibodies to, dot immunobinding assay, **121**, 489
Plasmids
interferon expression
construction, **119**, 368, 418, 437, 445, 471
expression of interferon γ gene, **119**, 380
rationale for design, **119**, 376
pACYC 184, cloning and orientation of mutant *lacY* gene, **125**, 227
pCV108, human genomic library, screening for apolipoprotein E, **128**, 816
pIN III, in construction of secretion cloning vectors, **125**, 139
applications, **125**, 143
for outer membrane localization, **125**, 143
for periplasmic localization, **125**, 140
pI.T7.α2.cop1, expression of recombinant human interferon α2, **119**, 167
Ti, *Agrobacterium tumefaciens*
introduction of foreign genes, **118**, 628
transformation of plants, **118**, 581, 634
vector, *Escherichia coli*-specific
application as broad host range vectors, **118**, 657
mobilization in gram-negative bacteria, **118**, 641, 654
transposon carrier
random mutagenesis of gram-negative bacteria, **118**, 644
site-specific insertion in gram-negative bacteria, **118**, 649, 656
viral DNA-containing
construction, rationale for, **118**, 661
introduction and expression in plants, **118**, 660, 663
preparation, **118**, 659
yeast, for secretion of human interferons
construction, **119**, 424
relevance and usefulness, **119**, 428

Plasmin
  in crystallized carbohydrate micro-
      spheres
    biological activity, assay, **112**, 121
    preparation, **112**, 120
    release studies, **112**, 122
  digestion of IgG, **116**, 20
Plasmodium sp.
  ingestion by macrophages, assay, **132**,
      642
  killing by macrophages, assay, **132**, 642
  maintenance, **132**, 641
Plastic surfaces
  macrophage separation on, **108**, 294
Plastids
  pea seedling, isolation of 3-hydroxy-3-
      methylglutaryl-CoA reductase, **110**,
      38
Plastoquinone
  biosynthesis, analytical techniques, **110**,
      313
  marker quinones and quinols
    isolation, **110**, 313
    purification, **110**, 315
Platelet-derived growth factor
  human
    assays, **109**, 752
    properties, **109**, 768
    purification, **109**, 759
    purity determination and standardiza-
        tion, **109**, 765
  interactions with
    interferon α in fibroblasts, assays, **119**,
        658
    interferon in Moloney sarcoma virus
        tumor cells *in vivo*, **119**, 664
  radioiodinated
    cell-associated
      determination, **109**, 74
      dissociation, **109**, 89
    preparation and characterization, **109**,
        70
    in receptor assay, **109**, 84
      binding conditions, **109**, 80
      binding solutions, **109**, 77
      Scatchard analysis, **109**, 92
    saturable binding, determination, **109**,
        74
Platelet-derived growth factor receptors
  evidence for mitogen receptor nature,
      109, 94

molecular properties, **109**, 97
  radioassay
    binding conditions, **109**, 80
    binding solutions, **109**, 77
    procedure and evaluation, **109**, 84
    Scatchard analysis, **109**, 92
Platelets
  human
    cell surface binding immunoassays,
        121, 714, 715
    purification of
      coagulation factor XIIIa, **113**, 366
      HLA-A and HLA-B antigens, **108**,
          593
    removal from phagocyte preparations,
        132, 228
Platinum
  binding to
    proteins, **114**, 152
    tRNA crystals, **114**, 160
PN 200-110
  in calcium channel assays
    basic binding study, **109**, 518
    differential labeling in brain, skeletal
        muscle, and heart, **109**,
        536
Poisson distribution
  in assessment of hybridoma monoclonal-
      ity, **121**, 412
Polarography
  dc, in analysis of quinones, **125**, 114
Polyacrolein
  -activated microspheres
    antibody binding, **112**, 158
    magnetically responsive, for cell
        separation, **112**, 162
    preparation, **112**, 155
  –agarose microspheres
    applications, **112**, 171
    synthesis and properties, **112**, 169
  microspheres
    antibody binding, **112**, 158
    applications, **112**, 171
    molecular structure and chemical
        analysis, **112**, 152
    preparation, **112**, 151, 164
  nanospheres, chromatographic charac-
      teristics, **112**, 156
Polyacrylamide gels
  (ADP-ribosyl)$_n$ proteins in, detection,
      106, 468

for high-resolution preparative isoelectric focusing
description, **104**, 262
load capacity, **104**, 265
protein detection techniques, **104**, 269
isoelectric focusing of lipoproteins, **128**, 434
optimal pH conditions, determination
by electrophoretic titration curves, **104**, 225, 229
by retention mapping, **104**, 227, 229
pantetheine-hydrolyzing enzyme in, visualization, **122**, 38
preparation for electrophoresis, **104**, 239, 241
in preparative isotachophoresis, **104**, 285, 296
radioactivity in, fluorographic detection, **104**, 461, 463
screening with spaghetti overlay technique, **121**, 453
urea-gradient, preparation, **131**, 166
Polyacrylhydrazidoagarose
characterization, **104**, 16
coupling of ligands, **104**, 17
derivatization, **104**, 15
preparation, **104**, 15
Polyadenylation
viral RNAs, methodology, **118**, 727
Poly(ADP-ribose)
characterization, **106**, 508
levels *in vivo*, quantification
fluorescence methods, **106**, 485
with poly(ADP-ribose) glycohydrolase and phosphodiesterases, **106**, 511
by radioimmunoassay, **106**, 498
tritium labeling method, **106**, 495
monomer unit and branched portion
preparation, **106**, 446
structure and properties, **106**, 447
preparation and purification, **106**, 441
properties, **106**, 443
protein-bound
chain length determination, **106**, 509
determination, **106**, 477
Poly(ADP-ribose) glycohydrolase
in analysis of poly(ADP-ribosyl)ated proteins, **106**, 508
assay, **106**, 505
Poly(ADP-ribose) synthetase
assay, **106**, 501
properties, **106**, 504
purification from bovine thymus and rat liver, **106**, 502
Poly(ADP-ribosyl)ation
nucleosomal chromatin, **106**, 513
analytical immunofractionation methods, **106**, 518
electrophoretic analysis, **106**, 517
overview, **106**, 438
Poly(alkyl cyanoacrylate) nanoparticles
preparation, **112**, 132
properties, **112**, 134
Polyclonal antibodies
calmodulin
for affinity chromatography, **102**, 107
for immunoprecipitation, **102**, 109
preparation, **102**, 105
L-glutamate decarboxylase
characterization, **113**, 8
production, **113**, 7
myosin
binding assays, **134**, 430
production and purification, **134**, 424
properties, **134**, 427
to recombinant interferon $\alpha 2$, in purification of interferon α, **119**, 451
site-directed, to *lac* permease of *Escherichia coli*
binding assays, **125**, 461
Fab fragment preparation, **125**, 458
preparation, **125**, 455
purification, **125**, 456
screening, **125**, 458
Polycyclic aromatic hydrocarbons
metabolism to carcinogenic derivatives during lipid peroxidation, detection, **105**, 347
Polyester cloth
immobilization of low-density lipoprotein mutants, **129**, 241
Polyethylene glycol
activation with tresyl chloride, **104**, 59
–ammonium sulfate, in concentration of human immune interferon, **119**, 58
in calmodulin crystallization, **102**, 146
in cell fusion for hybridoma production, **121**, 3
embedding protocol for immunofluorescence microscopy of cytoskeletal components, **134**, 580

in fusion procedure for production of antibodies to weak immunogens, **121**, 187
in macromolecule crystallization, **114**, 120
in osmotic stress measurement of membrane intermolecular forces, **127**, 407
–protein interactions, measurement, **114**, 73
in protein precipitation, **104**, 351
in protoplast fusion, **118**, 579
removal from solutions, **104**, 355, 363
stimulation of plasmid DNA uptake into protoplasts, **118**, 584
–triazine dye complexes, preparation, **104**, 357

Polygalacturonase
purification from *Colletotrichum lindemuthianum*, **118**, 10

Poly-γ-D-glutamate synthetase
Bacillus licheniformis
assay, **113**, 148
properties, **113**, 157
purification, **113**, 150

Poly-γ-D-glutamic acid
enzymatically synthesized
isolation and purification, **113**, 152
properties, **113**, 160
structural characterization, **113**, 153

Poly(glycolic acid) polymers
applications, **112**, 436
composition, **112**, 438
dose form, **112**, 443
drug loading, **112**, 441
drug release, controlled systems, **112**, 437
molecular weight and dispersity, **112**, 440
preparation, **112**, 443

Polyhydric alcohols
and nonmetabolizable analogs, preparation, **125**, 481
purification by scavenging, **125**, 481
transport by bacteria
assays, **125**, 479, 483
defective mutants, selection, **125**, 476, 477, 478

Poly ICL-CM dextran
interferon induction assays, **119**, 104
preparation, **119**, 103

Polyinosinic · polycytidylic acid
induction of interferons, **119**, 119, 132

Poly(I) · poly(C), *see* Polyinosinic · polycytidylic acid

Poly(lactic acid) polymers
applications, **112**, 436
composition, **112**, 438
dose form, **112**, 443
drug loading, **112**, 441
drug release, controlled systems, **112**, 437
molecular weight and dispersity, **112**, 440
preparation, **112**, 443

Poly(lactic/glycolic acid) polymers
applications, **112**, 436
composition, **112**, 438
dose form, **112**, 443
drug loading, **112**, 441
drug release, controlled systems, **112**, 437
molecular weight and dispersity, **112**, 440
preparation, **112**, 443

Poly(lactide-*co*-glycolide) resin
microcapsules containing norethisterone
characterization, **112**, 111
preparation, **112**, 102

Poly(L-lysine)
–6-aminonicotinamide conjugates
preparation, **112**, 272
properties, **112**, 282
bound monolayers
lymphocyte separation on, **108**, 130
preparation, **108**, 127
–daunomycin conjugates
preparation, **112**, 274
properties, **112**, 277
–doxorubicin conjugates
preparation, **112**, 274
properties, **112**, 277
–methotrexate conjugates
preparation, **112**, 273
properties, **112**, 279

Polylysine hydrobromide
–amatoxin conjugates
biological properties, **112**, 233
preparation by mixed anhydride coupling, **112**, 231

Polymerization
proteins
cooperative, in self-assembly, **130**, 48
differential scanning microcalorimetry, **130**, 73
Polymers
aprotinin-releasing hydrogel, structure, function, and release rates, **112**, 519
ethylene–vinyl acetate, as macromolecule delivery systems
kinetic analysis, **112**, 414
preparation, **112**, 400
glucose-sensitive, preparation and characterization, **112**, 489
hydron, as macromolecule delivery systems
kinetic analysis, **112**, 414
preparation, **112**, 400
linear, formation, small-angle light scattering measurements, **130**, 35
magnetically responsive, as macromolecule delivery systems
kinetic analysis, **112**, 414
preparation, **112**, 411
triggering system, **112**, 413
membranes
diffusion principles, **112**, 496
drug delivery rates and mechanisms, **112**, 500
nitroglycerine-releasing hydrogel, structure, function, and release rates, **112**, 517
poly(lactic/glycolic acid)
applications, **112**, 436
composition, **112**, 438
dose form, **112**, 443
drug loading, **112**, 441
drug release, controlled systems, **112**, 437
molecular weight and dispersity, **112**, 440
preparation, **112**, 443
poly(ortho ester), biodegradable
hydrolysis and drug release, **112**, 425
preparation, **112**, 423
random coil, size-exclusion chromatography, **117**, 142
urea-sensitive, preparation and characterization, **112**, 493
Poly(methyl methacrylate) nanoparticles
preparation, **112**, 130
properties, **112**, 134
Polymyxin B
in milk, bioluminescence test, **133**, 277
Polynucleotide kinase
T_4, in synthesis of (2′-5′)-oligoadenylic acid analog, **119**, 525
Polynucleotides
crystalline heavy-atom derivatives, preparation, **114**, 173
crystallization conditions, **114**, 168
Poly-L-ornithine
stimulation of plasmid DNA uptake into protoplasts, **118**, 584
Polyornithine hydrobromide
–amatoxin conjugates
biological properties, **112**, 233
preparation by mixed anhydride coupling, **112**, 231
Poly(ortho esters)
biodegradable polymer systems
hydrolysis and drug release, **112**, 425
preparation, **112**, 423
Polypeptide β_1
chemical characterization, **116**, 242
purification, **116**, 240
Polypeptides
calmodulin-binding, detection by gel overlay assay, **102**, 205
chloroplast
extraction from leaves, **118**, 364
immunodecoration, **118**, 367
cytochrome-*c* oxidase
composition, **126**, 36
topography, **126**, 25
cytochrome *d* and *o* terminal oxidase complex, composition, **126**, 99
mitochondrial uptake, role of protonmotive force, **125**, 130
in receptosomes, electrophoretic analysis, **109**, 268
regulatory, copurification with glutamyl-tRNA synthetase, **113**, 49
reversed-phase HPLC, **104**, 190, 208
Polyprenols
assays, **111**, 209
C_{15}–C_{60}
separation and chain length determination, **110**, 301

synthesis in *Myxococcus fulvus* cell-free system, **110**, 300
isolation from plant tissue, **111**, 208
radiolabeled, preparation, **111**, 203
Polyprenyl phosphates
assays, **111**, 214
isolation and analysis, **110**, 284
radiolabeled, preparation, **111**, 204
Polyprenyl pyrophosphates
enzymatic hydrolysis, **110**, 153
Polysaccharides
cell wall
analytical techniques, **118**, 25
extraction and isolation, **118**, 11
cell wall-like, isolation from sycamore extracellular medium, **118**, 19
in phagocytes, staining, **132**, 135, 179
Polysomes
bovine adrenal gland
isolation, **128**, 898
purification of low-density lipoprotein receptor mRNA, **128**, 899
Poly(vinylpyridine)
copolymer microspheres, preparation and properties, **112**, 186
metallic microbeads, preparation and properties, **112**, 185
microspheres, applications, **112**, 188
monomer microspheres
preparation, **112**, 177
properties, **112**, 176, 179
Poly(vinylpyrrolidone)
in osmotic stress measurement of membrane intermolecular forces, **127**, 408
Porin
Escherichia coli
characterization, **125**, 317
crystallization, **125**, 318
purification, **125**, 313
solubilization, **125**, 311
Porphobilinogen
stereospecifically labeled, preparation, **123**, 375, 439
Porphobilinogen deaminase
mammalian
in coupled-enzyme assay for porphobilinogen synthase, **123**, 343
purification, **123**, 340
Porphobilinogen synthase
bovine liver
activation, **123**, 429
assay, **123**, 428
properties, **123**, 433
purification, **123**, 429
mammalian, coupled-enzyme assay, **123**, 339
Porphyrin methyl esters
HPLC
with radial compression columns, **123**, 346
on silica column, **123**, 352
Porphyrins
and isomers, HPLC with radial compression columns, **123**, 346
precursors, isotachophoretic analysis, **123**, 370
Post-proline cleaving enzyme, *see* Prolyl endopeptidase
Posttranslational modifications
acylations–deacylations, **107**, 3
ADP-ribosylation, **106**, 399
alkylations and dealkylations, **106**, 265
at aminoacyl-tRNA level, **106**, 141
α-amino and α-carboxyl groups, **106**, 165
general aspects, **106**, 3
glycosylations, **106**, 399
halogenations, **107**, 281
hydroxylations, **107**, 281
miscellaneous derivatives, **107**, 503
nonenzymatic, **106**, 77
oxidations, **107**, 281
Potassium
gramicidin A channels, energy profiles, computational methods, **127**, 261
in solution, coordination number, **127**, 326
Potassium chloride
concentration, effect on glutamate carboxylase activity, **107**, 557
Potassium peroxodisulfate
in preparation of poly(methyl methacrylate) nanoparticles, **112**, 131
Potassium phosphate
^{18}O-labeled, synthesis, **126**, 631
Potato virus X
infectivity, effect of plant IVR, **119**, 733
Precipitants
in protein crystallization, **104**, 372, 373

Precipitation
affinity, *see* Affinity precipitation
calmodulin
with ammonium sulfate, **102**, 16
with ethanol, **102**, 16
cytochrome-*c* oxidase with ammonium sulfate, **126**, 164
glutamine synthetase
with polyethylene glycol, **113**, 225
with zinc, **113**, 221
IgG
with ammonium sulfate, **116**, 6
with caprylic acid, **116**, 13, 136
with Rivanol, **121**, 634
IgM
in agar, **121**, 627
with ammonium sulfate, **116**, 29
after dialysis, **116**, 29
immuno-, *see* Immunoprecipitation
interferon α with acidic ethanol, **119**, 28, 41
interferon δ with ammonium sulfate, **119**, 99
lipoproteins from blood samples, **129**, 83
mevalonate kinase with ammonium sulfate, **110**, 74
monoclonal IgE with ammonium sulfate, **116**, 79
myeloperoxidase with ammonium sulfate, **132**, 373
polyclonal IgE with ammonium sulfate, **116**, 81
proteins with polyethylene glycol, **104**, 351
RNase L with ammonium sulfate, **119**, 494
selective, in dissociation and isolation of glycoprotein hormone subunits, **109**, 743, 745
Pregnant mare serum gonadotropin
assembly
detection, **109**, 737
methods for, **109**, 747
deglycosylated
chromatographic purification, **109**, 729
preparation
with anhydrous HF, **109**, 726
with trifluoromethanesulfonic acid, **109**, 733
disassembly
detection, **109**, 737
methods for, **109**, 739
Pregnenolone
formation from cholesterol, determination, **111**, 354
Prenylation
chlorophyllide *a* in *Capsicum annuum* plastids, **110**, 274
Prenyl phosphates
ion-pair chromatography, **111**, 250
Prenyltransferase, *see* Dimethylallyl-*trans*transferase
Pressure
high, fluorescence polarization of 6-propionyl-2-dimethylaminonaphthalene, **130**, 516
surface, *see* Surface pressure
Pristane
effect on ascites tumor formation and monoclonal antibody production, **121**, 375
Prodrugs
acetaminophen, structure, effect on metabolism, **112**, 341
amide
formation by N-acylation, **112**, 357
N-Mannich bases
applications, **112**, 351
hydrolysis, **112**, 348
preparation, **112**, 348
amine
enamin derivatives, **112**, 359
N-Mannich bases
applications, **112**, 351
hydrolysis, **112**, 348
preparation, **112**, 348
β-amino alcohol, oxazolidine derivatives, **112**, 359
aspirin, structure, effect on metabolism, **112**, 343
assay specificity, **112**, 320
design considerations, **112**, 345
drug yields
assay specificity, **112**, 320
assessment, **112**, 317
evaluation, **112**, 309
gastrointestinal absorption, improvement by enzyme targeting, **112**, 360

imide, N-Mannich bases
applications, **112**, 351
hydrolysis, **112**, 348
preparation, **112**, 348
lability predictions, correlation analysis, **112**, 324, 333
objectives, **112**, 309
phenacetin, structure, effect on metabolism, **112**, 341
reversal kinetics
assessments, **112**, 312
in vitro screening, **112**, 316
reversal mechanisms
chemical, **112**, 311
enzymatic, **112**, 310
mixed, **112**, 312
reversible groups, **112**, 330
site specificity, **112**, 321
ureide, N-Mannich bases
applications, **112**, 351
hydrolysis, **112**, 348
preparation, **112**, 348
Proenkephalin A
input to neuronal areas, analytical techniques, **124**, 617
Proenkephalin B
input to neuronal areas, analytical techniques, **124**, 617
Proenkephalins
cDNA-containing vaccinia virus vectors
construction, **124**, 298
detection, **124**, 300
expression analysis, **124**, 305
purification, **124**, 301
structural analysis, **124**, 303
transcription, **124**, 303
precursor mRNA, measurement, **124**, 557
Proflavin
hydrated, X-ray and neutron diffraction analysis, **127**, 169
Progesterone
intrauterine device with, structure, function, and release kinetics, **112**, 505
plasma, chemiluminescence immunoassay, **133**, 400
Prohistidine decarboxylase
Lactobacillus 30a
assay, **122**, 135
properties, **122**, 137
purification, **122**, 136
Prolactin
binding in mammary gland samples
inhibition by antireceptor serum, **109**, 669
receptor assay for, **109**, 668
function in isolated luteal cells, **109**, 304
Prolactin receptors
antibodies to
assay, **109**, 668
inhibition of prolactin binding, **109**, 669
mono- and bivalent fragments, effects, **109**, 673
production, **109**, 668
from mammary gland and other tissues
binding assays, **109**, 157
electrophoretic analysis, **109**, 160
photoaffinity labeling, **109**, 161, 165
solubilization and purification, **109**, 158
Proline
isomerization, detection
isomeric specific proteolysis method, **131**, 109
in model peptides, **131**, 117
in proteins, **131**, 121
Prolyl endopeptidase
LHRH-degrading, assay, **103**, 546
Prolyl hydroxylase
plant
assays, **107**, 361, 365
cofactor requirements, **107**, 366
extraction for *in vitro* assay, **107**, 362
natural and synthetic substrates, **107**, 364
properties, **107**, 367
purification, **107**, 363
subcellular distribution, **107**, 367
substrate specificity, **107**, 364, 368
Pronase
in elution of absorbed cells from monolayers, **108**, 132
proteolysis of tubulin, **134**, 185
Pro-opiomelanocortin
gene expression, analysis by *in situ* hybridization, **124**, 497, 510
1,2-Propanediol, *see* Propylene glycol
Propane jet
in quick freezing of microtubules, **134**, 631

Propidium
as retrograde tracer for neuropeptide-specific pathways, **103**, 666
Propidium iodide
as marker reagent in FACS, **108**, 208
Propionyl-CoA carboxylase synthetase
assay, **107**, 265
properties, **107**, 262, 274
purification, **107**, 271
reaction mechanism, **107**, 275
sequence homology with other synthetases, **107**, 264
6-Propionyl-2-dimethylaminonaphthalene
fluorescence polarization at high pressure, **130**, 516
Propylene glycol
removal from reduced NAD preparations, **122**, 154
Prostaglandin E_1
induction of glucagon responsiveness in transformed MDCK cells, **109**, 357
regulation of cAMP levels in hybrid cells, testing procedures, **109**, 327
responsiveness of MDCK cells
evaluation, **109**, 363
measurement, **109**, 361
Prostaglandin $F_{2\alpha}$
function in isolated luteal cells, **109**, 304
Prostaglandin H synthase
generation of oxygen radicals, assays, **105**, 412
Prostaglandins
bradykinin-stimulated production by dermal fibroblasts
analytical chromatographic techniques, **109**, 493
effects of inhibitors of arachidonate release and metabolism, **109**, 486
radioimmunoassay, **109**, 485
effects on hyporesponsiveness to interferon, **119**, 707
Protease
ATP-dependent, reticulocyte, in protein ubiquitination, **106**, 255
Ca^{2+}-dependent, rat brain
assay, **102**, 283
properties, **102**, 287
purification, **102**, 285
in isolation of erythrocyte marginal bands, precautions, **134**, 250
localization in gels, **104**, 438
Staphylococcus aureus, proteolysis of tubulin, **134**, 186
Protein A
binding Ig, removal from fetal bovine serum, **121**, 301
in chromatographic separation of
IgG subclasses, **116**, 11, 131
lymphocytes, **108**, 132
electron-dense derivatives
in detection of surface IgG, **108**, 411
preparation, **108**, 408
in enrichment of IgM, **116**, 30
fluorescent
in detection of surface IgG, **108**, 409
preparation, **108**, 407
immobilized, crosslinking to IgG, **121**, 743
interactions with immunoglobulins, **108**, 470
iodination, **119**, 685
in isolation of lymphocyte antigens, **108**, 473
in large-scale purification of murine IgG subclasses, **116**, 11, 133
microparticle-bound
in detection and quantitation of surface IgG, **108**, 411
preparation, **108**, 408
in radioimmunoassay of interferon effects on antigen expression, **119**, 686
radiolabeled
in assay for antibodies to cellular and soluble antigens, **121**, 786
preparation, **108**, 407; **121**, 785
in quantitation of surface IgG, **108**, 410
in reverse plaque assay for Ig-secreting cells, **121**, 348
sensitized erythrocytes, in agglutination assay for immunoglobulins, **121**, 556
Protein-arginine methyltransferase
wheat germ
assay, **106**, 269
properties, **106**, 272
purification, **106**, 271
Proteinase
Staphylococcus aureus, in digestion of rat IgG subclasses, **121**, 669

Proteinase I
Dictyostelium discoideum, O^{β}-(*N*-acetyl-α-glucosamine-1-phosphoryl)serine in
chemical characterization, **107**, 173
spectroscopic characterization, **107**, 176
Proteinase K
proteolysis of tubulin, **134**, 186
Protein disulfide-isomerase
assays, **107**, 282
molecular properties, **107**, 291
purification from bovine liver, **107**, 288
specificity of action, **107**, 292
subcellular location, **107**, 287
tissue distribution, **107**, 286
Protein dynamics
analysis by hydrogen exchange kinetics, **127**, 643
atomic
biological function, **131**, 295
theoretical analysis, **131**, 292
time-dependence, local and collective effects, **131**, 294
breathing motions
analytical techniques, **131**, 436
measurement, **131**, 508
hinge-bending
analytical techniques, **131**, 301
biological function, **131**, 302
hydration effects, **127**, 207
internal
classification, **131**, 283, 391
in crystals, **131**, 394
detection by NMR in solution, **131**, 307
fluorescence studies, **131**, 518
hydrogen isotope exchange techniques for measurement, **131**, 448
Mössbauer spectroscopy, **131**, 568
neutron diffraction study, **131**, 433
NMR measurements and dynamic simulations, **131**, 362
role in protein function, **131**, 426
solid-state NMR, **131**, 327
theoretical analysis, **131**, 285
X-ray diffraction study, **131**, 396
regional melting, analytical techniques, **131**, 443
rigid body motions, description, **131**, 300
side chain motions
analytical techniques, **131**, 296, 317
biological function, **131**, 298
terminal methyl rotor groups
assignment of conformations, **131**, 437
confidence levels of assignments, **131**, 438
preferred orientations, **131**, 441
Protein folding
amide proton exchange as probe, **131**, 136
analytical techniques, historical development, **131**, 3
in aqueous methanol under equilibrium conditions, **131**, 175
in cryosolvents, **131**, 174
disulfide bonds as probes, **131**, 83
in fragment complementing system, kinetics and intermediate states, **131**, 204
intermediates in
amide proton exchange study, **131**, 146
circular dichroism study, **131**, 126
urea-gradient electrophoresis, **131**, 156
kinetic mechanisms
description, **131**, 53
determination, **131**, 58
mutational analysis, **131**, 250
pathway reconstruction, **131**, 89
stopped-flow circular dichroism, **130**, 526
at subzero temperatures, **131**, 179
transmembrane, analysis with pyridoxal phosphate, **125**, 650
Protein-glutamate methyltransferase
assays, **102**, 164; **106**, 296
properties, **106**, 297, 307
purification from brain and erythrocytes, **102**, 165; **106**, 301
Protein-glutamine γ-glutamyltransferase
guinea pig hair follicle
assay, **113**, 372
properties, **113**, 374
purification, **113**, 373
guinea pig liver
assay, **113**, 359
properties, **113**, 363
purification, **113**, 361
human epidermis
assay, **113**, 370
properties, **113**, 372
purification, **113**, 371

Protein kinase
Ca^{2+}-activated, properties and function, **107**, 89
Ca^{2+}/phospholipid-dependent, *see* Protein kinase C
catalyzed reaction, **107**, 81
classifications, **107**, 86
cyclic AMP-dependent
peptide substrates, preparation and assay, **107**, 138
phosphorylation of human interferon γ, **119**, 296
properties and function, **107**, 86
related mutations in Y1 adrenal tumor cells, characterization, **109**, 355
cyclic GMP-dependent
peptide substrates, preparation and assay, **107**, 139
properties and function, **107**, 88
distribution and isozymes, **107**, 82
double-stranded RNA-dependent
assay, **119**, 502
purification from fibroblasts, **119**, 502
effectors of, **107**, 100
epidermal growth factor-stimulable, identification as EGF receptor, **109**, 824
hormone receptor-associated, analysis with 5′-*p*-fluorosulfonylbenzoyl adenosine, **109**, 816
identification
with activators and inhibitors, **107**, 133
by addition of purified kinases, **107**, 134
by tissue distribution analysis, **107**, 134
natural substrates, identification, **107**, 134
nomenclature, **107**, 86
nuclear, properties and function, **107**, 97
phosphorylation reaction, evaluation, **107**, 99
polyamine-dependent
autophosphorylation, assay, **107**, 158
induction by interferon in Ehrlich ascites tumor cells, **107**, 160
regulatory properties, assays, **107**, 154
protein substrates, properties, **107**, 83, 145
regulation, **107**, 83
synthetic substrates, preparation and assay, **107**, 136
tyrosine-specific, *see* Protein-tyrosine kinase
Protein kinase C
mixed micelle assay, **124**, 353
in neuroendocrine tissue, activation by diacylglycerols, **124**, 60
rat brain
activated form, properties, **102**, 284
assay, **124**, 350
preparation for Ca^{2+}-protease I assay, **102**, 282
properties and function, **107**, 91
purification, **124**, 351
Protein methylase I, *see* Protein-arginine methyltransferase
Protein methylase II, *see* Protein-glutamate methyltransferase
Protein methylase III, *see* Histone-lysine methyltransferase
Protein phosphatase, *see* Phosphoprotein phosphatase
Proteins
acetylated, selective recovery, **107**, 230
acetylation
amino-terminal
inhibition, **106**, 181
overview, **106**, 165
side-chain, analytical techniques, **107**, 224
ADP–ATP carrier
activity, influencing factors, **125**, 617
atractyloside-binding site, photolabeling, **125**, 665
chemical modification, **125**, 659
detection with fluorescent analogs, **125**, 620
ESR studies, **125**, 638
fluorescent probes, **125**, 639
isolation, **125**, 610
^{31}P NMR studies in detergent micelles, **125**, 634
reconstitution in phospholipid vesicles, **125**, 613, 633
spin labeling, **125**, 635
transport assay, **125**, 615
–ADP-ribose conjugates
characterization, **106**, 466
purification, **106**, 462
ADP-ribosylation, arginine-dependent, **106**, 399

affinity chromatography with immobilized antibodies, **104**, 21, 47
affinity HPLC, **104**, 218
affinity partitioning, **104**, 356
alanine carrier, *Escherichia coli*
purification, **125**, 486
reconstitution into proteoliposomes, **125**, 490
transport assay, **125**, 490
amino acid derivatives in, ion-exchange chromatography, **106**, 17
amino acid reactivities with heavy atoms, **114**, 149
amino acid side chains
conformation, **127**, 185, 189
distributions, **127**, 192
interactions, computational methods, **127**, 187
stability, **127**, 194
amino-containing, coupling to activated agarose, **104**, 18
antibodies to, detection on blots with enzyme-conjugated anti-Ig antibodies, **121**, 502
arginine in, enzymatic deimination, **107**, 624
assembly, electrostatic stabilization, **130**, 452
asymptotic mass migration boundary shapes, calculation, **130**, 6
axial ratios, estimation from covolume measurements, **117**, 182
Bence-Jones, isolation and purification, **116**, 102
Bence-Jones Rhe
crystal structure, automated electron density map analysis, **115**, 219
structure refinement, resolution of phase ambiguity, **115**, 100
binding
active transport-related, in bacteria, **125**, 284, 289
affinity chromatography, **104**, 50, 54
blotted onto paper, identification, **104**, 458
bound ADP-ribose, determination, **106**, 472
bound lysine, enzymatic methylation and demethylation, **106**, 274
bound water, mobility analysis by Mössbauer spectroscopy, **127**, 197
calcium-binding, *see* Calcium-binding proteins
γ-carboxyglutamic acid residues
assays, **113**, 138
decarboxylation, **107**, 549
protection from decarboxylation, **107**, 550
quantitation, **113**, 140
tritium labeling, **107**, 545
cell-wall, Bradford assay, **118**, 26
chemiluminogenic conjugates
applications, **133**, 566
immunoreactivity, **133**, 561
properties, **133**, 561
synthesis, **133**, 558
chlorophyll-binding, isolation
from barley chlorina *f*2 mutant, **118**, 348
from pea leaves, **118**, 342
from spinach, **118**, 351
chloroplast
gene transcription *in vitro*, **118**, 232
in vitro synthesized, analytical techniques, **118**, 275, 293
cholesteryl ester transfer
assays, **111**, 275
purification from human plasma, **111**, 280
chromatography on hydroxyapatite columns, **117**, 370
citrulline in
identification and estimation, **107**, 626
occurrence, **107**, 625
coating of culture surfaces for phagocytosis studies, **132**, 212
cobalamin-binding, assay for assessment of phagocytic function, **132**, 124, 167
coupling to polyacrylhydrazidoagarose, **104**, 18
C-reactive, role in opsonization, **132**, 305
crystalline
density measurements
with aqueous Ficoll solutions, **114**, 187
in hydration determination, **114**, 176
methodology, **114**, 179

in molecular weight determination, **114**, 176
under salt-free conditions, **114**, 182
hydrogen–deuterium exchange, determination, **114**, 513
isomorphous heavy-atom derivatives, preparation, **114**, 147
neutron diffraction studies, **114**, 511
X-ray diffraction studies, **115**, 117
crystallization, **104**, 372, 378
in capillary tubes, **114**, 128
by free interface diffusion, **114**, 140
growth cessation phase, **114**, 108
growth phase, **114**, 101
influencing factors, **104**, 370
nucleation phase, **114**, 80
precipitating agents in, **104**, 373
seeding methods, **104**, 380
crystallography
diffraction methods in, **114**, 20
experimental procedures in, **114**, 41
historical review, **114**, 3
cuticle, enzymatic release of halogenated tyrosine, **107**, 429
cytosolic 1,25-dihydroxyvitamin D-binding
in assay for 1,25-dihydroxyvitamin D, **123**, 180, 187
preparation, **123**, 186
denaturation curves, urea- and guanidine hydrochloride-induced
analysis, **131**, 272
determination, **131**, 267
–detergent micelles
detergent removal by ultrafiltration, **104**, 327
dimensions, **104**, 320, 323
nature of, **124**, 145
in detergent solutions, molecular weight determination, **117**, 41
and detergents, separation by gel filtration of crude membrane extracts, **104**, 323
displacement chromatography, **104**, 113
with ampholytes as displacers, **104**, 115
with carboxymethyldextran displacers, **104**, 118
disulfide bonds in
conformational forces in formation, **131**, 104
enzymatic formation and isomerization, **107**, 281
enzymatic reduction–oxidation by thioredoxin, **107**, 295
formation, breakage, and rearrangement, **131**, 84
formation pathway
chemistry, **107**, 307
reconstruction, **107**, 328
intermediate species, trapping, **107**, 315
nonenzymatic formation and isomerization, **107**, 301
trapped intermediated species
analytical techniques, **107**, 318
identification and characterization, **107**, 323
trapping, **131**, 87
disulfide reagents
intermolecular, **131**, 95
intramolecular, **131**, 100
dopa-containing
colorimetric assays, **107**, 399
isolation techniques, **107**, 405
radiochemical assay, **107**, 404
ecdysteroid carrier, *see* Ecdysteroid carrier proteins
electron density maps, automated analyses, **115**, 206
electrostatic interactions, calculations, **130**, 143
enzymatic bromination, **107**, 441
erythrocyte carboxyl methylated
isolation of D-aspartic acid β-methyl ester, **106**, 332, 341
radiolabeled, preparation, **106**, 332
Escherichia coli, localization with secretion cloning vectors, **125**, 138
extracellular, occurrence of Glu–Lys crosslink, **107**, 251
folate transport, purification
from L1210 cells, **122**, 267
from *Lactobacillus casei*, **122**, 261
from transport-defective *Lactobacillus casei*, **122**, 266
fragment complementing systems
fragment exchange in, **131**, 207

preparation, **131**, 192
in gels
blotting onto paper, **104**, 455
silver staining, **104**, 441
staining with Coomassie Blue, **104**, 439
globular
for calibration of size-exclusion chromatographic columns, **117**, 147
structural domains, automatic recognition, **115**, 430
glucuronamide linkage, amino-terminal, detection, **106**, 210
glutamine-binding, from *Escherichia coli*
in analysis of glutamine transport in isolated membrane vesicles, **125**, 306, 308
assays, **125**, 304
purification, **125**, 303
H–D exchange ratios, determination, **131**, 444
heat-stable C-serum, stimulatory effect on 3-hydroxy-3-methylglutaryl-CoA reductase, **110**, 50
α-helices, schematic drawing methods, **115**, 364
heme-binding, *see* Heme-binding proteins
hinge, *see* Hinge protein
hydrated
calorimetric analysis, **127**, 148
NMR analysis, **127**, 154
hydration
measurement techniques, **117**, 219
solid-phase studies, **127**, 284
hydrogen bonds, side-chain, proton polarization, **127**, 445
hydrophobic, *see* Hydrophobic proteins
hydrophobic chromatography with alkylagarose supports, **104**, 69
20-hydroxyecdysone-binding
assay, **111**, 445
properties, **111**, 448
purification, **111**, 446
immunosorbent chromatography, **104**, 381
immunostaining on nitrocellulose, **121**, 848
inhibitory
for F_1F_0-ATPase
properties, **126**, 508
purification, **126**, 507
radiolabeling, **126**, 730
for protein phosphatase
assays, **107**, 116
purification, **107**, 124
interacting, sedimentation patterns
in ligand–ligand systems, **130**, 30
microheterogeneity effects, **130**, 19
intracellular, occurrence of Glu–Lys crosslink, **107**, 252
iodination by thyroid peroxidase, **107**, 445
iron–sulfur, *see* Iron–sulfur protein
juvenile hormone-binding
binding assays, **111**, 499, 506
demonstration, **111**, 500
photoaffinity labeling with epoxyfarnesyl diazoacetate, **111**, 509
juvenile hormone carrier
assay, **111**, 482
properties, **111**, 486
purification, **111**, 484
labeled sites, localization, **131**, 514
labile proton exchange, kinetics and mechanisms, **131**, 311
λ repressor
–DNA interactions, DNase footprint titrations, **130**, 132
folding and stability, genetic analysis, **131**, 265
large, diffraction data, X-ray diffractometry
background determination, **114**, 391
data collection strategies, **114**, 393
radiation decay corrections, **114**, 386
in X-ray tube, **114**, 392
ligand-binding, assay
electrophoretic techniques, **117**, 388
equilibrium gel penetration techniques, **117**, 342
by fluorescence spectroscopy, **117**, 400
by gel chromatography, **117**, 346
spectroscopic techniques, **117**, 381
–ligand interactions, differential scanning microcalorimetry, **130**, 67
lipid transfer I and II
activity assays, **129**, 807
cellular origin, **129**, 814
characterization, **129**, 802, 804

chemical inhibition, **129**, 813
isolation, **129**, 798
natural occurrence, **129**, 812
stability, **129**, 814
lipoprotein
apolipoprotein A-containing, composition, **129**, 142
staining, **128**, 418, 422
loops and miscellaneous features, schematic drawing methods, **115**, 372
luciferin-binding, dinoflagellate
assays, **133**, 324
properties, **133**, 326
purification, **133**, 324
lumazine, *see* Lumazine proteins
luminescence immunoassays, **133**, 354
major sperm, isolation from nematode, **134**, 418
mammalian, *in vivo* racemization
analysis, **106**, 107, 110
applications and implications, **106**, 113
detection, analytical methods, **106**, 99
kinetics and mechanism, **106**, 101
membrane, *see* Membrane proteins
methionine sulfoxide in, enzymatic reduction, **107**, 352
methylated
analytical techniques, **106**, 265
natural occurrence, **106**, 266
microtubule-associated, *see* Microtuble-associated proteins
molecular weight determination by polyacrylamide gel electrophoresis, **104**, 250
monoclonal antibodies to preselected regions, preparation and properties, **121**, 69, 88
myristyl blocking groups, amino-terminal, identification, **106**, 205
native and soluble, enzymatic hydrolysis, **106**, 69
N_s and N_i coupling
ADP-ribosylation by pertussis and cholera toxins, **109**, 566
analytical techniques and assays, **109**, 450
functional analysis with pertussis toxin as probe, **109**, 558
purification, **109**, 446, 454
nonenzymatic glycosylation, measurement, **106**, 77
nucleotide regulatory, transfer during cell fusion, effect on hormonal responsiveness, **109**, 374
oligomeric
denaturation–dissociation, **131**, 225
reconstitution studies, **131**, 218
renaturation–reassociation, **131**, 233
oligomycin sensitivity-conferring
interactions with F_0- and F_1-ATPase, **126**, 458
purification, **126**, 455
radiolabeling, **126**, 729
oxidatively modified
extent of modification, assessment, **107**, 373
preparation, **107**, 372
oxysterol-binding
assays, **110**, 15
properties, **110**, 18
partial specific volume, calculation
in concentrated salt and amino acid solutions, **117**, 60
in dilute urea solution, **117**, 53
phasing by isomorphous replacement, **115**, 15
O-phosphates in
chemical stability to acid, **107**, 4
detection, **107**, 13
lability to base, **107**, 5
linkage characterization, enzymatic methods, **107**, 7
quantification, **107**, 15
release by partial acid hydrolysis, **107**, 10
semipurification, **107**, 11
phosphate transport
isolation, **125**, 697, 702, 703
reconstitution, **125**, 699, 702, 703
phosphohistidyl, ^{31}P NMR studies, **107**, 77
5′-phosphopyridoxyl-containing
preparation, **122**, 121
in preparation of monoclonal antibodies to vitamin B_6, **122**, 122
phosphoramidate-containing, detection and characterization, **107**, 23
phosphorylated, ^{31}P NMR studies, **107**, 36

phosphorylation
detection in crude tissue extracts, **107**, 131
in intact cells and *in vivo*, **107**, 142
quantitative techniques, **107**, 139
phosphoseryl, ^{31}P NMR studies, **107**, 68
phosphotyrosine in, quantification, **107**, 18
–plant mtDNA complexes, preparation for electron microscopy, **118**, 485
plasma, vitamin K-dependent
functional properties, **107**, 511
isolation, **107**, 507
structure, **107**, 510
platelet, precipitating immunoassays, **121**, 704, 712
polyacrylamide gel electrophoresis, **104**, 244
poly(ADP-ribosyl)ated, analysis with poly(ADP-ribose) glycohydrolase and phosphodiesterases, **106**, 505
posttranslationally modified amino acids
HPLC probes, **106**, 22
mass spectrometric identification, **106**, 29
posttranslational modifications, *see* Posttranslational modifications
precipitation with polyethylene glycol
action mechanism, **104**, 352
advantages, **104**, 351
concentration requirements, **104**, 353
effect of protein–protein and protein–ligand interactions, **104**, 354
preparation for ion-exchange HPLC, **104**, 189
preparative isoelectric focusing, **104**, 272
detection techniques, **104**, 269
recovery, influencing factors, **104**, 271
resolution optimization, **104**, 266
preparative isotachophoresis, **104**, 281, 299
proline isomerization, analysis by isomeric specific proteolysis, **131**, 121
–protein interactions, differential scanning microcalorimetry, **130**, 59
proton conduction
application to ATPase proteolipids, **127**, 424
buffering capacity and refractory phase, **127**, 433
with diodes, **127**, 431
response to oscillating electrical fields, **127**, 436
stationary current, titration, **127**, 430
theoretical model, **127**, 421
proton translocation, measurement with laser-induced proton pulse, **127**, 522
regulatory P_{II}, in *Escherichia coli*
assays, **107**, 198; **113**, 227
properties, **113**, 231
purification, **113**, 228
retention on ion-exchange columns, **104**, 170, 227, 229
retinol-binding
interstitial, from eye
preparative-scale purification, **123**, 105
small-scale purification, **123**, 104
in serum, ELISA, **123**, 85
reversed-phase HPLC, **104**, 190, 208
riboflavin-binding, *see* Riboflavin-binding protein
ribosomal, from chloroplasts
analytical techniques, **118**, 209
purification, **118**, 207
S-100
antibody preparation, **102**, 256
enzyme immunoassay, **102**, 259
purification from bovine brain, **102**, 256
sedimentation equilibrium in air-turbine ultracentrifuge, **117**, 3
selective labeling
functional, **131**, 510
kinetic, **131**, 510
selenium-containing, *see* Selenoproteins
self-assembly, cooperative
critical concentration for, interpretation, **130**, 57
polymerization mechanism, **130**, 48
sedimentation measurements, **130**, 55
time-resolved X-ray scattering measurements, **130**, 58
turbidity measurements, **130**, 49
self-associations, *see* Self-associations
β-sheets, schematic drawing methods, **115**, 364
small molecules, interactions with membrane surfaces, **125**, 81

in solution
colorimetric assay, **104**, 415
FTIR spectroscopy, **130**, 300
hydrodynamic measurements, structural interpretation, **117**, 154
small-angle neutron scattering studies, **130**, 79
–solvent preferential interactions
measurement, **114**, 61
mechanisms, **114**, 74
stable tubule only polypeptide, purification, **134**, 166
sterol carrier, *see* Sterol carrier protein
striated flagellar root
analytical techniques, **134**, 411
antibody screening, **134**, 412
purification, **134**, 411
structural domains
definition and location, **115**, 422
determining principles, **115**, 427
SU45, from sea urchin egg
assays, **134**, 20, 22
fluorescence measurements, **134**, 22
isolation, **134**, 15, 18
sulfhydryl group oxidation by thiol/disulfide exchange, differential effects, **107**, 336
tau, *see* Tau protein
temperature-induced changes, scanning microcalorimetry, **131**, 15
thermochemiluminescent labeling, **133**, 538
transfer to bile salt micelles by gel filtration, **104**, 325
translocation across membranes, role of protonmotive force, **125**, 129
transport
analysis with *lacZ* gene fusions, **125**, 150
mutant isolation and characterization with transposons, **125**, 157
tricarboxylate carrier
incorporation into liposomes, **125**, 694
isolation, **125**, 692
reconstituted, properties, **125**, 695
tyrosine sulfate in, detection, **107**, 203
tyrosine sulfation stoichiometry, determination, **107**, 221
ubiquitination, **106**, 238
uncoupling, *see* Uncoupling protein
unfolded, fast-folding and slow-folding forms
assays, **131**, 73, 79
refolding kinetics, **131**, 72
vitamin B_{12}-binding
assay in human neutrophils, **132**, 273
purification, **123**, 28
Protein solubility
cavity theory, **114**, 59
electrostatic theory, **114**, 54
preferential interactions
measurement, **114**, 61
mechanisms, **114**, 74
thermodynamic definition, **114**, 51
Protein structure
in aqueous solutions, analysis by D_2O–H_2O neutron difference maps, **127**, 329
circular dichroism studies, **130**, 208, 519
comparisons
classification scheme, **115**, 399
equivalent residues, analysis, **115**, 413
general comparison criteria, **115**, 411
phylogenic tree construction, **115**, 417
quantitative methods, **115**, 400, 405
computer-generated illustrations
applications, **115**, 389
program operation, **115**, 382
prediction
by amino acid sequence homology test, **130**, 197
by comparison of residue characteristics, **130**, 200
membrane-bound proteins, **130**, 192
with Raman amide I and III spectra, **130**, 311
soluble proteins, **130**, 188
primary, definition, **131**, 218
quaternary, definition, **131**, 218
refinement
constrained–restrained least-squares, **115**, 282
crystallographic, **115**, 313
free atom insertion for extending and refining phases, **115**, 112
real space, **115**, 237, 305
rigid-body, **115**, 308
schematic drawings
reproduction, **115**, 378

shading and color, **115**, 376
specific methods, materials, and setup, **115**, 362
secondary, definition, **131**, 218
stability, electrostatic effects, **130**, 437
tertiary
antiparallel α, **115**, 344
antiparallel β, **115**, 349
definition, **131**, 218
parallel α/β, **115**, 346
small irregular proteins, **115**, 356
Protein synthesis
γ-carboxyglutamic acid-containing proteins *in vitro*, **113**, 136
high-molecular-weight proteins in wheat germ system, **118**, 154
initiation and elongation factors
assays, **118**, 114
functional properties, **118**, 127
physical properties, **118**, 124
purification from wheat germ, **118**, 110, 116
in mammalian cells, inhibition by cordycepin analog of 2,5-p_3A_n, **119**, 668
in wheat germ cell-free extracts, **118**, 128
yeast mitochondrial, Glu-tRNA identification as intermediate, **106**, 152
Protein translocators
biochemical isolation, **126**, 4
concepts of, **126**, 1
phosphate–triose phosphate–phosphoglycerate
identification, **125**, 717
inhibitory effect of amino acid reagents, measurement, **125**, 718
isolation, **125**, 723
labeling with amino acid reagents, **125**, 718
properties, **125**, 708
reconstitution, **125**, 724
reconstitution in proteoliposomes, **126**, 5
unifying principles, **126**, 7
Protein-tyrosine kinase
associated activity of platelet-derived growth factor receptors, **109**, 98
phosphotyrosine in reaction mixtures, quantification, **107**, 18
properties and function, **107**, 98
Proteolipids
ATPase, proton conduction, **127**, 424
chloroplast membrane
characterization, **126**, 523
purification, **126**, 522
radiolabeling, **126**, 521
reconstituted, proton conduction, **126**, 527
reconstitution, **126**, 526
Proteoliposomes
cytochrome-*o* oxidase-containing
characterization, **126**, 121
preparation, **126**, 118
preparation, **126**, 576
reconstitution of
bacterial alanine transport system, **125**, 490
bacterial *lac* permease, **125**, 440
bacterial melibiose carrier, **125**, 385
chloroplast CF_0-F_1, **126**, 516
cytochrome bc_1, **126**, 179
cytochrome *d* terminal oxidase complex, **126**, 91
cytochrome oxidase, **126**, 151
F_1F_0 and F_0 subunits of ATP synthase, **126**, 577, 578, 579
formate dehydrogenase, **126**, 392
fumarate reductase, **126**, 392
methylmalonyl-CoA decarboxylase Na^+ transport activity, **125**, 544
mitochondrial phosphate transport protein, **125**, 699, 702, 703
oxaloacetate decarboxylase catalyzing Na^+ uptake, **125**, 538
phosphate–hexose 6-phosphate antiport, **125**, 559
protein translocators, **126**, 5
Proteolysis
enzymatic, IgG subclasses, **121**, 654, 657, 659, 661, 664, 668
monitoring assays, **121**, 655, 665, 669
optimal conditions, **121**, 665, 669
IgA susceptibility to, **116**, 62
isomeric specific, in analysis of proline isomerization, **131**, 107
phospholamban, in removal of [^{32}P]phosphopeptides, **102**, 271
tubulin, in isolation of functional domains, **134**, 181
Prothymosin α
assay, **116**, 257
biological activities, **116**, 265

isolation, **116**, 261
properties, **116**, 262
tissue distribution, **116**, 264
Proton abstraction
during L-lysine 6-aminotransferase reaction, stereochemistry, **113**, 100
Proton conduction
in proteins
application to ATPase proteolipids, **127**, 424
buffering capacity and refractory phase, **127**, 433
with diodes, **127**, 431
response to oscillating electrical fields, **127**, 436
stationary current, titration, **127**, 430
theoretical model, **127**, 421
in reconstituted vesicles
ATP synthase subunit-containing, **126**, 582
chloroplast membrane proteolipid-containing, **126**, 526
Proton exchange
amide, as probe of protein folding pathways
competition method, **131**, 139, 150
medium resolution studies, **131**, 141
NMR studies, **131**, 143
pulse-labeling method, **131**, 141, 152
in proteins, kinetics
from folded state, **131**, 499
formalisms for, **131**, 479
at histidine C-2, **131**, 478
and mechanisms, **131**, 311
multiple-site, data analysis, **131**, 480
at surface, **131**, 498
Proton motion
diffusive versus tunneling, **127**, 267
in ferroelectrics, **127**, 279
in framework hydrates, **127**, 277
in insertion compounds, **127**, 281
in layered hydrates, **127**, 277
in nonaqueous conductors, **127**, 283
in particle hydrates, **127**, 274
tunneling, in biological systems
calculations, **127**, 81
coupling to molecular motion, **127**, 83
model, **127**, 79
solvent effects, **127**, 89
Protonmotive force
in bacterial flagellar motion
manipulation, **125**, 570
measurement, **125**, 565
reconstitution, **125**, 571
functional role, **125**, 134
generated by electron transport, estimation
with dual-inhibitor titrations, **127**, 553
from kinetics of secondary ion uptake, **127**, 546
in light-driven thylakoid membranes, **127**, 549
with microelectrodes, **127**, 547
during reconstitution of phosphorylation, **127**, 551
during respiration, **127**, 542
generation in anaerobic bacteria
by end product translocation
metabolic energy yield, **125**, 506
under nonphysiological conditions, **125**, 500
under physiological conditions, **125**, 503
role of specific transport proteins, **125**, 497
energy recycling model, **125**, 495
during solute transport, **125**, 493
in mitochondrial polypeptide uptake, **125**, 130
in protein export in bacteria, **125**, 133
occurrence in protein-translocating systems, **125**, 136
and secondary transport
historical perspectives, **126**, 1
unifying principles, **126**, 7
Proton polarizability
hydrogen bonds
in biological systems, **127**, 445
and hydrogen-bonded systems, **127**, 440
measurement by infrared spectroscopy, **127**, 455
role in enzyme function, **127**, 450
Proton pulse
laser-induced, in measurement of proton translocation in membranes and proteins, **127**, 523
Proton pumps
description, barrier models, **127**, 470

light-induced
efficiency measurements, **127**, 578
reaction energetics and dynamics, **127**, 580, 583
Proton translocation
in bacteriorhodopsin
FTIR studies, **127**, 649
resonance Raman spectroscopy, **127**, 587
cooperative, redox Bohr effects, **126**, 331
through F_0 vesicles
assay, **126**, 605
influencing factors, **126**, 607
in lipid bilayer membranes, effect of phase transitions, **127**, 484
across membranes
description, barrier models, **127**, 465
energetics, theoretical calculation, **127**, 456
measurement
by displacement currents, **127**, 610
with laser-induced proton pulse, **127**, 522
model systems for, **127**, 471
with nitroxide spin probes, **127**, 738
passive, through F_0 subunit of ATPase
assay, **126**, 570
reconstitution, **126**, 578
reconstitution with photosystem I reaction center and CF_1CF_0, **126**, 285
Proton transport
primary and secondary
energetics, **125**, 5, 9
energy conversion efficiency, **125**, 14
nonequilibrium thermodynamics, **125**, 12
stoichiometry, determination, **125**, 6, 11
secondary, and protonmotive force
historical perspectives, **126**, 1
unifying principles, **126**, 7
Protoplasts
plant
culture preparation
influencing factors, **118**, 562
methodology, **118**, 570
fusion methods, **118**, 579, 595
inhibitory potency of antiviral substances, assay, **119**, 732
inoculation with
double-stranded DNA virus, **118**, 701
single-stranded DNA virus, **118**, 704
tobacco mosaic virus, **119**, 747
isolation, **118**, 550, 570, 673; **119**, 731, 745
plant regeneration from
influencing factors, **118**, 551
methodology, **118**, 570
plasmid-transformed
characterization, **118**, 590
selection, **118**, 588, 675
proliferation, influencing factors, **118**, 551
transformation with foreign DNA, **118**, 581, 674
virus multiplication
assays, **119**, 748
interferon effects, assay, **119**, 751
Protozoa, *see also specific protozoa*
ciliated, as sources of spasmin, **134**, 406
Proximity criterion
in structural analysis of biomolecular hydration by computer simulation, **127**, 21
Pseudomonas
MA (ATCC 23819), purification of *N*-methyl-L-glutamate synthase, **113**, 37
P. fluorescens
ice nucleation in plant leaves, characteristics, **127**, 731
purification of aspartate ammonia-lyase, **113**, 624
P. putida, purification of
Fe-superoxide dismutase, **105**, 91
5-oxo-L-prolinase, **113**, 454
P. syringae, ice nucleation in plant leaves, characteristics, **127**, 731
P. viridiflava, ice nucleation in plant leaves, characteristics, **127**, 731
Pteridines
from blood cells and plasma
extraction, **122**, 283
rapid screening procedure, **122**, 292
reverse-phase HPLC, **122**, 275
Pterins
liquid chromatography with electrochemical detection, **122**, 300

Pteroylpentaglutamate
reduction, **122**, 321
synthesis, **122**, 320
Pteroylpolyglutamates
substrates and inhibitors, preparation and analysis, **122**, 333
Pulse generators
high-voltage, in transient electric birefringence system, **117**, 201
Pulse radiolysis
in determination of
free radical redox potentials, **105**, 179
reaction selectivity of secondary free radicals, **105**, 492
and dosimetry, **105**, 175
equipment for, **105**, 172, 178
optical and conductivity signal analysis, **105**, 176
in oxygen-centered radical formation, **105**, 169
in paraquat radical generation, **105**, 523
in primary radical formation, **105**, 168
Pulse width
axial light-loss, in flow cytometric analysis of neuroendocrine cell diameter, **103**, 233
Pumiliotoxin
interactions with ion channels in neuroendocrine tissue, analysis, **103**, 402
Pumpkin fruit
purification of prenyltransferases, **110**, 169
Pyranine
in measurement of pH changes across lipid bilayers and membrane proteins, **127**, 751
Pyridine nucleotides
intramitochondrial, redox state
determination, **105**, 437
hydroperoxide effects, **105**, 439
Pyridoxal dehydrogenase
human liver, assay, **122**, 115
Pyridoxal kinase
human liver, assay, **122**, 111
rat liver
purification, **122**, 100
in synthesis of [^{14}C]pyridoxal 5′-phosphate, **122**, 101
Pyridoxal oxidase
human liver, assay, **122**, 115
Pyridoxal 5′-phosphate
in analysis of ADP/ATP carrier protein transmembrane folding, **125**, 651
binding site in enzymes, isolation and sequence determination, **106**, 122, 135
^{14}C-labeled
preparation, **122**, 97
purification, **122**, 102
dependent enzymes, spectral properties, **106**, 117
in identification of chloroplast phosphate transport protein, **125**, 719
in labeling of chloroplast phosphate transport protein, **125**, 718
Pyridoxal 5′-phosphate phosphatase
human liver, assay, **122**, 114
Pyridoxamine-phosphate oxidase
human liver, assay, **122**, 113
rabbit liver
assay, **122**, 117
purification, **122**, 118
Pyridoxamine (pyridoxine) 5′-phosphate oxidase, *see* Pyridoxamine-phosphate oxidase
Pyridoxine
^{14}C-labeled
enzymatic phosphorylation, **122**, 101
preparation, **122**, 98
purification, **122**, 99
Pyridoxine (pyridoxamine) 5′-phosphate oxidase, *see* Pyridoxamine-phosphate oxidase
4-(2-Pyridylazo)resorcinol
in spectrophotometric determination of iron in heme proteins, **123**, 323
Pyrimethamine
loading in poly(lactic/glycolic acid) polymers, **112**, 438
Pyrocyctis lunula
bioluminescence
particulate activity, **133**, 310
soluble components, **133**, 312, 318, 324
cell culture and harvesting, **133**, 309
Pyroglutamic acid
protein N-terminal, mass spectrometric identification, **106**, 37
Pyrophosphate
allylic and homoallylic isoprenoid
purification, **110**, 139

synthesis, **110**, 133
allylic, photolabile analogs
absorption spectrum and photolability, **110**, 121
photolysis conditions, **110**, 120
as prenyltransferase substrates and inhibitors, **110**, 121
preparation, **110**, 117
nonaprenyl, preparation, **110**, 329
photoaffinity analogs
applications, **110**, 128
preparation, **110**, 125
properties, **110**, 128
polyprenyl, *see* Polyprenyl pyrophosphates
Δ^1-Pyrroline-5-carboxylate synthase
rat intestinal mucosa
assay, **113**, 113
properties, **113**, 119
purification, **113**, 118
Pyruvate
transport in chloroplast envelope, **125**, 713
Pyruvate carboxylase synthetase
assay, **107**, 265
properties, **107**, 262, 274
purification, **107**, 271
reaction mechanism, **107**, 275
sequence homology with other synthetases, **107**, 264
Pyruvate dehydrogenase kinase
properties and function, **107**, 94
Pyruvate–oxygen oxidoreductase chain
Escherichia coli, reconstitution in proteoliposomes, **126**, 93

Q

Quench-flow methods
chemical, for analysis of mechanochemical ATPases
data analysis, **134**, 699
data collection, **134**, 697
instrumentation, **134**, 693
Quick freezing
apparatus, **134**, 599
lung slices, **134**, 601
microtubules by propane jet, **134**, 631
Quin 2
as indicator of intracellular Ca^{2+}
general methodology, **124**, 118
specific applications, **124**, 125
Quinazoline derivatives
loading in poly(lactic/glycolic acid) polymers, **112**, 438
Quinones
electrochemical analysis
cyclic voltammetric techniques, **125**, 112
data analysis, **125**, 118
dc polarographic techniques, **125**, 114
practical considerations, **125**, 115
partition coefficients, HPLC determination, **125**, 121
redox states, **125**, 109
Quisqualic acid
excitotoxic properties in neuroendocrine research, **103**, 384

R

Racemization
in vivo, in mammalian proteins
analysis, **106**, 107, 110
applications and implications, **106**, 113
detection, analytical methods, **106**, 99
kinetics and mechanism, **106**, 101
Radiation inactivation
macromolecules, in molecular weight determination
activity measurements, **117**, 89
data analysis, **117**, 91
irradiation effects, **117**, 65
irradiation techniques, **117**, 73
sample preparation, **117**, 84
target analysis, **117**, 69
and target size analysis
applications to interferons, **119**, 258
principles and theory, **119**, 256
Radicals, *see also specific radicals*
oxygen-centered
absorption spectra and kinetics, **105**, 177
formation by pulse radiolysis, **105**, 169
primary
absorption spectra and kinetics, **105**, 177
formation by pulse radiolysis, **105**, 168

Radioallergosorbent test
IgE antibody, **116**, 92
Radiochemical assay
acetoacetyl-CoA synthetase, **110**, 5
[acyl-carrier-protein] acetyltransferase, **122**, 53
arginine deiminase, **107**, 628
cholesterol acyltransferase, **111**, 286
dihydrofolate reductase, **122**, 346
dolichol esterase, **111**, 480
dolichol kinase, **111**, 471
dolichyl-phosphate phosphatase, **111**, 477
dolichyl-pyrophosphate phosphatase, **111**, 548
dopa proteins, **107**, 404
ecdysone 3-epimerase, **111**, 439
ecdysone oxidase, **111**, 421
ferrochelatase, **123**, 402
folate oligoglutamate:amino acid transpeptidase, **122**, 367
20-hydroxyecdysone-binding protein, **111**, 445
3-hydroxy-3-methylglutaryl-CoA reductase, **110**, 52
3-hydroxy-3-methylglutaryl-CoA synthase, **110**, 21
25-hydroxyvitamin D 1α-hydroxylase, **123**, 161
isopentenyl-diphosphate Δ-isomerase, **110**, 93
juvenile hormone III biosynthesis *in vitro*, **111**, 530
juvenile hormone esterase, **111**, 490
lecithin–cholesterol acyltransferase, **129**, 776, 780
lysolecithin acyltransferase, **129**, 790
mevalonate kinase, **110**, 72
mevalonate 5-pyrophosphate decarboxylase, **110**, 87
pantetheine-hydrolyzing enzyme, **122**, 36
phosphomevalonate kinase, **110**, 79
plasma lipid transfer proteins, **129**, 809
prenyltransferases, **110**, 156, 167
pyruvoyl-dependent histidine decarboxylase, **122**, 129
thiamin-diphosphate kinase, **122**, 24
Radioenzymatic assay
acetylcarnitine, **123**, 259
acetyl-CoA, **123**, 259
catecholamines, **103**, 475
microscale
catecholamines, **103**, 484
serotonin, **103**, 488
pantothenic acid, **122**, 33
plasma mevalonate, **110**, 58
Radioimmunoassay
(ADP-ribosyl)$_n$ proteins on gels, **106**, 469
arginine vasopressin in neuroendocrine cells, **103**, 144
calcium-binding protein, **102**, 292
calmodulin-dependent protein phosphatase, **102**, 253
colony-stimulating factor 1, **116**, 579
competitive, Igs from bovine × murine hybridomas, **121**, 248, 249, 252
corticotropin-releasing factor, **103**, 569
culture-well solid-phase, for screening antibody-producing cells, **121**, 438
cyclic AMP, **109**, 362; **132**, 432
cyclic GMP, **109**, 827; **132**, 432
cytokinins, **110**, 355
in detection of nonenzymatic protein glycosylation, **106**, 83
1,25-dihydroxyvitamin D_3 receptors, **123**, 204
β-endorphin, **103**, 562
enkephalin peptides, **124**, 619
enkephalins, **103**, 561
glucosylated proteins, **128**, 359, 363
growth hormone-releasing factor, **124**, 379, 391
hybridoma-secreted antibodies, **121**, 204
IgE antibody, **116**, 93
IgG subclasses during peptic digestion, **121**, 655
IgM, **121**, 629
immunoglobulin allotypes, **116**, 163
immunoglobulin idiotypes, **116**, 180
immunoglobulin isotypes, **116**, 149
immunoglobulins in lymphocyte secretions, **108**, 432
interferon from amniotic fluid, **119**, 548
isotypic rheumatoid factor, **116**, 154
lactoferrin, for assessment of phagocytic function, **132**, 164
lecithin–cholesterol acyltransferase, **129**, 773
LHRH, **124**, 264
live-cell, tumor antigens, **119**, 682

α_2-macroglobulin antibodies in hybridoma supernatants, **121**, 64
multivalent antibody, for hybridoma screening, **121**, 425
myoglobin antibodies, **121**, 76
osteocalcin, **107**, 528
poly(ADP-ribose), in quantification of *in vivo* levels, **106**, 498
prostaglandins, **109**, 485
protein-bound ADP-ribose, **106**, 475, 479
prothymosin α, **116**, 257
serum amyloid A, **128**, 314
sandwich, interferon αA and γ, **119**, 582, 588
solid-phase
for epitope analysis, **121**, 461
monoclonal and site-directed polyclonal antibodies to *lac* permease, **125**, 458
transcobalamin II, **123**, 36
solution-phase
with avidin–biotin, for analysis of antibody affinity constants, **121**, 468
for hybridoma screening, **121**, 433
somatomedins, **109**, 775, 803
somatostatin, **124**, 419
small oligopeptides, preparation of haptens and substrates for, **103**, 435
substance P, **124**, 338
TGAL antigen-specific helper factor, **116**, 349
thymopoietin, **116**, 285
thymosin α_1, **116**, 239, 244
total IgE, **116**, 88
T-suppressor factor, **116**, 319
tuftsin, **132**, 323
ubiquitin, **116**, 285
Radioimmunosorbent test
total IgE, **116**, 90
Radioisotopes
in gels, fluorographic detection, **104**, 460
Radiolabeling
albumin microspheres with ^{99m}Tc, **112**, 43
in analysis of methylated proteins, **106**, 266
angiotensin II, **109**, 112
antibodies
to receptor-bound interferon with ^{125}I, **119**, 327
for spaghetti overlay technique, **121**, 455
with tritium, **121**, 670
antigens
for multivalent antibody radioimmunoassay, **121**, 427
for solution-phase radioimmunoassay, **121**, 434
anti-Ig antibodies for culture-well radioimmunoassay, **121**, 442
antiisotypic antibodies, **116**, 150
anti-mouse IgG, in analysis of interferon effects on antigen expression, **119**, 685
apolipoprotein B, **129**, 401
apolipoprotein E, **129**, 484
apolipoproteins, **129**, 255, 285, 425, 430
aspulvinone, **110**, 321
avidin, **109**, 443
bacteria, in assessment of phagocytic function, **132**, 103, 141
bile acids, **111**, 111
calcitonin, **109**, 42
calcium channels in brain, heart, and skeletal muscle, **109**, 515
calmodulin-dependent protein phosphatase, **102**, 251
γ-carboxyglutamic acid in proteins, **107**, 545; **113**, 143
cellular cholesterol, **129**, 630, 653
chloroplast 32-kDa membrane protein, **118**, 384
cholecystokinin, **109**, 65
colony-stimulating factor 1, **116**, 576
CRF receptors, **124**, 565
cultured hepatoma cells, for analysis of insulin receptor phosphorylation, **109**, 613
DNA, in analysis of mRNA concentration and half-life, **109**, 580
DNA probes by nick-translation, **128**, 868
dolichol, **111**, 203
dolichyl phosphate, **111**, 204
dual, in protein conformational analysis, **131**, 459
β-endorphin, **103**, 550
enkephalin peptides, **124**, 620
enkephalins, **103**, 550
erythrocyte carboxyl methylated proteins, **106**, 332

F_1-ATPase essential residues, **126**, 741
F_1-ATPase inhibitor peptide, **126**, 663
fibroblasts with [^{14}C]arachidonate, **109**, 488, 493
gastrin, **109**, 58
glucagon, **109**, 6
L-glutamate with radiolabeled α-ketoglutarate, **113**, 61
glutathione, **113**, 569
granulocyte colony-stimulating factor, **116**, 617
granulocyte–macrophage colony-stimulating factor, **116**, 598
growth hormone, **109**, 694
H-2 antigens, **108**, 453
hepatocyte phosphoinositides, **109**, 469
high-density lipoproteins, **128**, 652; **129**, 651
H^+–sugar transport proteins in *Escherichia coli*, **125**, 404, 409, 417
human chorionic gonadotropin derivatives, **109**, 204
20-hydroxyecdysone, **111**, 443
hypusine in growing lymphocytes, **106**, 346
Ia antigens, **108**, 453, 532
IgG, **121**, 247, 804
immunoglobulins, **116**, 203
inhibitory protein IF_1, **126**, 730
insulin receptor subunits, **109**, 594
interferon α, **119**, 267, 276
interferon α2, **119**, 327
interferon αA, **119**, 306
interferon β, **119**, 284, 400
interferon γ, **119**, 281
interferons, **119**, 263, 292, 322
interleukin 3, **116**, 551
lipoproteins, **129**, 285, 348, 425, 430, 500, 545
low-density lipoprotein receptors in cultured cells, **109**, 413
Lyt-1, Lyt-2, and Lyt-3 antigens, **108**, 670, 672
melibiose, **125**, 379
membrane antigens, **121**, 723
β_2-microglobulin, **108**, 497
mitochondrial membrane-bound enzymes with *N*,*N'*-dicyclohexylcarbodiimide, **125**, 93, 95, 97, 99, 105, 107
monoclonal antibodies, effect on immunoreactivity, **121**, 681
monoclonal and site-directed polyclonal antibodies to *lac* permease, **125**, 461
murine class I MHC antigens, **108**, 507
nerve growth factor, **109**, 22
neuropeptides, **103**, 34
(2'-5')-oligoadenylate, **119**, 353
oligomycin sensitivity-conferring protein, **126**, 729
parathyroid hormone, **109**, 49
peptide hormones for receptor photoaffinity labeling, **109**, 143
peptides
  comparative methods, **124**, 18
  direct methods, **124**, 21
platelet-derived growth factor, **109**, 70
polyprenols, **111**, 203
polyprenyl phosphate, **111**, 204
protein A, **108**, 407; **119**, 685; **121**, 785
pyridoxal 5'-phosphate, **122**, 97
Qa and TL antigens, **108**, 550
in quantification of poly(ADP-ribose) *in vivo*, **106**, 495
recombinant interferon αA, **119**, 286
RNA, in analysis of mRNA concentration and half-life, **109**, 583
selenocysteine in proteins, **107**, 581, 603
stereospecific, porphobilinogens
  at acid side chain, **123**, 376
  at aminomethyl side chain, **123**, 380
sterol precursors, **111**, 312
surface immunoglobulins, **108**, 426, 427
TGAL antigen-specific helper factor, **116**, 350
thymosin α_1 synthetic analog, **116**, 245
thyrotropin, in assay of thyroid-stimulating antibody, **109**, 679
toxins and A chains, **112**, 213
transcobalamin II, **123**, 42
tyrosine sulfate in proteins, **107**, 204
viral RNA, **118**, 719

Radioligand binding assay
  angiotensin II receptors, **109**, 110
  calcitonin receptors, **109**, 40
  cholecystokinin receptors, **109**, 64
  epidermal growth factor receptors, **109**, 101
  gastrin receptors, **109**, 56
  glucagon receptors, **109**, 3
  nerve growth factor receptors, **109**, 21
  parathyroid hormone receptors, **109**, 48
  prolactin receptors, **109**, 157

riboflavin-binding protein, **122**, 221
solubilized opiate receptors, **124**, 185
soluble neuropeptide receptors, **124**, 167
Radiolysis
in generation of selective oxidative free radicals, **105**, 491
pulse, *see* Pulse radiolysis
Radiometers
commercial
components, **133**, 588
suppliers, **133**, 591
Radiometric assay
S-adenosylmethionine:calmodulin (lysine) *N*-methyltransferase, **102**, 160
S-adenosylmethionine:protein carboxylmethyltransferase, **102**, 164
adenylate cyclase, **132**, 423
calmodulin carboxylmethylesterase, **102**, 166
coupled, for dihydrofolate reductase, **122**, 360
guanylate cyclase, **132**, 425
lactoferrin, for assessment of phagocytic function, **132**, 165
Radioreceptor assay
colony-stimulating factor 1, **116**, 581
1,25-dihydroxyvitamin D, **123**, 176
growth hormone-releasing factor, **124**, 400
interleukin 2, **116**, 499
interleukin 3, **116**, 551
platelet-derived growth factor, **109**, 756
somatomedin C, **109**, 803
Raman spectroscopy
amides I and III, in prediction of protein secondary structure, **130**, 311
in aqueous solutions, difference spectra from OH stretching region, **127**, 102
low-frequency, in determination of hydrogen bonding
in $NaClO_4$ aqueous solutions, **127**, 99
in pure water, **127**, 94, 95, 97
in sucrose aqueous solutions, **127**, 101
in measurement of proton translocation in bacteriorhodopsin, **127**, 587
resonance, *see* Resonance Raman spectroscopy
water in ice phases, **127**, 313
Rangia cuneata
isolation of histidinoalanine, **106**, 352
Rare earth compounds
binding to proteins, **114**, 152
Real space refinement
protein structure
description and applications, **115**, 237, 305
representative *R* factors, **115**, 250
Receptosomes, *see* Endosomes
Redox potential
Bohr effects, measurement in cytochrome system of mitochondria, **126**, 331
cytochrome *d* and *o* terminal oxidase complexes, determination, **126**, 105
free radical, determination by pulse radiolysis, **105**, 179
photosynthetic reaction center–cytochrome bc_1 hybrid system, regulation, **126**, 298
Redox reactions
interferon α, **119**, 251
membrane, relationship to surface potential, **125**, 58
quinones, **125**, 109
Redox state
nicotinamide nucleotides and flavins in intact tissues, measurement, **123**, 311
Reduction
acetylated cytochrome *c*, in assay for superoxide, **105**, 14, 376
chemical and enzymatic, generation of paraquat radical, **105**, 523
dioxygen, products, **105**, 4
electrochemical, pteroylpentaglutamate, **122**, 321
enzymatic, methionine sulfoxide in proteins and peptide, **107**, 352
ferricytochrome *c*, in assay for
superoxide anion, **105**, 14, 365, 372, 375, 394, 432
superoxide dismutase, **105**, 101
folylpolyglutamates, **113**, 180
γ-glutamylcysteine disulfide to γ-glutamylcysteine, **113**, 559
IgA, for monomer preparation, **116**, 139
IgG, for chain separation, **116**, 20
IgM
for chain separation, **116**, 35
for monomer preparation, **116**, 139

nitro blue tetrazolium, in assay for
superoxide anion, **105**, 15, 394
superoxide dismutase, **105**, 95
protein disulfides by thioredoxin, **107**, 296
selective, chloramines, **132**, 580
vitamin D metabolites, **123**, 138

Rejection
skin grafts, appraisal, **108**, 26

Replacing factor
T-cell, *see* T-cell replacing factor

Reservoirs
membrane-coated
applications, **112**, 504
drug release, **112**, 500

Resins
activated
coupling capacity, prediction, **104**, 11
coupling of ligands, **104**, 17
determination, **104**, 8, 14
activation
with carbamate derivatives, **104**, 11
with 1-cyano-4-dimethylaminopyridinium tetrafluoroborate, **104**, 7
cyanogen bromide method, **104**, 4
with *N*-cyanotriethylammonium, **104**, 7
with *p*-nitrophenyl cyanate, **104**, 7
ion-exchange
in isolation of thymopoietin, **116**, 287
in preparation of high-purity water, **104**, 395

Resonance Raman spectroscopy
in analysis of membrane surface charge density, **127**, 689
ligand binding, **130**, 350
UV, biopolymers, **130**, 331

Respiration
driven proton translocation, in estimation of protonmotive force, **127**, 542
mitochondrial
assay in tumor cells cocultivated with macrophages, **132**, 516
regulatory control in heart, **125**, 22

Respiratory burst
in phagocytosis, overview, **132**, 353

Restriction endonuclease analysis
chloroplast DNA, **118**, 174, 182, 605
plant mitochondrial DNA, **118**, 456, 609
plant nuclear DNA, **118**, 65
effect of cytosine methylation, **118**, 69
site polymorphisms, as genetic markers, **118**, 82

Reticulocyte lysates
rabbit
actin amino-terminal acetylation and processing, **106**, 179
translation of LHRH precursor mRNA, **124**, 322

Reticulocytes
ATP-dependent proteolytic enzymes, and ubiquitination, **106**, 255

Retina
specific α-bungarotoxin–luciferin binding, **133**, 47
whole, dispersed cells from
chemical dispersion technique, advantages and disadvantages, **103**, 367
identification and properties, **103**, 363
preparation, **103**, 363

Retinal
extraction as oxime, **123**, 53
HPLC, **123**, 55
applications, **123**, 58
comparison with bleaching analysis, **123**, 58

Retinoic acid
cis–trans isomerization, HPLC analysis, **123**, 84
extraction methods, **123**, 114
geometric isomers
purification, **123**, 76
radiolabeled, purification, **123**, 82
separation, **123**, 82
methylation, **123**, 116
physiological levels, quantification, **123**, 116

Retinoids
Fourier transform infrared spectroscopy, **123**, 92
purification, **123**, 95

Retinol, *see* Vitamin A

Retinyl esters
separation and quantitation, **123**, 68

Retinyl hexadecyl ether
biological applications, **129**, 834
synthesis, **129**, 832

Retinyl phosphate mannose
enzymatic synthesis, **123**, 63
separation, **123**, 65, 66

Retroviruses
induced cell fusion, interferon effects, **119**, 704
purification, **119**, 703
Reverse hemolytic plaque assay
secretion from neuroendocrine cells, **124**, 443
Reverse-phase chromatography
carotenoids, **111**, 193
C_5 to C_{20} isoprenoid benzoates and naphthoates, **111**, 252
cobalamins in biological material, **123**, 11
conjugated and free bile acids, **111**, 55, 59
corrinoids, **123**, 17
cytokeratin polypeptides, **134**, 369
GAT antigen-specific suppressor factors, **116**, 329, 334
glucagon iodination products, **109**, 6
granulocyte colony-stimulating factor, **116**, 610
granulocyte–macrophage colony-stimulating factor, **116**, 595, 596, 598
insulin-like growth factors, **109**, 788, 796
interleukin 2, **116**, 515, 533
interleukin-2 tryptic peptides, **116**, 520
lymphotoxin, **116**, 443
mast cell growth factor, **116**, 559
membrane proteins, **104**, 338
methanopterin and tetrahydromethanopterin derivatives, **122**, 422
methotrexate polyglutamate derivatives formed *in vivo*, **122**, 340
mobile- and stationary-phase parameters, selection, **104**, 199
neuropeptides, **103**, 535
nucleotides from NADH preparations, **122**, 155
polypeptides and proteins, **104**, 190, 208
prothymosin α, **116**, 262
pteridines from blood cells and plasma, **122**, 275
resolution optimization, **104**, 192
retention mechanisms and behavior, **104**, 191, 194
sterols, **111**, 10
sulfates, **111**, 62
α-thymosins, **116**, 256
in transacetylase assay, **106**, 195
tumor necrosis factor, **116**, 451
ubiquinones in biological samples, **105**, 150
Rheumatoid factor
isotypic, radioimmunoassay, **116**, 154
Rhizobium
broad host range vectors, application of *Escherichia coli* plasmids, **118**, 657
deletion creation with Tn*10*, **118**, 653
DNA isolation, **118**, 528
H_2 uptake negative mutants
characterization, **118**, 532
isolation, **118**, 529
in isolation of *hup* genes, **118**, 534
preparation, **118**, 529
screening, **118**, 531
nitrogen fixation genes, Tn*5* mapping, **118**, 519
site-specific insertion of selective markers, **118**, 649, 656
transposon mutagenesis, **118**, 644
vector plasmid mobilization in, **118**, 641, 654
Rhodamine
as staining reagent in FACS, **108**, 208
Rhodopseudomonas
R. blastica, ATP synthase subunit-encoding genes
cloning and DNA sequencing, **125**, 231
transcriptional mapping, **125**, 245
R. sphaeroides
cell growth and harvesting, **123**, 437
photosynthetic reaction center–cytochrome bc_1 hybrid system
construction, **126**, 296, 303
electron-transfer reactions, **126**, 300
redox poising, **126**, 298
solubilization, **126**, 300
photosynthetic reaction center, isolation and light energy conversion, **126**, 294
purification of
aa_3-type cytochrome-*c* oxidase, **126**, 138
5-aminolevulinate synthase, **123**, 435
cytochrome *b*, **126**, 326
cytochrome bc_1, **126**, 183, 186, 282
ferrochelatase, **123**, 410
3-hydroxybutyrate dehydrogenase from cell-free extracts, **104**, 111

malate dehydrogenase from cell-free extracts, **104**, 111

Rhodospirillum rubrum

ATP synthase subunit-encoding genes

cloning and DNA sequencing, **125**, 231

transcriptional mapping, **125**, 245

purification of

F_1-ATPase subunits, **126**, 528

membrane-bound inorganic pyrophosphatase, **126**, 538

Ribityl side chain

riboflavin and ring-substituted analogs, ^{1}H NMR, **122**, 241

Riboflavin

isotope dilution assay, **122**, 221

ribityl side chain, ^{1}H NMR, **122**, 241

urinary, fluorometric titration with apoflavoprotein, **122**, 234

Riboflavin-binding protein

avian

isolation, **122**, 227, 230, 231

properties, **122**, 232

radioligand binding assay, **122**, 224

Riboflavin kinase

radiosubstrate HPLC assay, **122**, 238

Riboflavin 5′-phosphate, *see* FMN (riboflavin 5′-phosphate)

Riboflavin phosphates

HPLC, **122**, 210, 212

ion-exchange chromatography, **122**, 213

preparation, **122**, 214, 216

stability, **122**, 217

structure and properties, **122**, 217

Riboflavin synthase

heavy, from *Bacillus subtilis*

assay, **122**, 193

crystallization, **122**, 195

properties, **122**, 196

purification, **122**, 194

subunit isolation, **122**, 196

Ribonuclease A

anion binding site

electrostatic effects on proton uptake and release, **130**, 433

pH-dependent development, **130**, 431

hydration, conformational effects, **127**, 382

native structure, cryosolvent effects, **131**, 174

refolding at subzero temperatures, **131**, 179

unfolding in methanol, **131**, 175

Ribonuclease L

assays, **119**, 491

crosslinking of (2′-5′)$(A)_3$[^{32}P]pCp to, **119**, 497

purification from murine EAT cells, **119**, 493

Ribonuclease S

anion binding site

electrostatic effects on proton uptake and release, **130**, 433

electrostatic free energy, **130**, 450

pH-dependent development, **130**, 431

crystal structure, automated electron density map analysis, **115**, 216

S-peptide, difference exchange behavior, **131**, 502

Ribonucleic acid, *see* RNA

Ribonucleotide reductase

Escherichia coli, in assay for glutaredoxins, **113**, 526

Ribonucleotides

UV resonance Raman spectroscopy, **130**, 341

Ribosomes

chloroplast

analytical techniques, **118**, 209

purification, **118**, 203

Ribulose-bisphosphate carboxylase

crystal intensity, multiwire area detector diffractometry, **114**, 470

duckweed, small subunit genes

cloning in phage λ, **118**, 397

expression, analytical techniques, **118**, 407

genomic clones

rapid gel analysis, **118**, 405

screening, **118**, 399

Rhodospirillum rubrum

assay, **118**, 427

expression, analytical techniques, **118**, 425, 427

gene cloning, **118**, 420

purification from *Escherichia coli*, **118**, 428

spinach, large and small subunits

measurement, **118**, 413

properties, **118**, 413

reassembly, **118**, 416
separation, **118**, 411
Synechococcus, large and small subunits
measurement, **118**, 413
properties, **118**, 413
reassembly, **118**, 416
separation, **118**, 411
Zea mays, large subunit genes
cloning, **118**, 429
expression in *Escherichia coli*, **118**, 431
Ribulose-bisphosphate carboxylase/oxygenase, *see* Ribulose-bisphosphate carboxylase
Rice
purification of elongation factor 1, **118**, 145
Richards optical comparator
refinements and methods for model coordinate measurement, **115**, 154
Ricin
A chain
–antibody conjugates
applications, **112**, 245
cytotoxicity, **112**, 243
preparation, **112**, 218, 241
storage and stability, **112**, 242
isolation, **112**, 210, 211, 239
radiolabeling, **112**, 213
–antibody conjugates
biological evaluation, **112**, 221
cytotoxicity, **112**, 222
physicochemical characterization, **112**, 220
preparation, **112**, 213
purification and storage, **112**, 219
handling, safety considerations, **112**, 224
–α_2-macroglobulin conjugates, preparation and applications, **112**, 303
–mannose 6-phosphate oligosaccharide conjugates, preparation and applications, **112**, 304
purification, **112**, 208, 239
radiolabeling, **112**, 213
Rickettsiae
biohazard, **125**, 254
cell space measurements, **125**, 257
cell water measurements, **125**, 256
membrane transport
bioenergetics, **125**, 259
microporous filtration analysis, **125**, 258
metabolic capacity, **125**, 257
purification, **125**, 254
taxonomy, **125**, 253
viability, **125**, 255
Rivanol
in precipitation of IgG, **121**, 634
RNA
chloroplast, from *in vitro* transcription reactions
primer extension analysis, **118**, 263
purification, **118**, 261, 275
coupling to polyacrylhydrazidoagarose, **104**, 18
firefly lantern, isolation, **133**, 5
hybridization techniques for endocrine and neuroendocrine studies, **124**, 269
hydrated, calorimetric analysis, **127**, 149
infectious viral, from cloned cDNA
plant infection with, **118**, 715
production and analysis, **118**, 712
messenger
apoaequorin
cDNA cloning, **133**, 301
cell-free translation, **133**, 300
purification, **133**, 299
apolipoprotein
cDNA probe synthesis, **128**, 672
hybridization assays, **128**, 680; **129**, 315
casein, phosphoserine codon usage, **106**, 158
concentration and half-life, measurement as function of hormonal treatment, **109**, 572
HLA antigen, FACS analysis, **108**, 240
Ia antigens
cDNA synthesis, **108**, 573
cell-free translation, **108**, 569
purification, **108**, 567
interferon, detection by *in situ* hybridization, **119**, 474
LHRH precursor
cell-free translation, **124**, 322
isolation, **124**, 320
low abundance red light-regulated, cDNA clones, identification, **118**, 369

low-density lipoprotein receptor
cDNA cloning, **128**, 901
immune purification, **128**, 899
translation, **128**, 900
luminescent *Escherichia coli* clone, analysis, **133**, 78
neuropeptide, measurement in discrete brain regions, **124**, 548
nucleotide precursors, specific activity, measurement, **109**, 592
phytochrome, cDNA clones, identification, **118**, 369
translation in wheat germ cell-free extracts, **118**, 128
T-suppressor factor
preparation, **116**, 323
translation in amphibian oocytes, **116**, 324
plant mitochondrial
analytical techniques, **118**, 482, 495
double-stranded species
characterization, **118**, 501, 504
isolation, **118**, 501
isolation, **118**, 472, 492, 497
single-stranded species
characterization, **118**, 501, 504
isolation, **118**, 503
polyadenylated, purification
from oats, **118**, 372
from rat spleen, **108**, 568
radiolabeled probes, in detection of neuropeptide mRNA, **124**, 497
ribosomal, from tobacco chloroplasts
analytical techniques, **118**, 209
purification, **118**, 207
transfer
binding to methionyl-tRNA formyltransferase, stoichiometry, **106**, 149
chloroplast and cytoplasmic
aminoacylation, **118**, 219
fractionation, **118**, 223
identification, **118**, 226
isolation, **118**, 215, 261
sequence analysis, **118**, 262
visualization and extraction, **118**, 225
crystalline, binding to heavy metal ions, **114**, 156
initiator, constrained–restrained least-squares structure refinement, **115**, 287
phenylalanine, constrained–restrained least-squares structure refinement, **115**, 282
viral
detection with
cDNA probes, **118**, 723
RNA probes, **118**, 717
double-stranded, isolation, **118**, 719
infectious cDNA clones
production, **118**, 707
properties, **118**, 705
in vitro transcription, **118**, 712
polyadenylation methods, **118**, 727
radiolabeled, preparation, **118**, 719
single-stranded, isolation, **118**, 718
translation in wheat germ system, **118**, 154
RNA polymerase
DNA-directed, *see* DNA-directed RNA polymerase
Ro 20-1724, *see* 4-(3-Butoxy-4-methoxybenzyl)-2-imidazolidinone
Rosetting techniques
for allotype-bearing cell enumeration, **116**, 172
for Fc receptor-bearing cell detection, **116**, 406
inhibitory, for IgG-binding factor detection, **116**, 410
for lymphocyte separation
indirect method with monoclonal antibody-labeled cells, **121**, 726
mixed method with soybean agglutinin, **108**, 176
for lymphoreticular cell detection
with antibody-coated latex spheres, **108**, 56
antigen-specific, **108**, 58
generalized methods, **108**, 51
with tumor cells, **108**, 60
for macrophage detection, applications of monoclonal antibodies, **108**, 322
for surface immunoglobulin detection, direct and mixed methods, **108**, 386
for thymulin detection, **116**, 271

Rotation function
in protein crystallography
angular variables, **115**, 59
applications, **115**, 69
calculation by fast Fourier transform, **115**, 63
computer programs, **115**, 63
origin removal, **115**, 59
sampling and resolution, **115**, 57
Rous sarcoma virus
as expression vehicle for human interferon β, **119**, 383
Rubredoxin
hydrated, X-ray and neutron diffraction analysis, **127**, 169
Ryegrass
isolation of protoplasts, **118**, 577

S

Saccharomyces cerevisiae, *see* Yeast
Salamander
isolation of erythrocyte marginal bands, **134**, 248
Salicylaldehyde
in assay for D-glutamate-D-amino acid transaminase, **113**, 109
Saliva
human, purification of vitamin B_{12}-binding proteins, **123**, 32
Salivary gland
Drosophila melanogaster, preparation of ecdysteroid receptors, **111**, 459
Salmonella typhimurium
active transport, binding protein-dependent
protein components, **125**, 281, 289
regulation, **125**, 288
system reconstitution
applicability for whole cells, **125**, 300
assays, **125**, 295
influencing factors, **125**, 296
in spheroplasts and treated cells, **125**, 291
system types, **125**, 279
chemotaxis-related carboxylmethylesterase, purification and properties, **106**, 321
chemotaxis-related carboxylmethyltransferase, properties and assays, **106**, 310
outer membrane, permeability measurement, **125**, 266
in oxygen free radical mutagenicity testing, **105**, 256
peptide transport
deficient mutants, isolation, **125**, 366
measurement, **125**, 370
tester strain TA102, for detection of oxidative mutagens, **105**, 249
Salts
concentrated solutions, for protein partial specific volume determinations, **117**, 60
effects on membrane surface potentials, **125**, 65, 68, 71
–protein interactions, measurement, **114**, 68
Samarium
binding to tRNA crystals, **114**, 157
Saponification
carotenoids, **111**, 122
Sarcoidosis
detection by cellular chemiluminescence, **133**, 501
Sarcolectins
human and animal
assays, **119**, 695
properties, **119**, 700
purification, **119**, 698
Sarcoplasmic reticulum
vesicles, canine cardiac
phospholamban extraction, **102**, 270
phosphorylation, **102**, 263
preparation, **102**, 262
properties, **102**, 263
Sarcosine dehydrogenase
rat liver
assay, **122**, 256
properties, **122**, 259
purification, **122**, 257
Schiff reagent
–periodic acid, in phagocyte staining, **132**, 135, 179
Schistosomula
killing by macrophages, assay, **132**, 656
preparation, **132**, 651

Scintillons
  dinoflagellate
    assays, **133**, 310
    properties, **133**, 311
    purification, **133**, 310
Scleroproteins
  halogenated tyrosines in
    effects of alkaline hydrolysis, **107**, 431
    enzymatic release, **107**, 429
    identification, **107**, 431
    isolation, **107**, 414
    separations, **107**, 423
    syntheses, **107**, 420, 432
Scopolamine
  delivery system, transdermal, structure, function, and release rates, **112**, 513
  transdermal application, rate-controlled, design, **112**, 451
Scopoletin
  oxidation, in measurement of hydrogen peroxide, **105**, 397
Scorpion venom
  interactions with ion channels in neuroendocrine tissue, analysis, **103**, 402
Sea urchin
  egg cytoplasmic dynein
    assays, **134**, 327, 339
    properties, **134**, 334, 350
    purification, **134**, 329, 344
  egg extracts
    fascin extraction, **134**, 17
    preparation, **134**, 15
    storage, **134**, 16
    SU45 isolation, **134**, 18
  eggs
    fertilization and embryo culturing, **134**, 206
    isolation of metaphase mitotic apparatus, **134**, 209
    native mitotic apparatus, characterization, **134**, 212
    preparation for isolation of tubulin, **134**, 131
  mariculture and aquarium maintenance, **134**, 201
Secretin
  regulation of cAMP levels, analytical techniques, **109**, 344
Secretion granules
  peptidergic, from brain
    isolation, **124**, 257
    lectin-binding sites, demonstration, **124**, 259
    as model system for intracellular analysis of peptide secretion, **124**, 254
    molecular aspects of peptide release, analytical techniques, **124**, 262
Sedimentation
  droplet, in cell surface tension determination, **132**, 54, 88
  interacting proteins
    in ligand–ligand systems, **130**, 30
    microheterogeneity effects on patterns, **130**, 19
  in measurement of
    nucleotide binding to uncoupling protein, **126**, 501
    protein self-assembly, **130**, 55
  microtubule–neurofilament networks, **134**, 553
  sperm dyneins, **134**, 322
Sedimentation coefficients
  macromolecules, determination by active enzyme centrifugation, **117**, 107
  proteins in solutions
    estimation from small-angle X-ray scattering data, **117**, 168
    structural calculations from, **117**, 174
Sedimentation displacement analysis
  1,25-dihydroxyvitamin D_3 receptors with monoclonal antibodies as probes, **123**, 201
Sedimentation equilibrium
  apolipoproteins in solution, **128**, 382
  in determination of
    cytochrome-*c* oxidase partial specific volume, **126**, 56
    protein molecular weight in detergent solutions, **117**, 41
  macromolecules, in air-turbine ultracentrifuge
    applications, **117**, 10
    description and methodology, **117**, 4, 24
    partitioning techniques, **117**, 25
    theoretical considerations, **117**, 7

Sedimentation velocity
apolipoproteins in solution, **128**, 382
in determination of cytochrome-*c* oxidase molecular weight, **126**, 57
difference, *see* Difference sedimentation velocity
Sedimentation volume
in cell surface tension determination, **132**, 54, 92
Seeding
crystals, **114**, 132
Selenium
incorporation into thiolase, **107**, 621
Selenocysteine
in proteins
analytical techniques, **107**, 579, 604
derivatization, **107**, 584
enzymatic degradation, **107**, 581
function, **107**, 578
identification, **107**, 585
occurrence, **107**, 576
radiolabeling, **107**, 581, 603
Selenocysteine lyase
selenocysteine degradation, **107**, 581
Selenomethionine
–IgG conjugate
applications, **121**, 776
preparation, **121**, 774
in proteins
characterization, **107**, 620
derivatization and identification, **107**, 623
Selenopeptides
glutathione peroxidase, preparation, **107**, 617
from plasma and renal selenoproteins, preparation, **107**, 617
Selenoproteins
chromatographic identification, **107**, 606
isotopic equilibration, **107**, 603
plasma P
purification, **107**, 612
selenopeptide preparation, **107**, 617
renal
selenopeptide preparation, **107**, 617
separation, **107**, 614
translational synthesis, **107**, 619
selenocysteine in
analytical techniques, **107**, 579
derivatization, **107**, 584
enzymatic degradation, **107**, 581
function, **107**, 578
identification, **107**, 585
occurrence, **107**, 576
radiolabeling, **107**, 581
selenomethionine in
characterization, **107**, 620
derivatization and identification, **107**, 623
Self-associations
ligand-induced
ligand binding isotherms, measurement and analysis, **117**, 496
velocity sedimentation study, **117**, 459
quantitative definition with migration patterns, **130**, 10
sedimentation patterns, microheterogeneity effects, **130**, 21
in velocity sedimentation, distinguishing criteria, **130**, 3
Sendai virus
induced cell fusion
in construction of hormonally responsive intact cell hybrids, **109**, 366
interferon effects, **119**, 704
induction of interferons, **119**, 36, 40, 137, 220
purification, **119**, 703
Sephadex G-10
in column separation of macrophages and lymphocytes, **108**, 303
Sepharose
–actin affinity resin, preparation, **134**, 463
–concanavalin A
in affinity purification of insulin receptor, **109**, 399, 402
immobilization of NAD glycohydrolase, **122**, 178
cytochrome *c*-linked, preparation, **126**, 65, 67
–2,2-diphenylpropylamine
applications, **133**, 105
binding to luciferase, **133**, 102
synthesis, **133**, 101
–*Helix pomatia* A hemagglutinin gels
in chromatographic separation of lymphocytes, **108**, 159
preparation, **108**, 155

iminodiacetic acid-activated, in nickel
  chelate chromatography of interferon γ, **119**, 201
immobilization of enzymes for continuous-flow bioluminescent assay, **133**,
  233
–insulin, in affinity purification of insulin
  receptor, **109**, 400, 403
–phenothiazine
  calmodulin-binding capacity, estima-
    tion, **102**, 29
  in calmodulin purification, **102**, 29
  coupling estimation, **102**, 24
  preparation, **102**, 18
–wheat germ agglutinin
  in affinity adsorption assay for soluble
    glucagon receptor, **109**, 13
  in affinity purification of insulin recep-
    tor, **109**, 399, 405
  in chromatographic separation of
    lymphocytes, **108**, 159
  preparation, **108**, 155
Sepharose 4B
–calmodulin, preparation, **102**, 211
in IgM chromatography, **116**, 31
Serine
  reactivity with heavy atoms, **114**, 152
Serotonin
  microradioenzymatic assay, **103**, 488
Serratia marcescens
  purification of anthranilate synthase and
    component ASI, **113**, 289, 290
Serum, *see also* Blood; Plasma
  avian, isolation of riboflavin-binding
    protein, **122**, 231
  fetal bovine, removal of protein A-
    binding Ig, **121**, 301
  human
    antibiotics, assay with luminescent
      bacteria, **133**, 281
    derived B-cell hybridoma growth
      promoters
      comparison with other growth
        promoters, **121**, 12
      overview, **121**, 11
      preparation, **121**, 12
    effect on hepatic lipase, **129**, 730
    purification of
      1,25-dihydroxyvitamin D, **123**, 179
      HLA antigens, **108**, 622

IgA, **116**, 44
IgD, **116**, 96
IgM, **116**, 29
insulin-like growth factors, **109**, 783
quantitation of
  1,25-dihydroxyvitamin D, **123**, 190
  retinol-binding protein, **123**, 85
  transcobalamin II, **123**, 44
  vitamin K, assay by HPLC with UV
    detection, **123**, 237, 243
immune, *see* Antiserum
lipid peroxide assay, **105**, 328
lipoprotein-deficient
  application in cholesterol efflux stud-
    ies, **129**, 637
  preparation, **129**, 634
murine, detection of thymulin, **116**, 279
porcine, purification of thymulin, **116**,
  272
rabbit, purification of
  heme-binding proteins on hemin-
    agarose, **123**, 336
  vitamin B₁₂-binding proteins, **123**, 32
rat, purification of
  insulin-like growth factor I and II,
    109, 776
  monoclonal antibodies, **121**, 646
  simultaneous determination of retinol
    and α-tocopherol by HPLC, **123**,
    215
Serum albumin
–amatoxin conjugates
  biological properties, **112**, 230, 233
  preparation, **112**, 227, 231
in assay for thyroid peroxidase, **107**, 449
coupling to phallotoxins
  with carbodiimides, **112**, 236
  with mixed anhydrides, **112**, 236
delipidation, **128**, 322
delivery systems
  polymeric and copolymeric
    kinetic analysis, **112**, 414
    preparation, **112**, 400
  magnetically responsive polymers
    kinetic analysis, **112**, 414
    preparation, **112**, 411
    triggering system, **112**, 413
density gradient media
  in equilibrium density separation of
    lymphoid cells, **108**, 109, 116

osmolarity and pH control, **108**, 103, 106
preparation and calibration, **108**, 107
encapsulated in crystallized carbohydrate magnetic microspheres, preparation for drug targeting, **112**, 126
fatty acid binding
data analysis, **128**, 333
measurement, **128**, 327
structure effects, **128**, 335
–fatty acid solutions, preparation, **128**, 322
lysine residues, nonenzymatic glucosylation, **106**, 88
detection and quantitation, **106**, 95
dissociation kinetics, measurement, **106**, 93
kinetics, determination, **106**, 90
measurement, **106**, 90
structural analysis by small-angle neutron scattering, **130**, 93, 111
structure and lipid transport, **128**, 336
as supplement in preparation of SFH medium, **121**, 284
Serum amyloid A
assay, **128**, 314
physical and chemical characteristics, **128**, 312
preparation, **128**, 312
Sesquiterpene cyclase
plant and fungal
assays, **110**, 389
extraction and purification, **110**, 398
properties, **110**, 384
Severin
assays, **134**, 12
purification, **134**, 10
Shape functions
in covolume measurement and protein axial ratio estimation, **117**, 184
Sialidase
in apolipoprotein E phenotyping, **128**, 281
Sialyltransferase
in hepatic Golgi subfractions, assay, **109**, 223
Silica, *see* Silicon dioxide
Silicic acid
in purification of interferon γ, **119**, 90, 146
Silicon dioxide
bound ligands, spectrophotometric determination, **104**, 217
in depletion of macrophages, **108**, 330
preparation for affinity HPLC, **104**, 212
tresylated, preparation, **104**, 58
Siliconization
chromatography columns for immunoaffinity separation of T-cell subsets, **121**, 743
Silkmoth
preparation of epoxyfarnesoic acid methyltransferase, **111**, 541
Silver staining
myosin, **102**, 71
proteins in gels, **104**, 441
Silylation
vitamin D metabolites, **123**, 137
Simian virus 40
temperature-sensitive mutant-transformed cells
mammalian, general properties, **109**, 387
rat fetal hepatocyte
establishment, **109**, 387
growth and maintenance, **109**, 395
metabolic and hormonal properties, **109**, 392
Sindbis virus
induction of guinea pig interferon, **119**, 127
Singlet oxygen
chemiluminescent, detection in biological systems, **105**, 221
chemiluminescent probes, requirements, **133**, 570
detection
kinetic methods, **105**, 41
luminescence assay, **105**, 42
traps for, **105**, 40
production by neutrophils, measurement, **105**, 396
sources, **105**, 37, 228
SITS
as retrograde tracer for neuropeptide-specific pathways, **103**, 665
Size-exclusion chromatography
applications, **104**, 169
column selection and maintenance, **104**, 163

equipment, **104**, 160
fundamental aspects, **104**, 155
interleukin 1, **116**, 463
interleukin 2, **116**, 515
neuropeptide precursors, **103**, 530
operating conditions, **104**, 168
random coil polymers, calibration, **117**, 144, 147
Skeletal muscle
α-bungarotoxin–luciferin binding, **133**, 47
calcium channels, assay, **109**, 513
cellular differentiation, interferon effects, measurement, **119**, 625
cultivation, **119**, 619
occurrence of Glu–Lys crosslink, **107**, 252
preparation from rat hindlimb, **102**, 76
quantitation of myosin P-light chain phosphorylation, **102**, 75
Skin
clonidine application, rate-controlled system design, **112**, 456
drug delivery systems
microsealed fabrication, **112**, 463
nitroglycerin-containing, structure, function, and release rates, **112**, 515
scopolamine-containing, structure, function, and release rates, **112**, 513
systemic, development, **112**, 449
17β-estradiol application, rate-controlled system design, **112**, 453
human, isolation of
Langerhans cells, **108**, 357, 686
transglutaminase, **113**, 371
murine
grafting, **108**, 20
isolation of Langerhans cells, **108**, 685
nitroglycerin application, rate-controlled system design, **112**, 456
occurrence of Glu–Lys crosslink, **107**, 256
scopolamine application, rate-controlled system design, **112**, 451
Slide immunoenzymatic assay
for antibacterial antibodies, **121**, 514, 521
for bacterial antigens, **121**, 516, 522
overview, **121**, 510
variations, **121**, 517
Small-angle light scattering
instrumentation for, **130**, 37
linear polymer formation, measurements, **130**, 44
maintenance procedures, **130**, 39
sample loading, **130**, 43
Small-angle neutron scattering
proteins in solution
experimental theory, **130**, 81
instrumentation for, **130**, 88
strongly interacting systems, **130**, 106
weakly interacting systems, **130**, 93
solvent around macromolecules, density and location measurement, **127**, 619
Small-angle X-ray scattering
in measurement of protein hydration, **117**, 247
proteins in solution
applications, **117**, 174
data correlations, **117**, 158
comparison with X-ray diffraction data, **117**, 171
measurement parameters, **117**, 155
Smooth muscle
calcium channels, assay, **109**, 513
isometric tension, measurement, **102**, 63
preparation from bovine trachea, **102**, 63
purification of
caldesmin, **134**, 38
metavinculin, **134**, 74
talin, **134**, 75
vinculin, **134**, 73
quantitation of myosin P-light chain phosphorylation, **102**, 65, 71
Sodium
counterion positions in hydrated DNA, computer simulation, **127**, 128
driven flagellar motors, in alkalophilic *Bacillus*, **125**, 582
gramicidin A channels, energy profiles, computational methods, **127**, 257
Sodium 9,10-anthraquinone-2-sulfonate
voltammetric measurements, **125**, 114
Sodium borohydride
in detection of N-terminal glucuronamide linkage in proteins, **106**, 211

Sodium borotritide
in quantitation of protein oxidative modification, **107**, 376
Sodium butyrate
induction of glucagon responsiveness in transformed MDCK cells, **109**, 357
Sodium carbonate
release from poly(ortho ester) polymer disks, **112**, 425
Sodium chloride
aqueous, neutron diffraction study, **127**, 13
release from poly(ortho ester) polymer disks, **112**, 428
Sodium cromoglycate
incorporation into hydrophilic albumin microspheres
assessment, **112**, 32
effect of protein concentration, **112**, 40
Sodium dodecyl sulfate
in gel electrophoresis, **104**, 241
Sodium iodohippurate
incorporation into hydrophilic albumin microspheres
assessment, **112**, 32
effect of protein concentration, **112**, 40
Sodium perchlorate
aqueous solutions, low-frequency Raman scattering, **127**, 99
Sodium sulfate
release from poly(ortho ester) polymer disks, **112**, 428
Sodium thiosulfate
as reducing agent in albumin microsphere labeling with ^{99m}Tc, **112**, 48
Sodium transport
bacterial
measurement, **125**, 333
reconstitution with biotin-dependent decarboxylases as pump, **125**, 547
with sugars, measurement, **125**, 381
by methylmalonyl-CoA decarboxylase
properties, **125**, 546
reconstitution in proteoliposomes, **125**, 544
by NADH oxidase in *Vibrio alginolyticus*, assay, **125**, 520
by oxaloacetate decarboxylase
assay, **125**, 536, 538
properties in reconstituted proteoliposomes, **125**, 539
in pyruvate carboxylation, **125**, 540
Software, *see also* Computer applications
for analysis of patch records, **124**, 235
for calculation of
protein conformation from circular dichroism, **130**, 256
proton translocation energetics in model membrane systems, **127**, 460
for computer-controlled scanning gel chromatography, **117**, 116
computer graphics, in molecular biology, **115**, 381
CORELS, for structure refinement of proteins and nucleic acids, **115**, 271
interactive graphics
crystallographic extensions, **115**, 167
in crystallographic refinement, **115**, 166
FRODO program, **115**, 158
as general molecular modeling system, **115**, 168
initial model building, **115**, 160
for ligand binding studies by ultrafiltration, **117**, 362
method of moments exponential fit program, **134**, 703
for minimum variance Fourier coefficient refinement, **115**, 15
for multiwire area detector diffractometer, **114**, 452
for rotation function calculation in protein crystallography, **115**, 63
for translation function calculation in protein crystallography, **115**, 67
Solanum melongena, *see* Eggplant
Solubility
proteins, *see* Protein solubility
Solubilization
apolipoproteins, **128**, 379
carotenogenic enzymes from *Neurospora*, **110**, 266
functional membrane proteins, **104**, 305
HLA-A and HLA-B antigens, **108**, 585
Ia antigens, **108**, 535, 538
integral membrane proteins, **104**, 332
leukocytes for myeloperoxidase purification, **132**, 372

membranes, **124**, 146
murine class I MHC antigens, **108**, 508
neuropeptide receptors, **124**, 164
opioid receptors, **124**, 177
Soluble immune response suppressor
antibodies, preparation and properties, **116**, 396
assay, **116**, 396
characterization, **116**, 401
detection, **116**, 397
overview, **116**, 368
oxidized
assay, **116**, 400
preparation, **116**, 400
producing lymphocytes, activation, **116**, 395
Solutions
aqueous
biomolecular equilibria
free energy calculations, **127**, 67, 76
role of entropy in, **127**, 65
hydrophobic interactions, theoretical and computational studies, **127**, 48
Monte Carlo simulations of hydration with proximity criterion analysis, **127**, 21
protein structure analysis with D_2O–H_2O neutron difference maps, **127**, 329
structure analysis by diffraction techniques, **127**, 13, 16, 325
superoxide generation by vacuum–UV photolysis, **105**, 81
water structure and ion positions in DNA, computer simulations, **127**, 128
ethanolic, superoxide generation by vacuum–UV photolysis, **105**, 81
microtubule assembly, X-ray scattering, **134**, 659
nonaqueous, superoxide anion generation
with crown ethers or cryptates, **105**, 75
electrochemical methods, **105**, 72
with tetramethylammonium superoxide, **105**, 78
Solvation
biopolymers, theoretical approaches, **127**, 369
Solvents
around macromolecules, density and location measurements, **127**, 619
–protein preferential interactions
measurement, **114**, 61
mechanisms, **114**, 74
Somatic cells
ATP extraction, considerations, **133**, 19
Somatomammotropin
human chorionic, *see* Human chorionic somatomammotropin
Somatomedin C
assays, **109**, 802
purification
from blood plasma, **109**, 814
from plasma Cohn fraction IV, **109**, 805
Somatomedins
assays, **109**, 802
chemistry, **109**, 799
nomenclature, **109**, 798
origin, **109**, 800
physical state in blood, **109**, 800
purification, **109**, 805
Somatostatin
antiserum, preparation, **124**, 337
in dissociated cerebral cortical cell cultures
content measurement, **124**, 419
release measurement, **124**, 417
dynamics, analysis by push–pull perfusion technique, **103**, 176
peptide biosynthesis, characteristics, **124**, 345
precursor mRNA, measurement, **124**, 560
processing products, identification, **124**, 335
regulation of cAMP levels, analytical techniques, **109**, 344
Somatostatin-28
microisolation from hypothalamus, **103**, 79
Somatostatin receptors
autoradiographic localization, **124**, 607
characteristics, **124**, 610
Somatotropin
release in perifused pituitary cells, **124**, 79

Sonication
for purification of recombinant human interferon γ, **119**, 206
for solubilization of opioid receptors, **124**, 182
Soybean agglutinin
in separation of lymphocytes
by mixed rosetting, **108**, 176
by selective agglutination, **108**, 175
Spaghetti overlay technique
for immunoautoradiographic screening of gels, **121**, 453
Spasmin
sources, **134**, 406
Spectinomycin
resistance in chloroplast mutants, selection for, **118**, 615
Spectrin
bovine and porcine brain, purification, **134**, 57
Spectrofluorometry
cytosolic free Ca^{2+} in hepatocytes, in assay of hormone effects, **109**, 555
dextran uptake by leukocytes, **108**, 343, 345
dityrosine, **107**, 380, 384
vitamin E, **105**, 144
Spectrophotometers
micro-, calibration for detection of fluorophores *in situ*, **105**, 342
Spectrophotometry
carotenoids, **111**, 147
chlorophylls, **110**, 306
cytochrome *b*-245, **132**, 385
cytochrome *d* terminal oxidase complex, **126**, 88, 101
cytochrome *o* terminal oxidase complex, **126**, 101
dehydroascorbate reductase, **122**, 10
diaphorase, **133**, 201
diode-array, 25-hydroxyvitamin D_3 metabolites, **123**, 147
dityrosine, **107**, 380, 384
ecdysone oxidase, **111**, 421
erythrocyte cytolysis, **132**, 493
fluorescent damage products of lipid peroxidation, **105**, 339
guanidine-specific ADP-ribosyltransferase, **106**, 405
H–D exchange in proteins, **131**, 460
in HPLC assay for
coumarin antagonists of vitamin K in blood, **123**, 225
vitamin K in tissues, **123**, 235
3-hydroxy-3-methylglutaryl-CoA synthase, **110**, 20
indoleamine 2,3-dioxygenase, **105**, 63
iron in heme proteins, **123**, 320
ligands in gel chromatographic columns, in concentration measurement, **117**, 351
lipid-conjugated dienes, **105**, 331
malonaldehyde in biological samples, **105**, 299
membrane potential in bioenergetic systems, **127**, 557
5-methyl-L-tetrahydrofolate, **122**, 383
mevalonate kinase, **110**, 60, 72
mevalonate 5-pyrophosphate decarboxylase, **110**, 88
mitochondrial pyridine nucleotides, in redox state determination, **105**, 437
NADase-catalyzed reactions, **122**, 169
NAD glycohydrolase, **122**, 173
NADH oxidase, **132**, 366
oxidatively modified proteins, **107**, 375
oxidoreductases, **133**, 201
phosphomevalonate kinase, **110**, 78
protein disulfide enzymatic reduction–oxidation by thioredoxin, **107**, 296
protein refolding at subzero temperatures, **131**, 179
protein unfolding in aqueous methanol, **131**, 175
reflectance, nicotinamide nucleotides and flavins in intact tissues, **123**, 311
sterol carrier protein, **111**, 294
sterols, **111**, 25
stopped-flow, protein folding, **131**, 62
vitamin E, **105**, 142
xanthophylls, **110**, 306
Spermatozoa
bovine
collection, **134**, 318
extraction of dynein ATPase, **134**, 321
washing and demembranization, **134**, 319
nematode
isolation, **134**, 418, 419

purification of major sperm protein, **134**, 418
Spheroplasts
bacterial, fusion with plant protoplasts, **118**, 585
yeast
isolation of mitotic spindles and spindle pole bodies, **134**, 228
lysis, **134**, 227
preparation, **134**, 226
Spinach
carotene synthesis in chloroplasts, **110**, 267
isolation of
acetyl-CoA:acyl carrier protein transacylase, **122**, 54
aminoacyl-tRNA synthetase, **118**, 218
chlorophyll-binding proteins, **118**, 351
chloroplast elongation factors, **118**, 296, 298
chloroplasts, **118**, 215, 255, 299
chloroplast transcription extracts, **118**, 234
chloroplast tRNAs, **118**, 215
dehydroascorbate reductase, **122**, 10
DNA polymerase α, **118**, 99
Spinacia oleracea, *see* Spinach
Spindle pole bodies
yeast
isolation, **134**, 228, 229
staining, **134**, 228
Spin labeling
ADP/ATP carrier protein, **125**, 635
cells, in determination of cell water viscosity, **127**, 365
Spin labels
in measurement of membrane surface charge density, **127**, 687
nitroxide, for transmembrane proton movement
electrical potential measurements, **127**, 744
gradient measurements, **127**, 742
volume measurements, **127**, 740
rotational motion, in analysis of intracellular viscosity, **127**, 239
Spin trapping
in assay for superoxide formation, **105**, 373, 377
free radicals, overview, **105**, 188
hydroxyl radicals, **105**, 18, 198
superoxide radicals, **105**, 198
Spleen
human, purification of HLA-A and HLA-B antigens, **108**, 594
murine
cell sedimentation profile, **108**, 82
collection of macrophages, **108**, 294
cultivation, **119**, 145
immunization for monoclonal antibody production, **121**, 33
preparation of
lymphocytes, **108**, 51
single-cell suspensions, **108**, 45, 75, 120
production of interferon γ, **119**, 146
removal methods, **108**, 43
rat
preparation of single-cell suspensions, **108**, 45
purification of
Ia antigens, **108**, 560
mRNA, **108**, 567
poly(A) RNA, **108**, 568
thymosin β_4, **116**, 268
thymosin β_{10}, **116**, 268
removal methods, **108**, 43
vitamin E levels, analytical techniques, **105**, 138
Squalene
radiolabeled
incubation with marine invertebrates and plant tissues, **111**, 320
preparation, **111**, 315
rat liver
assay, **111**, 294
properties, **111**, 298
purification, **111**, 296
Squalene epoxidase, *see* Squalene monooxygenase
Squalene monooxygenase
rat liver
assay, **110**, 376
properties, **110**, 380
purification, **110**, 377
Squalene synthetase
role in formation of 12-*cis*-dehydrosqualene, **110**, 373
yeast
assays, **110**, 363

catalytic characteristics, **110**, 360
isolation, **110**, 369
lipid depletion, **110**, 370
reconstitution, **110**, 371
solubilization, **110**, 370
Squid
optic lobe, isolation of neurofilaments, **134**, 373
Staining
chloroplast and cytoplasmic tRNAs, **118**, 225
enzymes in gels, **104**, 419
glycoproteins in gels, **104**, 447
^{3}H-labeled chromosomal G-bands, **128**, 874
immunoenzymatic, for detection of monoclonal antibodies and antigens, **121**, 855
immunofluorescent, *see* Immunofluorescent staining
immunohistochemical, neuropeptides, **103**, 458
intracellular, in dissociated tissue culture electrophysiological recording, **103**, 127
kinetochore in metaphase chromosomes, **134**, 271
lipids in lipoproteins, **128**, 418, 422
with monoclonal antibodies, **121**, 835
negative
Chlamydomonas basal bodies, **134**, 260
lipoproteins, **128**, 443
nitrocellulose blots with flagellar dyneins, **134**, 304
phagocytic cells, **132**, 132
phosphoproteins in gels, **104**, 451
proteins
with Amido Black, for antibody detection on blots, **121**, 502
in gels with Coomassie Blue, **104**, 439
in lipoproteins, **128**, 418, 422
silver, *see* Silver staining
T-cell subsets with fluorochromes, **121**, 751
Toxoplasma in macrophages, **132**, 634
yeast mitotic spindles and spindle pole bodies, **134**, 228
Stains-all, *see* 1-Ethyl-2-(3-(1-ethylnaphtho[1,2-*d*]thiazolin-2-ylidene)-2-methylpropenyl)naphthol[1,2-*d*]thiazolium bromide
Stannous chloride
as reducing agent in albumin microsphere labeling with ^{99m}Tc, **112**, 48
Starfish
purification of asteriarubin, **111**, 241
Steam distillation
isoprenoids, **111**, 150, 154
Stearic acid
^{3}H-labeled, in analysis of phospholipid turnover in cultured cells, **124**, 429
Stem cells
multipotential hematopoietic, assay, **116**, 555
Stereochemically restrained refinement
macromolecular structure
applications, **115**, 268
computer program design, **115**, 260
observational functions, **115**, 253
strategy, **115**, 264
weighting considerations, **115**, 263
Steroid hormones
and neuropeptides, simultaneous localization in brain, **103**, 631
receiving cells in CNS, immunocytochemistry, **103**, 639
Steroidogenesis
adrenocortical cell, activation by photoaffinity labeling, **103**, 65
luteal cells, regulation by luteinizing hormone, **109**, 302, 310
Steroids
antibody preparation, **133**, 393
surface chemiluminescent immunoassays, **133**, 387
Sterol carrier protein
rat liver
assay, **111**, 294
properties, **111**, 298
purification, **111**, 296
Sterologenesis
regulation and analysis with oxysterols, **110**, 9
Sterols
identification and quantitation
chromatographic techniques, **111**, 5, 37
spectral techniques, **111**, 25

metabolism in yeast, analysis with mutants, **111**, 333
oxygenated, *see* Oxysterols
in plants and marine invertebrates, analytical techniques, **111**, 311
Stigmatellin
types A and B, inhibition of cytochrome bc_1, **126**, 261
Stilbene oxide
cis and *trans*, in assay of epoxide hydrolases, **111**, 308
Stomach
rat, oxyntic mucosa, gastrin receptor assay, **109**, 56
Stopped-flow circular dichroism
protein conformational changes, **130**, 519
protein transient folding intermediates, **131**, 132
Stopped-flow methods
for analysis of mechanochemical ATPases
data collection and analysis, **134**, 681, 684
instrumentation, **134**, 679
Streptavidin
–agarose
biotin-binding activity, **121**, 722
in membrane antigen immunoaffinity isolation, **121**, 724
preparation, **121**, 721
stability, **121**, 723
bound ligands, displacement by biotin, **109**, 443
isolation on iminobiotinyl AH-Sepharose 4B, **109**, 425
Streptococcus
S. agalactiae, specific hybridomas, production, **121**, 263
S. lactis, hexose 6-phosphate antiport, reconstitution, **125**, 558
Streptomyces UC5319
purification of cyclases, **110**, 403
Streptomycin
resistance in chloroplast mutants, selection for, **118**, 615
Streptonigrin
mutagenicity testing with *Salmonella typhimurium* TA102, **105**, 250
Streptozotocin
in selection of bacterial mutants defective in polyhydric alcohol transport, **125**, 476
Strobilurin
types A, B, and C, inhibition of cytochrome bc_1, **126**, 260
Strongylocentrotus droebachiensis, *see* Sea urchin
Strongylocentrotus purpuratus, *see* Sea urchin
Subdermal implants
cylinder-shaped, fabrication, **112**, 465
Submitochondrial particles
simultaneous determination of oxygen consumption and NAD(P)H oxidation, **122**, 167
Substance P
antiserum preparation, **124**, 337
assay, **132**, 328
binding studies, **132**, 333
monoclonal antibodies, assay, **121**, 220
peptide biosynthesis, characteristics, **124**, 342
and peroxidase, bispecific monoclonal antibodies, assay, **121**, 217
processing products, identification, **124**, 335
properties, **132**, 326
synthesis, **103**, 28
Subtilisin
proteolysis of tubulin, **134**, 183
Succinate–cytochrome-*c* reductase
bovine cardiac mitochondria
iron–sulfur protein-depleted complex, purification, **126**, 216
isolation, **126**, 214
reconstitutively active iron–sulfur protein, purification, **126**, 214
Succinate dehydrogenase
Bacillus subtilis
biosynthesis, **126**, 413
genetics, **126**, 409
immunoprecipitation and composition, **126**, 403
cardiac mitochondrial, assay, **125**, 18
Escherichia coli
assays, **126**, 380
dimer purification, **126**, 381
holoenzyme purification, **126**, 385
Wolinella succinogenes
assays, **126**, 397

properties, **126**, 390
purification, **126**, 388
reconstitution, **126**, 392
N-Succinimidyl 3-(4-hydroxyphenyl)propionate, *see* Bolton–Hunter reagent
N-Succinimidylpropionate
^{3}H-labeled, antibody radiolabeling with, **121**, 670
N-Succinimidyl-3-(2-pyridyldithio)propionate
in antibody coupling to toxins, **112**, 215
Succinoylavidin
bound ligands, displacement by biotin, **109**, 443
Sucrose
aqueous solutions, low-frequency Raman scattering, **127**, 101
–cyanuric chloride, in assay for tissue sites of lipoprotein degradation, **129**, 619
Sucrose gradients
in purification of
chloroplast DNA, **118**, 167
plant mitochondria, **118**, 491
plasma membranes, **118**, 42
Sudan black
in phagocyte staining, **132**, 134, 178
Sugars
bacterial transport mutants, isolation and characterization with transposons, **125**, 157
dysopsonic function, overview, **132**, 308
reducing, in cell wall, colorimetric assay, **118**, 26
Sulfadiazine
loading in poly(lactic/glycolic acid) polymers, **112**, 438
Sulfates
reverse-phase liquid chromatography, **111**, 62
Sulfation
protein tyrosine, stoichiometry, determination, **107**, 221
Sulfhydryls
protein, oxidation by thiol/disulfide exchange, differential effects, **107**, 336
Sulfitolysis
oxidative, in cleavage of IgA disulfide bonds, **116**, 63
and reoxidation, human interferon αA, **119**, 252
Sulfonyl chlorides
organic, in immobilization of ligands, **104**, 56
Sulfosuccinimidobiotin
in biotinylation of monoclonal antibodies, **121**, 720
Sulfur
as ligand to metals in proteins, magnetic circular dichroism study, **130**, 283
Sulfuric acid
and thymol, in staining of glycoproteins in gels, **104**, 447
treatment of hexane extracts, in vitamin E determination, **105**, 141
Superfusion systems
for measurement of neuropeptide release, **124**, 469
Superoxide dismutase
action mechanism, **105**, 27
assays, **105**, 92, 95, 101, 457
cellular concentration, manipulation methods, **105**, 409
Cu,Zn-
electroimmunoassay, **105**, 97
electron density maps, interpretation, **115**, 189
isolation, **105**, 89, 90
mechanism, fine details, **105**, 30
in detection of
paraquat-mediated oxyradical formation, **105**, 525
superoxide anion release from monocytes and macrophages, **105**, 365
effect on chemiluminescence of probes for phagocyte oxygenation activity, **133**, 473
Fe-, isolation from bacteria, **105**, 91
general properties, **105**, 27
inhibition of indoleamine 2,3-dioxygenase reaction, **105**, 65
Mn-, preparation from solid tissues, **105**, 90
role in superoxide scavenging in chloroplasts, **105**, 427
subcellular distribution in rat liver, **105**, 105
Superoxide radicals
assays, **105**, 11, 14, 15

for assessment of phagocytic function, **132**, 112, 119, 151, 152, 158
biological sources, **105**, 59
chemistry in nonaqueous solutions, analytical techniques, **105**, 71
generation
in aqueous and ethanolic solutions by vacuum–UV photolysis, **105**, 81
drug-induced, detection and measurement, **105**, 505
in illuminated thylakoids, **105**, 423
by paraquat
detection and measurement, **105**, 525
methodology, **105**, 523
production
by liver cell fractions, measurement, **105**, 373
by macrophages, microassay, **132**, 410
by mitochondria, measurement, **105**, 429
by monocytes and macrophages, measurement, **105**, 365
by neutrophils, measurement, **105**, 361, 370, 394
systemic consequences, **105**, 386
reaction with bacterial luciferase flavin radical, **133**, 144
role in indoleamine 2,3-dioxygenase reaction, **105**, 65
scavenging in chloroplasts, **105**, 426
spin trapping, **105**, 198
Superoxygenase, *see* Indoleamine 2,3-dioxygenase
Suppressor factors
antigen-specific
detection and characterization, **116**, 316
preparation from suppressor T cells, **116**, 312
subunits, preparation and separation, **116**, 322
T cell-derived, overview, **116**, 299
from T-cell hybridomas, production and characterization, **116**, 306
GAT antigen-specific
assays, **116**, 328
characterization, **116**, 337
production, **116**, 327
purification, **116**, 329
soluble immune response, *see* Soluble immune response suppressor
T-cell, for mixed leukocyte response
bioassays, **116**, 420
immunochemical properties, **116**, 426
physicochemical properties, **116**, 424
production, **116**, 417
Surface area
molecular, calculations for known structures, **115**, 457
Surface potential
apolipoprotein monolayer adsorbed at air–water interface, measurement, **128**, 397
biological membranes
in analysis of membrane interaction with small protein molecules, **125**, 81
changes during energization, **125**, 78, 80
governed processes, characteristics, **125**, 63, 83
and intramembrane electric field, **125**, 76
and ion concentration, **125**, 58
measurement techniques, **125**, 64; **127**, 678
probes, **125**, 78
and redox potential, **125**, 75
salt effects, **125**, 65, 68
theoretical description, **125**, 60
electrokinetic
effect on phagocytosis, **132**, 68
measurement, **132**, 65
small cells or vesicles, estimation
with charged probes, **125**, 52
from electrophoretic mobility measurement, **125**, 55
Surface pressure
apolipoprotein monolayer adsorbed at air–water interface, measurement, **128**, 390, 392
Surface tension
phagocytes and bacteria, measurement, **132**, 22, 70
theoretical aspects, **132**, 19
Suspension cultures
anterior pituitary, preparation, **109**, 296
from friable callus cultures
establishment, **118**, 546

growth curves, establishment, **118**, 549
human cells, binding of interferon, analytical techniques, **119**, 305
monocytes, preparation, **132**, 207, 243
Sycamore
suspension-cultured
extraction and purification of cell wall components, **118**, 8
preparation of cell walls, **118**, 6
Synapsin I
bovine and porcine brain, purification, **134**, 67
Synchronization
plant cells in culture and in meristems by amphidicolin, **118**, 87
Synchrotron radiation
applications, **114**, 281; **134**, 666
in direct phase determination by anomalous scattering, **115**, 53
instrumentation, **134**, 662
for measuring fluorescence lifetimes, **130**, 484
properties, **134**, 657
scattering from solutions, **134**, 659
as X-ray source, **114**, 285
Synovial membranes
caprine
cultivation, **119**, 553
sensitivity to heterologous interferons, **119**, 554

T

Talin, *see* Thaumatin
Target size analysis
after radiation inactivation
applications to interferons, **119**, 258
principles and theory, **119**, 256
Tau protein
bovine brain
properties, **134**, 159
purification, **134**, 157
Taurine aminotransferase
Acromobacter superficialis
assay, **113**, 102
properties, **113**, 106
purification, **113**, 104
Taurine-glutamate transaminase, *see* Taurine aminotransferase
Taxol
induction of aster formation in eggs, **134**, 199
in purification of
microtubule-associated proteins, **134**, 116
microtubules, **134**, 106, 116
T-cell growth factor
in assay for thymosin fraction 5, **116**, 233
T-cell replacing factor
overview, **116**, 364
T cells
alloantigen-activated, preparation for IgG-binding factor production, **116**, 404
allo-Ia-restricted antigen-specific, generation, **108**, 267
antigen distribution and immunological function, **108**, 627
in assay for T-suppressor factor, **116**, 421
cytotoxic
applications, **132**, 474
assay with anti-erythrocyte hybridomas as targets, **121**, 761, 764
freezing, **132**, 473
generation with liposome-incorporated HLA-DR antigens, **108**, 607
hybridomas, generation, **132**, 470
preparation, **132**, 469
self-reactive, induction by allogeneic effect factor, **116**, 431
cytotoxicity, ^{51}Cr release assay, **132**, 440
depletion with 5-bromodeoxyuridine and light, **108**, 263, 268
epifluorescence microscopy, **121**, 748
functional analysis with nude mouse, **108**, 351
hybridoma preparation for IgG-binding factor production, **116**, 404
immunoaffinity chromatography, **121**, 744
induced specific macrophage cytotoxicity, assays, **132**, 531
for induction of graft-versus-host reactions, sources, **108**, 29
producing T-suppressor factor, preparation, **116**, 418

production of
allogeneic effect factor, **116**, 428
interleukin 2, **116**, 503, 530
in proliferation assay for interleukin 2, **116**, 495, 526
response to parathyroid hormone, assays, **109**, 631
separation
on cellular monolayers, **108**, 125
by indirect rosette technique, **121**, 734, 736
from large granular lymphocytes, **132**, 449
as source of TGAL antigen-specific helper factor, **116**, 341, 343, 346
supernatants containing B-cell stimulatory factor 1, preparation, **116**, 374
suppressor
antigen-specific
detection and characterization, **116**, 316
establishment of hybridomas, **116**, 314
preparation, **116**, 312
separation, **116**, 313
producing soluble immune response suppressor, activation, **116**, 395
production of hybridomas, **116**, 306
second order, induction by T-suppressor factor, **116**, 423
Technetium-99m
in labeling of albumin microspheres, **112**, 43
Teleocidin
induction of interferon δ production by leukocytes, **119**, 98
Tempamine
rotational motion in erythrocytes, in analysis of intracellular viscosity, **127**, 240
Temperature
effect on
chemiluminescent response of polymorphonuclear leukocytes, **133**, 485
citrate transport in tumor mitochondria, **125**, 684
phagocytosis, **132**, 97
optimum, for solubilization of membrane proteins, **104**, 310
small volumes, measurement and control, **124**, 190
subzero, protein refolding kinetics, **131**, 179
variation, in protein crystallization, **104**, 377; **114**, 125
Tempol
rotational motion in erythrocytes, in analysis of intracellular viscosity, **127**, 240
Tension
surface, *see* Surface tension
Terasaki plates
in phagocytosis microassays, **132**, 214
Terbutryne
resistance in chloroplast mutants, selection for, **118**, 619
Terminal deoxynucleotidyltransferase, *see* DNA nucleotidylexotransferase
Termite soldiers
isolation and identification of diterpenes, **110**, 417
Terpenoids
chromatographic analysis, **111**, 176
Testis
mouse, purification of Leydig cells, **109**, 275
rat, purification of
calmodulin, **102**, 3
glutathione transferase, **113**, 508
Leydig cells, **109**, 275
Testosterone
delivery systems, analytical techniques, **112**, 392
Testosterone nicotinate–*N*-methyl iodide
site-specific conversion to testosterone, *in vitro* determination, **112**, 395
Testosterone propionate
in hormonal bursectomy, **108**, 5
Tetanus toxoid
stimulation of B cells before fusion, **121**, 149
Tetracyanoethylene
radiolabeled, in detection of lipid-conjugated dienes, **105**, 336
12-*O*-Tetradecanoylphorbol-13-acetate
effects on interferon γ production
in plateletpheresis residues, **119**, 50
in T-cell lymphoma or hybridoma cell lines, **119**, 52

5,6,7,8-Tetrahydromethanopterin
identification and quantitation, **122**, 418, 422
properties, **122**, 420
purification, **122**, 419
Tetrahydropteroylpolyglutamate
preparation and analysis, **122**, 336
Tetrahymena thermophila
axonemal microtubules
complexing with dynein, **134**, 316
repolymerization, **134**, 314
cultivation, **134**, 307
isolation of dynein, **134**, 309
Tetramethylammonium chloride
elimination of base composition effects on DNA, **124**, 317
Tetramethylammonium superoxide
in preparation of superoxide in nonaqueous solutions, **105**, 78
2,2,5,5-Tetramethylpyrrolidine-*N*-oxyl-3-carboxylic acid
rotational motion in erythrocytes, in analysis of intracellular viscosity, **127**, 240
Tetramethylrhodamine isothiocyanate
coupling to calmodulin, **102**, 6
Tetranitromethane
in assay for superoxide, **105**, 11
modification of F_1-ATPase tyrosine residues, **126**, 727
Tetraselmis subcordiformis
cultivation, **134**, 410
purification of striated flagellar root protein, **134**, 411
Tetrodotoxin
interactions with ion channels in neuroendocrine tissue, **103**, 402
Texas Red
–enkephalin conjugate
preparation, **103**, 220
properties, **103**, 221
as staining reagent in FACS, **108**, 208
Thaumatin
chicken gizzard smooth muscle
assay, **134**, 76
purification, **134**, 74
gel overlays, in assay for protein–protein interactions, **134**, 557
Thawing
hybridoma cells, **103**, 467
Theophylline
oral delivery system, structure, function, and release rates, **112**, 512
Thermistors
in transient electric birefringence system, **117**, 204
Thermochemiluminescence
detection apparatus, **133**, 551
1,2-dioxetanes labels
applications, **133**, 538
synthesis, **133**, 534
energy transfer, **133**, 544
principles, **133**, 531
related immunoassay, **133**, 548
Thiamin
HPLC, **122**, 15
SP-Sephadex chromatography, **122**, 20
Thiamin diphosphate
HPLC, **122**, 15
Thiamin-diphosphate kinase
bovine brain
assay, **122**, 24
properties, **122**, 29
purification, **122**, 26
Thiamin monophosphate
HPLC, **122**, 15
SP-Sephadex chromatography, **122**, 20
Thiamin pyrophosphate
^{35}S-labeled, preparation, **122**, 25
SP-Sephadex chromatography, **122**, 20
Thiamin triphosphate
biosynthetic enzyme, bovine brain
assay, **122**, 24
properties, **122**, 29
purification, **122**, 26
HPLC, **122**, 15
SP-Sephadex chromatography, **122**, 20
Thin-layer chromatography
carbocyclase products, **110**, 410
carotenoids, **105**, 157; **111**, 130
5α-cholesterol and oxidation products, in singlet oxygen determination, **105**, 45
cobalamins in biological materials, **123**, 8
dityrosine, **107**, 379, 384
dolichols and polyprenols, **111**, 202
free and conjugated ecdysteroids, **111**, 391
hexane extracts, in vitamin E determination, **105**, 141

4-hydroxybenzoate polyprenyltransferase products, **110**, 330
isodityrosine, **107**, 393
in juvenile hormone esterase assay, **111**, 493
long-chain acylcarnitine esters, **123**, 286
mevalonate phosphates and isopentenyl pyrophosphate, **111**, 248
phosphatidylinositol, **109**, 508
phosphoinositides, **124**, 433
phospholipids, **124**, 431
prostaglandins, **109**, 497
radioactive acylcarnitines, **123**, 273
Thin-layer electrophoresis
tyrosine sulfate, **107**, 219
Thiobarbituric acid
in detection of
lipid peroxidation products, **105**, 290
nonenzymatic protein glycosylation, **106**, 85
Thiolase, *see* Acetyl-CoA acetyltransferase
Thiol–protein disulfide oxidoreductase
physical properties, **113**, 543
physiological role, **113**, 543
Thiols
–disulfide exchange
differential effects on enzyme activities, **107**, 336
kinetic competence, **107**, 345
in redox control of enzyme activities, **107**, 330
–disulfide ratio, variation *in vivo*, **107**, 332
Thioltransferase
human placenta
assay, **113**, 521
properties, **113**, 523
purification, **113**, 522
5-Thio-2-nitrobenzoic acid
in assay of chloramine-derived Cl^+, **132**, 578
Thiophane ring
variations in biotin analogs, HPLC, **122**, 66
Thiophosphates
diagnostic tests, **107**, 29
Thioredoxin
in enzymatic reduction–oxidation of protein disulfides, **107**, 295
Thoracic duct
murine, collection of lymphocytes, **108**, 270
rat, as lymphocyte source for hybridoma production, **121**, 56
Threonine
reactivity with heavy atoms, **114**, 152
Thylakoid membranes
isolation of chlorophyll-binding proteins, **118**, 338
light-dependent pH gradient, in estimation of proton motive force, **127**, 549
subfractionation
detergent treatment and centrifugation method, **118**, 335
mechanical disruption and phase partition method, **118**, 326
Thymectomy
neonatal mouse and rat
effect on immune functions, **108**, 13
reconstitution by engraftment of thymic tissue, **108**, 10
surgical procedure, **108**, 11
Thymic hormones
biological effects, **116**, 216
chemical properties, **116**, 216
overview, **116**, 213
Thymidine
^{3}H-labeled, in assay for
lymphocyte suicide, **108**, 259
macrophage cytotoxicity, **132**, 465, 518, 563, 567
monocyte/macrophage cytotoxicity, **132**, 464
platelet-derived growth factor, **109**, 752
Thymidylate synthase
in chain length determination of tissue folate polyglutamate, **122**, 313
Thymocytes
allogeneic, culture-conditioned medium, preparation, **121**, 20
as feeder cells in hybridoma production, **121**, 298, 312
in proliferation assay for interleukin 1, **116**, 461, 468
proliferative response to interleukin 2, inhibition by interferon, **119**, 662
Thymol
and sulfuric acid, in staining of glycoproteins in gels, **104**, 447

Thymopoietin
  assays, **116**, 280
  isolation, **116**, 286
  properties, **116**, 290
Thymosin α_1
  assays, **116**, 244
  chemical characterization, **116**, 242
  deacetylated, in assay for transacetylase, **106**, 195
  isolation, **116** 256
  purification, **116**, 235
Thymosin α_{11}
  isolation, **116**, 256
Thymosin β_4
  biological properties, **116**, 253
  chemical characterization, **116**, 252
  isolation, **116**, 266, 268, 269
  properties, **116**, 269
  purification, **116**, 248
Thymosin β_9
  isolation, **116**, 266
  properties, **116**, 269
Thymosin β_{10}
  isolation, **116**, 268, 269
  properties, **116**, 269
Thymosin β_{11}
  isolation, **116**, 268
  properties, **116**, 269
Thymosin fraction 5
  assays, **116**, 230
  biological properties, **116**, 225
  chemical characterization, **116**, 224
  component polypeptides, nomenclature, **116**, 223
  isolation of
    thymosin α_1 and related peptides, **116**, 235, 237, 256
    thymosin β_4, **116**, 250
  purification, **116**, 220
Thymosin fraction 5A
  assays, **116**, 230
  biological properties, **116**, 225
  chemical characterization, **116**, 224
  isolation of
    polypeptide β_1, **116**, 240
    thymosin β_4, **116**, 249, 250
  purification, **116**, 220
Thymulin
  bioassay, **116**, 271
  cellular receptors for, **116**, 275
  characterization, **116**, 274
  effects on T-cell functions, **116**, 270
  origin, **116**, 276
  purification, **116**, 272
  serum carriers, **116**, 279
  structure–activity relationships, **116**, 276
  zinc binding, **116**, 274
Thymulin receptors
  detection on lymphoblastoid T cells, **116**, 275
Thymus
  bovine, purification of
    cytosolic proteins binding 1,25-dihydroxyvitamin D, **123**, 178
    glutaredoxin, **113**, 537
    poly(ADP-ribose) synthetase, **106**, 502
    thymopoietin, **116**, 286
    thymosin β_4, **116**, 266
    thymosin β_9, **116**, 266
    thymosin fraction 5 and 5A, **116**, 220
  endocrine role, **116**, 213
  murine
    engraftment techniques, **108**, 16
    immune functions, ablation effects, **108**, 13
    localization of thymulin, **116**, 278
    preparation of single-cell suspensions, **108**, 45, 75
    purification of thymosin β_4, **116**, 250
    removal methods, **108**, 43
  porcine, purification of thymosin α_1, **116**, 237
  rat
    engraftment techniques, **108**, 16
    immune functions, ablation effects, **108**, 13
    preparation of single-cell suspensions, **108**, 45
    purification of
      prothymosin α, **116**, 261
      Thy-1 antigens, **108**, 656
    removal methods, **108**, 43
Thyroglobulin
  in assay for thyroid peroxidase, **107**, 449
  enzymatic iodination and coupling, **107**, 476
  nonenzymatic iodination and coupling, **107**, 483

Thyroid gland
bovine, purification of iodotyrosine deiodinase, **107**, 492
stimulating antibody, assays
receptor modulation procedures, **109**, 677
thyroid stimulation procedures, **109**, 686
Thyroid hormones
enzymatic iodination and coupling, **107**, 476
nonenzymatic iodination and coupling, **107**, 483
phenoxy ether formation, mechanism, **107**, 476
Thyroid peroxidase
assay, **107**, 448
iodination and coupling reactions, **107**, 476
dissociation, **107**, 483
product identification, **107**, 485
properties, **107**, 456
solubilization and purification, **107**, 450
Thyroid-stimulating hormone
assembly
detection, **109**, 737
methods for, **109**, 747
disassembly
detection, **109**, 737
methods for, **109**, 739
Thyrotropin
antibody dilution analysis, **121**, 476
binding to receptor, in assay for thyroid-stimulating antibody, **109**, 677
–daunomycin conjugates, preparation, **112**, 267
deglycosylated
chromatographic purification, **109**, 729
preparation, **109**, 726, 733
immunochemiluminometric assay, **133**, 382
iodination and purification, **109**, 679
Thyrotropin-releasing hormone
synthesis, **103**, 26
Thyroxine
free, immunochemiluminometric assay, **133**, 384
and iodothyronine, separation, **107**, 425
Tissue cultures
brain explants, preparation, **103**, 357
dissociated, intracellular recording techniques, **103**, 111
hypothalamic explants
for analysis of peptide synthesis and secretion, **124**, 359
in serum-supplemented media, techniques for, **103**, 315, 318
sympathetic neurons, techniques for, **103**, 334
Tissues
freezing and sectioning, **102**, 126
Titration
antibodies, for interferon neutralization, **119**, 559, 568
DNase footprint, for analysis of protein–DNA interactions, **130**, 132
dual-inhibitor, in estimation of proton motive force, critical assessment, **127**, 553
fluorometric, *see* Fluorometric titration
potentiometric, cytochrome *b*-245, **132**, 389
TLC, *see* Thin-layer chromatography
Tobacco
antiviral substances
assay, **119**, 732
production, **119**, 731
properties, **119**, 733, 734
purification, **119**, 731, 734
calli and suspension cultures, establishment, **118**, 545
crown gall tumors
growth techniques, **110**, 341
purification of dimethylallylpyrophosphate:5′-AMP transferase, **110**, 341
gene transfer into
by protoplast transformation, **118**, 668
with Ti plasmid vector, **118**, 627
isolation of
chloroplast ribosomal proteins and rRNA, **118**, 207
chloroplast ribosomes, **118**, 203
chloroplasts, **118**, 272
protoplasts, **118**, 571, 673
tobacco etch virus, **118**, 692
tobacco mosaic virus, **118**, 690
plasmid-transformed
gene expression, analysis and verification, **118**, 637

selection and regeneration, **118**, 635,
675
protoplasts
effects of human interferon, assay,
119, 744
preparation, **119**, 745
purification of
antiviral protein, **119**, 739
oligoadenylate synthetase, **119**, 754
regeneration from protoplasts, **118**, 573
replication of
tobacco etch virus, **118**, 692
tobacco mosaic virus, **118**, 690
effects of (2′-5′)-oligoadenylates and
analogs, **119**, 759
seeds, aseptic germination, **118**, 540
Tobacco etch virus
propagation in tobacco, **118**, 692
properties, **118**, 691
purification techniques, **118**, 692, 693
Tobacco hornworm
purification of
ecdysone 3-epimerase, **111**, 438
ecdysone 20-monooxygenase, **111**, 457
3-hydroxy-3-methylglutaryl-CoA
reductase, **110**, 53
juvenile hormone carrier protein from
hemolymph, **111**, 484
Tobacco mosaic virus
infection with, in analysis of plant antiviral protein, **119**, 734
multiplication in tobacco, **118**, 690
assays, **119**, 748
effects of (2′-5′)-oligoadenylates and
analogs, measurement, **119**, 759
interferon effects, assay, **119**, 751
properties, **118**, 689
purification techniques, **118**, 690, 693
RNA translation in wheat germ system,
118, 154
α-Tocopherol, *see* Vitamin E
Toluene 2,4-diisocyanate
neuropeptide conjugation methods, **103**,
453
p-Toluenesulfonyl chloride
activation of agarose, **104**, 59
Tomato
gene transfer with Ti plasmid vector,
118, 627

plasmid-transformed
gene expression, analysis and verification, **118**, 637
selection and regeneration, **118**, 635
plastids, purification of
isopentenyl pyrophosphate isomerase,
110, 217
phytoene synthetase, **110**, 216
prenyltransferase, **110**, 217
purification of mitochondrial DNA, **118**,
439, 446
Tomato golden mosaic virus
propagation in plants, **118**, 703
Toxins, *see also specific toxins*
bacterial antiphagocytic, overview, **132**,
311
handling, safety considerations, **112**, 224
Toxoplasma gondii
killing by macrophages, assay, **132**, 630
maintenance, **132**, 629
Trace elements
as supplements in preparation of SFH
medium, **121**, 285
Transacetylase, *see* N-Acetyltransferase
Transcarboxylase synthetase
assay, **107**, 265
properties, **107**, 262, 274
purification, **107**, 271
reaction mechanism, **107**, 275
sequence homology with other synthetases, **107**, 264
Transcobalamin II
human
antiserum preparation, **123**, 40
purification, **123**, 38
radiolabeling, **123**, 42
solid-phase radioimmunoassay, **123**, 44
Transcription
ATP synthase subunit genes, mapping,
125, 245
chloroplast genes *in vitro*, **118**, 232, 253
infectious viral cDNA clones, **118**, 712
lipoprotein genes
regulation, **128**, 53
required sequences, **128**, 51
neuropeptide cDNA insert in vaccinia
virus vector, **124**, 303
–translation reactions, coupled, in chloroplast lysates, **118**, 270

Transdermal dosage forms
  Nitrodisc and Nitro-Dur, structure, function, and release rates, **112**, 515
  rate-controlled
    design, **112**, 451
    development, **112**, 449
    fabrication, **112**, 463
  Transderm-Scop, structure, function, and release rates, **112**, 513
Transferase
  affinity chromatography, **104**, 21, 22
  localization in gels, **104**, 424
Transferrin
  –chemiluminescent protein conjugates
    applications, **133**, 566
    immunoreactivity, **133**, 561
    incorporation assay, **133**, 560
    properties, **133**, 561
    synthesis, **133**, 560
  seminal plasma, chemiluminescence immunoassay, **133**, 561
  as supplement in preparation of SFH medium, **121**, 280
Transferrin receptors
  antibodies to
    coupling to ricin A chain, **112**, 241
    modification for toxin conjugation, **112**, 240
    –ricin A chain conjugates
      applications, **112**, 245
      cytotoxicity, **112**, 243
      storage and stability, **112**, 242
Transformation
  B cells with Epstein–Barr virus
    methodology, **121**, 145, 170
    overview, **121**, 141
  NIH3T3 cells for enhanced sensitivity to human interferons, **119**, 599
  plant protoplasts
    confirmation, **118**, 675
    in direct gene transfer to plants, **118**, 668
    with foreign DNA, **118**, 581
  plants with Ti plasmid vectors, **118**, 634
Transforming growth factor
  interactions with interferons in fibroblasts, assays, **119**, 658
Transglutaminase, *see* Protein-glutamine γ-glutamyltransferase
Transglycosidation
  with immobilized NADase, **122**, 180
Transhydrogenase, *see* NAD(P)$^+$ transhydrogenase
Transient electric birefringence
  in determination of biopolymer structure, **117**, 198
Transition metals
  magnetic circular dichroism in proteins, **130**, 278
Translation
  chloroplast genes, initiation site determination, **118**, 309
  by isolated chloroplasts, **118**, 286
    monitoring, **118**, 292
    product analysis, **118**, 293
  LHRH precursor mRNA in cell-free system, **124**, 322
  transcription-coupled, in chloroplast lysates, **118**, 270
  T-suppressor factor mRNA in amphibian oocytes, **116**, 324
  viral RNAs in wheat germ system, **118**, 154
Translation function
  in protein crystallography
    applications, **115**, 69
    calculations, **115**, 66
    computer programs, **115**, 67
Transplantation
  normal and neoplastic tissues in nude mouse, **108**, 352
Transposons
  in isolation and characterization of bacterial sugar transport mutants, **125**, 157
  Tn*5*
    marker exchange in *Rhizobium*, **118**, 522
    mutagenesis techniques, **118**, 521
    in preparation of *Rhizobium* H_2 uptake negative mutants, **118**, 529
    random mutagenesis of gram-negative bacteria, **118**, 644, 654
  Tn*10*, in creation of deletions in *Rhizobium*, **118**, 653

Trehalose
  interaction with microsomal vesicle membranes
    differential scanning calorimetry, **127**, 698
    infrared spectroscopy, **127**, 698
Triacylglycerol
  lipoprotein, mass determination, **129**, 557
Trialkyl glyceryl ether
  biological applications, **129**, 842
  synthesis, **129**, 835
Triazine dyes
  on agarose gel, concentration determination, **104**, 105
  applications in biotechnology, **104**, 112
  coupling to support matrices, **104**, 102
  immobilized, in affinity chromatography, **104**, 109
  –polyethylene glycol complexes, preparation, **104**, 357
  purification and properties, **104**, 101
  structure and chemistry, **104**, 98
Tricarboxylate
  carrier protein, mitochondrial
    incorporation into liposomes, **125**, 694
    isolation, **125**, 692
    reconstituted, properties, **125**, 695
Trichothecium roseum
  purification of cyclases, **110**, 402
Triethylamine
  in activation of affinity resins, **104**, 6
Triethyl phosphate
  gas chromatography, **126**, 626
  mass spectrometry, **126**, 629
  synthesis, **126**, 619, 625
2,2,2-Trifluoroethanesulfonyl chloride
  activation of
    agarose and other chromatographic supports, **104**, 58
    polyethylene glycol, **104**, 59
Trifluoromethanesulfonic acid
  in deglycosylation of glycoprotein hormones, **109**, 733
Trifluorothymidine
  –asialofetuin conjugates, preparation and applications, **112**, 302
Triglyceride
  enzymatic quantification, **129**, 113
  lipoprotein metabolism, analytical techniques, **129**, 408, 418
  transfer assay for plasma lipid transfer proteins I and II, **129**, 812
Triiodothyronine
  and thyroxine, separation, **107**, 425
ε-*N*-Trimethyl-L-lysine
  chromatographic properties, **123**, 294
  synthesis, **123**, 292
Trimethyltempamine
  rotational motion in erythrocytes, in analysis of intracellular viscosity, **127**, 240
2,4,6-Trinitrobenzene sulfonate
  in identification of chloroplast phosphate transport protein, **125**, 719
  in labeling of chloroplast phosphate transport protein, **125**, 718
Triornithine
  in selection of bacterial mutants deficient in oligopeptide permease, **125**, 367
Triose phosphate
  transport protein in chloroplast envelope
    identification, **125**, 717
    inhibitory effect of amino acid reagents, measurement, **125**, 718
    isolation, **125**, 723
    labeling with amino acid reagents, **125**, 718
    properties, **125**, 708
    reconstitution, **125**, 724
Tripeptide permease
  bacterial, deficient mutants, isolation, **125**, 368
Tripneustes gratilla, *see* Sea urchin
Tritium
  antibody radiolabeling with, **121**, 670
  –hydrogen exchange in proteins
    analytical techniques, **131**, 459, 514
    applications, **131**, 502
    mechanism, **131**, 471
    methodology, **131**, 457, 511
    multiple-site, data analyses, **131**, 480
  labeling of γ-carboxyglutamic acid in proteins, **107**, 545
  as label in radioimmunoassay for cGMP, **109**, 827
  in quantification of poly(ADP-ribose) *in vivo*, **106**, 495

Triton N-101
in purification of cytochrome *b*-245, **132**, 384
Triton X-100
mixed micelles, in protein kinase C assay, **124**, 353
in preparation of cytochrome-*c* oxidase two-dimensional crystals, **126**, 22
solubilization of
opioid receptors, **124**, 182
thyroid peroxidase, **107**, 453
in subfractionation of thylakoid membranes, **118**, 335
Trityrosine
and dityrosine, halogenated, separation, **107**, 427
Troponin C
interaction with phosphorylase kinase, **102**, 225
True Blue
as retrograde tracer for neuropeptide-specific pathways, **103**, 664
Trypan blue
in neutrophil viability assessment, **132**, 96, 135
in phagocytic assay by fluorescence quenching, **132**, 201
Trypsin
–concanavalin A conjugates
activity assays, **112**, 250
characterization, **112**, 253
preparation, **112**, 251
purification, **112**, 252
coupling to sulfonated support, **104**, 61
crystals, neutron structure, determination, **114**, 545
digestion of
chloroplast 32-kDa membrane protein, **118**, 394
IgG and subclasses, **116**, 18; **121**, 664
tubulin, **134**, 181
in isolation of anterior pituitary cells, **109**, 293
solvent structure analysis with D_2O–H_2O neutron difference maps, **127**, 338
Trypsin inhibitor
coupling to sulfonated supports, **104**, 60, 61
delivery systems, polymeric and copolymeric
kinetic analysis, **112**, 414
preparation, **112**, 400
Tryptophan
ADP–ATP carrier protein, as intrinsic fluorescent probes, **125**, 648
in proteins
fluorescence studies, **131**, 524, 533
torsional motion, **131**, 296
radiolabeled, in assay for indoleamine 2,3-dioxygenase, **105**, 63
reactivity with heavy atoms, **114**, 151
Tryptophan dimethylallyltransferase
Claviceps SD58
assays, **110**, 336
properties, **110**, 338
purification, **110**, 336
Tryptophan synthase
Escherichia coli, α subunit folding and stability, genetic analysis, **131**, 264
Tubericidin
analog of (2′-5′)-oligoadenylic acid, synthesis, **119**, 526
Tubulin
antibodies
production, **134**, 455
specificity, **134**, 458
bovine brain, separation from microtubule-associated proteins, **134**, 94
chromatography on hydroxyapatite columns, **117**, 379
content in erythrocyte marginal bands, **134**, 241
detyrosinolated
preparation, **106**, 231, 233
properties, **106**, 237
proportions, determination, **106**, 232
fluorescent analogs
characterization, **134**, 524, 525
preparation, **134**, 522, 526
porcine brain
assays, **134**, 180
limited proteolysis, **134**, 181
sea urchin egg, isolation, **134**, 128
self-association patterns in velocity sedimentation, **130**, 4
tyrosinolated
preparation, **106**, 233

properties, **106**, 237
proportions, determination, **106**, 232
Tubulin-tyrosine ligase
applications, **106**, 231
assays, **106**, 225; **134**, 171
properties, **106**, 231
purification from porcine brain, **106**, 227; **134**, 173, 176
Tuftsin
antiserum preparation, **132**, 323
deficiency, effect on opsonization, **132**, 316
overview, **132**, 318
radioactive, synthesis, **132**, 321
radioimmunoassay, **132**, 325
role in opsonization, **132**, 304
Tumor cells, *see also* Cell lines
adenocarcinoma of cecum, in mouse, effects of interferon treatment, **119**, 650
adenoma, tissue dispersion, **103**, 291
ascites, pristane effects on formation, **121**, 375
CCRF-CEM, cytotoxic effects of chloramines, **132**, 592
colorectal carcinoma, antibody-dependent lysis by macrophages, assay, **132**, 567
Ehrlich ascites
extract preparation, **119**, 518
interferon induction of polyamine-dependent protein kinase, **107**, 160
purification of
$(2'\text{-}5')(A)_n$ synthetase, **119**, 518
RNase L, **119**, 493
epidermoid carcinoma, cultured, purification of low-density lipoprotein receptors, **109**, 409, 413
fibrosarcoma, lysis by macrophages, assay, **132**, 563
hepatoma
cocultivation with macrophages, **132**, 511
cytotoxic effect of macrophages, assays, **132**, 516
insulin receptor phosphorylation, **109**, 613, 616
mitochondrial citrate transport, measurement, **125**, 671
separation from macrophages, **132**, 514
immunocytoma
nonsecreting, fusion with spleen cells, **121**, 237
as source of monoclonal IgE, **116**, 77
leukemic
extraction of corrinoids, **123**, 16
interferon effects on growth, **119**, 632
purification of folate transport proteins, **122**, 267
Leydig, isolation from rodents, **109**, 280
lymphoblastic leukemia
cocultivation with macrophages, **132**, 511
cytotoxic effect of macrophages, assays, **132**, 516
separation from macrophages, **132**, 514
lymphoblastoid, centrosomes
assay, **134**, 263
isolation, **134**, 267
lymphoid, fluorescence-activated cell sorting, **108**, 235
lymphoma
antibody-dependent binding and lysis by macrophages, assay, **132**, 565
comparative feeder cell effects, **121**, 298
interferon-mediated cytolysis, microassay, **119**, 574
production of interferon γ, **119**, 53
in proliferation assay for interleukin 3, **116**, 546
purification of interleukin 2, **116**, 531
mammary
effect of chemoimmunoliposomes, **121**, 827
human, IgG-based localization in athymic mice, **121**, 808, 811
viability assay, **132**, 488
mastocytoma, binding and lysis by macrophages, assay, **132**, 559
melanoma, human and murine
cultivation, **119**, 614
effects of interferon, measurement, **119**, 617
melanin assays, **119**, 614
Moloney sarcoma virus-induced, antagonistic effect of platelet-derived

growth factor and murine interferon, **119**, 664
mucoid carcinoma of breast, in mouse, effects of interferon treatment, **119**, 650
myeloid leukemic, histocompatibility antigen expression, interferon-induced changes, **119**, 688
myeloma
cell line selection for hybridoma production, **121**, 312
cloning, **121**, 315
comparative feeder cell effects, **121**, 298
cultivation in SFH medium, **121**, 286
fusion with
chronic lymphocytic leukemia cells, **121**, 114
lymphocytes from various sources, **121**, 57
spleen cells, **121**, 12, 31, 46, 62, 241, 315
fusion protocol for preparation of
antibodies to weak immunogens, **121**, 185
antigen-specific hybridomas, **121**, 197
harvesting for hybridoma production, **121**, 30
maintenance in culture, **121**, 74
NS-1 and derivatives, cultivation in KSLM medium, **121**, 268
overview of human cell lines, **121**, 112
parental lines in production of reactive monoclonal antibodies, **103**, 462
preparation for fusion, **121**, 61
screening with culture-well solid-phase radioimmunoassay, **121**, 448
myelomonocytic leukemic, purification of mast cell growth factor, **116**, 557, 563
neuroblastoma, centrosomes
assay, **134**, 263
isolation, **134**, 264
neuroblastoma × glioma hybrid
culture and handling, **109**, 321
generation and characteristics, **109**, 317
(neuroblastoma × glioma) × neuroblastoma hybrid–hybrid
culture and handling, **109**, 321
generation, **109**, 321
hormonal regulation of
adenylate cyclase in homogenates, testing procedures, **109**, 333
guanidinium ion uptake, testing procedures, **109**, 338
intracellular cAMP levels, testing procedures, **109**, 327
intracellular cGMP levels, testing procedures, **109**, 337
morphological differentiation, **109**, 325
osteogenic sarcoma, in cytoreceptor assay for 1,25-dihydroxyvitamin D, **123**, 191
pituitary
cytosolic Ca^{2+} changes, monitoring, **124**, 129
membrane potential changes, monitoring, **124**, 129
phospholipid turnover, measurement, **124**, 424
plasmacytoma
as human hybridoma fusion partners, **121**, 126
morphology, **121**, 123
non-Ig-secreting, selection, **121**, 136
preparation of ascites fluid, **116**, 127
purification of monoclonal Igs, **116**, 128
sources, **116**, 128
stability, **116**, 128
ultrastructural analysis, **121**, 125
primary, from cancer patients
collection and cultivation, **119**, 638
colony growth, evaluation, **119**, 639
interferon effects, testing, **119**, 640
prostatic
adenine nucleotides
assay, **133**, 36
extraction, **133**, 32
cultivation, **133**, 29
QT35, expression of human interferon β with Rous sarcoma viral genome, **119**, 385
recognition by lymphocytes, assay, **108**, 60
superoxide dismutase assay, **105**, 457
Swarm chondrosarcoma, inhibition of *in vivo* growth by 2,5-3′dA_3 core, **119**, 671

transplantation and growth in nude mouse, **108**, 352
tuberculosis, intracellular viscosity measurement, **127**, 249
Tumor necrosis factor
amino acid sequence homology with lymphotoxin, **116**, 456
bioassay, **116**, 450
characterization, **116**, 453
purification, **116**, 451
supernatants with, production, **116**, 449
Tumors
crown gall, tobacco
growth techniques, **110**, 341
purification of dimethylallylpyrophosphate:5′-AMP transferase, **110**, 341
Tunicamycin
inhibition of glucagon receptor induction in transformed MDCK cells, **109**, 359
Turbidity
in measurement of protein self-assembly, **130**, 49
Turnip
introduction and expression of cloned viral DNA, **118**, 660, 663
propagation of cauliflower mosaic virus, **118**, 697
Turnip yellow mosaic virus
RNA translation in wheat germ system, **118**, 154
Tyramine
–cellobiose, in assay for tissue sites of lipoprotein degradation, **129**, 622
Tyrosinase
Neurospora, S^{β}-(2-histidyl)cysteine properties and assay, **106**, 355
Tyrosine
binding to hemoproteins, resonance Raman spectroscopy, **130**, 405
F_1-ATPase
chemical modification, **126**, 724
radiolabeling, **126**, 746, 749
halogenated derivatives
effects of alkaline hydrolysis, **107**, 431
enzymatic release from cuticle protein, **107**, 429
identification, **107**, 431
isolation from invertebrate scleroproteins, **107**, 414
separation, **107**, 423
syntheses, **107**, 420, 432
incorporation in tubulin, **106**, 223
neuronal
assay, **103**, 508
concentration determination, **103**, 508
specific activity, determination, **103**, 508
in proteins
aromatic ring flips and rotational motions, NMR spectroscopy, **131**, 317
fluorescence studies, **131**, 533
torsional motion, **131**, 296
release from tubulin, **106**, 231, 233
reactivity with heavy atoms, **114**, 151
Tyrosine ligase
tubulin-specific, *see* Tubulin-tyrosine ligase
Tyrosine protein kinase, *see* Protein-tyrosine kinase
Tyrosine *O*-sulfate
chemical synthesis, **107**, 202
detection in proteins, **107**, 203
properties, **107**, 203

U

Ubiquinol
in assay for cytochrome *d* and *o* terminal oxidase complexes, **126**, 88, 106
in biological samples
chromatographic detection, **105**, 150
extraction, **105**, 149
preparation, **105**, 148
formation in *Vibrio alginolyticus* membranes, assay, **125**, 529
Ubiquinol–cytochrome-*c* oxidoreductase, *see* Ubiquinol–cytochrome-*c* reductase
Ubiquinol–cytochrome-*c* reductase, *see also* Cytochrome bc_1
bovine cardiac mitochondria
isolation and properties, **126**, 295
–photosynthetic reaction center hybrid system
construction, **126**, 296, 303
electron-transfer reactions, **126**, 300

redox poising, **126**, 298
solubilization, **126**, 300
Neurospora crassa
cleavage products
isolation, **126**, 203
structural properties, **126**, 206
electron-transfer activities, **126**, 207
isolation, **126**, 202
membrane crystals
formation, influencing factors, **126**, 193
preparation, **126**, 192
structural analysis, **126**, 194, 197
subunit topography, **126**, 197
reconstitution, **126**, 206
structural properties, **126**, 206
Ubiquinol oxidase
Paracoccus denitrificans
assays, **126**, 306
properties, **126**, 313
purification, **126**, 307
Ubiquinone-50
NADPH oxidase-associated
characterization, **132**, 364
preparation, **132**, 358
Ubiquinones
in biological samples
chromatographic detection, **105**, 150
extraction, **105**, 149
preparation, **105**, 148
partition coefficients, HPLC determination, **125**, 125
Ubiquitin
active form, detection of terminal glycine residues, **106**, 258
assays, **116**, 280
and ATP-dependent proteolytic enzymes of reticulocytes, **106**, 255
–histone 2A conjugate
branched structure, **106**, 248
chromosomal localization, **106**, 253
deconjugation, **106**, 260
functional role, **106**, 261
isolation from bovine chromatin, **106**, 241
primary structure, **106**, 241
properties, **106**, 239
terminal amino acid sequences, **106**, 245
isolation, **116**, 288
properties, **116**, 290
structure, **106**, 238
Ubiquitination
proteins, **106**, 238
UDP-*N*-acetylglucosamine–lysosomal enzyme *N*-acetylglucosaminephosphotransferase
assays, **107**, 163
rat liver
properties, **107**, 170
purification, **107**, 168
Ultracentrifugation
air-turbine, macromolecule sedimentation equilibrium
applications, **117**, 10
description and methodology, **117**, 4, 24
partitioning techniques, **117**, 25
theoretical considerations, **117**, 7
analytical
application to lipoproteins, **129**, 29, 37
calibration, **129**, 32
cell construction and alignment, **129**, 27
apolipoprotein E-containing lipoproteins, **129**, 146
lipoproteins
density gradient method, **128**, 170
sequential flotation method, **128**, 155; **129**, 70
vertical spin density gradient method, **128**, 181
macromolecules, by difference sedimentation velocity technique
applications, **117**, 38
practical aspects, **117**, 28
sources of variation, **117**, 35
nascent lipoprotein particles, **129**, 291
protein complexes in detergent solution, **117**, 41
zonal, *see* Zonal ultracentrifugation
Ultrafiltration
chloramines, **132**, 584
in concentration of human immune interferon, **119**, 69
in detergent removal from protein-containing micelles, **104**, 327
interleukin 1, **116**, 471
in ligand binding studies, **117**, 354

Millipore-membrane, in purification of
  interferon γ, **119**, 85
in preparation of high-purity water, **104**,
  395
thymopoietin, **116**, 286
thymosin fraction 5 and 5A, **116**, 222
Ultraviolet irradiation
in preparation of high-purity water, **104**,
  399
Uncoupling protein
from brown adipose tissue
  assays, **126**, 495
  hydrodynamic characterization, **126**,
    497
  nucleotide binding
    general properties, **126**, 499
    measurement techniques, **126**, 501
  purification, **126**, 491
mitochondrial, detection with fluorescent
  nucleotide analogs, **125**, 627
reconstitution in phospholipid vesicles,
  127, 772
Undecaprenylpyrophosphate synthetase
Lactobacillus plantarum
  assays, **110**, 120, 282
  photoinactivation, **110**, 122
  photolabile pyrophosphate analog
    activity as substrates and inhibi-
      tors, **110**, 121
  photolysis, **110**, 120
  properties, **110**, 294
  purification, **110**, 120, 289
  separation and identification in mix-
    tures of prenyltransferases, **110**,
    287
Undecylhydroxydioxobenzothiazole
inhibition of cytochrome *bc*₁, **126**, 264
Undecylhydroxynaphthoquinone
inhibition of cytochrome *bc*₁, **126**,
  265
Unspecific monooxygenase
inducible, FACS analysis, **108**, 239
Uracil
³H-labeled, uptake by *Toxoplasma* in
  macrophages, assay, **132**, 636
Uranyl salts
binding to proteins, **114**, 152
Urea
as buffer for determination of protein
  partial specific volume, **117**, 53

protein denaturation curves
  analysis, **131**, 272
  determination, **131**, 267
  stock solutions, preparation, **131**, 269
Urease
controlled polymer system, preparation
  and characterization, **112**, 493
Ureides
N-Mannich bases
  applications as prodrugs, **112**, 351
  hydrolysis, **112**, 348
  preparation, **112**, 348
Ureido ring
variations in biotin analogs, HPLC, **122**,
  66
Uric acid
determination, **105**, 164
formation, **105**, 163
function in biological systems, **105**, 162
interference in dichlorofluorescein assay
  for lipid hydroperoxides, **105**, 355
Uridine 5'-triphosphate
biotinylation, **109**, 439
Uridylyl-removing enzyme
Escherichia coli
  assays, **113**, 231
  properties, **113**, 234
  purification, **113**, 232
Uridylyltransferase
assay, **107**, 198
Urinary tract
human
  bacteriurea detection, **133**, 23
  bioluminescence screening, **133**, 24
  sources of infection, **133**, 22
Urine
excretion of 4-carboxyglutamate, **107**,
  537
fluorometric titration of riboflavin, **122**,
  234
human, purification of
  colony-stimulating factor 1, **116**,
    574
  monoclonal Ig light chains, **116**, 102
  short-chain acylcarnitines, **123**, 265,
    271
mevalonate assay, **110**, 63
pterin detection by liquid chromatogra-
  phy/electrochemistry system, **122**,
  301

Uronic acids
cell wall, colorimetric assays, **118**, 26
Uroporphyrin I
in porphobilinogen synthase-coupled enzyme assay, **123**, 339
Uroporphyrinogen
and coproporphyrin isomers, simultaneous separation, **123**, 387
isomers, HPLC separation, **123**, 385
Uroporphyrinogen decarboxylase
chicken erythrocyte
assay, **123**, 416
properties, **123**, 420
purification, **123**, 417
Uterus
microsealed drug delivery system, fabrication, **112**, 463
UTP–glucose-1-phosphate uridylyltransferase
localization in gels by simultaneous capture technique, **104**, 437
UTP:P_{II} uridylyltransferase
Escherichia coli
assays, **113**, 231
properties, **113**, 234
purification, **113**, 232

V

Vaccinia virus
recombination plasmid vectors
construction, **124**, 298
detection, **124**, 300
expression analysis, **124**, 305
purification, **124**, 301
structural analysis, **124**, 303
transcription, **124**, 303
Vagina
microsealed drug delivery system, fabrication, **112**, 463
Valine
dipeptides with, in selection of bacterial mutants deficient in dipeptide permease, **125**, 370
Valporic acid
rate-controlled delivery by osmotic system, **112**, 474
Vanadate
application as probe for cytoskeletal mechanochemistry, **134**, 486
effects on cells, **134**, 484
inhibition of
actomyosin ATPase, **134**, 485
dynein ATPase, **134**, 485
Vapor diffusion
in crystallization of
bacterial porin, **125**, 320
proteins, **104**, 375
Vasoactive intestinal peptide
regulation of cAMP levels in primary glial cultures, analytical techniques, **109**, 344
Vasopressin
arginine, *see* Arginine vasopressin
biosynthesis, *in vivo* studies, **103**, 512
effects on
Ca^{2+} fluxes in perfused liver and isolated hepatocytes, **109**, 553
phosphorylase in hepatocytes, **109**, 557
hypothalamic, measurement in tissue culture, **124**, 362
responsiveness of MDCK cells
evaluation, **109**, 363
measurement, **109**, 361
Veillonella alcalescens
purification of methylmalonyl-CoA decarboxylase, **125**, 540
Velocity sedimentation
in analysis of ligand-induced protein self-association
advantages and limitations, **117**, 460
at multiple equilibria, **117**, 490
qualitative analysis, **117**, 467
quantitative analysis, **117**, 485
in cell separation
applications, **108**, 80
limitations, **108**, 73
methodology, **108**, 67
theoretical description, **108**, 66
for determination of asymptotic boundary shapes, **130**, 12
in measurement of protein hydration, **117**, 247
self-associations in, distinguishing criteria, **130**, 3
in separation of
anterior pituitary cells, **103**, 260
Celsep technique, **103**, 266
elutriation technique, **103**, 271

unit gravity method, **103**, 260
rosettes, **108**, 62
(±)-Verapamil
^{3}H-labeled, labeling of calcium channels with, **109**, 543
Veratridine
interactions with ion channels in neuroendocrine tissue, **103**, 402
Veratrine, *see* Cevadine
Very-low-density lipoproteins
apolipoproteins B and E
characterization, **129**, 340
protease processing, **129**, 339
delipidation, **128**, 345, 349, 351
interactions with low-density lipoprotein receptor, **129**, 331
isolation, **129**, 324
metabolic conversions, analysis with specific radioactivity data, **129**, 385
metabolism, enzymatic assay in whole plasma, **129**, 704
subfractionation, **129**, 328
synthesis, analysis in cultured hepatocytes, **129**, 278
Vesicles
asolectin, reconstitution of mitochondrial F_0F_1-ATPase, **126**, 474
envelope, bacterial
chloride transport assays, **125**, 515, 518
halorhodopsin content estimation, **125**, 519
light-generated chloride gradients, **125**, 516
membrane potential, measurement, **125**, 47, 511, 514
pH changes, measurement, **125**, 513, 514
preparation, **125**, 511, 513
reconstitution of proton-driven flagellar motor, **125**, 571
surface potential, measurement, **125**, 52
volume changes, **125**, 515
inside-out, preparation from human erythrocytes, **102**, 49
lipid, preparation with alkyl glycoside detergents, **125**, 34
membrane, from *Escherichia coli*
Ca^{2+} uptake assay, **125**, 335
glutamine transport, assays and properties, **125**, 306
preparation, **125**, 305, 334
nonlysosomal non-Golgi, isolation by Percoll density gradient centrifugation, **109**, 227
phospholipid, *see* Phospholipid vesicles
right-side out subcellular, measurement of sugar transport, **125**, 399
sarcoplasmic reticulum
phospholamban extraction, **102**, 270
phosphorylation, **102**, 263
physicochemical properties, **102**, 263
preparation, **102**, 262
submitochondrial
preparation, **126**, 429
reconstitution of F_0F_1-ATPase, **126**, 432
Vesicular stomatitis virus
in antiviral assay of interferons, **119**, 534
defective interfering particles, induction of chicken interferon, **119**, 119
induction of guinea pig interferon, **119**, 127
ts mutants, induction of chicken interferon, **119**, 119
Vibration
in dissociated tissue culture electrophysiological recording, control, **103**, 117
molecular, coupling to proton tunneling in biological systems, **127**, 83
Vibrio
V. alginolyticus
growth in culture, **125**, 521
and Na^+ pump-defective mutants
generation of membrane potential and pH gradients, analysis, **125**, 524
H^+ fluxes induced by oxygen pulse, measurement, **125**, 525
isolated membranes, preparation and respiratory activities, **125**, 527
isolation, **125**, 521
Na^+ extrusion, assay, **125**, 523
V. harveyi
aldehyde-requiring mutant
induction, **133**, 122

luciferase purification by HPLC, **133**, 123
bioluminescence-related acyltransferases
assay, **133**, 183
purification, **133**, 185
properties, **133**, 187
V. succinogenes, purification of asparaginase, **113**, 610
Video light microscopy
single microtubules, **134**, 561
Video tracking device
for assay of myosin movement, **134**, 538
Villin
intestinal microvillus, purification, **134**, 26, 31
Vinca rosea, *see* Periwinkle
Vinculin
chicken gizzard smooth muscle
assay, **134**, 76
purification, **134**, 73, 510
fluorescently labeled
assay, **134**, 511
microinjection, **134**, 513
preparation, **134**, 510
gel overlays, in assay for protein–protein interactions, **134**, 557
Vinyl acetate
–ethylene copolymers, as macromolecule delivery systems
kinetic analysis, **112**, 414
preparation, **112**, 400
Vinylpyrenes
as chemiluminescent probes of singlet oxygen
emission spectra, **133**, 577
in hepatic microsomal extracts, **133**, 580
properties, **133**, 572
Violaxanthin cycle
in isolated chloroplasts, **110**, 308
spectrophotometric analyses, **110**, 306
Violaxanthin deepoxidase
lettuce chloroplast
assay, **110**, 310
properties, **110**, 311
purification, **110**, 311
Viral particles
interferon-inducing, titer calculations, **119**, 113
Viruses, *see also specific viruses*
as inducers of chicken interferon, **119**, 118
replication, inhibitory potency of plant IVR, assay, **119**, 732
Viscoelastometry
in molecular weight determination for elongated molecules, **117**, 94
Viscometry
falling-ball, *see* Falling-ball viscometry
Viscosity
proteins in solutions, estimation from small-angle X-ray scattering data, **117**, 170
solvent, effect on fluorescence polarization at high pressure, **130**, 515
Vitamin A
binding protein
interstitial, purification from eye, **123**, 102
in serum, ELISA, **123**, 85
geometric isomers
extraction from tissues, **123**, 76
handling and storage, **123**, 76
purification, **123**, 76
radiolabeled, purification, **123**, 81
separation, **123**, 78
separation and quantitation by HPLC, **123**, 68
and α-tocopherol, in plasma or serum, simultaneous determination by HPLC, **123**, 215
Vitamin B_6
metabolic forms, cation-exchange HPLC, **122**, 102
metabolizing enzymes, assays, **122**, 110
monoclonal antibodies
applications, **122**, 124
preparation, **122**, 122
Vitamin B_{12}
binding protein
assay in human neutrophils, **132**, 273
in assessment of phagocytic function, **132,** 124, 167
purification, **123**, 28
in biological material, assay, **123**, 3
extraction from cells, **123**, 16
hydrated, X-ray and neutron diffraction analysis, **127**, 169

–intrinsic factor receptor, canine ileal
assay, **123**, 23
purification, **123**, 24
resolution, **123**, 17
synthesis, **123**, 16
Vitamin D
induced Ca^{2+}-binding proteins, ELISA, **123**, 154
and metabolites, isolation and identification, **123**, 127
Vitamin D_2
quantitation in human milk, **123**, 167
Vitamin D_3
quantitation in human milk, **123**, 167
Vitamin E
extraction from
blood, **105**, 139
tissues, **105**, 136, 140
HPLC determination, **105**, 134
interference in dichlorofluorescein assay for lipid hydroperoxides, **105**, 355
and retinol, simultaneous determination in plasma or serum by HPLC, **123**, 215
spectrofluorometric assay, **105**, 144
spectrophotometric assay, **105**, 142
Vitamin K
assay in tissues by HPLC with UV detection, **123**, 235
coumarin antagonists in blood, assay by HPLC, **123**, 223
dependent plasma proteins
functional properties, **107**, 511
isolation, **107**, 507
structure, **107**, 510
Vitamin K_1
assay in tissues by HPLC with UV detection, **123**, 242
Vitamin K_2
bacterial, HPLC analysis, **123**, 251
partition coefficients, determination by HPLC, **125**, 125
Vitrification
definition, **134**, 613
microtubules, **134**, 627
Volatilization
chloramines, **132**, 583
Voltage clamp
in dissociated tissue culture electrophysiological recording, **103**, 128
Voltammetry
catecholamines *in situ*, **103**, 479
cyclic, *see* Cyclic voltammetry
Volume
halorhodopsin-containing envelope vesicles, chloride-induced changes, **125**, 515
molecular, calculations for known structures, **115**, 444
partial specific, *see* Partial specific volume
Von Willebrand factor
fragment, rocket inhibition assay, **121**, 708

W

Waldenström diseases
IgM purification from serum of patients, **116**, 29
Warfarin
in plasma, HPLC with UV detection, **123**, 225
Water
assay for ^{18}O, **126**, 637
–biomolecule interactions, calorimetric analysis, **127**, 141
cellular
in *Artemia* cysts
control, **127**, 234
properties, **127**, 236
viscosity analysis with spin labels, **127**, 239, 360
common impurities, **104**, 391
conformational forces between membrane and surfaces, direct measurements, **127**, 353
–DNA solutions, structure, computer simulation, **127**, 128
–D_2O neutron difference maps, in solvent structure analysis of proteins, **127**, 329
freezing and thawing rates in plant tissue, **127**, 764
frozen
bacterial nucleation in plant leaves
characteristics, **127**, 731
quantitation, **127**, 731, 732
tube nucleation test, **127**, 733

interactions with antifreeze proteins in cold-water fish, **127**, 297, 301
intracellular damage evaluation by freeze-fracture methods, **127**, 704
phase studies, **127**, 303
heavy, structure, neutron difference analysis, **127**, 325
high-purity, preparation, **104**, 391
interactions with antifreeze peptides in cold-water fish, **127**, 302
intracellular, measurement
in mycoplasmas, **125**, 264
in rickettsiae, **125**, 256
liquid
interactions in biological systems, NMR study, **127**, 151
low-frequency Raman scattering analysis, **127**, 91
in nonpolar solvents, vibrational spectroscopy, **127**, 106
in partially frozen plant tissue, NMR analysis, **127**, 762
structure
characterization, **127**, 9
neutron diffraction analysis, **127**, 162
X-ray diffraction analysis, **127**, 5, 162
membrane
quantitation by neutron diffraction and X-ray techniques, **127**, 217
role in structure and function, analysis by deuteration, **127**, 662
–membrane interactions
differential scanning calorimetry, **127**, 698
infrared spectroscopy, **127**, 698
mobility in proteins, analysis by Mössbauer spectroscopy and microwave absorption, **127**, 196
penetration depth into lipid bilayers
capacitance measurements, **127**, 514
X-ray diffraction measurements, **127**, 512
permeability of lipid bilayer membranes, effect of phase transitions, **127**, 482
protein content
dynamic effects, **127**, 207
measurement, **127**, 287
pure, structural analysis
method of difference, **127**, 323
neutron and X-ray diffraction, **127**, 320
solutions, structural analysis
method of difference, **127**, 323
neutron and X-ray diffraction, **127**, 320
supercooling in plant tissues, **127**, 768
treatment systems
description, **104**, 392
selection considerations, **104**, 401
Water channels
in erythrocytes
chemical modifications, **127**, 607
diffusional permeability measurements, **127**, 599, 601
osmotic permeability measurements, **127**, 603
volume change measurements, **127**, 605
Western blotting
for detection of apolipoprotein glycosylation, **128**, 360
proteins onto paper, **104**, 455
Wheat germ
cell-free extracts
mRNA translation, **118**, 129
preparation, **118**, 129, 156
storage, **118**, 157
synthesis of high-molecular-weight proteins, **118**, 154
translation assay, **118**, 157
purification of
elongation factor 1, **118**, 148
initiation and elongation factors, **118**, 109, 130
protein methylase I, **106**, 271
transacetylase, **106**, 195
Wheat germ agglutinin
–Sepharose conjugates
in affinity adsorption assay for soluble glucagon receptor, **109**, 13
in affinity purification of insulin receptor, **109**, 399, 405
in chromatographic separation of lymphocytes, **108**, 159
preparation, **108**, 155
Wolinella succinogenes
purification of
formate dehydrogenase, **126**, 389
fumarate reductase, **126**, 388

Woodward's reagent K
in coupling monoclonal antibodies to microspheres, **112**, 80
modification of F_1-ATPase carboxylic groups, **126**, 722
Wool
and morphological components
enzymatic hydrolysis, **106**, 61
extent and distribution of diagnostic components, **106**, 62
inhibitors originating from substrate preparations, **106**, 67
isolation, **106**, 60
separation, **106**, 61

X

Xanthophyll cycles
chromatographic analyses, **110**, 305
in isolated chloroplasts, **110**, 308
spectrophotometric analyses, **110**, 306
X-ray absorption spectrometer
for EXAFS measurements, design, **117**, 424
X-ray diffraction
crystals
data collection
by microdensitometry, **114**, 200
with multiwire area detector diffractometer, **114**, 416, 452
oscillation method with large unit cells, **114**, 211
with rotation camera using synchrotron radiation source, **114**, 281
with television area detector diffractometers, **114**, 472, 486
with X-ray film, **114**, 204
dynamic information from, **115**, 117
cytoskeletal fibers, **134**, 633
with double-focusing mirrors, alignment method, **114**, 316
electron density maps, calculation by fast Fourier transform
algebra for, **115**, 328
organization of, **115**, 333
immunoglobulins and fragments, preparative methods, **116**, 190
from large proteins, measurement by X-ray diffractometry
background determination, **114**, 391
data collection strategies, **114**, 393
radiation decay corrections, **114**, 386
in X-ray tube, **114**, 392
macromolecules, structure refinement
definition, **115**, 227
limits, **115**, 229
model accuracy, **115**, 232
monitoring, **115**, 231
number of observations, **115**, 230
in protein crystallography, **114**, 20
protein internal dynamics, **131**, 389
proteins in solution
data correlations, **117**, 158
geometric parameters, **117**, 174
measurement parameters, **117**, 155
water
around biomolecules, structural study, **127**, 162
in measurement of penetration depth into lipid bilayers, **127**, 512, 515
in membranes, quantitative study, **127**, 217
pure and in solution, structural study, **127**, 5, 320, 325
X-ray films
in crystal diffraction data collection, **114**, 204
processing for determination of Bragg reflection intensity from oscillation photographs, **114**, 241
X rays
biological effects, **134**, 665
mutagenicity testing with *Salmonella typhimurium* TA102, **105**, 250
rotation anode generators
characteristics and design considerations, **114**, 300
operation and maintenance, **114**, 307
sealed-tube generators, characteristics and design considerations, **114**, 300
synchrotron sources, **114**, 285
X-ray scattering
anomolous, in direct phase determination
applications, **115**, 49
multiple-wavelength analysis, **115**, 53
phase ambiguity, resolution, **115**, 106
phasing power, **115**, 48
scattering factors, **115**, 43
structure factors, **115**, 44

theoretical foundation, **115**, 43
time-resolved
in analysis of protein self-assembly, **130**, 58
microtubule assembly, **134**, 657
X-ray tubes
for X-ray crystal diffractometry of large proteins, **114**, 392
Xyloglucans
extraction and isolation, **118**, 17
Xylose
transport system in *Escherichia coli*
assays, **125**, 388, 399
associated membrane proteins, identification, **125**, 404, 417, 419
genetics, **125**, 404
properties, **125**, 389

Y

Yeast
cell-free extracts
preparation, **123**, 365
production of [^{14}C]coproporphyrin III, **123**, 366
growth in culture, **123**, 364
interferon expression vectors, construction, **119**, 418, 424
mitochondrial protein synthesis, Glu-tRNA identification as intermediate, **106**, 152
particle preparation for phagosome–lysosome fusion assay, **132**, 259
purification of
cytochrome bc_1, **126**, 173, 183, 185
diphthamide in elongation factor 2, **106**, 381
farnesylpyrophosphate synthetase, **110**, 183
glutamate synthase, **113**, 334
hexokinase, **104**, 110
L-kynurenine transaminase, **113**, 92
NAD synthetase, **113**, 300
prenyltransferase, **110**, 147
squalene synthetase, **110**, 369
spheroplasts
isolation of mitotic spindles and spindle pole bodies, **134**, 228
lysis, **134**, 227
preparation, **134**, 226
sterol auxotrophs
in analysis of sterol metabolism, **111**, 341
description, **111**, 334
isolation, **111**, 339
sterol metabolism, HPLC, **111**, 48
sterol mutants
in analysis of sterol metabolism, **111**, 343
characterization, **111**, 336
description, **111**, 333
isolation, **111**, 334
synthesis of interferon, **119**, 420
Yeda press
in subfractionation of thylakoid membranes, **118**, 326

Z

Zea mays, *see* Maize
Zinc
binding to thymulin, **116**, 274
Zinostatin
magnetic circular dichroism, **130**, 284
mutagenicity testing with *Salmonella typhimurium* TA102, **105**, 250
Zonal electrophoresis
monoclonal Ig light chains, **116**, 102
IgG, **116**, 7
Zonal ultracentrifugation
advantages and disadvantages, **129**, 13
applications to lipoproteins, **129**, 14
methodology, **129**, 5
Zymosan
in chemiluminescent macrophage activation assay, **132**, 545
opsonized
in assessment of phagocytic function, **132**, 138
stimulation of neutrophils, effect on chemiluminigenic response, **133**, 476
radiolabeled particles
phagocytosis by macrophages, assay, **132**, 329
preparation, **132**, 328

Contributor Index

Boldface numerals indicate volume number.

A

Aamodt, Eric J., **134**, 544
Abakerli, R. B., **105**, 36
Abbott, William, **113**, 567
Abraham, Carmela R., **134**, 388
Abrams, Paul G., **121**, 107
Absolom, Darryl R., **132**, 16, 95, 281
Achtstaetter, Thomas, **134**, 355
Ackerman, Sharon H., **126**, 667
Ackers, Gary K., **130**, 132
Acuto, Oreste, **108**, 624
Adair, W. Lee, **111**, 201
Adair, W. Steven, **121**, 453
Adamietz, Peter, **106**, 461
Adams, Dolph O., **132**, 555
Aebersold, Paul, **119**, 579
Aebi, Hugo, **105**, 121
Aebig, K. W., **134**, 408
Aeschbach, Robert, **107**, 377
Agarwal, R. C., **115**, 112
Aggarwal, Bharat B., **116**, 441, 448
Aggerbeck, Lawrence P., **128**, 457
Agnew, William S., **110**, 359
Aguet, Michel, **119**, 321
Aguilera, Greti, **109**, 110; **124**, 590
Ahl, Reinhard, **119**, 211
Ahlquist, Paul, **118**, 704
Ahmad, Mushtaq, **133**, 109
Ailhaud, G., **109**, 377
Airas, R. Kalervo, **122**, 33
Aitken, Alastair, **106**, 205
Akerboom, Theo P. M., **105**, 445
Akey, Christopher W., **126**, 434
Akhtar, M., **123**, 375
Akiyoshi, D. E., **110**, 340
Alaupovic, P., **128**, 297
Albers, John J., **129**, 78, 763, 797
Albersheim, Peter, **118**, 3
Alberta, Julia A., **126**, 353
Aldred, P., **124**, 534
Ålin, Per, **113**, 504, 507
Allen, Charles M., **110**, 117, 281
Allen, Robert C., **133**, 449
Allfrey, Vincent G., **107**, 224
Allison, J. P., **108**, 614
Allison, R. Donald, **113**, 419
Allison, William S., **126**, 741
Almond, Janice R., **123**, 159
Alpers, David H., **123**, 23
Altendorf, Karlheinz, **126**, 569
Alvarez-Gonzalez, Rafael, **106**, 483
Amadò, Renato, **107**, 377
Ambudkar, Suresh V., **125**, 558
Ames, Bruce N., **105**, 249, 352
Amidon, Gordon L., **112**, 360
Amir-Zaltsman, Y., **133**, 284
Amos, W. Brad, **134**, 404
Anantharamaiah, G. M., **128**, 627
Anderson, Bruce M., **122**, 156, 169, 173
Anderson, Constance D., **122**, 156, 173
Anderson, James Michael, **102**, 9
Anderson, Mary E., **113**, 548, 555
Anderson, Paul, **119**, 281
Anderson, Richard G. W., **129**, 201
Andersson, Bertil, **118**, 325
Andersson, Stefan, **111**, 364
Andondonskaja-Renz, Berta, **122**, 273
Andreu, José Manuel, **117**, 346; **130**, 47
Andrews, David W., **108**, 600
Andrews, Patricia C., **105**, 358; **132**, 369
Andrews, T. John, **118**, 410
Andrews, William W., **126**, 741
Anraku, Yashuhiro, **126**, 94
Antonelli, Guido, **119**, 69
Antoni, Guido, **121**, 193
Appella, Ettore, **108**, 494
Aquila, Heinrich, **125**, 650
Arakawa, Tsutomu, **114**, 49; **117**, 60
Archinard, Philippe, **126**, 455

Argiro, Vincent, **103**, 334
Argos, Patrick, **130**, 185
Arhelger, Chris, **116**, 416
Arita, Jun, **103**, 607
Armellini, Doretta, **121**, 193
Arnaud, Claude D., **109**, 40, 48
Arndt, U. W., **114**, 472
Arnheiter, Heinz, **119**, 312, 326, 332
Arnold, Lyle J., Jr., **112**, 270
Artymiuk, P. J., **114**, 397
Asada, Kozi, **105**, 422
Asada, Yasuhiko, **113**, 90
Ashman, Leonie K., **121**, 497
Ashwell, Gilbert, **109**, 232
Asmus, Klaus-Dieter, **105**, 167
Atassi, M. Zouhair, **121**, 69
Atkins, Robert C., **121**, 828
Atkinson, David, **128**, 582
Atmar, Valerie J., **107**, 154
Atwater, Jonathan A., **119**, 499
Audhya, Tapan, **116**, 279
Auerbach, Anthony, **103**, 147; **124**, 190
Aune, Thomas M., **116**, 395
Avila, Jesús, **134**, 179
Aviv, Dvora, **118**, 595
Aw, T. C., **123**, 185
Azzi, Angelo, **125**, 86; **126**, 45, 64, 72, 78, 138, 159, 520

B

Baba, Tsuneo, **110**, 117
Babior, Bernard M., **105**, 358
Babiss, Lee E., **119**, 611, 619
Bach, Jean-François, **116**, 269
Bacher, Adelbert, **122**, 192, 209
Bachorik, Paul S., **129**, 78
Bada, Jeffrey L., **106**, 98
Badger, Thomas M., **124**, 79
Badwey, John A., **132**, 353, 365
Baggiolini, Marco, **132**, 267, 395
Baines, Anthony J., **134**, 55
Balaban, Alexandru T., **127**, 738
Baldwin, Robert L., **131**, 3, 51
Baldwin, Thomas O., **133**, 98, 248
Balkwill, Frances R., **119**, 649
Ball, G. D., **119**, 35
Baltscheffsky, Margareta, **126**, 538
Bamberger, Mark, **129**, 628
Bancroft, Carter, **124**, 269
Banerjee, Debendranath, **129**, 283
Bannister, J. V., **105**, 88
Bannister, W. H., **105**, 88
Banquerigo, M. L., **121**, 322
Baralle, F. E., **128**, 727
Bard, J. D. J., **118**, 270
Bar-Joseph, Moshe, **118**, 742
Barker, H. A., **113**, 121
Barker, Jeffery L., **103**, 93
Barkley, Robert M., **107**, 563
Barnard, G., **133**, 284
Barnard, Graham F., **110**, 155
Barnea, Ayalla, **124**, 254
Baron, Samuel, **119**, 574
Bar-Shavit, Zvi, **132**, 326
Bartlett, Desiree L., **110**, 171
Bartolome, Jorge, **103**, 590
Bashor, Mark M., **122**, 234
Basse, Per, **119**, 688
Basten, A., **108**, 254
Bates, Michael D., **102**, 195
Battenberg, Elena L. F., **103**, 670
Battersby, Rüdiger V., **104**, 281
Battye, Frank, **121**, 748
Baukal, Albert, **109**, 110
Bayer, E. A., **133**, 284
Bayer, Ernst, **109**, 316
Baynes, John W., **106**, 88
Bazaes, Sergio, **110**, 78
Bazin, Hervé, **121**, 234, 622, 638, 663
Beattie, Diana S., **126**, 173
Beaudrie, Ken, **128**, 181
Beck, William S., **123**, 3
Beckner, Suzanne K., **109**, 356, 360
Beckwith, Margaret, **116**, 416
Behrman, Harold R., **109**, 298
Bekesi, Eva, **119**, 204, 296
Bell, Robert M., **124**, 57, 353
Bello, Jake, **119**, 103
Benbow, Uli, **125**, 119
Bendedouch, Dalila, **130**, 79
Bender, Timothy P., **109**, 704
Benditt, Earl P., **128**, 311
Benedict, Christine V., **107**, 397
Ben-Jonathan, Nira, **103**, 249
Bennett, Barbara A., **103**, 493
Bennett, John, **118**, 338
Bennett, Vann, **134**, 55
Bennink, Jack R., **108**, 270

Benson, Douglas M., **129**, 857
Ber, Rosalie, **121**, 381
Berg, Kurt, **119**, 688
Bergeron, John J. M., **109**, 219
Bergstresser, Paul R., **108**, 683
Bergstrom, James D., **110**, 3
Berlin, Richard D., **108**, 336
Berry, Edward A., **126**, 305
Berry, Marla J., **119**, 499
Bethea, Cynthia L., **103**, 287
Beveridge, David L., **127**, 21
Beychok, Sherman, **130**, 519
Beyer, Klaus, **125**, 631
Beyer, Peter, **110**, 267, 299; **111**, 248
Bhat, Pangala V., **123**, 75
Bibb, William F., **121**, 417
Bieber, L. L., **123**, 264, 276
Bielski, Benon H. J., **105**, 81
Biemann, Klaus, **106**, 29
Bill, Kurt, **126**, 64, 72
Billheimer, Jeffrey T., **111**, 286
Billiau, Alfons, **119**, 145, 88
Biltonen, Rodney L., **130**, 534
Bino, Tamar, **119**, 136
Bird, R. P., **105**, 299
Birnbaumer, Lutz, **109**, 3, 446, 566
Bishop, Cheri, **128**, 223
Bishop, David F., **123**, 339
Bishop, James E., **122**, 50
Bjercke, Robert J., **121**, 562
Blackmore, P. F., **109**, 550
Blackshear, Perry J., **104**, 237; **112**, 520, 530
Blanchard, John S., **104**, 404
Blanchard, Steven G., **103**, 219
Blanquet, Sylvain, **106**, 141
Bleackley, R. Chris, **116**, 526
Blessinger, Clare, **121**, 438
Block, Marc R., **125**, 639, 658
Block, Steven M., **134**, 531
Block, Timothy, **121**, 771
Bloom, Floyd E., **103**, 670
Bloom, Mark, **118**, 309
Blume, Alfred, **127**, 480
Blumenstock, Frank A., **132**, 334
Blumenthal, Robert, **128**, 657
Bodley, James W., **106**, 378
Bodmer, Judith L., **112**, 298
Bodor, Nicholas, **112**, 381
Boekema, Egbert J., **126**, 344
Boeshore, Maury L., **118**, 437
Boettcher, Brian, **113**, 458
Boeyé, Albert, **121**, 332
Bognar, Andrew L., **122**, 349
Bogner, Werner, **125**, 650
Boguski, Mark S., **128**, 753
Bohemier, Denise, **107**, 563
Böhlen, Peter, **103**, 72
Bolli, Reinhard, **126**, 45
Bollon, Arthur P., **119**, 359
Bolt, D. J., **121**, 433
Bolton, A. E., **124**, 18
Bolton, Hilary, **104**, 318
Bonner, William M., **104**, 460
Bono, M. Rosa, **108**, 600
Borchardt, Roy A., **129**, 536
Bordas, Joan, **134**, 657
Borek, Carmia, **105**, 464
Borle, André B., **124**, 90
Borthwick, Iain A., **123**, 395
Bosin, Ejal, **123**, 85
Boss, Barbara D., **121**, 27
Boston, Raymond C., **129**, 366
Boström, Hans, **111**, 364
Bosworth, John M., Jr., **121**, 541
Bothwell, Mark, **121**, 771
Bottema, Cynthia D. K., **111**, 333
Boucheix, Claude, **121**, 580
Boulay, François, **125**, 639, 658
Bourque, Don P., **118**, 201, 270
Boussioux, Anne Marie, **133**, 209
Boveris, Alberto, **105**, 429
Bowen-Pope, Daniel F., **109**, 69
Bowers-Komro, Delores M., **122**, 63, 116
Boxer, Steven G., **130**, 484
Boyer, Paul D., **126**, 618
Boylan, Michael, **133**, 70
Bøyum, Arne, **108**, 88
Brabson, John S., **113**, 627
Bradley, William A., **129**, 319
Brady, R. C., **134**, 217
Braiman, Mark S., **127**, 587
Brandolin, Gérard, **125**, 639, 658
Brandts, John F., **131**, 107
Brass, Johann M., **125**, 289
Braude, Irwin A., **119**, 72, 193
Brauer, Manfred, **107**, 36
Braun, Barry S., **125**, 119
Braun, Jörg, **133**, 354
Braun, Rüdiger W., **121**, 737

Bredehorst, Reinhard, **106**, 472
Brenowitz, Michael, **130**, 132
Breslow, Jan L., **128**, 690, 712, 773, 788
Bretscher, Anthony, **134**, 24, 37
Brewer, H. Bryan, Jr., **128**, 223; **129**, 482
Bridges, C. D. B., **123**, 102
Bridges, Richard J., **113**, 567
Brightman, Milton W., **103**, 187
Briles, David E., **116**, 174
Brinkley, B. R., **134**, 268
Brisson, N., **118**, 659
Britton, George, **111**, 113
Broadwell, Richard D., **103**, 187
Brodin, Thomas, **121**, 295
Brody, Tom, **107**, 552; **122**, 367
Broger, Clemens, **126**, 64, 72
Brooks, Charles L, III, **127**, 369
Brooks, Kathryn, **116**, 372
Brooks, Timothy L., **121**, 615
Broquist, Harry P., **123**, 303
Brot, Nathan, **107**, 352; **118**, 309
Brown, Barry L., **109**, 293
Brown, Dale C., **116**, 467
Brown, John J., **111**, 509
Brown, Kenneth L., **123**, 14
Brown, Larry, **112**, 399
Brown, McKay, **121**, 327
Brown, Michael S., **109**, 405
Brown, Richard C., **133**, 366
Brown, Susan S., **134**, 9
Brown, W. Virgil, **129**, 186, 384, 457
Browning, Karen S., **118**, 109
Bruce, J. Malcolm, **125**, 109, 119
Bruck, Claudine, **121**, 587
Brugh-Collins, Michele, **123**, 61
Bruhn, Thomas, **103**, 565; **124**, 389
Brunengraber, Henri, **110**, 58, 100
Brunner, K. Theodor, **132**, 437
Brunori, M., **105**, 22
Bruns, Gail A. P., **128**, 788
Brunzell, John D., **128**, 181
Bryan, Joseph, **129**, 857; **134**, 13
Buckel, Wolfgang, **125**, 547
Bucks, Rodney R., **130**, 484
Buenafe, Abigail, **121**, 327
Büge, Ursula, **126**, 32
Buhrow, Susan A., **109**, 816
Builder, Stuart E., **119**, 77
Bukberg, Phillip R., **129**, 457
Bulinski, Jeannette Chloë, **134**, 147, 453
Bullough, David A., **126**, 741
Bundgaard, Hans, **112**, 347
Bunn, Paul A., Jr., **121**, 678
Burakoff, Steven J., **108**, 607
Burchiel, Scott W., **121**, 596
Burger, J. J., **112**, 27, 43
Burgess, Antony W., **116**, 588
Burghaus, Petra A., **127**, 746
Burke, M. J., **127**, 761
Burman, C. J., **119**, 35
Burridge, K., **134**, 69
Busch, Harris, **106**, 238
Buttriss, Judith L., **105**, 131
Byers, David, **133**, 183
Byler, D. Michael, **130**, 290

C

Cabib, Enrico, **104**, 415
Cabral, F., **134**, 217
Cadenas, Enrique, **105**, 221
Caerteling, Gerda C., **122**, 412
Cafiso, David S., **127**, 502
Caldini, A. L., **133**, 557
Caliguri, Edward J., **124**, 402
Calton, Gary J., **104**, 381
Camara, Bilal, **110**, 244, 274; **111**, 544
Cambier, John C., **103**, 227
Cande, W. Zacheus, **134**, 473
Cane, David E., **110**, 383
Canfield, Robert E., **109**, 638
Cann, John R., **130**, 19
Cantell, Kari, **119**, 27, 54
Capaldi, Roderick A., **126**, 22
Capella, Peter, **124**, 402
Capobianchi, Maria R., **119**, 69
Capponi, Alessandro M., **124**, 116
Card, J. Patrick, **103**, 619
Cardarelli, Pina, **132**, 334
Cardemil, Emilio, **110**, 86
Carey, Luc, **133**, 183
Carlberg, Inger, **113**, 484
Caron, Marc G., **103**, 577
Carpenter, Graham, **109**, 101
Carr, Steven A., **106**, 29
Carrasco, Nancy, **125**, 453
Carrea, Giacomo, **133**, 238
Casey, Robert P., **125**, 86; **126**, 13
Caspers, Martin, **119**, 441, 453
Castle, Lawrence, **105**, 293
Castleman, S., **125**, 27

Castranova, Vincent, **132**, 498
Cathcart, Richard, **105**, 352
Catignani, G. L., **123**, 215
Catt, Kevin J., **109**, 110; **124**, 590
Catterall, James R., **132**, 626
Cederbaum, Arthur I., **105**, 516
Cerdá-Olmedo, Enrique, **110**, 220
Cerottini, Jean-Charles, **132**, 437
Chafouleas, James G., **102**, 104
Chan, Lawrence, **128**, 745, 851, 863; **129**, 297
Chandler, David, **127**, 48
Chang, Chung Nan, **119**, 424
Chang, David J., **128**, 811
Chang, Ernest S., **111**, 494, 509
Chang, K.-P., **132**, 603
Chang, Kwen-Jen, **103**, 219
Chang, M., **112**, 150
Chang, May, **119**, 204
Chang, Thomas Ming Swi, **112**, 195
Chany, Charles, **119**, 541, 694
Chany-Fournier, Françoise, **119**, 694
Chapman, M. John, **128**, 70
Charbonneau, Harry, **102**, 17
Charton, M., **112**, 323
Chastain, Jane L., **122**, 63
Chazotte, Brad, **125**, 35
Chen, Chien-Hwa, **119**, 177
Chen, Ching-Hong, **129**, 763
Chen, Christina Y., **119**, 424
Chen, G. Chi, **128**, 519
Chen, Jeng-Chyh, **121**, 615
Chen, Sow-Hsin, **130**, 79
Cheney, Richard, **134**, 42
Cheung, Marian C., **129**, 130
Cheung, Wai Yiu, **102**, 39, 244
Chevalier, Pamela, **124**, 402
Chien, Yie W., **112**, 461
Chin, Christopher C. Q., **106**, 17, 69
Chock, P. Boon, **113**, 213
Chothia, Cyrus, **115**, 420
Chou, Janice Yang, **109**, 385
Chrambach, Andreas, **104**, 305
Chrispeels, Maarten J., **107**, 361
Christman, Michael F., **105**, 249
Christy, M. Robert, **107**, 563
Chu, B., **117**, 256
Chua, Nam-Hai, **118**, 232
Chuman, Hiroshi, **124**, 3
Chung, Byung Hong, **128**, 181; **129**, 704
Chung, Soo Il, **113**, 358
Cianfriglia, Maurizio, **121**, 193
Cidlowski, John A., **122**, 120
Ciferri, Orio, **118**, 296
Ciocca, Daniel R., **121**, 562
Cirillo, Vincent P., **125**, 259
Civin, Curt I., **121**, 322
Clark, Brian R., **121**, 459
Clark, Joan E., **122**, 385, 392
Clark, M. R., **121**, 548
Clark, Michael F., **118**, 742
Clarke, Steven, **106**, 310, 330
Clayton, Richard N., **103**, 32
Clegg, James S., **127**, 230
Clejan, Liviu, **126**, 173
Clementi, Enrico, **127**, 114
Cleveland, Patrick H., **121**, 525
Cleveland, W. Louis, **121**, 95
Clifford, Dennis P., **105**, 393
Coburn, Stephen P., **122**, 102
Cocking, E. C., **118**, 578
Codina, Juan, **109**, 446, 566
Coghlan, J. P., **124**, 534
Cohen, Alfred M., **112**, 530
Cohen, Bennett N., **118**, 309
Cohen, G., **122**, 72
Cohen, Georges N., **113**, 596, 600
Cohen, Gerald, **105**, 305, 510, 516
Cohen, Philip, **102**, 219, 227; **106**, 205
Cohen, Sara, **119**, 136, 464
Cohen, William D., **134**, 232
Cohn, Waldo E., **106**, 3
Colcher, David, **119**, 682; **121**, 802
Cole, Susan P. C., **121**, 120, 140
Coleman, Peter S., **125**, 671; **126**, 667
Coleman, Timothy, **107**, 352
Coligan, John E., **108**, 437
Coling, D. E., **134**, 408
Coller, Barry S., **121**, 412
Coller, Hilary A., **121**, 412
Collier, Kenneth J., **119**, 481
Collins, Anne, **125**, 697
Collins, Christine A., **134**, 116
Colman, Paul D., **122**, 309
Colwell, Dawn E., **121**, 42
Comb, Michael, **124**, 278
Cone, John T., **128**, 181
Conn, P. Michael, **102**, 195; **103**, 49, 401; **124**, 36, 57
Consler, Thomas G., **117**, 27
Conway de Macario, Everly, **121**, 509
Cook, Robert J., **122**, 251, 255

Cook, William J., **102**, 143
Cooper, Arthur J. L., **113**, 63, 66, 69, 71, 73, 76, 80, 338, 344, 350, 602
Cooper, Herbert L., **106**, 344
Cooper, Jack R., **122**, 20, 24
Cooper, Philip H., **132**, 395
Cooper, Ralph L., **103**, 483
Coppenhaver, Dorian H., **119**, 199
Corbley, Michael J., **126**, 72
Cori, Osvaldo, **110**, 406
Cormier, Milton J., **102**, 17, 195; **133**, 288
Cormont, Françoise, **121**, 234, 622, 638
Costello, M. Joseph, **127**, 704
Cowsar, Donald R., **112**, 101
Craig, Susan W., **134**, 78, 460
Crastes de Paulet, André, **133**, 209
Creek, Kim E., **123**, 61
Creighton, Thomas E., **107**, 305; **131**, 83, 156
Creswick, P., **108**, 254
Crissman, Robert S., **121**, 183
Croce, Carlo M., **121**, 120
Cronin, Michael J., **103**, 294
Cross, Andrew R., **132**, 378
Croteau, Rodney, **110**, 383
Crowe, John H., **127**, 696
Crowe, Lois M., **127**, 696
Crowl, Robert, **119**, 376
Cséplo, Agnes, **118**, 611
Cuatrecasas, Pedro, **103**, 219; **109**, 399
Cuello, A. C., **121**, 210
Cuff, George W., **112**, 84
Cullen, Susan E., **108**, 467, 523
Culler, Michael D., **124**, 67
Cumber, A. J., **112**, 207
Curtis, Ann, **124**, 318
Curtiss, Linda K., **128**, 354
Cushing, Gary L., **129**, 167
Cutting, John A., **104**, 451
Czapski, Gidon, **105**, 209
Czech, Michael P., **109**, 179

D

Daëron, M., **116**, 403
Dailey, Harry A., **123**, 401, 408
Dalton, David, **103**, 565
Danho, Waleed, **125**, 453
Danielsson, Henry, **111**, 364
Dânşoreanu, Manase M., **122**, 161
Darfler, Frederick J., **109**, 356, 360
D'Ari, Linda, **113**, 169
Darszon, Alberto, **127**, 486
Darvill, Alan G., **118**, 3
Daughaday, William H., **109**, 773
Daugherty, Bruce L., **119**, 434
Davey, M. R., **118**, 578
Davies, David R., **104**, 370
Davies, Glenn E., **128**, 811
Davies, Hugh A., **119**, 533
Davis, Bruce H., **108**, 336
Davis, John M., **121**, 307
Davis, Jonathan, **134**, 55
Davis, Lawrence C., **117**, 116
Davis, Roger A., **129**, 272, 536
Davisson, V. Jo, **110**, 130
Dawson, Adam J., **118**, 470
Dawson, Paul A., **129**, 254
Dayton, Claire H., **132**, 334
Deamer, David W., **127**, 471
Dean, C. J., **121**, 52
Dean, Roger T., **112**, 298
De Baetselier, Patrick, **133**, 507
de Boer, Martin, **132**, 225
De Boever, J., **133**, 387
De Brabander, M., **134**, 592
DeChiara, Thomas M., **119**, 403, 416
Decker, Karl, **122**, 185
De Clercq, Lieve, **121**, 234, 622, 638
Dedman, John R., **102**, 1
de Graan, Pierre N. E., **109**, 129
Deisenhofer, J., **115**, 303
de la Torre, Javier, **134**, 179
Deléage, Gilbert, **126**, 417
De Ley, Marc, **119**, 88, 145
Dell, Ralph B., **129**, 384
Del Maestro, Rolando, **105**, 386
De Luca, Luigi M., **123**, 61
DeLuca, Marlene, **133**, 3, 198, 215
De Maeyer, Edward, **119**, 474
De Maeyer-Guignard, Jaqueline, **119**, 474
Demaille, Jacques G., **102**, 261
Demel, Rudy A., **129**, 738
De Mey, J., **134**, 592
Dempsey, Mary E., **111**, 293
Dencher, Norbert A., **127**, 746
Denner, Larry, **113**, 3
Den Otter, Willem, **132**, 458, 531
de Petris, Stefanello, **108**, 413

de Roo, J. E., **112**, 43
Desai, Indrajit D., **105**, 138
Desnick, Robert J., **123**, 339
De Souza, Errol B., **124**, 560
Dessen, Philippe, **106**, 141
Detrich, H. William, III, **134**, 128
Deutch, Charles E., **106**, 198
Devash, Yair, **119**, 752, 759
DeVault, Don, **127**, 79
DeVries, Arthur L., **127**, 293
deVries, Simon, **126**, 293
de Vrij, Wim, **126**, 159
Dewald, Beatrice, **132**, 267
De Weger, Roel A., **132**, 458, 531
de Wet, Jeffrey R., **133**, 3
Dhadwal, H. S., **117**, 256
Diamond, R., **115**, 237
Dianoux, Anne-Christine, **126**, 712
Dianzani, Ferdinando, **119**, 69
Dickson, Robert B., **109**, 257
Dijkema, Rein, **119**, 441, 453
Dillard, Cora J., **105**, 337
DiMaria, Peter, **106**, 274
Dimroth, Peter, **125**, 530, 540
Di Paola, Eugene A., **107**, 224
Diplock, Anthony T., **105**, 131
Di Sabato, Giovanni, **108**, 43
Ditta, G., **118**, 519
Dix, Thomas A., **105**, 347
Dixon, Jack E., **106**, 170; **113**, 541
Djiane, Jean, **109**, 156, 667
Dobson, C. M., **131**, 362
Dobson, Pauline R. M., **109**, 293, 827
Doig, M. T., **122**, 313
Dollinger, Gavin, **127**, 649
Domingo, Derrick L., **112**, 238
Doner, Landis W., **122**, 3
Dory, Ladislav, **129**, 660
Douglas, Carolyn, **103**, 565
Douglass, James, **124**, 309
Downs, Thomas R., **124**, 371
Dragutan, Ileana, **127**, 738
Draper, H. H., **105**, 299
Dray, Fernand, **103**, 409
Drebin, Jeffrey A., **121**, 587
Dresel, Hans A., **129**, 216
Drezner, Marc K., **123**, 159
Driscoll, Charles F., **112**, 56
Driscoll, Donna M., **128**, 41
Drubin, David, **134**, 156
Drugeon, G., **118**, 154
Drummond, Robert J., **119**, 183
Drysdale, Beth-Ellen, **132**, 549
Dubbeld, Martin, **119**, 453
Dubochet, Jacques, **127**, 719
Duc-Goiran, Paulette, **119**, 541
Duffel, Michael W., **113**, 516
Duffy, Thomas E., **113**, 350
Dufy, Bernard, **103**, 93
Dufy-Barbe, Luce, **103**, 93
Dukovich, Mitchell, **116**, 480
Duncan, W. E., **123**, 185
Dunlap, Claud E., III, **124**, 617
Dunlap, Jay C., **133**, 307
Dunlop, Patricia C., **106**, 378
Dunn, Stanley D., **126**, 434
Dupont, Yves, **125**, 639
Dupuis, Alain, **126**, 712
Durbin, Richard M., **114**, 211
Durda, Paul J., **108**, 666
Durkin, Tracy A., **124**, 402
Dusanter-Fourt, Isabelle, **109**, 667
Dutton, P. Leslie, **125**, 109, 119; **126**, 293

E

Eager, Kendra B., **121**, 59
Ebeling, James, **124**, 57
Eberle, Alex N., **109**, 129
Ebsworth, Nicolette, **119**, 643
Eby, William C., **121**, 783
Edelman, Elazer, **112**, 399
Edelman, Marvin, **118**, 384
Edelstein, Celina, **128**, 151, 339
Edelstein, Stuart J., **126**, 434
Edmond, John, **110**, 3
Edmondson, Dale E., **122**, 240
Edwards, Carol A., **126**, 211
Ehnholm, Christian, **129**, 716
Ehrenberg, Benjamin, **127**, 678
Ehrig, Karin, **126**, 770
Ehrlich, Paul H., **109**, 638; **121**, 695
Eichberg, Joseph, **109**, 504
Eichele, G., **114**, 132
Eid, Pierre, **119**, 347
Eisenbarth, George S., **103**, 459
Eisenberg, Shlomo, **129**, 347
Eisenstein, Laura, **127**, 649
Ejiri, Shin-ichiro, **118**, 140

Ekman, Pia, **107**, 130
Elder, Michael, **114**, 199
Elliott, Bruce E., **108**, 49
Elliott, William H., **111**, 51; **123**, 395
Elshourbagy, Nabil A., **128**, 753
Elson, E. L., **130**, 454
Enderby, J. E., **127**, 319
Endo, Toyoshi, **102**, 256
Engebrecht, JoAnne, **133**, 83
Engel, W. D., **126**, 224
Engers, Howard D., **132**, 437
Englander, Joan J., **131**, 508
Englander, S. W., **131**, 508
Engleman, Edgar G., **121**, 168
English, James P., **112**, 101
Engström, Lorentz, **107**, 130
Erickson-Viitanen, Susan, **116**, 265
Eriksen, Nils, **128**, 311
Eriksson, Véronique, **113**, 520
Erlanger, Bernard F., **121**, 95
Erlitz, Fran, **119**, 403
Ernster, L., **126**, 428
Esch, Fred S., **103**, 72
Eskay, Robert L., **124**, 359
Eskinazi, Daniel P., **121**, 783
Esko, Jeffrey D., **129**, 237
Esteban, Jose, **121**, 802
Esterbauer, Hermann, **105**, 319
Evans, W. H., **109**, 246
Evans, William S., **103**, 294
Evenson, Debra A., **103**, 393
Everse, Johannes, **132**, 491, 537
Exton, J. H., **109**, 550
Ey, Peter L., **121**, 497
Eyzaguirre, Jaime, **110**, 78

F

Fahey, Diana, **119**, 284
Fain, J. N., **109**, 469
Faivre-Bauman, Annie, **103**, 313
Falk, Gunnar, **125**, 230
Familletti, Philip C., **119**, 63
Fantes, Karl H., **119**, 233
Fara, John W., **112**, 470
Faulstich, Heinz, **112**, 225
Faynor, Steven M., **122**, 152
Fazekas de St. Groth, S., **121**, 360
Feher, G., **114**, 77
Feldman, Richard I., **126**, 643
Felix, Arthur M., **119**, 242
Ferguson-Miller, Shelagh M., **125**, 16, 27
Fernandez, Rafael, **116**, 372
Ferrara, Napoleone, **124**, 245
Ferry, David R., **109**, 513
Fetter, Richard D., **127**, 704
Feyereisen, R., **111**, 442, 530
Fidge, Noel H., **129**, 443
Fidler, Isaiah J., **132**, 594
Field, James B., **109**, 566
Fielding, Christopher J., **111**, 267, 274; **129**, 783
Fiers, Walter, **119**, 366
Fillingame, Robert H., **126**, 545, 558
Fink, Anthony L., **131**, 173
Fink, George, **124**, 318
Finkelman, Fred D., **116**, 121
Finn, Frances M., **109**, 418
Finney, John L., **127**, 284
Finter, N. B., **119**, 35
Fiol, C., **123**, 276
Fish, Leonard E., **118**, 282
Fisher, Harvey F., **113**, 16
Fisher, M. R., **127**, 91
Fisher, Paul B., **119**, 611, 619, 629
Fisher, Ronald R., **126**, 353
Fisher, Waldo R., **128**, 247
Fitch, Walter M., **128**, 773
Fitchen, J., **118**, 419
FitzGerald, David, **119**, 332
Fiume, Luigi, **112**, 225
Flaherty, Lorraine, **108**, 481
Flavin, Martin, **106**, 223
Fleischmann, W. Robert, Jr., **119**, 574
Fleisher, David, **112**, 360
Fleming, Jennie E., **123**, 401
Fletterick, Robert J., **114**, 386
Fliss, Henry, **107**, 352
Flohé, Leopold, **105**, 93, 114
Floyd, Robert A., **105**, 231
Floyd-Smith, Georgia, **119**, 489
Flückiger, Rudolf, **106**, 77
Flügge, U. I., **125**, 705, 716
Fluhr, Robert, **118**, 611
Flygare, Susanne, **104**, 364
Föhles, J., **106**, 58
Folk, J. E., **106**, 344; **113**, 358
Fonda, Margaret L., **113**, 11
Fong, S.-L., **123**, 102

Foon, K. A., **121**, 107
Foote, Christopher S., **105**, 36
Foran, Judith, **108**, 607
Forbes, Gayle M., **128**, 690
Ford, J. R., **117**, 256
Forest, C., **109**, 377
Forni, L. G., **105**, 179
Forni, Luciana, **108**, 413
Forrester, J. A., **112**, 207
Forte, Trudy M., **128**, 442
Forti, Robert L., **119**, 533
Foster, David L., **125**, 429; **126**, 545
Foster, David M., **129**, 366, 395
Foster, S. J., **107**, 582
Foung, Steven K. H., **121**, 168
Fourme, R., **114**, 281
Foxall-VanAken, S., **125**, 27
Foxwell, B. M. J., **112**, 207
Fraley, Robert T., **118**, 627
Franěk, František, **121**, 277, 631
Frangione, Blas, **116**, 3
Frank, Nancy E., **123**, 167
Franke, Christine A., **124**, 295
Franke, Werner W., **134**, 355
Frasier, Susan G., **117**, 301
Fràter-Schröder, Marijke, **123**, 36
Frauenfelder, Hans, **127**, 207
Frazer, Ian H., **121**, 748
Freedman, Robert B., **107**, 281
Freeman, J. P., **118**, 578
Freer, Stephan T., **115**, 235
Freisheim, James H., **122**, 293
Fridman, W. H., **116**, 403
Fridovich, Irwin, **105**, 59
Friedl, Peter, **126**, 579, 770
Friedman, Daniel L., **108**, 43
Friedman, Harold L., **127**, 3
Friesen, H. G., **109**, 692
Frohman, Lawrence A., **124**, 371
Fry, Stephen C., **107**, 388
Fujii, Dennis, **124**, 245
Fujii, Hiroshi, **110**, 153, 192, 199
Fujimoto, Yoshinori, **111**, 346
Fujitaki, James M., **107**, 23
Fukazawa, Chikafusa, **128**, 801
Fuke, Motohiro, **119**, 359
Fukui, Yoshio, **134**, 573
Fuller, N. L., **127**, 400
Fung, B. M., **127**, 151
Furlong, Clement E., **125**, 279
Furtek, Cheryl A., **105**, 263
Furth, Anna J., **104**, 318
Futai, Masamitsu, **126**, 588, 595

G

Gabig, Theodore G., **132**, 355
Gabriel, Othmar, **104**, 416
Gagnon, Claude, **134**, 318
Galli, Maria Grazia, **118**, 87
Gallop, Paul M., **106**, 77; **107**, 516; **113**, 133
Galun, Esra, **118**, 595
Gander, John E., **104**, 447; **107**, 172
Ganong, Barry R., **124**, 57
Ganther, H. E., **107**, 582, 593
Ganzi, Gary C., **104**, 391
Garavito, R. Michael, **125**, 309
Gardner, Jerry D., **109**, 64, 288
Garger, S. J., **118**, 717
Garvey, Justine S., **108**, 285
Gastinel, Louis Noel, **116**, 269
Gaubatz, John W., **129**, 167
Gautheron, Danièle C., **126**, 417, 455, 761
Gee, Connie E., **124**, 510
Gelabert, M. J., **116**, 403
Geller, Bruce L., **105**, 105
Gennis, Robert B., **126**, 87, 138
Georgiades, Jerzy A., **119**, 83
Gera, Abdullah, **119**, 729
Gery, Igal, **116**, 456
Geuens, G., **134**, 592
Getz, Godfrey S., **128**, 41
Ghetie, Victor, **108**, 132, 405
Gheuens, J., **121**, 425
Ghisla, S., **133**, 140
Giangiacomo, Kathleen M., **126**, 293
Gianturco, Sandra H., **129**, 319
Gibson, Joyce Corey, **129**, 186, 457
Gibson, Margaret M., **125**, 365
Gilbert, Hiram F., **107**, 330
Gill, Thomas J., III, **121**, 386
Gilles-Gonzalez, Marie A., **125**, 190
Gilley, Richard M., **112**, 101
Gilliland, Gary L., **104**, 370
Gimsing, Peter, **123**, 3
Ginn, Stephen, **121**, 183
Ginsberg, Henry N., **129**, 186, 384, 457
Ginsburg, Mary F., **134**, 232

Giovanella, Beppino C., **108**, 348
Girotti, Stefano, **133**, 238
Glaser, Thomas, **109**, 316
Glassy, Mark C., **121**, 525
Glazer, Alexander N., **106**, 359
Glende, Eric A., Jr., **105**, 331
Glenney, John R., Jr., **102**, 204
Glick, Bruce, **108**, 3
Glickman, Robert M., **129**, 519
Glines, Laura A., **129**, 26
Glineur, Corrine, **121**, 587
Glode, L. Michael, **121**, 702
Glossmann, Hartmut, **109**, 110, 513
Go, Masayoshi, **124**, 349
Goad, L. J., **111**, 311
Godinot, Catherine, **126**, 417, 455, 761
Goff, C. G., **106**, 418
Goldberg, Eugene P., **112**, 18
Goldman, David, **104**, 441
Goldman, Rachel, **132**, 326
Goldsby, Richard A., **121**, 244, 433, 797
Goldsmith, Paul, **124**, 245
Goldstein, Allan L., **116**, 213, 219, 233, 248
Goldstein, Gideon, **116**, 279
Goldstein, Joseph L., **109**, 405
Goldstein, L., **122**, 72
Golub, S. H., **108**, 363
Goodall, Gregory J., **116**, 255
Goodenough, J. B., **127**, 263
Gorbunoff, Marina J., **117**, 370
Gordon, David A., **128**, 671
Gordon, Jeffrey I., **128**, 753
Gorevic, Peter D., **116**, 3
Gorga, Joan C., **108**, 607
Goroff, Diana K., **108**, 274
Gospodarowicz, Denis J., **103**, 287
Goss, Neil H., **107**, 261
Goswami, Ajit, **107**, 488
Goto, Akio, **112**, 139
Gottlieb, Paul D., **108**, 666
Gotto, Antonio M., Jr., **128**, 3; **129**, 216, 857
Gottschalk, M., **119**, 211
Grady, Richard K., Jr., **124**, 402
Graham, Angus, **133**, 70
Granados, Edward N., **119**, 103
Grand, Roger, **102**, 219
Grant, Gary, **133**, 189
Grant, Steven, **119**, 629
Gratton, Enrico, **127**, 207
Graves, Donald J., **106**, 403
Gray, J. Dixon, **108**, 363
Green, Monika J., **105**, 3
Green, Ralph, **123**, 14
Greenberg, Bruce M., **118**, 253
Greenwood, Robert S., **103**, 132
Greer, Jonathan, **115**, 206
Gregg, Richard E., **129**, 482
Gregory, Roger B., **131**, 448
Greiner, John W., **119**, 629, 682
Griffith, Owen W., **113**, 445, 461
Grimaldi, P., **109**, 377
Grisham, Matthew B., **132**, 569, 585
Gromet-Elhanan, Zippora, **126**, 528
Grosfeld, Haim, **119**, 464
Gruber, Kenneth A., **103**, 524
Gruissem, Wilhelm, **118**, 253
Grumet, F. Carl, **121**, 168
Grzesiek, Stephan, **127**, 746
Guerrieri, F., **126**, 331
Guffanti, A. A., **125**, 352
Guidry, Albert J., **121**, 244
Guilfoyle, Tom J., **118**, 696
Guirard, Beverly M., **122**, 139
Gulik-Krzywicki, Tadeusz, **128**, 457
Gundberg, Caren M., **107**, 516; **113**, 133
Gundersen, Gregg G., **134**, 453
Günzler, Wolfgang A., **105**, 114
Gupta, Arun, **110**, 327
Gupta, Sohan L., **119**, 263, 340
Gurd, Frank R. N., **130**, 413, 437
Gurevitz, M., **118**, 419
Gustafson, Gary L., **107**, 172
Guthenberg, Claes, **113**, 499, 507
Gutknecht, John, **127**, 471
Gutman, Menachem, **127**, 522
Gutteridge, John M. C., **105**, 47
Guynn, Robert W., **122**, 381
Gwynne, John T., **129**, 679
Gyure, L. A., **121**, 52

H

Haas, Werner, **132**, 467
Hackenbrock, Charles R., **125**, 35
Hackett, Neil R., **125**, 190
Haddad, J. G., **123**, 185
Haegert, David G., **108**, 386

Haenni, A. L., **118**, 154
Hageman, R. V., **122**, 399
Hagen, Tory M., **122**, 116
Hager, L. P., **107**, 439
Hagerman, Paul J., **117**, 198
Hajra, Amiya K., **122**, 50
Halk, Edward L., **118**, 766
Hall, J. G., **121**, 52
Halley, Bruce A., **123**, 92
Hallick, Richard B., **118**, 253
Halliwell, Barry, **105**, 47
Halperin, Gideon, **129**, 816
Hamamoto, Toshiro, **126**, 123, 640
Hamilton, James A., **128**, 472
Hamlin, Ronald, **114**, 416
Hammarström, M.-L., **108**, 153
Hammarström, S., **108**, 153
Hammock, Bruce D., **111**, 303, 487
Hamprecht, Bernd, **109**, 316, 341
Hancock, Wayne W., **121**, 828
Hand, P. Horan, **119**, 682
Hanley-Bowdoin, Linda, **118**, 232
Hanna, Bruce A., **133**, 22
Hannappel, Ewald, **116**, 265
Hannig, Kurt, **108**, 180
Hannun, Yusuf, **124**, 353
Hanover, John A., **109**, 257
Hansen, Ernil, **108**, 180
Hansen, R. Gaurth, **122**, 36
Hanson, Maureen R., **118**, 437
Hantke, Uwe, **133**, 354
Hara, Ichiro, **129**, 57
Haralambidis, J., **124**, 534
Harbin, Bertille M., **123**, 401
Hardman, David A., **128**, 262
Hardman, Karl D., **115**, 381
Harford, Joe, **109**, 232
Haritos, A. A., **116**, 255
Harmar, Anthony J., **124**, 335
Harmon, J. T., **117**, 65
Harper, Angela M., **132**, 378
Harper, Jeffrey F., **102**, 122
Harper, Mary E., **128**, 863
Harrington, Charles A., **109**, 504
Harris, Curtis C., **103**, 409
Harrison, Stephen C., **114**, 211
Hart, Russell C., **102**, 17, 195
Hartline, Richard A., **113**, 639, 664
Hartmann, Dan-Paul, **108**, 325
Hasenson, Michael, **133**, 27
Hashimoto, Tadao, **126**, 504
Hashimoto, Yoshiyuki, **121**, 817
Hassan, Hosni M., **105**, 254, 404, 523
Hassinen, Ilmo E., **123**, 311
Hastings, J. Woodland, **133**, 140, 307
Hatch, Linda, **105**, 162
Hatefi, Youssef, **126**, 360
Hatfield, J. Michael, **103**, 257
Hatzfeld, Mechthild, **134**, 355
Hauschka, Peter V., **107**, 516; **113**, 133
Hausinger, Robert P., **122**, 199
Hauska, Günter, **126**, 271, 285
Haussler, Mark R., **123**, 199
Havel, Richard J., **128**, 3; **129**, 591
Hawkes, Richard, **121**, 484
Hawkes, Wayne C., **107**, 602
Hayaishi, Osamu, **105**, 61; **106**, 450, 500
Hayward, James N., **103**, 132
Hayzer, David J., **116**, 26
Hazum, Eli, **103**, 58, **124**, 47
Hearn, Milton T. W., **104**, 190
Hed, Jan, **132**, 198
Hederstedt, Lars, **126**, 399
Hedo, Jose A., **109**, 593
Heeb, Mary J., **104**, 416
Heikkila, Richard E., **105**, 510
Heil, Matthew F., **108**, 285
Heimer, Edgar P., **119**, 242
Heldt, H. W., **125**, 705, 716
Helinski, Donald R., **133**, 3
Heller, J., **112**, 422
Hellström, U., **108**, 153
Henderson, G. B., **122**, 260
Henderson, Louis, **116**, 540
Henderson, Peter J. F., **125**, 157, 387
Hendren, R. Wayne, **103**, 547
Hendrickson, Wayne A., **115**, 41, 252
Heney, Gayle C., **122**, 83
Henslee, Jerry G., **113**, 113
Herberg, John T., **109**, 13, 207
Herbert, Edward, **124**, 278, 295, 309
Heremans, Hubertine, **119**, 145
Herlt, Maria, **126**, 498
Hermans, Jan, **115**, 171
Heron, Iver, **119**, 688
Herscowitz, Herbert B., **108**, 274
Hersh, Louis B., **113**, 36
Hershey, Howard P., **118**, 369
Herzenberg, L. A., **108**, 197
Herzlinger, Doris, **125**, 453

Hesterberg, Lyndal K., **117**, 97
Hetru, C., **111**, 411
Hewlett, N. Gail, **118**, 201
Hibbs, John B., Jr., **132**, 508
Hice, Rita, **102**, 17
Hicks, Kevin B., **122**, 3
Hidaka, Hiroyoshi, **102**, 185, 256
Higashi, Naoki, **119**, 93
Higgins, Christopher F., **125**, 365
Hildebrandt, John D., **109**, 446, 566
Hill, H. Allen O., **105**, 3
Hillson, David A., **107**, 281
Hilpert, Wilhelm, **125**, 540
Hilz, Helmuth, **106**, 461, 472
Himmelstein, K. J., **112**, 422
Hinds, Thomas R., **102**, 47
Hinkkanen, Ari, **122**, 185
Hinz, Hans-Jürgen, **130**, 59
Hirano, Susan S., **127**, 730
Hirata, H., **125**, 485
Hirokawa, Nobutaka, **134**, 598
Hirvonen, Sinikka, **119**, 27, 54
Hisanaga, Shin-ichi, **134**, 337
Hitzeman, Ronald A., **119**, 424
Hjelmeland, Leonard M., **104**, 305; **124**, 135
Ho, May-Kin, **108**, 313
Hoben, Patricia, **113**, 55
Hochstein, Paul, **105**, 162
Hodge, Tony P., **118**, 453
Hodges, Thomas K., **118**, 41
Hoefer, Michael T., **103**, 249
Hoffmann, J. A., **111**, 411
Hofmann, Klaus H., **109**, 418
Hohn, T., **118**, 659
Hokanson, John E., **128**, 181
Hokland, Marianne, **119**, 688
Hokland, Peter, **119**, 688
Holbrook, Stephen R., **114**, 167
Holcenberg, John S., **113**, 257
Holdsworth, George, **128**, 288
Hollis, Bruce W., **123**, 167, 176
Holloway, Christopher J., **104**, 281
Hollstein, Monica, **105**, 249
Holmgren, Arne, **107**, 295; **113**, 525
Holmquist, Barton, **130**, 270
Holzman, Rita B., **133**, 98, 248
Holzman, Thomas F., **133**, 98, 248
Hom, S. S. M., **118**, 528
Hommes, N. G., **110**, 340
Honda, Susumu, **119**, 204
Honek, John F., **122**, 199
Hong, Jau-Shyong, **103**, 547
Hong, Jen-shiang, **125**, 180, 302
Hoogenraad, Nicholas J., **121**, 375
Horbett, T. A., **112**, 484
Horecker, B. L., **106**, 193; **116**, 255, 265; **117**, 342
Hořejší, Václav, **104**, 275
Horne, Donald W., **122**, 269
Horowitz, Bernard, **119**, 39
Horsch, Robert B., **118**, 627
Horst, Ronald L., **123**, 127
Hotta, Kunimoto, **119**, 481
Houdebine, Louis-Marie, **109**, 667
Hovmöller, Sven, **126**, 191
Howard, A. J., **114**, 452
Hruby, Dennis E., **124**, 295
Hsu, Tingsen, **113**, 468
Hubbard, Roger A., **108**, 139
Hudson, Bruce, **130**, 331
Huennekens, F. M., **122**, 260; **123**, 28
Humbel, René E., **109**, 782
Humbert, James R., **132**, 520
Humble, Elisabet, **107**, 130
Hummelen, Jan C., **133**, 531
Hundal, T., **126**, 428
Hunt, Arthur G., **122**, 43; **125**, 302
Hunt, S., **107**, 413
Hunter, James T., **108**, 348
Hunter, Kenneth W., Jr., **121**, 541
Huszar, Gabor, **106**, 287
Huttner, Wieland B., **107**, 200
Huynh, Quang Khai, **122**, 135
Hyams, Jeremy S., **134**, 225
Hymer, Wesley C., **103**, 257

I

Ihle, James N., **116**, 540
Iida, Shozo, **102**, 135
Ikekawa, Nobuo, **111**, 346, 352
Ikenoya, Satoru, **105**, 147
Illum, Lisbeth, **112**, 67
Imae, Yasuo, **125**, 582
Imai, Jiro, **119**, 522
Imai, Kazuhiro, **133**, 435
Imai, Yoh, **110**,375
Ingebritsen, T. S., **107**, 102

Ingham, Kenneth C., **104**, 351
Inglot, Anna D., **119**, 657
Ingram, C. D., **124**, 207
Innerarity, Thomas L., **129**, 542
Innis, Michael A., **119**, 397
Inouye, Masayori, **125**, 138
Iqbal, M. Perwaiz, **122**, 346
Isaac, R. E., **111**, 377
Isaacs, N. W., **115**, 112
Ishii, Koichi, **110**, 184
Ishizaka, Kimishige, **116**, 76
Israelachvili, Jacob, **127**, 353
Issartel, Jean Paul, **126**, 712
Itoh, Shigeru, **125**, 58
Itoku, Kendall A., **105**, 341
Iverius, Per-Henrik, **129**, 691
Iwamoto, Machiko, **132**, 481
Iyengar, Ravi, **109**, 13, 207
Izzo, G., **126**, 331

J

Jabalquinto, Ana María, **110**, 86
Jackson, Cynthia L., **128**, 788
Jackson, J. Baz., **127**, 557
Jackson, Richard L., **128**, 288; **129**, 738
Jackson, Susan, **116**, 157
Jacob, Harry S., **105**, 378
Jacobs, Steven, **109**, 399
Jacobsen, Donald W., **123**, 14, 28
Jacobson, Elaine L., **106**, 483
Jacobson, Myron K., **106**, 483
Jaenicke, Rainer, **131**, 218
Jagendorf, Andre T., **118**, 282
Jakoby, William B., **113**, 495, 510, 516
Jakway, James P., **121**, 481
James, Stephanie L., **132**, 649
Jamoulle, Jean-Claude, **119**, 522
Janin, Joël, **115**, 420
Jansonius, J. N., **114**, 132
Janzen, Edward G., **105**, 188
Jaquet, Philippe, **103**, 287
Jarausch, Jochen, **126**, 32
Jarett, Leonard, **109**, 187
Jaton, Jean-Claude, **116**, 26
Jefferson, M. Margaret, **132**, 569, 585
Jelsema, Carole L., **109**, 480
Jennes, Lothar, **103**, 448; **124**, 36
Jennings, Walter, **111**, 149
Jensen, Lyle H., **115**, 227
Jensen, Robert T., **109**, 64, 288
Jensson, Helgi, **113**, 504
Ji, Inhae, **109**, 203
Ji, Tae H., **109**, 203
Jiang, Pan Hong, **119**, 694
Job, Didier, **134**, 160
Johansson, Göte, **104**, 356
Johnson, Fred L., **129**, 45
Johnson, Kenneth A., **134**, 306, 677
Johnson, Leonard R., **109**, 56
Johnson, Mark L., **109**, 572
Johnson, Mary I., **103**, 334
Johnson, Michael L., **117**, 301; **130**, 534
Johnson, William J., **132**, 555
Johnston, Richard B., Jr., **105**, 365
Jolley, Diane, **124**, 389
Jonas, Ana, **128**, 553
Jones, Benjamin L., **110**, 209
Jones, C., **123**, 375
Jones, Glenville, **123**, 141
Jones, Harold P., **105**, 389
Jones, Mary Ellen, **113**, 113
Jones, Owen T. G., **132**, 378
Jones, Patricia P., **108**, 452
Jones, Philip D. E., **112**, 67
Jones, Simon J., **125**, 190
Jones, T. Alwyn, **115**, 157
Jones, Valerie P., **118**, 470
Jones, Wanda M., **113**, 108
Jones-Mortimer, Maurice C., **125**, 157
Jordan, Peter M., **123**, 427, 435
Jovell, Robert J., **121**, 509
Juarez, Dolores V., **108**, 683
Juarez-Salinas, Hector, **106**, 483; **121**, 615
Juillerat, Marcel A., **131**, 185
Jung, Vincent, **119**, 597

K

Kaback, H. Ronald, **125**, 214, 429, 453; **126**, 113, 370
Kadenbach, Bernhard, **126**, 32
Kaetzel, Marcia A., **102**, 1
Kagamiyama, Hiroyuki, **113**, 83
Kagan, Elliott, **108**, 325
Kagawa, Yasuo, **126**, 1, 604, 640
Kahlon, Talwinder S., **129**, 26
Kahn, C. Ronald, **109**, 593, 609

Kahn, Daniel, **106**, 141
Kahn, Maria, **119**, 383
Kahn, R., **114**, 281
Kaiser, Donald A., **134**, 423
Kajikawa, Norio, **102**, 279
Kalata, Kenneth, **114**, 486
Kam, Z., **114**, 77
Kamen, B. A., **122**, 339
Kamiya, Toshio, **121**, 817
Kamo, Naoki, **125**, 46
Kanazawa, Hiroshi, **126**, 588, 595
Kandutsch, Andrew A., **110**, 9
Kane, John P., **128**, 262, 519; **129**, 123
Kanno, Masamoto, **116**, 311
Kao, Fa-Ten, **128**, 851
Kaplan, Ronald S., **125**, 671
Kapp, Judith A., **116**, 295, 303, 325
Karathanasis, Sotirios K., **128**, 690, 712
Karlsson, R., **114**, 132
Karnovsky, Manfred L., **132**, 353, 365
Karp, Matti T., **122**, 147
Karplus, Martin, **127**, 369; **131**, 283, 362
Kaseman, Deborah S., **113**, 305
Kasting, Norman W., **103**, 176
Kasuga, Masato, **109**, 609
Katayama, Kouichi, **105**, 147
Katinger, H. W. D., **121**, 352
Kato, Kanefusa, **102**, 256
Kato, Tetsuro, **112**, 139
Katoh, Masao, **109**, 156, 667
Katona, Eva, **127**, 662
Katsuki, Hirohiko, **110**, 373
Kattchee, Philip A., **122**, 381
Katz, David H., **116**, 428
Kaufman, Bennett, **121**, 797
Kaufman, James F., **108**, 600
Kauppinen, Hanna-Leena, **119**, 27, 54
Kawade, Yoshimi, **119**, 558
Kawamoto, Tomoyuki, **121**, 266
Kawanishi, Shosuke, **123**, 415
Kawasaki, Takashi, **122**, 15
Kearney, John F., **116**, 174
Keegstra, Kenneth, **118**, 316
Keen, Peter M., **124**, 335
Keith, Alec D., **127**, 360
Kelder, Bruce, **119**, 383, 582, 588
Kell, Douglas B., **127**, 538
Keller, Jonathan, **116**, 540
Keller, R. Kennedy, **111**, 201
Kelley, Jim L., **128**, 170
Kelly, Paul A., **109**, 156, 667
Keltjens, Jan T., **122**, 412
Kempf, Christoph, **104**, 329
Kempner, Ellis S., **117**, 65; **119**, 255
Kennett, Roger H., **121**, 59
Kern, Daniel, **113**, 42
Kerner, J., **123**, 264
Keszthelyi, Lajos, **127**, 610
Keyomarsi, Khandan, **122**, 309
Khan, Masood N., **109**, 219
Khananshvili, Daniel, **126**, 528
Khorana, H. Gobind, **125**, 190
Kido, Hiroshi, **106**, 193; **117**, 342
Kidwell, William R., **106**, 505
Kiehart, Daniel P., **134**, 423
Kielian, Margaret, **132**, 257
Kikkawa, Ushio, **124**, 349
Kilian, Mogens, **116**, 37
Kilpatrick, Brian F., **103**, 577
Kim, Chong H., **126**, 238
Kim, J. B., **133**, 387
Kim, Peter S., **131**, 136
Kim, Sangduk, **106**, 295
Kim, Sung-Hou, **114**, 156, 167
Kindt, Thomas J., **116**, 157
King, Chi-Hsin Richard, **110**, 171
King, Jonathan, **131**, 250
King, Stephen M., **134**, 225, 291
King, Tsoo E., **126**, 238
Kirsch, Wolff M., **107**, 563
Kirschner, Marc, **134**, 156, 261
Kirshner, Norman, **103**, 305
Kishimoto, Akira, **102**, 279
Kissinger, Peter T., **122**, 300
Kita, Kiyoshi, **126**, 94
Kitano, Tatsuro, **124**, 349
Kitchell, Judith P., **112**, 436
Kittler, Jason M., **122**, 120
Kizer, John Stephen, **103**, 435
Klass, Michael R., **134**, 414
Klausner, Richard D., **104**, 340; **128**, 657
Klebanoff, Seymour J., **105**, 399
Klee, C. B., **102**, 227
Kleiman, Norman J., **108**, 43
Klein, Gérard, **126**, 712
Kleinig, Hans, **110**, 267, 299; **111**, 248
Klevit, Rachel E., **102**, 82
Klingenberg, Martin, **125**, 618; **126**, 490, 498; **127**, 772
Klostermeyer, H., **106**, 58; **107**, 258

Knight, Ernest, Jr., **119**, 284
Knott, R. B., **127**, 217
Knudsen, Peter, **108**, 600
Knutson, Grace S., **119**, 499
Kobatake, Yonosuke, **125**, 46
Kobayasi, Syoyu, **125**, 582
Koch, Tad H., **107**, 563
Koeppe, John K., **111**, 509
Kohen, F., **133**, 284, 387
Kohn, Joachim, **104**, 3
Kohr, William J., **116**, 448; **119**, 424
Kolattukudy, P. E., **106**, 210
Kolbe, Hanno V. J., **125**, 697
Konings, Wil N., **125**, 492; **126**, 159
Konishi, Kiyoshi, **126**, 94
Konopinska, Danuta, **132**, 318
Koolman, J., **111**, 419, 429
Koopman, William J., **116**, 146
Kopito, Ron R., **110**, 58
Kornfeld, Stuart, **107**, 163
Korngold, Robert, **108**, 270
Koshland, Daniel E., Jr., **106**, 310, 321
Kossiakoff, Anthony A., **114**, 530; **131**, 433
Koszewski, Nick J., **123**, 127
Kovalick, Gae E., **111**, 509
Kovář, Jan, **121**, 277
Koyama, Tanetoshi, **110**, 153, 192, 199
Kozbor, Danuta, **121**, 120, 140
Kramer, James J., **113**, 113
Krämer, Reinhard, **125**, 610, 692
Kramer, Richard A., **119**, 416
Kraus, R. J., **107**, 582, 593
Krause, James E., **103**, 511, 539
Krauss, Ronald M., **128**, 181, 417
Krebs, Keith E., **128**, 387
Kreil, Günther, **106**, 218
Kreis, Thomas E., **134**, 507
Kresheck, Gordon C., **127**, 141
Kreuter, Jörg, **112**, 129
Kreuz, Klaus, **111**, 248
Kricka, Larry J., **133**, 229, 331, 404
Krief, Patricia, **121**, 580
Krieger, Monty, **128**, 608; **129**, 227
Krinks, M. H., **102**, 227
Krinsky, Norman I., **105**, 155; **132**, 369
Kröger, Achim, **126**, 387
Krul, Elaine S., **128**, 527
Krulwich, T. A., **125**, 352
Krupen, K., **116**, 325
Kruski, Arthur W., **128**, 170
Kuehn, Glenn D., **107**, 154
Kuhar, Michael J., **124**, 560
Kuhn-Nentwig, Lucia, **126**, 32
Kuhs, W. F., **127**, 303
Kulaev, I. S., **126**, 447
Kumar, A. Ashok, **122**, 293
Kümel, Gunther, **121**, 737
Kummer, Udo, **121**, 670
Kumosinski, Thomas F., **117**, 154, 219
Kung, Hsiang-fu, **119**, 204, 292, 296, 315
Kunicki, Thomas J., **121**, 702
Kunz, Heinz W., **121**, 386
Kurfürst, M., **133**, 140
Kuriyama, Ryoko, **134**, 190
Kuusi, Timo, **129**, 716
Kuwata, Tsuguo, **119**, 702
Kvamme, Elling, **113**, 241

L

Laakso, Simo, **105**, 126
Labbe, Pierre, **123**, 363
Labes, M., **118**, 640
Labhardt, A. M., **131**, 126
Lachman, Lawrence B., **116**, 467
Lacko, Andras G., **129**, 763
Lacroix, Andre, **123**, 75
LaFata, Margherita, **121**, 193
Lagacé, Lisette, **102**, 104
Laghai-Newton, Aghdas, **123**, 435
Lakowicz, Joseph R., **131**, 518
Lambert, Nigel, **107**, 281
Lamoyi, Edmundo, **121**, 652
Lampen, J. Oliver, **106**, 365
Lamport, Derek T. A., **106**, 523
Landau, Bernard R., **110**, 100
Landon, Roxanne M., **124**, 412
Lane, Leslie C., **118**, 687
Lane, Richard D., **121**, 183
Lang, Johanna, **105**, 319
Lang, Les, **107**, 163
Langanger, G., **134**, 592
Langer, Jerome A., **119**, 248, 305
Langer, Robert, **112**, 399
Langone, John J., **121**, 871
Lansdorp, P. M., **121**, 855
Lanyi, Janos K., **125**, 510
Lapointe, Jacques, **113**, 42, 50
Larson, Kerstin, **113**, 520

Larsson, Per-Olof, **104**, 212, 364
Lasky, Stephen R., **119**, 499
Lattman, Eaton, **115**, 55
Lau, Yun-Fai, **128**, 811
Lauer, Stephen J., **118**, 109
Läuger, P., **127**, 465
Lauquin, Guy J. M., **125**, 639, 658
Lawrence, Glen D., **105**, 305
Lawson, Alexander M., **111**, 63
Lax, Sandra R., **118**, 109
Le, Junming, **119**, 48
Le, Ngoc-Anh, **129**, 384, 457
Leach, Franklin R., **133**, 51
Leaver, Christopher J., **118**, 470
LeBoeuf, Renee C., **128**, 877
Lebon, Pierre, **119**, 541
Lee, Choongeun, **119**, 667
Lee, Dae-Sil, **125**, 190
Lee, David K., **119**, 153
Lee, James, **132**, 318
Lee, James C., **117**, 27, 97, 381
Lee, John, **133**, 109, 149
Lee, Nancy S., **128**, 375
Lee, Sang-Sun, **122**, 237
Lefker, Bruce A., **132**, 355
Lefkowitz, Doris L., **132**, 537
Lefkowitz, Stanley S., **132**, 537
Lehnert, Bruce E., **132**, 183
Lemire, Bernard D., **126**, 377
Lengeler, Joseph W., **125**, 473
Lengyel, Peter, **119**, 489, 516
Leonard, Kevin, **126**, 191
Lepault, Jean, **127**, 719
Le Peuch, Christian J., **102**, 261
Le Peuch, Danielle A.-M., **102**, 261
Lerch, Konrad, **106**, 355
Leroux, P., **124**, 607
Lesiak, Krystyna, **119**, 522
Lesk, Arthur M., **115**, 381
Lesko, Stephen A., **105**, 539
Letarte, Michelle, **108**, 642, 654
Levasseur, Sylvain, **113**, 42
Levin, David E., **105**, 249
Levine, J. E., **124**, 466
Levine, Joel, **134**, 42
Levine, Rodney L., **107**, 370
Levings, C. S., III, **118**, 485
Lew, P. Daniel, **124**, 116
Lewis, C. Ann, **105**, 231
Lewis, Garry P., **122**, 319
Lewis, Thomas A., **111**, 333
Li, Bo-Liang, **119**, 351
Li, Luyuan, **113**, 451
Li, Shi Wu, **119**, 351
Li, Ting-Kai, **122**, 97
Lian, Jane B., **107**, 516; **113**, 133
Liebes, Leonard F., **122**, 10
Liem, Ronald K. H., **134**, 380
Lilius, Esa-Matti, **105**, 126
Lim, C. K., **123**, 383, 389
Lin, C. Shirley, **109**, 625
Lin, Chi-Shui, **126**, 490
Lin, Chin-Tarng, **113**, 3; **129**, 297
Lin, E. C. C., **125**, 467
Lin, Leo S., **119**, 183
Lin, Lung-Nan, **131**, 107
Lin, Michael C., **109**, 356, 360
Lin, Shuo-Liang, **127**, 649
Lindgren, Frank T., **129**, 26
Lindmo, Tore, **121**, 678
Ling, Nicholas C., **103**, 72
Link, Th. A., **126**, 224, 253
Linke, Petra, **126**, 201
Liou, G. I., **123**, 102
Liptak, Raymond A., **119**, 153
Lister, Richard M., **118**, 742
Litvin, David A., **108**, 298
Liu, Xin-Yuan, **119**, 351, 676
Ljungdahl, Lars G., **122**, 385, 392
Ljungdahl, Per O., **126**, 181
Lloyd-Williams, Paul, **125**, 109, 119
Lo, Kin-Ming, **125**, 190
Lobaugh, Bruce, **123**, 159
Locker, Daniel, **121**, 491
Loebenstein, Gad, **119**, 729
Loewen, Peter C., **122**, 330
Loewy, Ariel G., **107**, 241
Löffler, Fridolin, **109**, 341
Lohmann, W., **105**, 451
Longo, William E., **112**, 18
Lonsdale, David M., **118**, 453
Lonsdale, H. K., **112**, 495
Loomis, Carson, **124**, 353
Loor, F., **108**, 371
López de Castro, José A. , **108**, 582
Lou, Lillian L., **106**, 330
Loudes, Catherine, **103**, 313
Low, Teresa L. K., **116**, 213, 219, 233, 248
Lowe, Christopher R., **104**, 97
Lown, J. William, **105**, 532

Löwy, I., **116**, 403
Lübbem, Mathias, **126**, 682
Lubiniecki, Anthony S., **119**, 77
Luborsky, Judith L., **109**, 298
Luchins, Jeremy, **130**, 519
Lücken, Uwe, **126**, 682, 733
Ludwig, Bernd, **126**, 153
Lufkin, Thomas, **124**, 269
Luhr, Jordan E., **108**, 274
Lui, Alec, **122**, 97
Luider, Theo M., **133**, 531
Lumeng, Lawrence, **122**, 97
Lunardi, Joël, **126**, 712
Lundin, Arne, **133**, 27
Lunn, Charles A., **125**, 138
Lünsdorf, Heinrich, **126**, 770
Lunte, Craig E., **122**, 300
Luque, Enrique H., **124**, 443
Lusis, Aldons J., **128**, 877
Luu, B., **111**, 411
Lynch, William, **134**, 37
Lynes, Michael A., **108**, 481

M

Macario, Alberto J. L., **121**, 509
MacDonald, E. M. S., **110**, 347
Macey, Robert I., **127**, 598, 738
Mackay, Ian R., **121**, 748
Macnab, Robert M., **125**, 563
Macpherson, Andrew J. S., **125**, 387
Mage, Michael G., **108**, 118, 125
Magun, Arthur M., **129**, 519
Mahaffee, Darien D., **129**, 679
Mahley, Robert W., **128**, 273, 801, 811; **129**, 145, 542
Mahuren, J. Dennis, **122**, 102
Maier, R. J., **118**, 528
Makino-Tasaka, Momoyo, **123**, 53
Male, David, **121**, 556
Maloney, Peter C., **125**, 558
Maloof, Farahe, **107**, 445
Maloy, W. Lee, **108**, 437
Manalan, A. S., **102**, 227
Mandelkow, Eckhard, **134**, 612, 633, 657
Mandelkow, Eva-Maria, **134**, 612
Maneckjee, Rhoda, **124**, 172
Manganiello, Vincent C., **109**, 480
Manjunath, P., **109**, 725
Mannella, Carmen A., **125**, 595
Mannering, Gilbert J., **119**, 718
Mannervik, Bengt, **113**, 484, 490, 499, 504, 507, 520
Manning, David R., **102**, 74
Manning, James M., **113**, 108
Manolagas, Stavros C., **123**, 190
Manouvriez, Patrick, **121**, 622
Mansurova, S. E., **126**, 447
Manthey, J. A., **107**, 439
Marbrook, J., **121**, 759
Marcel, Yves L., **128**, 432
Marchalonis, John J., **108**, 139
Marcus, A., **118**, 128
Marcus, Philip I., **119**, 106, 115
Marcus-Samuels, Bernice, **109**, 656
Marder, Jonathan B., **118**, 384
Margel, Shlomo, **112**, 164
Margolis, Robert L., **134**, 160
Mariani, Massimo, **121**, 193
Mariz, Ida K., **109**, 773
Markert, Michèle, **105**, 358
Marnett, Lawrence J., **105**, 347, 412
Marra, Johan, **127**, 353
Marsh, Julian B., **129**, 498
Marsh, Mary E., **106**, 351
Martensen, Todd M., **107**, 3
Martin, Joseph B., **103**, 176
Martin, Mitchel, **124**, 278
Martin, Nancy C., **106**, 152
Martin, Thomas F. J., **124**, 424
Martinez, Hugo M., **130**, 208
Marzetta, Carol A., **129**, 45
Mason, Ronald P., **105**, 416
Mason, W. T., **124**, 207
Massagué, Joan, **109**, 179
Massey, John B., **128**, 403, 515
Massone, Annalisa, **121**, 193
Mastro, Andrea M., **127**, 360
Matheson, Iain B. C., **133**, 109
Matsuda, Toshio, **122**, 20
Matsukura, Hiroshi, **125**, 582
Matsumura, Toshiharu, **132**, 481
Matsushita, Kazunobu, **126**, 113
Matsuura, Katsumi, **126**, 293
Mattera, Rafael, **109**, 566
Matteucci, Mark, **119**, 424
Matthew, James B., **130**, 413, 437
Matthews, Brian W., **114**, 176; **115**, 397
Matthews, Rowena G., **122**, 333, 372

Mattoo, Autar K., **118**, 384
Maurizi, Michael R., **113**, 627
May, Brian K., **123**, 395
Mayne, Leland, **130**, 331
McCann, Richard O., **133**, 288
McCarron, Richard M., **108**, 274
McCaslin, Darrell R., **117**, 41
McClean, Phillip E., **118**, 437
McClure, Don B., **121**, 266
McComas, Warren, **125**, 214
McConathy, W. J., **128**, 297
McCord, Joe M., **105**, 389
McCormick, Donald B., **122**, 63, 116, 237
McCormick, Frank, **119**, 397
McCoy, John M., **125**, 190
McElvany, K. D., **107**, 439
McEnery, Maureen W., **126**, 470
McFadden, Philip N., **106**, 330
McFarlin, D. E., **121**, 425
McFaul, Steve J., **132**, 491
McGarrigle, Richard, **124**, 190
McGarry, Michael, **119**, 103
McGhee, Jerry R., **121**, 42
McGinity, James W., **112**, 84
McIntire, William S., **106**, 369
McIntosh, L., **118**, 419
McIntosh, T. J., **127**, 511
McKelvy, Jeffrey F., **103**, 511, 539; **124**, 548
McKenzie, J. M., **109**, 677
McLean, Larry R., **129**, 738
McMaster, W. Robert, **108**, 558
McNeil, Michael, **118**, 3
McPherson, Alexander, **114**, 112, 120, 125
Means, Anthony R., **102**, 104
Mednis, Alma D., **116**, 379
Meeker, Rick B., **103**, 132
Mefford, Ivan N., **124**, 402
Mehlhorn, Rolf J., **105**, 215; **127**, 738
Meighen, Edward, **133**, 70, 172, 183, 189
Meints, R. H., **118**, 485
Meisler, Natalie T., **122**, 120
Meister, Alton, **113**, 59, 185, 305, 327, 338, 344, 350, 379, 390, 393, 400, 438, 445, 451, 458, 468, 555, 567, 571, 602
Mendelsohn, Frederick A. O., **124**, 590
Meng, Martha, **128**, 223
Merril, Carl R., **104**, 441
Merrill, Alfred H., Jr., **122**, 110
Messeri, G., **133**, 557
Mestecky, Jiri, **116**, 37
Mezei, Mihaly, **127**, 21
Michalek, Suzanne M., **121**, 42
Miksztal, Andrew R., **105**, 71
Millan, Monica, **124**, 590
Miller, Barbara E., **102**, 291; **123**, 154
Miller, Gregory V., **105**, 341
Miller, Joyce F., **106**, 403
Miller, K. F., **121**, 433
Miller, Mark S., **122**, 227
Miller, Michael J., **126**, 87
Miller, Paul J., **132**, 243
Miller, Richard G., **108**, 64
Mills, David, **118**, 41
Mills, Kingston, **121**, 726
Milstein, C., **121**, 210
Miranda, Armand F., **119**, 611, 619
Miron, Talia, **104**. 3
Mirshahi, Massoud, **121**, 580
Mishell, Barbara B., **108**, 303
Mishell, Robert I., **108**, 303
Misra, Dhirendra N., **121**, 386
Mitchell, William M., **119**, 533
Mitchison, T. J., **134**, 261
Mittal, Chandra K., **132**, 422, 428
Mitzka-Schnabel, Ursula, **110**, 253
Miwa, Masanao, **106**, 441, 495
Miyamoto, Carol, **133**, 70
Mizel, Steven B., **116**, 480
Miziorko, Henry M., **110**, 19
Moehring, J. M., **106**, 388
Moehring, T. J., **106**, 388
Mogensen, Knud Erik, **119**, 267, 347
MohanaRao, J. K., **130**, 185
Moldow, Charles F., **105**, 378
Molenaar, R., **109**, 275
Molinaro, Giuseppe A., **121**, 783
Molony, L., **134**, 69
Molteni, Luigi, **112**, 285
Momany, Frank A., **124**, 3
Moncuit, J., **116**, 403
Monger, Daniel J., **110**, 51
Mongini, Patricia K. A., **121**, 438
Monji, Nobuo, **123**, 85
Monroe, John G., **103**, 227
Montal, Mauricio, **126**, 123
Montgomery, Robert R., **121**, 702
Moody, Carmella S., **105**, 254
Moody, David E., **111**, 303
Moore, Linda L., **132**, 520

Moore, Marsha N., **128**, 745
Moore, Robert Y., **103**, 619
Moradi-Ameli, Mahnaz, **126**, 761
Moran, Richard G., **122**, 309
Morch, M. D., **118**, 154
Morel, G., **124**, 607
Morhenn, Vera B., **108**, 357
Mori, Hiroshi, **134**, 573
Morin, John E., **113**, 541
Morino, Yoshimasa, **106**, 116
Morisaki, Masuo, **111**, 346, 352
Mornex, Françoise, **121**, 802
Morrell, Joan I., **103**, 639
Morris, Alan G., **119**, 96
Morris, Mariana, **103**, 493, 524; **124**, 359
Morris, R. O., **110**, 340, 347
Morrisett, Joel D., **128**, 472; **129**, 167
Morse, David, **133**, 189
Morse, Philip D., II, **127**, 239
Mosbach, Klaus, **104**, 56, 364; **121**, 352
Moschera, John A., **119**, 177, 204
Moser, Christopher C., **126**, 293
Mosher, Mary E., **126**, 558
Mosier, D. E., **108**, 294
Moss, Joel, **106**, 411, 430; **109**, 480
Motsenbocker, Marvin A., **107**, 602
Motta, Geneviève, **121**, 491
Moura, Teresa, **127**, 598
Moyle, William R., **109**, 638; **121**, 695
Mulcahy, Linda, **119**, 383
Müller, Armin, **105**, 311
Müller, Michele, **126**, 78
Mullet, John E., **118**, 232
Mulligan, B. J., **118**, 578
Munding, Anton, **125**, 631
Munkres, Kenneth D., **105**, 263
Murata, Kousaku, **113**, 567
Murofushi, Hiromu, **106**, 223
Murphy, Mary A., **108**, 274
Murtaugh, Timothy J., **102**, 158
Murtiashaw, Martha H., **106**, 88
Musliner, Thomas A., **128**, 417

N

Na, George C., **117**, 459, 496
Nacy, Carol Ann, **132**, 603
Nadler, J. Victor, **103**, 393
Nagai, Yoshitaka, **132**, 481
Nagashima, Fujio, **106**, 116
Nairn, Roderick, **108**, 505
Najjar, Victor A., **132**, 318
Nakanishi, Koji, **127**, 649
Nałęcz, Katarzyna A., **126**, 45
Nałęcz, Maceij J., **125**, 86
Napoli, Joseph L., **123**, 112, 127
Nargessi, R. D., **122**, 67
Nayar, Rajiv, **132**, 594
Neary, Joseph T., **107**, 445
Néauport-Sautès, C., **116**, 403
Négrel, R., **109**, 377
Negro-Vilar, Andres, **124**, 67
Neil, James, **124**, 190
Neill, Jimmy D., **124**, 443
Neilson, G. W., **127**, 319
Nelsestuen, Gary L., **107**, 503, 507
Nelson, Eldon C., **123**, 92
Nelson, Nathan, **118**, 352; **126**, 285, 520
Nelson, Phillip G., **103**, 111
Nelson, Sidney D., **112**, 340
Némethy, George, **127**, 183
Nerland, Donald E., **119**, 715
Nes, William R., **111**, 3
Nestel, Paul J., **129**, 443
Neubacher, H., **105**, 451
Neukom, Hans, **107**, 377
Newell, D. G., **108**, 393
Newman, Michael J., **125**, 429
Newman, Thomas C., **128**, 671
Newton, Colin J., **122**, 152, 155
Newton, Kathleen J., **118**, 488
Nice, Edouard C., **116**, 588
Nichol, Lawrence W., **117**, 182; **130**, 6
Nicholls, David G., **127**, 557
Nichols, Alex V., **128**, 417
Nicola, Nicos A., **116**, 600
Nicolas, Jean-Claude, **133**, 209
Nicolson, Garth L., **121**, 717
Niedel, James E., **124**, 57
Nielsen, C., **114**, 452
Nielsen, Jennifer B. K., **106**, 365
Nielsen, Peter, **122**, 209
Nielsen, T. B., **117**, 65
Niezgodka, Malgorzata, **108**, 666
Nikaido, Hiroshi, **125**, 265
Nilsson, Kurt, **104**, 56; **121**, 352
Ning, Robert Y., **119**, 153
Nishimura, Mitsuo, **125**, 58
Nishino, Kohsuke, **122**, 24

Nishino, Tokuzo, **110**, 167, 373
Nishizuka, Yasutomi, **102**, 279; **124**, 349
Nissenson, Robert A., **109**, 40, 48
Nivison, Helen T., **118**, 282, 437
Noeman, Saad A., **121**, 386
Noguchi, Masato, **133**, 298
Nordhausen, Robert W., **128**, 442
Norman, Anthony W., **102**, 291; **123**, 154
Northrop, Dexter B., **122**, 152, 155
Notari, Robert E., **112**, 309
Noumi, Takato, **126**, 595
Nozaki, Mitsuhiro, **113**, 83
Nunez, J., **107**, 476
Nussbaum, Samuel R., **109**, 625
Nuydens, R., **134**, 592
Nyrén, Pål, **126**, 538

O

Oberfelder, Robert W., **117**, 27, 381
Oberley, Larry W., **105**, 457
O'Brien, Peter J., **105**, 370
Ockenhouse, Christian F., **132**, 637
O'Connell, M., **118**, 640
O'Connell, Mary A., **118**, 437
O'Connor, Clare M., **106**, 330
O'Connor, John D., **111**, 458
Odashima, Kazunori, **127**, 649
Odessey, Richard, **123**, 284
Ogawa, Yasushi, **111**, 274
Ogura, Kyozo, **110**, 153, 167, 184, 188, 192, 199, 206, 320
O'Halloran, T., **134**, 69
Ojima, Nobutoshi, **110**, 320
Okajima, Masako, **132**, 549
O'Kane, Dennis J., **133**, 109, 149
Okazaki, Mitsuyo, **129**, 57
Oláh, Imre, **108**, 3
Oliver, Janet M., **108**, 336
Olmsted, J. B., **134**, 467
Olney, John W., **103**, 379
Olsen, Kenneth W., **123**, 324
Olsson, Lennart, **121**, 295
Olwin, Bradley B., **102**, 148
O'Malley, Judith A., **119**, 103
Ono, Teruo, **110**, 375
Opella, Stanley J., **131**, 327
Oppenheim, Joost J., **116**, 357
Oppenheimer, Norman J., **106**, 399
Oram, John F., **129**, 645
Ordovas, Jose M., **129**, 420
Orlando, C., **133**, 557
Orozco, Emil M., Jr., **118**, 232
Orr, George A., **122**, 83, 87
Orrenius, Sten, **105**, 505
Ortaldo, John R., **132**, 445
Osborne, James C., Jr., **128**, 213, 375
O'Shea, Paul, **126**, 78
Östberg, Lars, **121**, 228
Östlund-Lindqvist, Ann-Margaret, **129**, 691
Ott, Gary S., **121**, 615
Otter, Tim, **134**, 291
Ötting, F., **105**, 93
Otto, Joann J., **134**, 555

P

Pace, C. N., **131**, 266
Packard, C. J., **129**, 566
Packer, Lester, **105**, 215; **127**, 738
Padan, Etana, **125**, 214, 337
Paetkau, Verner, **116**, 526
Paik, Woon Ki, **106**, 265, 268, 274
Paik, Young-Ki, **128**, 811
Pajtasz, Elżbieta, **119**, 657
Paladini, Alejandro A., Jr., **130**, 493
Palaszynski, Edmund, **116**, 540
Palkovits, Miklós, **103**, 368
Palmer, J., **112**, 43
Palmer, Jeffrey D., **118**, 167
Palmieri, Ferdinando, **125**, 692
Palukaitis, Peter, **118**, 723
Pan, Yu-ching E., **119**, 204
Panayotatos, Nikos, **119**, 166
Pande, Shri V., **123**, 259, 297
Papa, S., **126**, 331
Parak, Fritz, **127**, 196; **131**, 568
Pardo, José V., **134**, 460
Park, Myung Hee, **106**, 344
Parker, Thomas S., **110**, 58
Parks, D. R., **108**, 197
Parks, Leo W., **111**, 37, 333
Parlo, Risa A., **125**, 671
Parmelee, David C., **134**, 355
Parr, Gary R., **131**, 185
Parsegian, V. A., **127**, 400
Parsons, Thomas F., **109**, 736
Pastan, Ira, **109**, 257

Paszkowski, Jerzy, **118**, 668
Patel, Lekha, **126**, 113
Patsch, Josef R., **129**, 3
Patsch, Wolfgang, **129**, 3
Payne, D. Michael, **106**, 483
Payne, William E., **126**, 325
Pazzagli, Mario, **133**, 420
Pearson, James C., **104**, 97
Pearson, Richard D., **132**, 603
Pedersen, Peter L., **126**, 470, 477, 484, 660
Pedrali-Noy, Guido, **118**, 87
Peleg, Edna, **103**, 249
Pellegrino, M. A., **108**, 614
Pelletier, G., **124**, 607
Penefsky, Harvey S., **126**, 608
Penin, François, **126**, 417, 455
Penningroth, Stephen M., **134**, 477
Pennock, J. F., **110**, 313
Pennoyer, Jeffrey D., **126**, 181
Penschow, J. D., **124**, 534
Pepe, Michael G., **128**, 354
Perlmann, P., **108**, 153
Perrin, Marilyn H., **124**, 164, 389
Perrot, Jean-Yves, **121**, 580
Perry, L. Jeanne, **119**, 424
Persson, Jörgen, **133**, 27
Perutz, Max, **114**, 3
Pessen, Helmut, **117**, 154, 219
Pestka, Sidney, **119**, 3, 14, 248, 255, 286, 305, 315, 383, 434, 481, 582, 588, 597
Peterfreund, Robert A., **103**, 347
Peters, Frens, **126**, 733
Peters, T. J., **123**, 383, 389
Petersen, N. O., **130**, 454
Peterson, Elbert A., **104**, 113
Peterson, Ronald C., **111**, 482
Pethig, Ronald, **105**, 238
Petsko, Gregory A., **114**, 141, 147; **131**, 389
Pfaff, Donald W., **103**, 639; **124**, 497
Pfendt, Eva A., **108**, 357
Pflumm, Mollie, **130**, 519
Phillips, A. W., **119**, 35
Phillips, D. C., **114**, 397; **115**, 117
Phillips, George N., Jr., **114**, 128
Phillips, Michael C., **128**, 387; **129**, 628
Phillips, Walter C., **114**, 300, 316
Pick, Edgar, **132**, 407
Pick, Uri, **126**, 512
Picton, Colin, **102**, 219
Pierce, Carl W., **116**, 295, 303, 395
Pierce, John G., **109**, 736
Pike, Beverley L., **121**, 341
Pike, J. Wesley, **123**, 199
Pirola, Byron A., **123**, 395
Pitas, Robert E., **129**, 542
Pittman, Ray C., **129**, 612
Plate, Charles A., **125**, 187
Pleau, Jean-Marie, **116**, 269
Plotsky, Paul M., **103**, 469
Poapst, Mary E., **129**, 395
Pohorille, Andrew, **127**, 64
Polacheck, Itzhack, **104**, 415
Poljak, Roberto J., **116**, 190
Pollard, Thomas D., **134**, 423
Pollet, Robert J., **117**, 3
Poole, Philip L., **127**, 284
Poolman, Bert, **126**, 159
Poonian, Mohindar S., **125**, 214
Porter, John C., **103**, 607
Porter, John W., **110**, 71, 209
Porter, Ned A., **105**, 273
Portetelle, Daniel, **121**, 587
Posner, Barry I., **109**, 219
Posner, Gary H., **133**, 569
Potekhin, Sergey A., **131**, 4
Potrykus, Ingo, **118**, 549
Potter, James D., **102**, 135
Potter, Jennifer, **104**, 318
Potter, Michael, **116**, 121
Potts, John T., Jr., **109**, 625
Pougeois, Richard, **126**, 712
Poulter, C. Dale, **110**, 130, 171; **111**, 252
Pousette, Åke, **133**, 27
Powell, Grace M., **128**, 375
Power, J. B., **118**, 578
Powers-Lee, Susan G., **113**, 27
Pownall, Henry J., **128**, 3, 403, 515
Pozzan, Tullio, **124**, 116
Prakash, V., **117**, 53, **130**, 3
Prasher, Douglas C., **133**, 288
Pratt, Lawrence R., **127**, 48, 64
Pratt, M. M., **134**, 325
Prelli, Frances C., **116**, 3
Presentini, Rivo, **121**, 193
Preston, Gregory M., **124**, 269
Prestwich, Glenn D., **110**, 417; **111**, 509
Prezioso, Girolamo, **125**, 692
Price, Madelon T., **103**, 379
Price, Paul A., **107**, 544, 548

Priddle, John D., **104**, 318
Priest, D. G., **122**, 313
Prince, Roger C., **125**, 109
Privalov, Peter L., **131**, 4
Propst, Friedrich, **109**, 316
Pross, Hugh F., **108**, 49
Proulx, Mario, **113**, 50
Pryce, Gareth, **121**, 556
Pryor, William A., **105**, 293
Pühler, A., **118**, 640
Pullman, Alberte, **127**, 250
Pullman, Maynard D., **126**, 460
Puppione, Donald L., **128**, 155
Purnell, Michael R., **106**, 505

Q

Quail, Peter H., **118**, 369
Quinlan, Roy A., **134**, 355

R

Rabinowitz, Jesse C., **113**, 169
Rabinowitz, Murray, **106**, 152
Radke, Gary A., **117**, 116
Radola, Bertold J., **104**, 256
Ragan, C. Ian, **126**, 360
Ragnarsson, Ulf, **107**, 130
Raines, Elaine W., **109**, 749
Rajagopalan, K. V., **122**, 399
Rajashekar, C. B., **127**, 761
Rall, Stanley C., Jr., **128**, 273
Ramakrishnan, Rajasekhar, **129**, 384
Ramirez, V. D., **124**, 466
Ramsdell, John S., **103**, 287
Rand, R. P., **127**, 400
Randall, Linda L., **125**, 129
Ransom, Janet H., **121**, 293
Rashidbaigi, Abbas, **119**, 286, 315, 582
Rathod, Pradipsinh K., **122**, 360
Ratner, B. D., **112**, 484
Rau, D. C., **127**, 400
Rau, Werner, **110**, 253
Rauch, Charles T., **134**, 160
Rauckman, Elmer J., **105**, 198
Rauschenbach, Peter, **122**, 209
Ravel, Joanne M., **118**, 109
Ray, Marjorie J., **128**, 181
Rayment, Ivan, **114**, 136, 316
Raziuddin, Arati, **119**, 340
Reading, Christopher L., **121**, 18
Reardon, Catherine A., **128**, 811
Reaves, Troy A., Jr., **103**, 132
Rebouche, Charles J., **123**, 290
Recknagel, Richard O., **105**, 331
Redman, Kent L., **106**, 179
Redpath, J. Leslie, **105**, 491
Rees, H. H., **111**, 377
Regnier, Fred E., **104**, 170
Rehberg, Edward, **119**, 383
Reilly, Edward B., **108**, 666
Reinhardt, Timothy A., **123**, 176
Reinisch, Lou, **131**, 568
Reiser, Georg, **109**, 316
Reisner, A. H., **104**, 439
Reisner, Yair, **108**, 168
Reitman, Marc L., **107**, 163
Remaut, Erik, **119**, 366
Rembaum, A., **112**, 150, 175
Remington, Jack S., **132**, 626
Remington, S. J., **115**, 303
Remold, Heinz G., **116**, 379
Ren, Wen Hua, **119**, 676
Renart, Jaime, **104**, 455
Repine, John E., **105**, 393
Reyes, Philip, **122**, 360
Reymond, Marianne J., **103**, 607
Reynolds, Jacqueline A., **117**, 41
Rhee, Sue Goo, **107**, 183; **113**, 213
Ribeiro-Neto, Fernando A. P., **109**, 566
Rich, Susan, **116**, 416
Richards, Frederic M., **115**, 145, 440
Richards, G., **112**, 150
Richardson, David C., **115**, 189
Richardson, Jane S., **115**, 189, 341, 359
Richey, James S., **104**, 223
Richter, Christoph, **105**, 435
Riddle, Vicki A., **133**, 98
Riendeau, Denis, **116**, 526; **133**, 172
Rilling, Hans C., **110**, 125, 145, 335
Rimoldi, Donata, **123**, 61
Ringe, Dagmar, **131**, 389
Riser, Mary E., **102**, 104
Rittenberg, Marvin B., **121**, 327
Rivier, Catherine, **103**, 565
Rivier, Jean, **103**, 565; **124**, 389
Rivin, Carol, **118**, 75
Rivnay, Benjamin, **104**, 340

Roach, Peter J., **107**, 81
Robb, Richard J., **116**, 493
Robbins, Richard J., **124**, 412
Roberts, James L., **124**, 510
Rockley, Mark G., **123**, 92
Rockley, Natalie L., **123**, 92
Roda, Aldo, **133**, 215, 238
Roder, John C., **121**, 120, 140
Rodgers, John R., **109**, 572
Rodriguez, Angel, **133**, 172, 183
Rodriguez, Russell J., **111**, 37, 333
Roe, Richard M., **111**, 487
Rogero, Jose R., **131**, 508
Rogers, G. E., **107**, 624
Rogers, Stephen G., **118**, 627
Roheim, Paul S., **129**, 660
Rojas, Francisco J., **109**, 3
Rojas, María Cecilia, **110**, 406
Romain, Paul L., **108**, 148, 624
Rommerts, F. F. G., **109**, 275
Ronan, Rosemary, **128**, 223
Roos, Dirk, **132**, 225, 250
Röper, Karin, **106**, 58
Rose, George D., **115**, 430
Rosen, Barry P., **125**, 328
Rosen, Gerald M., **102**, 195; **105**, 198
Rosen, Henry, **105**, 399
Rosen, Jeffrey M., **109**, 572
Rosenberg, Andreas, **127**, 630; **131**, 448
Rosenberg, Hagai, **119**, 136
Rosenberg, Isadore N., **107**, 488
Rosenblatt, Michael, **109**, 625
Rosenbusch, Jürg P., **125**, 309
Rosenfeld, Claude, **121**, 580
Rosenstreich, David L., **108**, 298
Rosenthal, Alan S., **109**, 625
Rosenthal, Walter, **109**, 446
Ross, A. Catharine, **123**, 68
Ross, Russell, **109**, 69, 749
Ross, W. C. J., **112**, 207
Rosseneu, M., **128**, 365
Rossiakoff, Anthony A., **127**, 329
Rossio, Jeffrey L., **121**, 107
Rossmann, Michael G., **114**, 237; **115**, 397
Rothblat, George H., **129**, 628
Rothenberg, Sheldon P., **122**, 346
Rothnagel, J. A., **107**, 624
Rothschild, Kenneth J., **127**, 343
Rotilio, G., **105**, 22
Rottem, Shlomo, **125**, 259
Rottenberg, Hagai, **125**, 3
Rougeot, Catherine, **103**, 409
Rousseaux, J., **121**, 663
Rousseaux-Prevost, R., **121**, 663
Roussell, Anne M., **112**, 530
Rowe, Paul M., **102**, 158
Rowe, Peter B., **122**, 319
Rowe, W. Bruce, **113**, 199
Roy, Swapan K., **119**, 153
Rozengurt, Enrique, **119**, 643
Rubenstein, Peter A., **106**, 179
Rubin, Byron, **115**, 391
Rubinstein, Ardon, **129**, 186
Ruch, Walter, **132**, 395
Ruddat, Manfred, **111**, 189
Rudel, Lawrence L., **129**, 45
Rudney, Harry, **110**, 327
Rudolph, Rainer, **131**, 218
Runhaar, Bert A., **132**, 458
Russell, David W., **128**, 895
Russell, David Warwick, **110**, 26
Russo, C., **108**, 614

S

Saba, Thomas M., **132**, 334
Sachs, Frederick, **103**, 147; **124**, 190
Sack, John S., **102**, 143
Sagami, Hiroshi, **110**, 184, 188, 206
Sagami, Ikuko, **110**, 199, 320
Sage, Becky A., **111**, 458
Sairam, M. R., **109**, 725
Saito, Takashi, **116**, 311
Sakai, Hikoichi, **134**, 337
Sakai, Senkiti, **109**, 156
Sakaki, Yoshiyuki, **133**, 298
Sakiyama, Fumio, **106**, 165
Sala, Francesco, **118**, 87
Salemme, F. R., **114**, 140; **115**, 154
Salisbury, J. L., **134**, 408
Salmon, D. M., **109**, 366
Salmon, E. D., **134**, 519
Salmon, Sydney, E., **119**, 635
Samanta, Himadri, **119**, 516
Sample, Sally, **119**, 579
Samuel, Charles E., **119**, 499
Sanchez-Madrid, Francisco, **121**, 239
Sanders, Sheila K., **134**, 460
Sandoval, Ignacio V., **104**, 455

Sandri, G., **126**, 428
Sano, Seiyo, **123**, 415
Santos, Eugenio, **126**, 370
Sar, Madhabananda, **103**, 631
Sarai, Akinori, **127**, 79
Sarkar, Fazlul H., **119**, 263
Sarkar, Hemanta K., **125**, 214
Sasa, Tsutomu, **123**, 370, 421
Sasagawa, Tatsuru, **106**, 22
Sass, Ronald L., **106**, 351
Sato, Gordon H., **121**, 266
Sato, J. Denry, **121**, 266
Satre, Michel, **126**, 712
Satterwhite, Dennis M., **110**, 92
Saul, Michael W., **118**, 668
Savage, Hugh, **127**, 162
Sawicki, Janet A., **108**, 494
Sawyer, Donald T., **105**, 71
Scanu, Angelo M., **128**, 144, 151, 339
Scearce, Richard M., **103**, 459
Schaber, Michael D., **119**, 416
Schachter, Beth S., **124**, 497
Schaefer, Ernst J., **129**, 420
Schaeffer, James M., **103**, 362; **124**, 29; **133**, 43
Schäfer, Günter, **126**, 682
Schäfer, Hans-Jochen, **126**, 649
Schägger, H., **126**, 224
Schairer, Hans Ulrich, **126**, 579, 770
Scheiner, Steve, **127**, 456
Scheirer, W., **121**, 352
Schellekens, Huub, **119**, 220, 441, 453
Scher, Malka G., **111**, 547
Schibler, M. J., **134**, 217
Schimmer, Bernard P., **109**, 350
Schlegel, Werner, **124**, 116
Schlom, Jeffrey, **119**, 629
Schlossman, Stuart F., **108**, 148, 624
Schmid, Franz X., **131**, 70
Schmidt, A., **118**, 128
Schmidt, John A., **116**, 456
Schnaper, H. William, **116**, 395
Schnapp, Bruce J., **134**, 561
Schneider, Erwin, **126**, 569
Schneider, Wolfgang J., **109**, 405
Schobert, Brigitte, **125**, 510
Schoelmerich, J., **133**, 215
Schoenborn, B. P., **114**, 510; **127**, 217
Schonfeld, Gustav, **128**, 527
Schram, Eric, **133**, 507
Schramm, Michael, **109**, 346
Schröder, Heinz C., **134**, 170
Schröder, Ulf, **112**, 116
Schroeder, H. R., **133**, 557
Schroer, Joyce A., **109**, 704
Schroit, Alan J., **132**, 594
Schuffman, Shirley S., **119**, 533
Schuldiner, Shimon, **125**, 337
Schulster, D., **109**, 366
Schulten, K., **127**, 419
Schulten, Z., **127**, 419
Schulter, Samuel F., **108**, 139
Schumaker, Verne N., **128**, 155, 247
Schuster, Anne M., **118**, 497
Schutt, Clarence E., **114**, 211
Schwartz, A., **112**, 175
Schwerzmann, Klaus, **126**, 660
Schwiers, Elizabeth, **105**, 352
Scott, Robert A., **117**, 414
Seal, S. N., **118**, 128
Sealy, Roger C., **105**, 479
Seasholtz, Audrey, **124**, 278
Seddon, Andrew P., **113**, 451
Seehra, Jasbir S., **123**, 427
Seelig, Gail Fouré, **113**, 379, 390
Seetharam, Bellur, **123**, 23
Segal, Anthony W., **132**, 378
Segrest, Jere P., **128**, 181; **129**, 704
Seifert, Hans, **124**, 389
Sekar, Vaithilingam, **107**, 154
Sekellick, Margaret J., **119**, 115
Seki, Yoshihiko, **123**, 415
Sekura, Ronald D., **109**, 446, 558, 566
Sela, Ilan, **119**, 734, 744, 752, 759
Seliger, Howard H., **133**, 569
Selkoe, Dennis J., **134**, 388
Senear, Donald F., **130**, 132
Senyei, Andrew E., **112**, 56
Serio, Mario, **133**, 420
Serrano, Luis, **134**, 179
Serwer, Philip, **130**, 116
Setchell, Kenneth D. R., **111**, 63
Seto, Shuichi, **110**, 167, 320
Seubert, Andreas, **123**, 346
Seubert, Sigrid, **123**, 346
Sevanian, Alex, **105**, 162; **111**, 303
Shafferman, Avigdor, **119**, 136, 464
Shah, Vinod K., **118**, 511
Shakhov, Yu. A., **126**, 447
Shaltiel, Shmuel, **104**, 69

Shane, Barry, **122**, 323, 349
Sharma, Rajendra K., **102**, 210
Sharma, Somesh D., **132**, 626
Sharon, Nathan, **108**, 168
Sharp, Stephen, **106**, 157
Shaw, Elizabeth K., **118**, 338
Shaw, Jane E., **112**, 448
Shaw, Roger, **111**, 51
Shea, Madeline A., **130**, 132
Shear, Hannah Lustig, **132**, 637
Shearer, Martin J., **123**, 223, 235
Sheetz, Michael P., **134**, 531
Shelness, Gregory S., **128**, 671
Shen, Fung-Win, **108**, 249
Shenolikar, S., **107**, 102
Shenolikar, Shirish, **102**, 219
Shepherd, J., **129**, 566
Sheriff, Steven, **115**, 41
Shevach, Ethan M., **108**, 262
Shier, W. Thomas, **112**, 248
Shih, Li-Chen Nancy, **116**, 467
Shikita, Mikio, **111**, 352
Shillito, Raymond D., **118**, 549
Shimakata, Takashi, **122**, 53
Shin, Hyun S., **132**, 549
Shin, Whan-Chul, **114**, 156
Shinka, Toshihiro, **110**, 167
Shinomiya, Masaki, **129**, 738
Shioi, Yuzo, **123**, 370, 421
Shiota, Makoto, **102**, 279
Shively, John E., **121**, 459
Shivers, Brenda D., **124**, 497
Shoemaker, William J., **103**, 347
Shook, F. C., **105**, 36
Shooter, Eric M., **109**, 21
Shortman, Ken, **108**, 102
Shoulders, Carol C., **128**, 727
Shpungin, Joseph, **127**, 329
Shumak, Steven L., **129**, 395
Shuman, Howard A., **125**, 150
Sibley, David R., **103**, 577
Sieburth, Leslie E., **118**, 338
Siedlik, Paul H., **105**, 412
Siegel, Frank L., **102**, 158
Sierzega, Michael E., **121**, 120
Sies, Helmut, **105**, 221, 311, 445
Sigrist-Nelson, Kristine, **126**, 520
Silber, Robert, **122**, 10
Siliciano, Janet D'Angelo, **134**, 78
Silver, Paul J., **102**, 62
Silver, Robert B., **134**, 200
Silverman, Michael, **133**, 83
Simon, John R., **134**, 487
Simon, R., **118**, 640
Simon, S. A., **127**, 511
Simons, Guus, **119**, 366
Simpson, J. S. A., **109**, 692
Sims, James L., **106**, 483
Singer, Thomas P., **106**, 369
Singh, Ambarish K., **111**, 509
Singh, Arjun, **119**, 424
Sipat, Abdullah, **110**, 40
Sisco, Paul H., **118**, 497
Sissom, Janice F., **103**, 607
Sjögren, Hans Olov, **121**, 295
Sjöquist, John, **108**, 132, 405
Sjövall, Jan, **111**, 63
Slater, Trevor F., **105**, 283, 319
Sliwkowski, Mark X., **107**, 620
Sloop, Charles H., **129**, 660
Slotkin, Theodore A., **103**, 590
Sluski, Richard, **107**, 563
Smith, D. S., **122**, 67
Smith, Janet L., **115**, 41
Smith, K. L., **112**, 495, 504
Smith, Louis C., **129**, 848, 857
Smith, Martyn T., **105**, 505
Smith, Philip F., **124**, 443
Smith, Robert M., **109**, 187
Smith, Roberta H., **118**, 539
Smith, Roberts A., **107**, 23
Smith, Temple, **128**, 773
Smith-Gill, Sandra J., **116**, 121
Smulson, Mark E., **106**, 438, 512
Snell, Esmond E., **122**, 128, 135, 139
Snell, William J., **134**, 252
Snowdowne, Kenneth W., **124**, 90
Snyder, Mark A., **106**, 321
Socolofsky, Theodore J., **117**, 116
Socorro, Lilian, **129**, 738
Soda, Kenji, **113**, 83, 90, 96, 102
Sohal, R. S., **105**, 484
Söll, Dieter, **113**, 55
Solomon, Alan, **116**, 101
Solomon, Frank, **134**, 139
Soloski, Mark J., **108**, 549
Soman, Gopalan, **106**, 403
Somerville, C. R., **118**, 419
Sone, Nobuhito, **126**, 145, 604
Song, Guoxing, **113**, 3

Sonnenfeld, Gerald, **119**, 715
Soodak, Morris, **107**, 445
Soper, Thomas S., **113**, 108
Sophianopoulos, Alkis J., **117**, 354
Sophianopoulos, Judith A., **117**, 354
Sorensen, Craig M., **116**, 303
Spadari, Silvio, **118**, 87
Spalding, David M., **116**, 146
Sparks, Robert A., **115**, 23
Spector, Arthur A., **128**, 320
Spencer, Steven A., **114**, 530
Spiegel, Sara, **119**, 729
Spiegelberg, Hans L., **116**, 95
Spitsberg, Vitaly, **126**, 434
Spitz, Douglas R., **105**, 457
Spitz, M., **121**, 33
Springer, Timothy A., **108**, 313; **121**, 239
Spudich, James A., **134**, 531
Squire, Phil G., **117**, 142
Srikumaran, S., **121**, 244
Srivastava, Gopesh, **123**, 395
Stacey, Dennis W., **119**, 383
Stadtman, Earl R., **113**, 213
Stadtman, Thressa C., **107**, 576
Stahl, Robert L., **122**, 10
Staley, Daphne, **124**, 57
Stanker, Larry H., **121**, 615
Stanley, E. Richard, **116**, 564
Stanley, Philip E., **133**, 14, 587
Stanssens, Patrick, **119**, 366
Stark, William S., **105**, 341
Staros, James V., **109**, 816
Steigemann, W., **115**, 303
Stein, Olga, **129**, 816
Stein, S., **116**, 325
Stein, Yechezkiel, **129**, 816
Steiner, Alton L., **102**, 122
Steiner, George, **129**, 395
Steiner, Sonia, **109**, 346
Steinkamp, John A., **132**, 183
Steinmetz, André, **118**, 212
Steinmuller, David, **108**, 20, 28, 35
Stempel, Kerstin E., **126**, 618
Stern, David B., **118**, 488
Sterner, Richard, **107**, 224
Stevens, James L., **113**, 510
Stevenson, Henry C., **108**, 242; **121**, 107; **132**, 243
Stevenson, Thomas T., **118**, 3
Stewart, A. A., **102**, 227
Stewart, Barbara H., **112**, 360
Stewart, Carleton C., **132**, 183
Stewart, John M., **103**, 3
Stewart, Kim D., **124**, 548
Stewart, Thomas S., **106**, 157
Stipani, Italo, **125**, 692
Stock, Jeffrey B., **106**, 310, 321
Stoehr, Peter J., **118**, 453
Stokstad, E. L. R., **122**, 367
Storm, Daniel R., **102**, 148
Stossel, Thomas P., **132**, 192
Straka, James G., **123**, 352
Strang-Brown, Priscilla, **102**, 135
Strange, Philip G., **109**, 827
Strauss, Herbert L., **127**, 106
Stremlo, Donna, **119**, 63
Strickland, Thomas W., **109**, 736
Stringfello, Dale A., **119**, 707
Stroh, Annemarie, **126**, 32
Strominger, Jack L., **108**, 600, 607
Stuart, D. I., **115**, 117
Stull, James T., **102**, 62, 74
Stumpf, Paul K., **122**, 53
Stumpf, Walter E., **103**, 448, 631; **124**, 36
Sturgess, Maria, **133**, 366
Styles, J. M., **121**, 52
Subbaiah, P. V., **129**, 790
Suelter, Clarence H., **125**, 16
Sugawara, Minoru, **121**, 817
Sugimura, Takashi, **106**, 438, 441, 495
Sugino, Hiromu, **119**, 204
Suhadolnik, Robert J., **119**, 667, 752, 759
Sumbilla, Carlota, **111**, 471
Sundberg, David K., **103**, 493, 524; **124**, 359, 617
Suresh, M. R., **121**, 210
Susi, Heino, **130**, 290
Sussman, Joel L., **115**, 271
Suttie, J. W., **107**, 552
Suzuki, Shinya, **121**, 817
Suzuki, Tatsuo, **123**, 53
Svenneby, Gerd, **113**, 241
Svoboda, James A., **111**, 437
Svoboda, Majorie E., **109**, 798
Swaney, John B., **128**, 613
Swanson, Larry W., **103**, 663
Swartz, James R., **119**, 424
Sweet, Robert M., **114**, 19

Switzer, Robert L., **113**, 627
Sygusch, Jurgen, **114**, 386; **115**, 15
Sykes, Brian D., **107**, 36
Szelke, Michael, **124**, 318
Szittner, Rose, **133**, 189

T

Tabas, Ira, **128**, 647
Tagawa, Kunio, **126**, 504
Taintor, Read R., **132**, 508
Takada, Masahiro, **105**, 147
Takahara, Masayasu, **125**, 138
Takeoka, Gary, **111**, 149
Tall, Alan R., **128**, 647; **129**, 469
Tallant, E. Ann, **102**, 39, 244
Tamaoka, Jin, **123**, 251
Tanaka, Toshio, **102**, 185
Taniguchi, Masaru, **116**, 311
Taniuchi, Hiroshi, **131**, 185
Tanizawa, Katsuyuki, **113**, 90, 96
Tappel, Al L., **105**, 337; **107**, 602
Tarnowski, S. Joseph, **119**, 153, 177, 242, 403, 588
Tarr, Criss, **119**, 72
Tate, Suresh S., **113**, 400, 471
Taussig, Michael J., **116**, 340
Taylor, Clinton, A., Jr., **129**, 612
Taylor, D. Lansing, **134**, 487
Taylor, Frederick R., **110**, 9
Taylor, John M., **128**, 753, 801, 811
Taylor, Simeon I., **109**, 656
Taylor, William A., **102**, 210
Taylor-Papadimitriou, Joyce, **119**, 643
Teitelbaum, Anne P., **109**, 40, 48
Telia, Marius I., **122**, 161
ten Brink, Bart, **125**, 492
Ten Eyck, Lynn F., **115**, 324
Termini, John, **127**, 649
Test, Samuel T., **132**, 401
Tewari, Krishna K., **118**, 186
Thaller, C., **114**, 132
Thanassi, John W., **122**, 120
Thatcher, David R., **119**, 166
Thelen, Marcus, **126**, 78
Thèze, J., **116**, 403
Thomas, Edwin L., **132**, 569, 585
Thomas, Gary, **124**, 295
Thompson, Ambler, **133**, 569
Thompson, Malcolm J., **111**, 437
Thompson, William F., **118**, 57
Thor, Hjördis, **105**, 505
Thorner, Michael O., **103**, 294; **124**, 389
Thorpe, Gary H. G., **133**, 331, 404
Thorpe, P. E., **112**, 207
Thorpe, Suzanne R., **106**, 88
Tiboni, Orsola, **118**, 296
Tice, Thomas R., **112**, 101
Tillotson, Jerry Ann, **122**, 234
Timasheff, Serge N., **114**, 49; **117**, 53, 60, 459, 496; **130**, 3, 47
Tixier-Vidal, Andrée, **103**, 313
Tobin, E. M., **118**, 396
Tokuda, Hajime, **125**, 520
Tokushige, Masanobu, **113**, 618
Tollefson, John H., **129**, 797
Tomita, Yoshimi, **119**, 702
Tomlinson, E., **112**, 27, 43
Torczynski, Richard M., **119**, 359
Torgner, Ingeborg AA., **113**, 241
Torrence, Paul F., **119**, 522
Torres, Anthony R., **104**, 113
Toth, Peter P., **125**, 16
Toyama, Seizen, **113**, 102
Tregear, G. W., **124**, 534
Trivers, Glennwood E., **103**, 409
Trowbridge, Ian S., **112**, 238
Troy, Frederic A., II, **113**, 146
Trumble, William R., **125**, 214
Trumpower, Bernard L., **126**, 181, 211, 305, 316, 325
Tsai, Kelly Peng, **119**, 177, 204
Tsao, Yuan-Kai, **128**, 745
Tsuchiya, Tomofusa, **125**, 377
Tsuji, Frederick I., **133**, 298
Tsujimoto, Masafumi, **119**, 93
Tsunasawa, Susumu, **106**, 165
Tsutsui, Ken, **123**, 331
Tu, Shiao-Chun, **133**, 128
Tuck, Martin, **106**, 268
Tulinsky, A., **115**, 77
Turck, C., **116**, 325
Turner, Bryan M., **121**, 848
Turpen, T. H., **118**, 717
Turunen, Pekka, **105**, 126
Tybulewicz, Victor L. J., **125**, 230
Tyring, Stephen, **119**, 574

U

Ueda, Kunihiro, **106**, 450, 500
Ueno, Hiroshi, **113**, 108
Ulitzur, S., **133**, 264, 275
Unden, Gottfried, **126**, 387
Underhill, Lisa H., **109**, 656
Underwood, P. A., **121**, 301
Unger, Klaus, **104**, 154
Ungibauer, Margit, **126**, 32
Unno, Katsuo, **112**, 139
Updyke, Timothy V., **121**, 717
Upper, Christen D., **127**, 730
Utiyama, Hiroyasu, **131**, 51
Uzé, Gilles, **119**, 267

V

Valdivia, Manuel M., **134**, 268
Vale, Ronald D., **109**, 21
Vale, Wylie, **103**, 347, 565; **124**, 389
Valentine, Joan S., **105**, 71
Vallee, Richard B., **134**, 89, 104, 116
VanAken, T., **125**, 27
Van Alten, Pierson J., **108**, 10
Vanaman, Thomas C., **102**, 296
Van Bruggen, Ernst F. J., **126**, 344
Van Buskirk, John J., **107**, 563
Van Damme, Jozef, **119**, 88
van der Meide, Peter H., **119**, 220, 441, 453
van der Molen, H. J., **109**, 275
Van Dyke, Chris, **133**, 493
Van Dyke, Knox, **132**, 498; **133**, 493
Van Heel, Marin G., **126**, 344
van Heyningen, Veronica, **121**, 472
Van Keuren, Margaret L., **104**, 441
van Mourik, P., **121**, 174
VanNess, Brian G., **106**, 378
van Osdol, William W., **130**, 534
van Oss, Carel J., **132**, 3
van Reis, Robert, **119**, 77
van Renswoude, Jos, **104**, 329, 340
Van Scott, Michael R., **132**, 498
Van Vunakis, Helen, **121**, 871
Van Wyk, Judson J., **109**, 798
Varesio, Luigi, **108**, 307
Varga, J. M., **112**, 259
Vasilescu, V., **127**, 662
Vaughan, Joan, **103**, 565; **124**, 389
Vaughan, Martha, **106**, 411, 430
Velan, Baruch, **119**, 136, 464
Vellom, Daniel C., **133**, 229
Via, David P., **129**, 216, 848
Viceps-Madore, Dace, **122**, 120
Vignais, Pierre V., **125**, 639, 658; **126**, 712
Viitanen, Paul V., **125**, 214, 429
Vijverberg, Kitty, **119**, 220, 441
Vilček, Jan, **119**, 48
Vincent, Pamela L., **102**, 158
Vincenzi, Frank F., **102**, 47
Vinogradov, Serge N., **123**, 320
Vita, Alberto, **106**, 193; **117**, 342
Vitetta, Ellen S., **108**, 426, 549; **116**, 372
Voetman, Alwin A., **132**, 250
Vogels, Godfried D., **122**, 412
von Bahr, Christer, **113**, 499
von Boehmer, Harald, **132**, 467
von Jagow, G., **126**, 224, 253
Vuorinen, Pauli I., **122**, 147

W

Wadsworth, Patricia, **134**, 519
Waechter, Charles J., **111**, 471, 547
Wagener, Christoph, **121**, 459
Wagner, Conrad, **122**, 251, 255
Wagner, Gerhard, **131**, 307
Waite, J. Herbert, **107**, 397
Wakabayashi, Yasuo, **113**, 113
Waldmann, Thomas A., **116**, 201
Walker, John E., **125**, 230
Walker, Patricia L, **102**, 135
Walker, Richard F., **103**, 483
Wall, Lee, **133**, 172
Wallace, M. A., **109**, 469
Wallace, Robert W., **102**, 39, 244
Walrafen, G. E., **127**, 91
Walsh, Christopher, **122**, 199
Walsh, Kenneth A., **106**, 22
Walsh, Mary T., **128**, 582
Waltersdorph, Ann M., **105**, 399
Wandosell, Francisco, **134**, 179
Wang, Bi-Cheng, **115**, 90
Wang, Elaine, **122**, 110
Wang, Jerry H., **102**, 210
Wang, Ning, **119**, 676
Wang, T. P., **119**, 676

Ward, Cheryl, **118**, 97
Ward, Larry D., **117**, 400
Ward, Samuel, **134**, 414
Warholm, Margareta, **113**, 499
Warnick, G. Russell, **129**, 101
Warrant, R. W., **114**, 156
Warshel, A., **127**, 578
Watenpaugh, Keith D., **115**, 3
Watson, John C., **118**, 57
Weaver, L. H., **114**, 132
Webb, D. R., **116**, 295, 325
Weber, Klaus, **102**, 204; **134**, 170
Webster, JoAnn J., **133**, 51
Weech, Philip K., **128**, 432
Weeks, Ian, **133**, 366
Wehland, Jürgen, **134**, 170
Wehr, C. Timothy, **104**, 133
Wei, Ru-Qi, **132**, 537
Weil, Jacques-Henry, **118**, 212
Weiner, Joel H., **126**, 377
Weiner, Richard I., **103**, 287; **124**, 245
Weinstein, John N., **128**, 657
Weinstein, Yacob, **116**, 540
Weirich, Gunter F., **111**, 437, 454, 540
Weisgraber, Karl H., **128**, 273; **129**, 145
Weiss, Benjamin, **102**, 171
Weiss, Hanns, **126**, 191, 201
Weiss, Stephen J., **132**, 401
Weissbach, Arthur, **118**, 97
Weissbach, Herbert, **107**, 352; **118**, 309
Welankiwar, Sudhakar, **105**, 155
Wellner, Daniel, **113**, 564
Wells, David E., **121**, 417
Welsh, Michael J., **102**, 110
Welte, Karl, **116**, 553
Westbrook, Edwin M., **114**, 187
Westbrook, Gary L., **103**, 111
Westerwoudt, Regine J., **121**, 3
Wetlaufer, Donald B., **107**, 301
White, Bruce A., **124**, 269
White, Harold B., III, **122**, 221, 227
White, Jeffrey D., **103**, 511; **124**, 548
White, Morris F., **109**, 609
Widder, Kenneth J., **112**, 56
Wielckens, Klaus, **106**, 472
Wienhausen, G., **133**, 198
Wigness, Bruce D., **112**, 530
Wikvall, Kjell, **111**, 364
Wilchek, M., **104**, 3; **133**, 284
Wilcox, Josiah N., **124**, 510
Wilhelmsen, Eric C., **107**, 602
Wilkinson, Miles F., **119**, 96
Will, Oscar H., III, **111**, 189
Willard, Mark, **134**, 42
Williams, David L., **128**, 671; **129**, 254
Williams, Kevin J., **128**, 647
Williams, Noreen, **126**, 477, 484, 667
Williams, Richard S., **118**, 338
Williams, Robert W., **130**, 311
Williams, Robley C., Jr., **130**, 35; **134**, 544
Williamson, Gary, **122**, 240
Williamson, Joanne M., **113**, 445, 589
Willingham, Mark C., **109**, 257
Willis, David H., Jr., **119**, 667
Willson, R. L., **105**, 179
Wilson, Dorothy M., **125**, 377
Wilson, E., **114**, 132
Wilson, Steven P., **103**, 305
Wilson, Susan D., **122**, 269
Wilson, T. Hastings, **125**, 377, 429
Wimpee, C. F., **118**, 396
Winge, Dennis R., **105**, 105
Winick, N., **122**, 339
Winkler, Edith, **126**, 498; **127**, 772
Winkler, Fritz K., **114**, 211
Winkler, Herbert H., **125**, 253
Winship, Timothy R., **119**, 125
Winzor, Donald J., **117**, 182; **130**, 6
Wise, Donald L., **112**, 436
Witman, George B., **134**, 280, 291
Witte, Pamela L., **121**, 381
Wittwer, Carl T., **122**, 36
Witztum, Joseph L., **128**, 354
Wlodawer, Alexander, **114**, 551; **127**, 162
Woehle, Diana, **119**, 177
Wognum, A. W., **121**, 855
Wohlrab, Hartmut, **125**, 697
Wold, Finn, **106**, 69
Wolosewick, John J., **134**, 580
Wong, Dennis M., **108**, 307
Wong, Peter K., **105**, 231
Wood, Harland G., **107**, 261
Wood, Keith V., **133**, 3
Wood, W. Graham, **133**, 354
Woodford, Terry A., **106**, 170
Woodhead, J. Stuart, **133**, 366
Woodside, A. B., **110**, 130
Wraight, Christopher J., **121**, 375
Wright, Lynda S., **102**, 158
Wright, Samuel D., **132**, 204

Wriston, John C., Jr., **113**, 608
Wu, Chuen-Shang C., **130**, 208
Wu, Jang-Yen, **113**, 3
Wu, Licia N. Y., **126**, 353
Wubben, Jacqueline, **119**, 220, 453
Wulf, John J., **119**, 424
Wunderlich, D., **119**, 682
Wüthrich, Kurt, **131**, 307
Wyckoff, Harold W., **114**, 330
Wynberg, Hans, **133**, 531
Wynn, Peter C., **124**, 590
Wyse, Bonita, W., **122**, 36

X

Xu, Tingsen, **113**, 59
Xuong, Ng. H., **114**, 452

Y

Yagi, Kunio, **105**, 328
Yagi, Toshiharu, **113**, 83
Yamamoto, Gayle, **103**, 565; **124**, 389
Yamamoto, Harry Y., **110**, 303
Yamamoto, Ralph, **119**, 183
Yamamoto, Tokuo, **128**, 895
Yang, Jen Tsi, **130**, 208
Yang, Tsan-Hsi, **111**, 252
Yang, Xiaohang, **126**, 316
Yankofsky, S. A., **122**, 72
Yapel, Anthony F., Jr., **112**, 3
Yeung, Clement W. T., **109**, 170
Yilma, Tilahun, **119**, 130, 551
Yin, H. L., **134**, 3
Yip, Cecil C., **109**, 170
Yip, Y. K., **119**, 48
Yonaha, Kazuo, **113**, 102
York, William S., **118**, 3
Yoshida, Ryotaro, **105**, 61
Yoshida, Yukuo, **126**, 504
Yoshikawa, Kazuaki, **103**, 547
Yoshimura, Tohru, **113**, 96
Yost, David A., **122**, 169, 173
Youngblood, William W., **103**, 435
Yousif, Abdullah E., **118**, 316
Yu, Myeong-Hee, **131**, 250
Yu, Nai-Tend, **130**, 350
Yuan, Dorothy, **108**, 426
Yumura, Shigehiko, **134**, 573
Yumura, Toshiko K., **134**, 573
Yung, Yee-Pang, **116**, 553
Yuzuriha, Teruaki, **105**, 147

Z

Zabriskie, T. Mark, **111**, 252
Zaccai, Giuseppe, **127**, 619
Zacharchuk, Charles M., **132**, 549
Zackroff, Robert V., **134**, 371
Zadravec, Sylvia, **105**, 319
Zagalsky, P. F., **111**, 216
Zagorski, W., **118**, 154
Zaitlin, D., **118**, 270
Zakarija, M., **109**, 677
Zalkin, Howard, **113**, 263, 264, 273, 278, 282, 287, 293, 297, 303
Zannis, Vassilis I., **128**, 690, 712, 823
Zawatzky, Rainer, **119**, 474
Zech, Loren A., **129**, 366
Zeheb, Ron, **122**, 83, 87
Zeijlemaker, W. P., **121**, 174, 855
Zeitler, Hans-Jörg, **122**, 273
Zetterqvist, Örjan, **107**, 130
Zhang, Jingyuan, **106**, 500
Zhang, Yu-Zhong, **126**, 22
Zheng, Hong Da, **119**, 676
Zimm, Bruno H., **117**, 94
Zoon, Kathryn C., **119**, 276, 312, 326, 332
Zuberi, Riaz I., **116**, 428
Zukin, R. Suzanne, **124**, 172
Zumstein, Peter P., **109**, 782
Zundel, Georg, **127**, 439
Zurawski, Gerard, **118**, 253
Zur Nedden, Dorothy L., **119**, 276, 312